Biotechnology

A Laboratory Skills Course

Second Edition

J. Kirk Brown

BIO·RAD

Developmental editor: *Taylor Page*

Coordinating editor: *Laura Lenz*

Technical editors: *Ashleigh Miller, Jeannie Spagnolo*

Contributing editor: *Yolanda Kowalewski*

Art Director: *Ken Shiokari*

Design and Illustrations: *KSD-CA*

Printer: *Walsworth Print Group*

ISBN 978-0-9832396-3-5

Printing number
1 2 3 4 5 6 7 8 9 10   18

Disclaimer

Students, educators, and researchers using experimental procedures outlined in this book do so at their own risk. Users of this book are solely responsible for obtaining the necessary licenses, certificates, and permissions to conduct any experimental procedures that are subject to national or local legislation or restrictions. Users of this book are responsible for the humane treatment of animals according to local and national guidelines. The author and Bio-Rad Laboratories, Inc. do not assume responsibility for failure of a user to do so.

**Bio-Rad
Laboratories, Inc.**

*Life Science
Group*

Web site bio-rad.com **USA** 1 800 424 6723
Australia 61 2 9914 2800 **Austria** 43 01 877 89019
Belgium 32 03 710 53 00 **Brazil** 55 11 3065 7550
Canada 1 905 364 3435 **China** 86 21 6169 8500
Czech Republic 36 01 459 6192 **Denmark** 45 04 452 10 00
Finland 35 08 980 422 00 **France** 33 01 479 593 00
Germany 49 089 3188 4393 **Hong Kong** 852 2789 3300
Hungary 36 01 459 6190 **India** 91 124 4029300
Israel 972 03 963 6050 **Italy** 39 02 49486600
Japan 81 3 6361 7000 **Korea** 82 2 3473 4460
Mexico 52 555 488 7670 **The Netherlands** 310 318 540 666
New Zealand 64 9 415 2280 **Norway** 47 0 233 841 30
Poland 36 01 459 6191 **Portugal** 351 21 4727717
Russia 7 495 721 14 04 **Singapore** 65 6415 3188
South Africa 36 01 459 6193 **Spain** 34 091 49 06 580
Sweden 46 08 555 127 00 **Switzerland** 41 0617 17 9555
Taiwan 886 2 2578 7189 **Thailand** 66 2 651 8311
United Arab Emirates 971 4 8187300
United Kingdom 44 01923 47 1301

10000099934 Ver B (12008528) US/EG   17-1161 0918 Sig 0118

Dedication

This book represents a lifetime of teaching experience and sacrifice. I would like to dedicate it to a number of people.

To my immediate family: my father and mother, Jim and Janice Brown, and my sister Lisa and her family, who fostered my natural curiosity into a lifetime of wonder about the world around me. To my wife, Lisa, and her family, whose love and support have always encouraged me to pursue my dreams and to grow, and to my daughter Lynae and son Ryan and their families who mean everything to me, I dedicate this book.

To my Bio-Rad family, especially to my friend and colleague Stan Hitomi, to whom I am so very grateful for all of the adventures and collaboration. To my late friend Ron Mardigian for believing enough in two teachers to listen to our ideas about what teachers actually want and need to successfully teach biotechnology to their students. To Bryony Ruegg, whose continued support, I owe so very much. To all of the Bio-Rad Explorer team members with whom I have traveled and taught with over the years, I dedicate this book.

To my Tracy Unified family, especially Dr. Jim Franco and Dr. Sheila Harrison, whose outside the box thinking enabled me to live my life as a teacher in two worlds, one in the classroom and the other partially in the business world. To my current supervisors, James Mousalimas and Jane Steinkamp, who tirelessly support STEM education for the students in San Joaquin County and to my former students and current users of this text, I dedicate this book.

About the Author

 *J. Kirk Brown is currently the Director of STEM Programs at the San Joaquin County Office of Education, in Stockton, CA. He is a National Board–certified teacher and the former Science Department Chair at Tracy High School, Tracy, CA, where he taught for 25 years. He facilitated engaging hands-on laboratory experiences in his International Baccalaureate Biology and Biotechnology courses.*

Kirk's passion for education extended into and beyond the postsecondary level. As an adjunct associate professor at San Joaquin Delta College, Stockton, CA, he taught courses in Core Biology and Fundamentals of Biotechnology, and was the lead instructor at the Edward Teller Education Center at the Lawrence Livermore National Laboratory (LLNL), in Livermore, CA. Currently he leads a team of professionals that conducts teacher professional learning programs and develops STEM-related opportunities for students in central California. He is on numerous leadership teams that help bring the Next Generation Science Standards to students in CA.

Kirk's dedication as a teacher has earned him numerous honors, recognition by students, and peer-nominated awards, including:

> *Genentech's Access Excellence Award (1996)*
>
> *Milken National Educator Award (1999)*
>
> *Outstanding Biology Teacher Award from the National Association of Biology Teachers (2003)*
>
> *Carlston Family Foundation Outstanding Teachers of America Award (2006)*
>
> *Biotechnology Educator DiNA Award from BayBio (inducted into the 2007 Pantheon)*
>
> *Cortopassi Family Foundation Outstanding Science Teacher Award (2008)*
>
> *Teacher of the Year for San Joaquin County (2012)*

Kirk has inspired generations of students and has seen his students become leaders in their fields. Many of Kirk's former students have attended high-profile universities, received science, technology, engineering, and math (STEM) degrees at all levels, become science teachers, and pursued a wide variety of careers. Many have been selected for prestigious honors themselves. As a lifelong mentor, Kirk maintains connections with his former students, building bridges among current and past students.

Contents

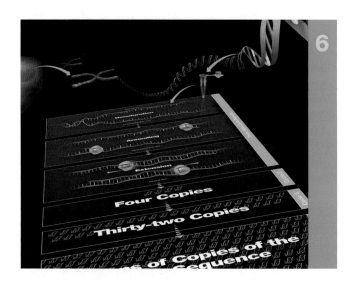

Chapter 8: Immunological Applications — 294

Chapter 8 Vignettes

Chapter 9: Research Projects — 330

Chapter 9 Vignettes

Appendices — 366

Preface

Dear Student,

As a science educator, I have always found that the key to student engagement is providing relevance and context. Biotechnology, with its direct application to student lives — from disease treatment and environmental monitoring to the snack food they eat — resonates with them and inspires them to learn. I hope this text does the same for you.

In addition, we know that the best way to learn about science is to do science. And when it comes to developing the particular skills associated with a specific field of science, the best way to learn those skills is do them often and repetitively the correct way. *Biotechnology: A Laboratory Skills Course* was written to help you learn about biotechnology by performing laboratory activities. Whether you plan to pursue a career in science or not, this course will give you an understanding of biotechnology and its implications in the real world that will serve as a foundation through your entire life. Moreover, as my own students have experienced, the hands-on nature and critical thinking required for these experiences will help you develop college- and career-ready skills that are necessary in a globally competitive world. The skills gained in the course also prepare you for pursuing laboratory science and may help you secure part-time work in a laboratory while studying at college or university.

The activities in this book cross many subjects and can be used to supplement a wide variety of courses that you might be taking now or will take in the future. The activities also bring together science, technology, engineering, and mathematics (STEM) in an integrated set of tools and processes. Whether you use this book to integrate a few concepts or utilize them all, you will gain valuable experience and a deeper appreciation of science and its role in the world today.

The goal of this book is to provide you with sufficient background for the laboratory activities without being overwhelming. As such, the background and activities are included together in each chapter so that the relevant background information is located in the same place as the activities.

This book is divided into nine chapters that cover different biotechnology concepts. Each chapter contains a background section that introduces you to the biology theory behind the techniques and provides background to the techniques themselves. Chapters 2–8 include a series of activities that directly link to the chapter topic.

Chapter 1 is a general introduction to biotechnology, the industry, its regulations, and the types of careers available in the field.

Chapter 2 focuses on basic skills that are essential to conducting biotechnology research, such as safety, maintaining a laboratory notebook, measuring volumes and mass, and making solutions.

Chapter 3 introduces microbiological and cell culturing techniques.

Chapters 4–6 introduce techniques for DNA manipulation, such as restriction digestion, transformation, CRISPR-Cas9, and the polymerase chain reaction (PCR).

Chapter 7 covers basic protein analysis techniques, including protein quantitation, chromatography, and polyacrylamide gel electrophoresis (PAGE), and the basics of protein production in industry.

Chapter 8 extends concepts learned in Chapter 7 to the use of antibodies in protein research.

Chapter 9 is the culmination of the course, where students take the skills developed over the duration of the course and apply them to an independent research project.

The activities are designed to build progressively over time and develop the laboratory skills and technical understanding that will give you the grounding necessary to perform basic independent research. To this end it is expected that you use a laboratory notebook to support your development so you learn how to document your work in the manner that practicing scientists would also use and can follow. You should take responsibility to write up your own work and utilize the rubric provided in Appendix F to help clarify the requirements for write-ups. I also encourage you to utilize Appendix E to evaluate your specific skills as they develop over time. I always suggest that you keep a running list of skills that you believe you have mastered in the back of your notebook. It is always nice to celebrate your achievements and this is one way to do so.

I hope this is just your first step into your exploration of biotechnology and you are inspired to engage in the process further. When you are taking courses, think about diving deeper into subjects that interest you and continue to develop your laboratory skills. It is in the use of these skills to answer questions or solve problems that you will really appreciate their magnitude. I hope you will be biotech-savvy citizens better able to make informed decisions in your own life and help others do the same.

The Polymerase Chain Reaction

6

Chapter 6

Chapter 6: Overview

6.1 Invention of PCR
6.2 What Is PCR?
6.3 Thermal Cyclers
6.4 Types of PCR
6.5 PCR Optimization
6.6 Techniques Based on PCR
6.7 Real-World Applications of PCR

Chapter 6: Laboratory Activities

6.A STR PCR Analysis
6.B GMO Detection by PCR
6.C Detection of the Human PV92 Alu Insertion
6.D Fish DNA Barcoding

Summary

The polymerase chain reaction (PCR) has revolutionized the study of living things. Invented by Kary Mullis in 1983, PCR has been a springboard for molecular biology research. It has been the basis of the Human Genome Project, modern forensic analysis, and genetic engineering. Using PCR, a small DNA sequence consisting of just a few hundred base pairs can be found within a genome of billions of base pairs. Billions of copies of the sequence can be generated, making the DNA sequences available for study and manipulation. Agriculture has been transformed by PCR with the advent of genetically modified crops. Cows and goats have been genetically engineered to produce pharmaceutical drugs in their milk, creating an industry called biopharming. PCR has made forensic analysis cheap, fast, and extremely accurate. Today's DNA profiles have less than a one in a trillion chance of matching another random individual, providing law enforcement with a powerful tool to fight crime. PCR has also been used to compare Neanderthal and human DNA to provide insights into how these populations interacted tens of thousands of years ago. The activities in this chapter use PCR to investigate DNA profiling, to detect genetic modifications in food, and to study human ancestry.

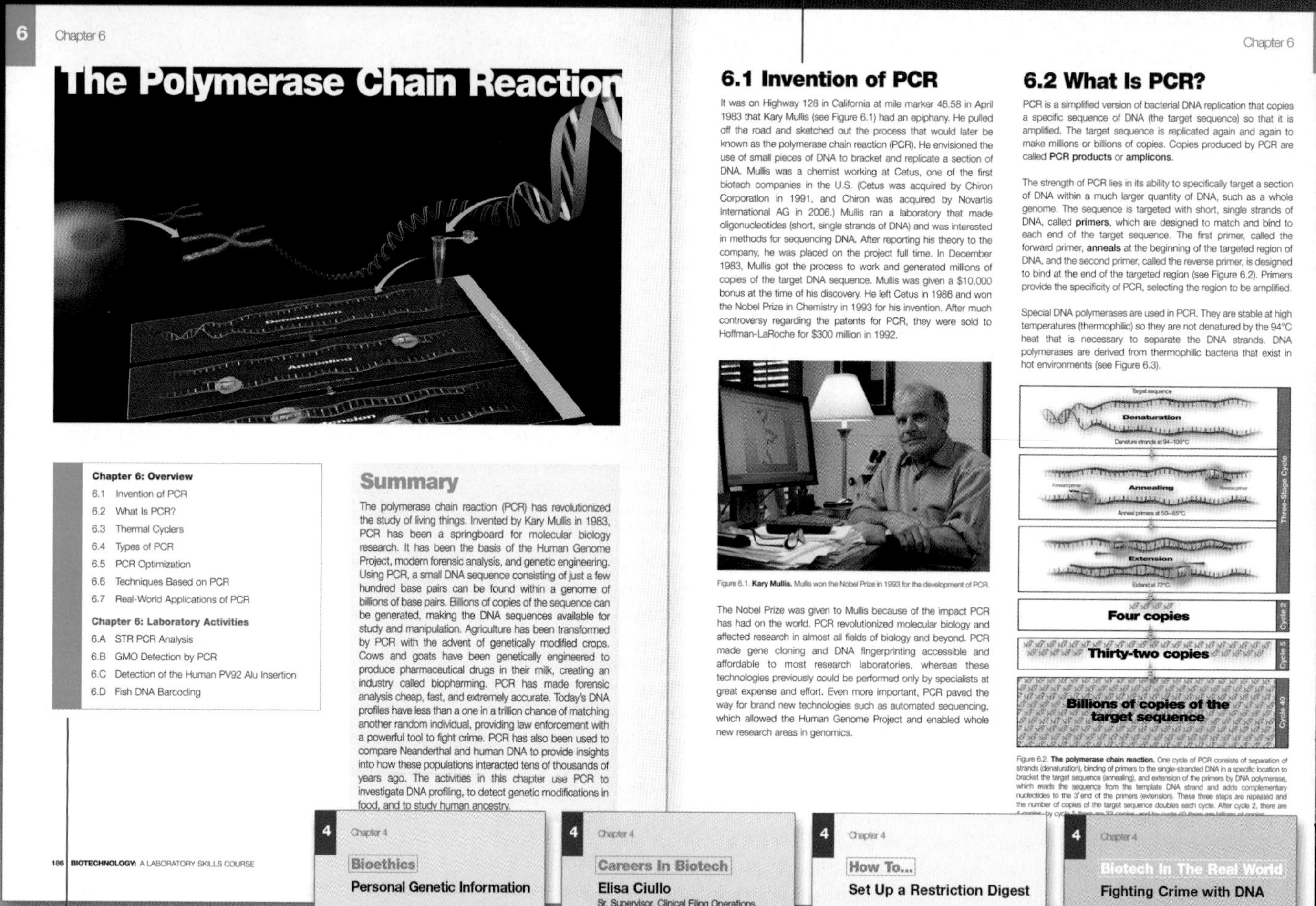

186 **BIOTECHNOLOGY:** A LABORATORY SKILLS COURSE

6.1 Invention of PCR

It was on Highway 128 in California at mile marker 46.58 in April 1983 that Kary Mullis (see Figure 6.1) had an epiphany. He pulled off the road and sketched out the process that would later be known as the polymerase chain reaction (PCR). He envisioned the use of small pieces of DNA to bracket and replicate a section of DNA. Mullis was a chemist working at Cetus, one of the first biotech companies in the U.S. (Cetus was acquired by Chiron Corporation in 1991, and Chiron was acquired by Novartis International AG in 2006.) Mullis ran a laboratory that made oligonucleotides (short, single strands of DNA) and was interested in methods for sequencing DNA. After reporting his theory to the company, he was placed on the project full time. In December 1983, Mullis got the process to work and generated millions of copies of the target DNA sequence. Mullis was given a $10,000 bonus at the time of his discovery. He left Cetus in 1986 and won the Nobel Prize in Chemistry in 1993 for his invention. After much controversy regarding the patents for PCR, they were sold to Hoffman-LaRoche for $300 million in 1992.

Figure 6.1. **Kary Mullis.** Mullis won the Nobel Prize in 1993 for the development of PCR.

The Nobel Prize was given to Mullis because of the impact PCR has had on the world. PCR revolutionized molecular biology and affected research in almost all fields of biology and beyond. PCR made gene cloning and DNA fingerprinting accessible and affordable to most research laboratories, whereas these technologies previously could be performed only by specialists at great expense and effort. Even more important, PCR paved the way for brand new technologies such as automated sequencing, which allowed the Human Genome Project and enabled whole new research areas in genomics.

6.2 What Is PCR?

PCR is a simplified version of bacterial DNA replication that copies a specific sequence of DNA (the target sequence) so that it is amplified. The target sequence is replicated again and again to make millions or billions of copies. Copies produced by PCR are called **PCR products** or **amplicons**.

The strength of PCR lies in its ability to specifically target a section of DNA within a much larger quantity of DNA, such as a whole genome. The sequence is targeted with short, single strands of DNA, called **primers**, which are designed to match and bind to each end of the target sequence. The first primer, called the forward primer, **anneals** at the beginning of the targeted region of DNA, and the second primer, called the reverse primer, is designed to bind at the end of the targeted region (see Figure 6.2). Primers provide the specificity of PCR, selecting the region to be amplified.

Special DNA polymerases are used in PCR. They are stable at high temperatures (thermophilic) so they are not denatured by the 94°C heat that is necessary to separate the DNA strands. DNA polymerases are derived from thermophilic bacteria that exist in hot environments (see Figure 6.3).

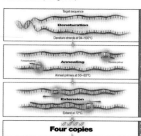

Figure 6.2. **The polymerase chain reaction.** One cycle of PCR consists of separation of strands (denaturation), binding of primers to the single-stranded DNA in a specific location to bracket the target sequence (annealing), and extension of the primers by DNA polymerase, which reads the sequence from the template DNA strand and adds complementary nucleotides to the 3' end of the primers (extension). These three steps are repeated and the number of copies of the target sequence doubles each cycle. After cycle 2, there are 4 copies; by cycle 5, there are 32 copies; and by cycle 40, there are billions of copies.

Chapter 6

Chapter overview
gives a roadmap of subject matter covered in the book.

Four types of vignettes
show how biotechnology concepts covered in the chapter play a role in our daily lives. Vignette topics include careers, real-life case studies, discussions about bioethics, and spotlights on key skills.

4

Chapter 4

Bioethics

Personal Genetic Information

As the cost of DNA sequencing continues to drop, the genetic testing industry is expanding rapidly. Nearly 100 companies, like 23andMe, Ancestry, and MyHeritage, now provide some form of direct-to-consumer genetic testing, and personal genetic tests are available online.

In most cases, these companies sell kits with materials and instructions for customers to send saliva samples to a commercial laboratory where the DNA is extracted and analyzed. These sequence data are analyzed to determine the customer's risk and carrier status for conditions like Alzheimer's disease, breast or lung cancer, and diabetes as well as many other traits.

These genetic tests pose no physical risk. But as private companies store and share personal genetic information, ethical questions arise. How might people interpret disease risk results and react to news that they are at higher risk for disease? Would they understand the correlation between a positive test result and the probability of developing the disease? Some companies require doctor approval prior to testing, but others, like 23andMe, do not, and offer contacts to genetic counselors instead. Could identifying risk inspire customers to change their lifestyle? If a woman learns she is at high risk for breast cancer, for example, how would she know whether to take radical action, such as a double mastectomy to eliminate the risk altogether, or to get more frequent mammograms instead?

Genetic information applies not only to the customer but also to their relatives. If someone shares information about their personal cancer risk with others, they may also be sharing information about the cancer risk of their siblings, children, or even grandchildren. In 2008, President George W. Bush signed into law the Genetic Information Nondiscrimination Act (GINA), which prohibits discrimination on the basis of information derived from genetics tests. Nevertheless, what types of consent should someone be required to get from family members before sharing that type of information online?

114 **BIOTECHNOLOGY:** A LABORATORY SKILLS COURSE

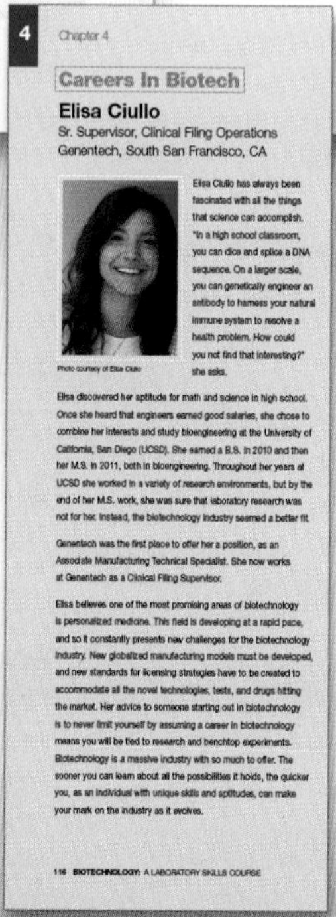

4

Chapter 4

Careers In Biotech

Elisa Ciullo
Sr. Supervisor, Clinical Filing Operations
Genentech, South San Francisco, CA

Photo courtesy of Elisa Ciullo

Elisa Ciullo has always been fascinated with all the things that science can accomplish. "In a high school classroom, you can dice and splice a DNA sequence. On a larger scale, you can genetically engineer an antibody to harness your natural immune system to resolve a health problem. How could you not find that interesting?" she asks.

Elisa discovered her aptitude for math and science in high school. Once she heard that engineers earned good salaries, she chose to combine her interests and study bioengineering at the University of California, San Diego (UCSD). She earned a B.S. in 2010 and then her M.S. in 2011, both in bioengineering. Throughout her years at UCSD she worked in a variety of research environments, but by the end of her M.S. work, she was sure that laboratory research was not for her. Instead, the biotechnology industry seemed a better fit.

Genentech was the first place to offer her a position, as an Associate Manufacturing Technical Specialist. She now works at Genentech as a Clinical Filing Supervisor.

Elisa believes one of the most promising areas of biotechnology is personalized medicine. This field is developing at a rapid pace, and so it constantly presents new challenges for the biotechnology industry. New globalized manufacturing models must be developed, and new standards for licensing strategies have to be created to accommodate all the novel technologies, tests, and drugs hitting the market. Her advice to someone starting out in biotechnology is to never limit yourself by assuming a career in biotechnology means you will be tied to research and benchtop experiments. Biotechnology is a massive industry with so much to offer. The sooner you can learn about all the possibilities it holds, the quicker you, as an individual with unique skills and aptitudes, can make your mark on the industry as it evolves.

116 **BIOTECHNOLOGY:** A LABORATORY SKILLS COURSE

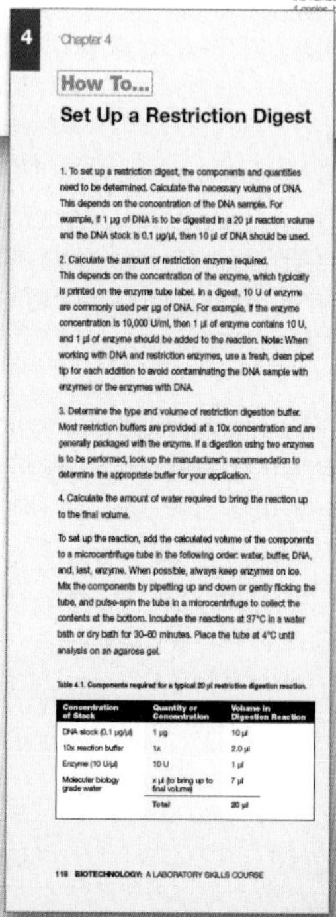

4

Chapter 4

How To...

Set Up a Restriction Digest

1. To set up a restriction digest, the components and quantities need to be determined. Calculate the necessary volume of DNA. This depends on the concentration of the DNA sample. For example, if 1 µg of DNA is to be digested in a 20 µl reaction volume and the DNA stock is 0.1 µg/µl, then 10 µl of DNA should be used.

2. Calculate the amount of restriction enzyme required. This depends on the concentration of the enzyme, which typically is printed on the enzyme tube label. In a digest, 10 U of enzyme are commonly used per µg of DNA. For example, if the enzyme concentration is 10,000 U/ml, then 1 µl of enzyme contains 10 U, and 1 µl of enzyme should be added to the reaction. Note: When working with DNA and restriction enzymes, use a fresh, clean pipet tip for each addition to avoid contaminating the DNA sample with enzymes or the enzymes with DNA.

3. Determine the type and volume of restriction digestion buffer. Most restriction buffers are provided at a 10x concentration and are generally packaged with the enzyme. If a digestion using two enzymes is to be performed, look on the manufacturer's recommendation to determine the appropriate buffer for your application.

4. Calculate the amount of water required to bring the reaction up to the final volume.

To set up the reaction, add the calculated volume of the components to a microcentrifuge tube in the following order: water, buffer, DNA, and, last, enzyme. When possible, always keep enzymes on ice. Mix the components by pipetting up and down or gently flicking the tube, and pulse-spin the tube in a microcentrifuge to collect the contents at the bottom. Incubate the reactions at 37°C in a water bath or dry bath for 30–60 minutes. Place the tube at 4°C until analysis on an agarose gel.

Table 4.1. Components required for a typical 20 µl restriction digestion reaction.

Concentration of Stock	Quantity or Concentration	Volume in Digestion Reaction
DNA stock (0.1 µg/µl)	1 µg	10 µl
10x reaction buffer	1x	2.0 µl
Enzyme (10 U/µl)	10 U	1 µl
Molecular biology grade water	x µl (to bring up to final volume)	7 µl
	Total	20 µl

118 **BIOTECHNOLOGY:** A LABORATORY SKILLS COURSE

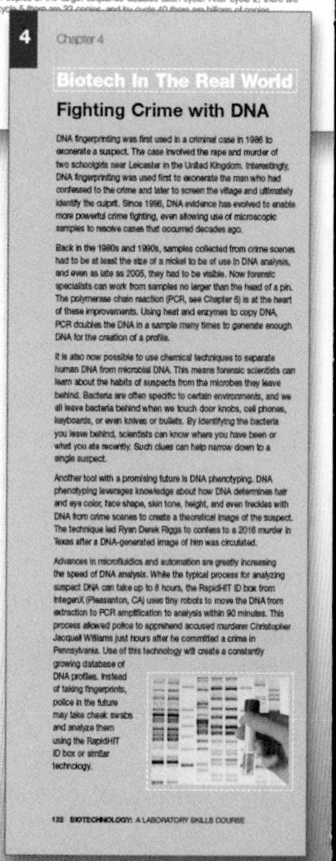

4

Chapter 4

Biotech In The Real World

Fighting Crime with DNA

DNA fingerprinting was first used in a criminal case in 1986 to exonerate a suspect. The case involved the rape and murder of two schoolgirls near Leicester in the United Kingdom. Interestingly, DNA fingerprinting was used first to exonerate the man who had confessed to the crime and later to screen the village and ultimately identify the culprit. Since 1986, DNA evidence has evolved to enable more powerful crime fighting, even allowing use of microscopic samples to resolve cases that occurred decades ago.

Back in the 1980s and 1990s, samples collected from crime scenes had to be at least the size of a nickel to be of use in DNA analysis, and even as late as 2005, they had to be visible. Now forensic specialists can work from samples no larger than the head of a pin. The polymerase chain reaction (PCR, see Chapter 6) is at the heart of these improvements. Using heat and enzymes to copy DNA, PCR doubles the DNA in a sample many times to generate enough DNA for the creation of a profile.

It is also now possible to use chemical techniques to separate human DNA from microbial DNA. This means forensic scientists can learn about the habits of suspects from the microbes they leave behind. Bacteria are often specific to certain environments, and we all leave bacteria behind when we touch door knobs, cell phones, keyboards, or even knives or bullets. By identifying the bacteria you leave behind, scientists can know when you have been or what you ate recently. Such clues can help narrow down to a single suspect.

Another tool with a promising future is DNA phenotyping. DNA phenotyping leverages knowledge about how DNA determines hair and eye color, face shape, skin tone, height, and even freckles with DNA from crime scenes to create a theoretical image of the suspect. The technique led Ryan Derek Riggs to confess to a 2010 murder in Texas after a DNA-generated image of him was circulated.

Advances in microfluidics and automation are greatly increasing the speed of DNA analysis. While the typical process for analyzing suspect DNA can take up to 8 hours, the RapidHIT ID box from IntegenX (Pleasanton, CA) uses tiny robots to move the DNA from extraction to PCR amplification to analysis within 90 minutes. This process allowed police to apprehend accused murderer Christopher Jacquel Williams just hours after he committed a crime in Pennsylvania. Use of this technology will create a constantly growing database of DNA profiles. Instead of taking fingerprints, police in the future may take cheek swabs and analyze them using the RapidHIT ID box or similar technology.

122 **BIOTECHNOLOGY:** A LABORATORY SKILLS COURSE

Activities implement the techniques described in the background information. Early activities focus on building basic skills, while later activities use those basic skills as a foundation for more advanced techniques.

Laboratory skills are acquired by performing the activity. The requirements necessary to claim proficiency in those skills are described in the Laboratory Skills Assessment Rubric in Appendix E.

Graphics illustrate the protocol steps of the activities.

Step-by-step protocols simplify procedures and provide guidance on results analysis

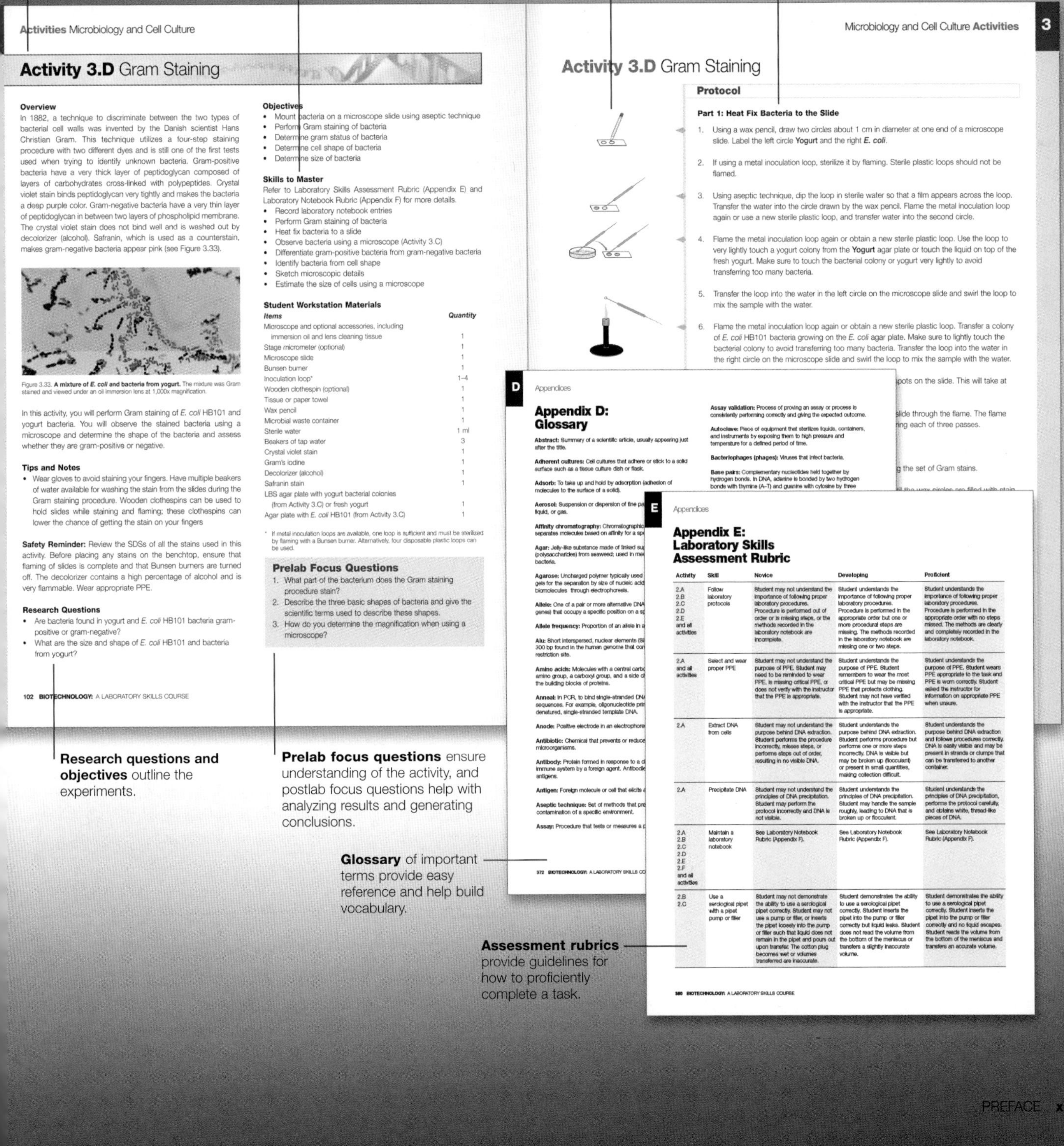

Activity 3.D Gram Staining

Overview

In 1882, a technique to discriminate between the two types of bacterial cell walls was invented by the Danish scientist Hans Christian Gram. This technique utilizes a four-step staining procedure with two different dyes and is still one of the first tests used when trying to identify unknown bacteria. Gram-positive bacteria have a very thick layer of peptidoglycan composed of layers of carbohydrates cross-linked with polypeptides. Crystal violet stain binds peptidoglycan very tightly and makes the bacteria a deep purple color. Gram-negative bacteria have a very thin layer of peptidoglycan in between two layers of phospholipid membrane. The crystal violet stain does not bind well and is washed out by decolorizer (alcohol). Safranin, which is used as a counterstain, makes gram-negative bacteria appear pink (see Figure 3.33).

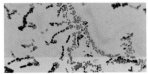

Figure 3.33. **A mixture of E. coli and bacteria from yogurt.** The mixture was Gram stained and viewed under an oil immersion lens at 1,000x magnification.

In this activity, you will perform Gram staining of E. coli HB101 and yogurt bacteria. You will observe the stained bacteria using a microscope and determine the shape of the bacteria and assess whether they are gram-positive or negative.

Tips and Notes

- Wear gloves to avoid staining your fingers. Have multiple beakers of water available for washing the stain from the slides during the Gram staining procedure. Wooden clothespins can be used to hold slides while staining and flaming; these clothespins can lower the chance of getting the stain on your fingers.

Safety Reminder: Review the SDSs of all the stains used in this activity. Before placing any stains on the benchtop, ensure that flaming of slides is complete and that Bunsen burners are turned off. The decolorizer contains a high percentage of alcohol and is very flammable. Wear appropriate PPE.

Research Questions

- Are bacteria found in yogurt and E. coli HB101 bacteria gram-positive or gram-negative?
- What are the size and shape of E. coli HB101 and bacteria from yogurt?

Objectives

- Mount bacteria on a microscope slide using aseptic technique
- Perform Gram staining of bacteria
- Determine gram status of bacteria
- Determine cell shape of bacteria
- Determine size of bacteria

Skills to Master

Refer to Laboratory Skills Assessment Rubric (Appendix E) and Laboratory Notebook Rubric (Appendix F) for more details.
- Record laboratory notebook entries
- Perform Gram staining of bacteria
- Heat fix bacteria to a slide
- Observe bacteria using a microscope (Activity 3.C)
- Differentiate gram-positive bacteria from gram-negative bacteria
- Identify bacteria from cell shape
- Sketch microscopic details
- Estimate the size of cells using a microscope

Student Workstation Materials

Items	Quantity
Microscope and optional accessories, including immersion oil and lens cleaning tissue	1
Stage micrometer (optional)	1
Microscope slide	1
Bunsen burner	1
Inoculation loop*	1–4
Wooden clothespin (optional)	1
Tissue or paper towel	1
Wax pencil	1
Microbial waste container	1
Sterile water	1 ml
Beakers of tap water	3
Crystal violet stain	1
Gram's iodine	1
Decolorizer (alcohol)	1
Safranin stain	1
LBS agar plate with yogurt bacterial colonies (from Activity 3.C) or fresh yogurt	1
Agar plate with E. coli HB101 (from Activity 3.C)	1

* If metal inoculation loops are available, one loop is sufficient and must be sterilized by flaming with a Bunsen burner. Alternatively, four disposable plastic loops can be used.

Prelab Focus Questions

1. What part of the bacterium does the Gram staining procedure stain?
2. Describe the three basic shapes of bacteria and give the scientific terms used to describe these shapes.
3. How do you determine the magnification when using a microscope?

Research questions and objectives outline the experiments.

Prelab focus questions ensure understanding of the activity, and postlab focus questions help with analyzing results and generating conclusions.

Activity 3.D Gram Staining

Protocol

Part 1: Heat Fix Bacteria to the Slide

1. Using a wax pencil, draw two circles about 1 cm in diameter at one end of a microscope slide. Label the left circle Yogurt and the right E. coli.

2. If using a metal inoculation loop, sterilize it by flaming. Sterile plastic loops should not be flamed.

3. Using aseptic technique, dip the loop in sterile water so that a film appears across the loop. Transfer the water into the circle drawn by the wax pencil. Flame the metal inoculation loop again or use a new sterile plastic loop, and transfer water into the second circle.

4. Flame the metal inoculation loop again or obtain a new sterile plastic loop. Use the loop to very lightly touch a yogurt colony from the **Yogurt** agar plate or touch the liquid on top of the fresh yogurt. Make sure to touch the bacterial colony or yogurt very lightly to avoid transferring too many bacteria.

5. Transfer the loop into the water in the left circle on the microscope slide and swirl the loop to mix the sample with the water.

6. Flame the metal inoculation loop again or obtain a new sterile plastic loop. Transfer a colony of E. coli HB101 bacteria growing on the E. coli agar plate. Make sure to lightly touch the bacterial colony to avoid transferring too many bacteria. Transfer the loop into the water in the right circle on the microscope slide and swirl the loop to mix the sample with the water.

Appendix D: Glossary

Abstract: Summary of a scientific article, usually appearing just after the title.

Adherent cultures: Cell cultures that adhere or stick to a solid surface such as a tissue culture dish or flask.

Adsorb: To take up and hold by adsorption (adhesion of molecules to the surface of a solid).

Aerosol: Suspension or dispersion of fine particles in a liquid, or gas.

Affinity chromatography: Chromatographic method used to separate molecules based on affinity for a spe...

Agar: Jelly-like substance made of linked sug... (polysaccharides) from seaweed; used in med... bacteria.

Agarose: Uncharged polymer typically used ... gels for the separation by size of nucleic acid... biomolecules through electrophoresis.

Allele: One of a pair or more alternative DNA ... genes) that occupy a specific position on a sp...

Allele frequency: Proportion of an allele in a ...

Alu: Short interspersed nuclear elements (SI... 300 bp found in the human genome that con... restriction site.

Amino acids: Molecules with a central carbo... amino group, a carboxyl group, and a side ch... the building blocks of proteins.

Anneal: In PCR, to bind single-stranded DNA... sequences. For example, oligonucleotide prim... denatured, single-stranded template DNA.

Anode: Positive electrode in an electrophore...

Antibiotic: Chemical that prevents or reduce... microorganisms.

Antibody: Protein formed in response to a ... immune system by a foreign agent. Antibodie... antigens.

Antigen: Foreign molecule or cell that elicits a...

Aseptic technique: Set of methods that pre... contamination of a specific environment.

Assay: Procedure that tests or measures a p...

Assay validation: Process of proving an assay or process is consistently performing correctly and giving the expected outcome.

Autoclave: Piece of equipment that sterilizes liquids, containers, and instruments by exposing them to high pressure and temperature for a defined period of time.

Bacteriophages (phages): Viruses that infect bacteria.

Base pairs: Complementary nucleotides held together by hydrogen bonds. In DNA, adenine is bonded by two hydrogen bonds with thymine (A–T) and guanine with cytosine by three

Glossary of important terms provide easy reference and help build vocabulary.

Appendix E: Laboratory Skills Assessment Rubric

Activity	Skill	Novice	Developing	Proficient
2.A 2.B 2.C 2.D 2.E and all activities	Follow laboratory protocols	Student may not understand the importance of following proper laboratory procedures. Procedure is performed out of order or is missing steps, or the methods recorded in the laboratory notebook are incomplete.	Student understands the importance of following proper laboratory procedures. Procedure is performed in the appropriate order but one or more procedural steps are missing. The methods recorded in the laboratory notebook are missing one or two steps.	Student understands the importance of following proper laboratory procedures. Procedure is performed in the appropriate order with no steps missed. The methods are clearly and completely recorded in the laboratory notebook.
2.A and all activities	Select and wear proper PPE	Student may not understand the purpose of PPE. Student may need to be reminded to wear PPE, is missing critical PPE, or does not verify with the instructor that the PPE is appropriate.	Student understands the purpose of PPE. Student remembers to wear the most critical PPE but may be missing PPE that protects clothing. Student may not have verified with the instructor that the PPE is appropriate.	Student understands the purpose of PPE. Student wears PPE appropriate to the task and PPE is worn correctly. Student asked the instructor for information on appropriate PPE when unsure.
2.A	Extract DNA from cells	Student may not understand the purpose behind DNA extraction. Student performs the procedure incorrectly, misses steps, or performs steps out of order, resulting in no visible DNA.	Student understands the purpose behind DNA extraction. Student performs procedure but performs one or more steps incorrectly. DNA is visible but may be broken up (flocculent) or present in small quantities, making collection difficult.	Student understands the purpose behind DNA extraction and follows procedures correctly. DNA is easily visible and may be present in strands or clumps that can be transferred to another container.
2.A	Precipitate DNA	Student may not understand the principles of DNA precipitation. Student may perform the protocol incorrectly and DNA is not visible.	Student understands the principles of DNA precipitation. Student may handle the sample roughly, leading to DNA that is broken up or flocculent.	Student understands the principles of DNA precipitation, performs the protocol carefully, and obtains visible, thread-like pieces of DNA.
2.A 2.B 2.C 2.D 2.E 2.F and all activities	Maintain a laboratory notebook	See Laboratory Notebook Rubric (Appendix F).	See Laboratory Notebook Rubric (Appendix F).	See Laboratory Notebook Rubric (Appendix F).
2.B 2.C	Use a serological pipet with a pipet pump or filler	Student may not demonstrate the ability to use a serological pipet correctly. Student may not use a pump or filler, or inserts the pipet loosely into the pump or filler such that liquid does not remain in the pipet and pours out upon transfer. The cotton plug becomes wet or volumes transferred are inaccurate.	Student demonstrates the ability to use a serological pipet correctly. Student inserts the pipet into the pump or filler correctly but liquid leaks. Student does not read the volume from the bottom of the meniscus or transfers a slightly inaccurate volume.	Student demonstrates the ability to use a serological pipet correctly. Student inserts the pipet into the pump or filler correctly and no liquid escapes. Student reads the volume from the bottom of the meniscus and transfers an accurate volume.

Assessment rubrics provide guidelines for how to proficiently complete a task.

Acknowledgements

Many people have contributed to this book and I appreciate all their help. Thank you to my students and colleagues, past and present, from whom I have learned so much and who have all contributed in some way to this book. Thank you to all the scientists and educators who reviewed the book. Thank you to all the people and organizations that directly contributed to the book including *Agdia Inc., Andreas Bolzer, AquaBounty Technologies, Asha Miles, CDC Public Health Image Library, Corning Life Science, Dave Conley, Denise Gangadharan, Dora Barbosa, Doug Boyd, Elisa Ciullo, Greg Hampikan, Greg Hoff, Ivica Letunić, Joanna Philips, Joshua Moore, Kary Mullis, Katie Dalpozzo, Keith Schuetz, Kevin McLaughlin, Kristi DeCourcy, Linda Strausbaugh, Matthew Betts, Miguel Esteban, Naturalnews.com, New England Biolabs, OHAUS Corporation, Science Buddies, Scott Chilton, Shimadzu Corporation, Sophy Wong, Sunny Choe, Tina Lanese, USDA ARS.*

Second Edition Reviewers

Alex Ward, *Bio-Rad Laboratories*

Bryony Ruegg, *Bio-Rad Laboratories*

James Hewlett, *Fingerlakes Community College*

Kenneth Oh, *Bio-Rad Laboratories*

Laura Moriarty, *Bio-Rad Laboratories*

Leala Thomas, *Bio-Rad Laboratories*

Lori Wojciechowski, *University of Florida*

Lucy Smith, *Bio-Rad Laboratories*

Sherry Annee, *Brebeuf Jesuit Preparatory School*

Tamara Mandell, *University of Florida*

Thuy Lau, *Bio-Rad Laboratories*

Previous Reviewers

Amanda Grimes, *Mesa High School Biotechnology Academy*

Amy Cote, *Science Consultant*

Annett Hahn-Windgassen, *Bio-Rad Laboratories*

Angela DeaMude, *Tech Council of Maryland*

Brad Williamson, *University of Kansas*

Bridgette L. Kirkpatrick, *Collin College*

Cinda Herndon-King, *Georgia Bio*

Damon Tighe, *Bio-Rad Laboratories*

David Palmer, *Bio-Rad Laboratories*

Dawn Reed, *Bakersfield Christian High School*

Dennis C. Yee, *Bio-Rad Laboratories*

Diane Sweeney, *Punahou School*

Doug Kain, *Merced College*

Henri Wintz, *Bio-Rad Laboratories*

Ingrid Miller, *Bio-Rad Laboratories*

Jay Chugh, *Acalanes High School*

Jeff Rapp, *Athens Technical College*

Joann Lau, *Bellarmine University*

Joe Siino, *Bio-Rad Laboratories*

John Walker, *Bio-Rad Laboratories*

Jose Lopez, *Bio-Rad Laboratories*

Joy Killough, *Westwood High School*

Julie Mathern, *Bio-Rad Laboratories*

Kathy Silvey, *Bio-Rad Laboratories*

Ken Gracz, *T. Wingate Andrews High School*

Larisa Haupt, *Griffith University*

Laurie Usinger, *Bio-Rad Laboratories*

Lee Olech, *Bio-Rad Laboratories*

Leigh Brown, *Bio-Rad Laboratories*

Linda Lingelbach, *Bio-Rad Laboratories*

Luanne J. Wolfgram, *Johnson County Community College*

Michele Gilbert, *Bio-Rad Laboratories*

Nik Chmiel, *Bio-Rad Laboratories*

Pat Jensen, *Bio-Rad Laboratories*

Paul Liu, *Bio-Rad Laboratories*

Sherri Andrews, *Bio-Rad Laboratories*

Stan Hitomi, *San Ramon School District*

Tamara Mandell, *University of Florida*

Tiffany Stevens, *Bakersfield Christian High School*

Xan Simonson, *Mesa High School Biotechnology Academy*

Xuemei He, *Bio-Rad Laboratories*

The Biotechnology Industry

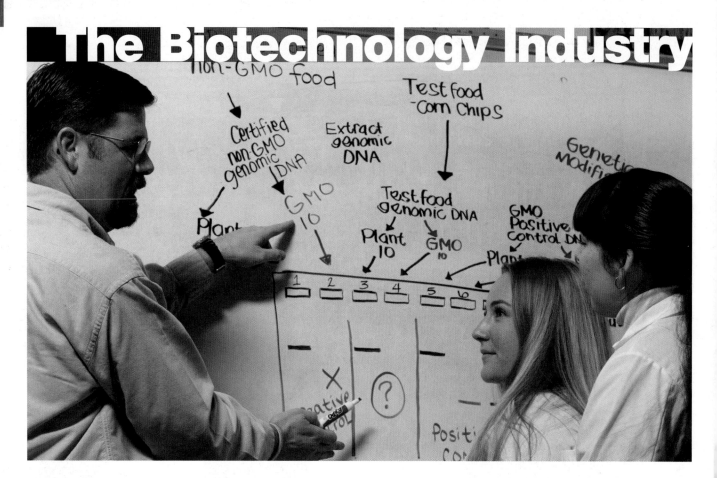

Summary

Have you ever eaten yogurt, done laundry, worn contact lenses, written on paper, or eaten a puffed corn snack? If so, you have used products of biotechnology. The products of biotechnology range from cutting-edge therapeutic drugs that target cancer to virus-resistant papaya to the enzymes added to laundry detergents to remove fat stains. The biotechnology industry is growing rapidly after its beginnings in the late 1970s. Career opportunities in the biotechnology sector are increasing as a result, and qualified job applicants will find their skills and knowledge in high demand.

Biotechnology is a major part of the global economy with companies in both the private and public sectors. Biotechnology is applied worldwide to produce a variety of products in biological research, healthcare, agriculture and food, forensics, and manufacturing. Standard practices are, therefore, required to ensure that the same products made in different countries meet the same quality and safety standards. Because it can be used to change the genetic makeup of organisms, including humans, biotechnology also raises ethical questions and requires public oversight and governmental regulation. This chapter provides a general outline of biotechnology — the industry, its regulations, and the types of careers that are available in the field.

1.1 What Is Biotechnology?

Biotechnology can be difficult to define. In fact, each country around the world has its own unique definition. Because biotechnology is based on many rapidly changing techniques, its definition is also constantly changing.

In its broadest sense, biotechnology is technology based on biology. Though we tend to think of biotechnology as a modern phenomenon, humans have been applying technology to living things for thousands of years. Examples include selectively breeding crops to improve yield, and using microbes to make yogurt, bread, and wine (see Figure 1.1). More specifically, **biotechnology** is the use of modern molecular and microbial techniques to make useful products or processes. As such, it is distinct from traditional animal breeding, plant selection, and fermentation practices.

Biotechnology is not a pure scientific discipline. It draws on knowledge and techniques from many of the biological sciences, such as genetics, molecular biology, biochemistry, cell biology, and microbiology. Biotechnology also draws on many nonbiological fields of study, such as engineering, chemistry, physics, and information technology. Conversely, many of these disciplines also draw on the methods that biotechnology has developed (see Figure 1.1).

Biotechnology distinguishes itself from other biological disciplines by its purpose: to develop products and processes. Because commercial products are created by the biotechnology industry, biotechnology makes a direct contribution to the world's economy.

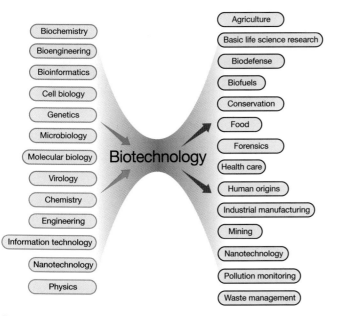

Figure 1.1. **The contributors and applications of biotechnology.** A number of scientific disciplines contribute to biotechnology (left; orange text represents biological sciences, and gray text represents other scientific disciplines). In turn, biotechnology is applied in numerous other fields and industries.

Biotechnology: Good or Bad?

Biotechnology is powerful. It can change the genetic makeup of organisms, including humans. Because of this power, some biotechnology is controversial and anti-biotechnology sentiment is not uncommon. Opponents of biotechnology express concern that biotechnology companies do not consider the impact of their technologies on the planet and on human life. Some concerns are unfounded and based on fear rather than scientific data, for example that genetically modified food will genetically modify the person eating it. However, biotechnology has real ethical implications that should be considered in the context of real scientific data.

Biotechnologists themselves appreciate the power of the technology. When the implications of DNA technology first became evident, the National Academy of the Sciences organized a conference in Asilomar, CA, in 1975. The participants wrote principles and guidelines for conducting recombinant DNA experiments to minimize biohazards that are generated during the experiments. The conference organizers also brought the implications of the technology to the public's attention to encourage discussion. Today, the ethical implications of new technologies are still central to all scientific discussion. These ethical implications range widely from the impacts of altering genetic traits in humans using CRISPR technology to the effects on human health and the environment of selling genetically engineered salmon.

Where would you draw the line? Would you accept cancer therapy with a drug made from a genetically engineered virus? Would you drink milk from a cow treated with a recombinant growth hormone? Would you eat genetically modified fish? Would you clone your pet? Would you genetically modify a pre-implantation embryo to fix a genetic disease? Would you clone yourself? Consider these questions as you learn more about the power of biotechnology. Use the knowledge you gain in this course to make informed decisions in your personal and professional life.

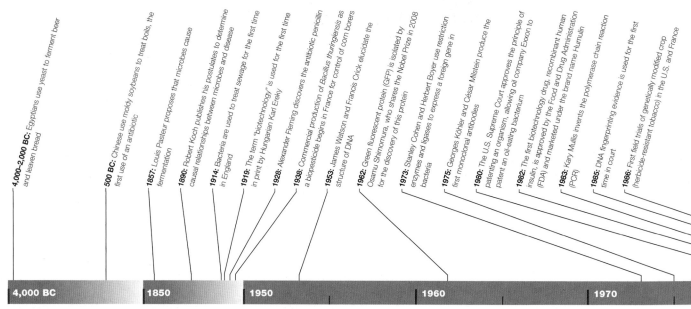

Figure 1.2. **Timeline of biotechnology**. Note the rapid expansion of biotechnology in the last 150 years.

The Biotechnology Toolkit

Biotechnology incorporates a set of tools and techniques to help solve a multitude of problems. These tools can be biological, chemical, instrumentation, or software. Some biotechnology companies, such as Bio-Rad Laboratories, are in the business of making new tools and techniques for researchers and businesses. Advances in biotechnological techniques can also be made by finding new ways to use or improve existing tools. The activities in this book introduce many of the tools and techniques used in biotechnology.

The biotechnology toolkit can include whole cells or molecules, such as DNA, RNA, and proteins that are obtained from nature. Scientists modify these cells and molecules to perform new tasks. Enzymes that cut, reattach, and edit DNA allow scientists to transfer DNA from one organism to another so that the recipient organism can perform a new and useful task. For example, the human insulin gene was inserted into bacteria to give the bacteria the ability to make human insulin. The insulin was subsequently purified. The process of manipulating genes is called **genetic engineering**. The genetic engineering and purification of insulin is used as therapy for diabetes (see Chapter 7). Genetically engineered bacteria and eukaryotic cells are biological tools that are often used as factories to produce novel proteins (**recombinant proteins**).

Chemistry is important in biotechnology, as many tools rely on chemical interactions. Chemical tools include **chromatography** resins that bind proteins based on particular properties of the proteins. This allows them to be purified from other cellular components. For example, pharmaceutical companies use chromatography resins to purify biological drugs, like insulin, produced in bacteria or other organisms.

Biotechnology also relies on laboratory instruments to perform and analyze biotechnological procedures. Some instruments measure, while others perform other functions. For example, **spectrophotometers** measure the amount of light absorbed by a solution to allow accurate quantitation of proteins, DNA, and even cells like bacteria. **Thermal cyclers** rapidly heat and cool tubes of DNA and enzymes, thereby enabling rapid DNA replication by a technique called the polymerase chain reaction (PCR; see Chapter 6). Among many other applications, PCR is used in forensic laboratories to fingerprint DNA.

Biotechnological tools and techniques are constantly being improved and applied in new and exciting ways to help solve the problems of humanity (see Figure 1.2).

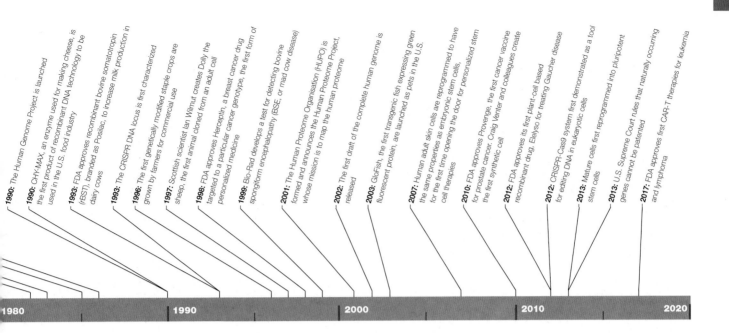

1990: The Human Genome Project is launched

1990: CHY-MAX, an enzyme used for making cheese, is the first product of recombinant DNA technology to be used in the U.S. food industry

1993: FDA approves recombinant bovine somatotropin (rBST), branded as Posilac, to increase milk production in dairy cows

1993: The CRISPR DNA locus is first characterized

1996: The first genetically modified staple crops are grown by farmers for commercial use

1997: Scottish scientist Ian Wilmut creates Dolly the sheep, the first animal cloned from an adult cell

1998: FDA approves Herceptin, a breast cancer drug targeted to a particular cancer genotype, the first form of personalized medicine

1999: Bio-Rad develops a test for detecting bovine spongiform encephalopathy (BSE, or mad cow disease)

2001: The Human Proteome Organisation (HUPO) is formed and announces the Human Proteome Project, whose mission is to map the human proteome

2002: The first draft of the complete human genome is released

2003: GloFish, the first transgenic fish expressing green fluorescent protein, are launched as pets in the U.S.

2007: Human adult skin cells are reprogrammed to have the same properties as embryonic stem cells, for the first time opening the door for personalized stem cell therapies

2010: FDA approves Provenge, the first cancer vaccine for prostate cancer. Craig Venter and colleagues create the first synthetic cell

2012: FDA approves its first plant-cell based recombinant drug, Elelyso for treating Gaucher disease

2012: CRISPR-Cas9 system first demonstrated as a tool for editing DNA in eukaryotic cells

2013: Mature cells first reprogrammed into pluripotent stem cells

2013: U.S. Supreme Court rules that naturally occuring genes cannot be patented

2017: FDA approves first CAR-T therapies for leukemia and lymphoma

1980　　　1990　　　2000　　　2010　　　2020

1.2 Who Uses Biotechnology?

Many industries and scientific disciplines use biotechnology (see Figure 1.1). Currently the main users of biotechnology are the life science research, healthcare, agricultural, food, and manufacturing industries. Biotechnology is also important in national defense, research into human origins (anthropology), forensics, and nanotechnology.

Life Science Research

The dividing line between basic life science research and biotechnology research is hazy. The goal of any basic research is to understand how a system works, and basic life science research is driven by curiosity to understand more about life. The goal of biotechnology research is to develop solutions that help solve problems; this type of research is called applied research. For example, if a new virus that infects humans is discovered, basic research scientists investigate the normal mechanism of action of the virus, its nucleic acid sequence, and the proteins that are important for its virulence. Applied scientists take that information and develop drugs and a vaccine to combat the virus. Basic life science researchers use many biotechnological tools and techniques in their investigations, and many biotechnological tools and techniques are based on their discoveries. These tools include **cell culture** and cell expression techniques to understand how cells function, grow, divide, and die. Researchers also use recombinant DNA techniques (see Chapter 4) to discover how genes and proteins function.

-Omics and Systems Biology

Life science research has shifted from studying how a single gene or protein functions to investigating how whole cells, organisms, populations, or ecosystems function at the molecular level. The term **systems biology** describes this approach. This shift to systems biology at the molecular level has become possible only because of advances in the biotechnology toolkit available to researchers. The disciplines using this approach have been termed "-omics":

- **Genomics** investigates the whole genome, the full complement of DNA in a cell

- **Proteomics** studies the proteome, the entire protein complement of a cell or organism

- **Transcriptomics** is the study of the transcriptome, every RNA transcript expressed in a cell or organism

- **Metabolomics** investigates the metabolome, all metabolites in a cell or organism present at a specific time

- **Microbiomics** investigates the microbiome, all microorganisms living in a place at the same time

The theory behind systems biology is that the reaction of a cell, organism, or ecosystem to an input (for example, a drug therapy) can be predicted only when all these factors have been considered.

Each of these -omics generates vast amounts of data, and one of the major challenges of the 21st century is finding ways to sort and analyze these data. Using information technology for biological applications is called **bioinformatics** (see Chapters 6 and 7), a rapidly growing area of biotechnology. Bioinformaticians find ways to gather, store, sort, and analyze data. They use the data to make computer models to predict cellular and organismal behavior.

Healthcare

Healthcare related products make up the largest segment of the biotechnology industry. Currently the biotechnology companies that generate the most revenue are those that develop biopharmaceuticals, medical devices, and diagnostic tools and tests.

Biotechnology and pharmaceutical companies use biotechnology methods and techniques to discover new drugs in a process called **drug discovery**. They use those same tools to develop and manufacture drugs, an expensive and time-consuming process. Very few drug candidates make it to the market. Most candidates are rejected before they are tested on people in clinical trials. Many more are rejected during the phases of clinical trials as companies narrow down the best drug to combat a particular disease. From start to finish, **drug development** can take 10–15 years and cost over $2.5 billion for each approved drug. This cost includes the research into all the drug candidates that were rejected.

Drug Discovery, Development, and Testing

Drug discovery begins with the hunt for molecules that could treat a disease. This often involves first identifying a target that has a role in causing or affecting the development of a disease or its symptoms. Most often, these targets are genes or proteins. Drugs are then developed to inhibit a particular gene or protein. Alternatively, an entire network of proteins may be involved in a disease, in which case all the proteins involved become targets for drug therapy. To minimize the risk of spending years and millions of dollars pursuing a drug that won't work, scientists first work to confirm the role of the target in the disease process.

Once a target has been identified, many biotechnology and pharmaceutical companies use automated high-throughput tools and techniques to screen drug candidate molecules. High throughput means that hundreds or thousands of candidates can be investigated all at once, rather than one by one. For example, **microarrays** (see Chapter 6) and microfluidic arrays can simultaneously screen thousands of drug candidate molecules. The candidates that bind to the target can be quickly identified and investigated further to determine which bind best or interfere with the target's activity. This helps narrow the candidates to find a lead compound with the best potential for becoming a drug.

Once a lead candidate has been identified, it goes through extensive testing and development. Scientists use a variety of techniques to analyze and modify the original drug candidate through chemical engineering to tweak its design and improve its performance. They also use various techniques to determine whether the drug candidate is likely to cause dangerous side effects. Finally, lead candidates are tested in humans for safety and efficacy in a series of clinical trials (see Figure 1.17 and Regulation of Products in Healthcare, later in this chapter).

Protein-Based Drug Production

Most biological drugs, sometimes called biopharmaceuticals, are proteins which are larger and more complex than traditional small molecule drugs. Because they are made using living cells, they can be quite expensive to produce. When developing the production process for a biological drug, companies optimize the production conditions to produce the most active protein in the most cost-effective way.

Some proteins, like insulin, can be made in bacteria, such as *Escherichia coli*. Though bacteria grow quickly, they have some limitations. Bacterial cells do not produce a lot of protein, and growing bacterial cell cultures in large production tanks called **bioreactors** can be expensive. Using bacterial cells, therefore, may not be cost-effective. In addition, proteins may have structural features that bacterial cells cannot create. For these proteins, companies must use yeast, plant, or animal cell cultures, or even whole plants or animals (see Figure 1.3).

When the growth process is done, the protein is extracted from the cells and purified using a combination of techniques (see Chapter 7). Finally, the protein is tested one last time, and then packaged and shipped to the customer. Regardless of the methods used, producing any drug involves quality control activities at each step of the way. This ensures that both the drug and production process meet the specifications that were reviewed and approved by the U.S. Food and Drug Administration (FDA) for safety and effectiveness.

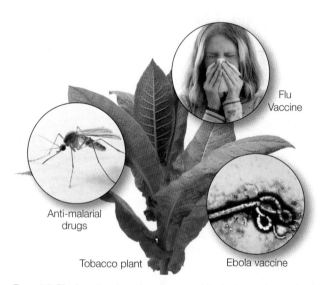

Figure 1.3. **Biopharming.** A number of crops and livestock are being used as living factories for a variety of drugs. As an example, the experimental Ebola drug ZMapp is being made in tobacco plants, as are the precursor molecules for anti-malarial drugs and the flu vaccine.

Personalized Medicine

A new era of **personalized medicine**, or pharmacogenomics (pharmaceuticals based on genomics), is revolutionizing healthcare. Like having a dentist make a cast of a patient's teeth for a retainer designed specifically for that individual, biotechnological advancements will enable patients to submit tissue samples and receive a treatment regimen designed specifically for them.

Individual characteristics, such as race/ethnicity, gender, and family history all affect responses to a drug and determine how well drugs and therapies work. Some drugs have been developed and are prescribed depending on these factors. But as sequencing individual human genomes becomes more cost-effective, there will soon come a time when a drug regimen is determined by the genotype of the patient or the tumor being targeted.

One of the first such drugs is trastuzumab (trade name Herceptin, see Figure 1.4). Herceptin works only in cancers that have a mutation in a specific gene, *HER2*, that causes the HER2 protein to be produced at levels that are too high. Cancer cells are screened to determine whether they are producing higher levels of HER2 protein before patients are given the drug. The decision to administer Herceptin is based on the presence of this single gene. Soon, though, treatments will be based on the entire genome of a patient.

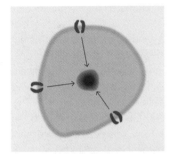

Normal Cell
A normal number of HER2 receptors send signals to the cell to grow and divide normally.

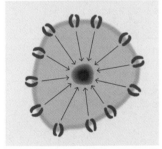

Cell with *HER2* Mutation
Too many HER2 receptors send signals causing the cell to grow and divide too quickly.

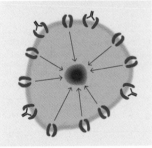

Cell Treated with Herceptin
When Herceptin binds to HER2 receptors it blocks them from sending signals to the cell to grow and divide.

Figure 1.4. **Herceptin, a cancer-specific drug.** The HER2 receptor signals a cell to grow and divide. Some people have mutations in the *HER2* gene that cause cells to create too many HER2 receptors. which can cause cells to grow and divide too quickly and develop into cancer. Herceptin attaches to HER2 receptors and prevents them from sending signals to the cell to grow and divide.

Many diseases and disorders are caused by genetic mutations. Gene editing techniques like CRISPR (clustered regularly interspaced short palindromic repeats, see Chapter 4) are in use now and may one day allow scientists to edit a patient's DNA to correct the mutations that cause diseases such as hemophilia, sickle-cell anemia, blindness, and cystic fibrosis.

Immunotherapy is another type of personalized medicine. It uses parts of a person's immune system to fight diseases such as cancer. One particularly promising approach, chimeric antigen receptor (CAR) T-cell therapy, involves collecting and using a patient's own immune cells to treat their cancer (see Chapter 8, Biotech in the Real World). The patient's T-cells are engineered to recognize and target specific types of cancers. The therapy is specific to the patient's cancer cells, so theoretically does not kill off healthy cells the way traditional chemotherapies and radiation therapies do. In 2017, two CAR T-cell therapies were approved by the FDA, one for children with acute lymphoblastic leukemia (ALL), and the other for adults with advanced lymphomas.

In addition, research is being conducted to explore using animals to produce organs for transplant into humans. The donor animal's genome would be altered to produce human proteins personalized to the individual needing the transplant. This would prevent the recipient from rejecting the organ, as well as eliminate the need for immunosuppressant drugs. With gene-editing tools such as CRISPR, scientists can remove immune-stimulating sugars from the surface of cells, which has been done using pig cells, introduce human genes that regulate blood coagulation to prevent dangerous clots, and edit out viral sequences that could infect the patient.

Clinical Diagnostics
In addition to developing drugs, biotechnology is used to diagnose diseases and other medical conditions. Doctors' offices, hospitals, and medical laboratories often use diagnostic tests developed by biotechnology companies for these purposes.

Many diseases are caused by **microorganisms** such as bacteria. Microbiological techniques can be used to identify the cause of an infection, and this allows the appropriate therapy to be prescribed. But traditional diagnosis of microbial infections takes days or even weeks, since bacteria need to be cultured before they can be identified. Biotechnology has helped to speed up diagnoses. In some cases, specialized microbial media have been developed to identify disease-causing bacteria quickly. In other cases, entirely new tests, often based on PCR or antibodies, have been developed that can detect and identify bacteria within hours or even minutes.

Antibodies are proteins that specifically recognize and bind to other proteins. They are commonly used in diagnostic tests to identify proteins that are indicators of medical conditions. Examples of antibody-based diagnostic tests include dipstick tests that are used in laboratories, doctors' offices, and occasionally by the general public (for example, pregnancy tests). Dipstick tests show the presence or absence of a protein involved in a medical condition using a urine, saliva, or blood sample. Analysis of antibody-based tests may also require a medical laboratory to have special instrumentation. Laboratory tests can determine the actual level of a protein (or multiple proteins) that may change when someone has a certain medical condition, which helps diagnose the stage, or seriousness, of the condition. Figure 1.5 shows a dipstick test that is used in a laboratory to test for HIV.

Figure 1.5. **Bio-Rad's Genie II HIV1/HIV2™ assay.** A serum sample is added to the circular area. If the serum is HIV positive, proteins will bind to the reagents in the strip, and colored stripes will develop in the viewing zone, depending on which proteins are present. This assay works similarly to a pregnancy test.

Agriculture

Agriculture uses biotechnology to enhance the growth and health of crops and animals. Since 1996, **genetically modified crops (GM crops)** with properties such as pest-resistance, herbicide-resistance, and extra nutritional content have been farmed to increase yields, decrease pesticide use, and increase food value (see Figure 1.6). Many of these traits are being "stacked" on top of one another by interbreeding plants to produce varieties with multiple genetically engineered properties. For example,

SmartStax corn has eight different genetically engineered traits. Whereas in 1997 only 4% of cotton farmed in the U.S. (by acreage) was genetically modified, by 2017 this number had risen to 92%, most of it modified for herbicide resistance.

Biotechnology has also affected how farmers raise livestock. Agricultural scientists found that, by treating dairy cows with an injectable growth hormone called recombinant bovine somatotropin (rBST; trade name Posilac), they could increase milk production, which increases farmers' yields and reduces the price of milk. Advances in technology that affect food production have yet to gain full consumer acceptance. For example, though scientific tests have shown no significant differences between milk from cows treated with rBST and non-rBST-treated cows, many consumers choose to buy milk from untreated cows.

Biotechnological practices often go hand in hand with government legislation. Antibiotics were developed to treat bacterial infections, but antibiotic-resistant bacteria are now a major healthcare concern. Fifty years ago, adding low levels of antibiotics to livestock feed was found to increase the animals' growth. This generated higher yields and reduced the price of meat, and the practice was approved by the FDA. Now, however, there are fears this practice is increasing the emergence of antibiotic-resistant

	Genetic Trait			
	Herbicide tolerance	Insect resistance	Other	Uses
Field corn	✔	✔	Drought tolerance	Animal feed, fuel, high-fructose corn syrup, cooking oil, starch, cereal and other food ingredients, alcohol, industrial uses
Sweet corn	✔	✔		Food
Canola	✔			Animal feed, cooking oil, biodiesel
Soybean	✔	✔		Animal feed, aquaculture, oil, fatty acid, biodiesel, food, lecithin, pet food, adhesives and building materials, printing ink, industrial uses
Sugar beet	✔			Sugar, animal feed
Cotton	✔	✔		Fiber, animal feed, cooking oil
Alfalfa	✔			Animal feed
Rainbow papaya			Disease resistance	Food
Summer squash			Disease resistance	Food
Apple			Non-browning	Food
Potato			Reduced bruising and black spot, blight resistance, non-browning, low acrylamide	Food

Figure 1.6. **Commercially available genetically modified crops.** In the U.S., eleven genetically modified crops are available. These crops have been engineered primarily for pest resistance and herbicide and drought tolerance. The two newest additions, apple and potato, were engineered for non-browning and other traits.

bacteria. The concern is that if animals are treated with an antibiotic over time, the bacteria living in them will become resistant to that drug. In addition, if people ingest the resistant bacteria through improperly cooked meat, they may not respond to antibiotic treatment if they become ill.

The use of subtherapeutic doses in animal feed or water to promote growth and improve feed efficiency has been banned in Europe since 2006. In 2007 and 2009, the Preservation of Antibiotics for Medical Treatment Act bill was introduced by U.S. senators proposing to phase out nontherapeutic use of antibiotics in animal feed. The practice was eliminated in the United States effective January 1, 2017.

Fisheries and the aquaculture industry are also using biotechnology. In 2010, controversy surrounded the FDA when it was deciding whether genetically engineered salmon could be sold to U.S. consumers. The genetically engineered salmon express growth hormones from different fish species that make them grow faster (see Figure 1.7). Many people are concerned about the impact of these fish when eaten or if they gain access to the wild.

Biotechnological research is also being conducted on the environments and populations of wild and farmed fish and shellfish to help conservation, management, and breeding.

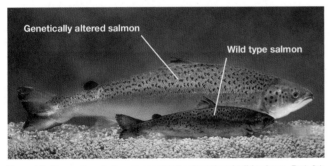

Figure 1.7. **Comparison between genetically altered and wild salmon.** Both fish are shown at 18 months of age. Image courtesy of AquaBounty.

Food

Biotechnology was first used in the food industry. For thousands of years, humans have been selecting microorganisms for fermentation to brew beer, bake bread, ferment soybeans, and curdle milk for yogurt and cheese (see Figure 1.8). In the modern era, yeasts and bacteria are genetically engineered to produce exactly what is required to increase the efficiency and yield of food processing. For example, bacteria important in the fermentation of dairy products used to be susceptible to infection by a virus. Infected batches would need to be discarded, which decreased yields. Scientists have now genetically engineered bacteria that are resistant to the virus, increasing yields of dairy products.

The food industry also uses biotechnology to produce key enzymes involved in food production. As an example, rennet is a complex of chymosin and other enzymes used to coagulate milk in the manufacture of cheese. Traditionally, rennet was extracted from the

Figure 1.8. **Ancient bakery in Pompeii.** Yeast would have been used here in the production of bread and other baked goods.

fourth stomach chamber of young, unweaned calves as part of veal production. As veal production declined in the mid-twentieth century, however, cheese makers sought other ways to coagulate milk. Though several sources of enzymes, including plants, fungi, and microbes, can substitute for animal rennet, the chymosin from these sources does not have the exact same properties of animal rennet, and so do not produce cheeses of the same quality.

In the late 1980s, scientists at the biotechnology company Pfizer isolated the calf chymosin gene and inserted it into the bacterium *Escherichia coli* (*E. coli*). The genetically modified *E. coli* strain produced large amounts of the mammalian enzyme. In 1990, the FDA approved Pfizer's chymosin for human consumption, as it was identical to the chymosin found in animal rennet and free of genetically modified bacteria. It was the first artificially produced enzyme to be registered and allowed by the FDA. This genetically engineered chymosin has a higher purity than animal rennet and produces more consistent cheese with less waste than an equivalent amount of crude calf stomach extract. By 2008, 80% to 90% of commercially made cheeses in the U.S. and Britain were made using genetically engineered chymosin.

The food industry also uses biotechnology to improve the quality or nutritional content of basic food staples (see Table 1.1). In this respect, the food industry is intimately linked to agricultural biotechnology. For example, golden rice is a type of rice that has been genetically engineered to express beta-carotene, a precursor to vitamin A, to help reduce vitamin A deficiency. Vitamin A deficiency leads to blindness in more than 250,000 children each year in the developing world. Although golden rice may seem to be a great solution to a global problem, it needs governmental approval before it can be sold for human consumption, and approval can take a long time. Golden rice was developed in the laboratory in 1999 but was still navigating governmental regulatory systems as of 2017.

Genetically modified animals with added nutritional value are also being developed, such as pigs with high levels of omega-3 fatty acids. However, as of 2017, no genetically modified animals have been approved for sale as food in the U.S. Though the AquaBounty salmon (see Figure 1.7) was approved by the FDA in 2015, within weeks the FDA issued a ban on the import and sale of genetically engineered fish until labeling guidelines were established (the salmon is commercially available in Canada).

Industrial Manufacturing

Biotechnology has helped to improve efficiency, increase yields, and reduce environmental impacts from manufacturing industries. Enzymes and microorganisms are used in place of chemicals and energy-intensive processes. For example, in the textile industry, stone-washed jeans used to be washed with stones and acid, which was expensive because it damaged the machinery. Now an enzyme called cellulase softens the denim. Using the enzyme reduces the environmental impact of the harsh chemicals and reduces the cost of the machinery.

Many manufacturing processes require conditions that are not optimal for most enzymes, such as high or low heat, high pressure, or highly acidic, alkaline, or salty conditions. However, some microorganisms thrive in harsh conditions like hot springs or salt marshes. Research into these organisms helps identify enzymes that can be used in traditionally suboptimal conditions. A familiar

example is the development of laundry detergents that work in the cold wash cycle (see Table 1.1). These detergents contain enzymes derived from microorganisms that thrive in cold water. As a result, the enzymes break down fat and protein stains without requiring heat, thereby reducing the demand for energy to heat the water. Petrochemicals are used to manufacture plastics, and plastic waste creates an environmental problem because plastics do not degrade. Plant-based plastics are now being produced as a more environmentally friendly alternative. Many large chemical companies have joined forces with smaller biotechnology companies to find enzymes that convert plant sugars into useful polymers to replace traditional plastics. Biological plastics have become competitive in price and performance with plastics made from petrochemicals. In addition, biological plastics take fewer resources and energy to produce, and reduce landfill waste because they are biodegradable.

Biofuels

Biotechnology is being used to develop alternative energy sources. Fossil fuels are limited and are having a detrimental effect on the environment. **Biofuels** are fuels that can be derived from living organisms, such as plants, algae, certain fungi, and even animal fats, which are renewable sources of energy.

Bioethanol is one type of fuel, and it can be produced by alcoholic fermentation of the sugars stored within plants and algae using microbes like yeast. Yeast can produce large amounts of bioethanol, has a high ethanol tolerance, and can ferment (break down) a wide range of sugars. However, it cannot completely break down some of the sugars stored in plant materials, for example, those stored in the cell walls or as starch. This means bioethanol production must involve commercially available enzymes, which are often produced using other genetically modified microbes. Research continues on developing yeast strains that can produce these enzymes themselves.

Biodiesel is another biofuel produced from vegetable oils. In the U.S., most biodiesel is made from soybean oil; however, canola oil, sunflower oil, recycled cooking oils, oils from algae, and even animal fats are also used. Genetically modified soybean and canola crops are often used for this purpose (see Figure 1.6) because they offer higher yields. Most diesel engines can run on biodiesel without needing any modifications.

As part of the Energy Independence and Security Act, the U.S. government dedicated billions of dollars to biofuel development. Many government laboratories are trying to find natural enzymes and to genetically engineer enzymes that can make a cost-effective biofuel. The biofuel industry has many other hurdles to overcome. For example, the amount of land and water that would be required to grow enough plants or algae to meet the world's energy needs using current biofuel technology is a major concern.

Table 1.1. **Industrial enzymes produced with biotechnology.**

Business	Applications	Examples
Detergents	Household cleaning and laundry	Dental cleansers, cold water detergents, stain removers
Textiles	Treatments for denim, silk, leather, etc.	Stonewashing denim, polishing silks, softening leather
Food processing	Baking, brewing, juice processing, milk coagulation	Maintaining moisture in bread, clarification of juices and beer, fermentation of beer and wine, syrup making, cheese and yogurt making
Pulp and paper	Reduction of costs and improvement of quality	Starch conversion, pitch control, bleaching, de-inking, sticky notes, and slime control

Mining

Biotechnology has also improved yields in mineral mining. Microorganisms such as *Thiobacillus ferrooxidans* that get energy by oxidizing sulfur compounds and iron are being used to leach minerals such as copper from mine waste piles (tailings) or low-grade ore in a process called bioprocessing. The copper can be collected in solution and purified. Mining companies now generate 25% of the world's copper through bioprocessing.

Pollution Monitoring and Waste Management

Controlling pollution is essential to maintaining our health and the health of the environment. Monitoring pollution levels allows appropriate personnel to be alerted and action taken quickly to control the spread of pollutants when needed. **Biosensors** are biotechnological instruments that convert the action of a biological molecule or organism into an electrical signal. Biosensors have been developed that use antibodies, enzymes, and PCR to detect pollutants in the air, soil, and water.

If pollutants are released into the environment, biotechnological tools can help with cleanup efforts. The use of organisms to convert hazardous waste into a less hazardous form is called **bioremediation** (or phytoremediation, if plants are used) (see Figure 1.9). This is not a new idea. Bacteria were first used to treat sewage in the early part of the 20th century. However, modern techniques are helping to identify new microorganisms and enzymes that have either new functions or can complete the same task more efficiently. The first patent issued for any organism was for an oil-eating bacterium (*Pseudomonas putida*) that could be used to help clean up oil spills.

Two forms of bioremediation can be used. First, new organisms (that may be genetically engineered) can be introduced to the site for cleanup. These would be designed to die off once all the waste has been consumed. Second, if naturally occurring microorganisms can break down the pollutant, their growth can be stimulated by providing special nutrients to a contaminated site. When applied to oil spills, bioremediation that encourages the growth of natural oil-eating bacteria in the environment appears to be more successful than introducing foreign or genetically engineered bacteria. In response to the BP oil spill in the Gulf of Mexico in 2010, a bioremediation company called Evolugate used naturally occurring bacteria from the gulf, actively selecting for mutations that improved their oil-eating abilities (see Figure 1.9). Since then, scientists have been actively trying to figure out how microbes gobble up the >1,000 types of chemical compounds found in oil. They have found that some microbes break down certain compounds better than others, but all can be made to work together as a bacterial community to consume oil.

Microorganisms have also been genetically engineered to help detoxify radioactive waste. For example, the rod-shaped bacterium *Deinococcus radiodurans* has been genetically engineered for bioremediation of solvents and heavy metals in a radioactive environment. It can survive all sorts of extremes,

Affecting Government Policy

Lake Apopka in Florida was once a popular spot for fishing. Following decades of pollution, however, the lake was known as one of Florida's most polluted bodies of water. In 1980, Tower Chemical Company, a local pesticide manufacturer, improperly dumped DDE, a breakdown product of DDT, into the lake, which caused the fish and alligator populations to plummet. In 1981, the Environmental Protection Agency (EPA) opened an investigation into the pollution of the lake, which was then designated as a Superfund cleanup site.

During the 1980s Dr. Louis Guillette, a University of Florida professor and scientist, was studying the reproductive biology of lizards, and was invited to join the effort in understanding the reproductive biology of alligators. His research brought him to Lake Apopka, where he discovered that, despite years of cleanup, male alligators in the lake showed a 25% reduction in the size of their genitals and

very low levels of testosterone when compared to alligators from nearby Lake Woodruff.

After learning that some environmental contaminants, like DDE, can act like hormones, Guillette demonstrated through laboratory experiments that even trace amounts of DDE, amounts not high enough to cause general toxic effects, could affect sexual development. He found DDE mimics the female reproductive hormone estrogen and causes male alligators to become "feminized." Such chemicals are called endocrine disruptors because they interfere with the body's endocrine system, the system that produces hormones that regulate metabolism, growth and development, sexual function, and reproduction.

Armed with this knowledge, Guillette and other scientists argued for changes to government policy. Since endocrine disruptors can also be present in food, plastic water bottles, and cosmetics, they argued, these products should be screened for trace levels of the disruptors. After all, humans (like alligators) are at the top of the food chain and so are exposed to cumulative effects. Guillette testified before congress in 1993 that, "Every man sitting in this room today is half the man his grandfather was. And the question is: are our children going to be just half the men we are?" referring to studies indicating that the sperm count in some populations of modern men is 50% lower than two generations ago.

In 1996, congress passed the Food Quality Protection Act, largely as a result of these lobbying efforts. The law requires the Environmental Protection Agency to test pesticides for endocrine disrupting chemicals. As for the alligators of Lake Apopka, their populations have still not fully recovered.

Figure 1.9. **Bioremediation of oil spills.** Many marine microbes that feed on oil and refined oil products are abundant where oil seeps naturally through fissures in the oceanic floor into seawater, as happens in the Gulf of Mexico. After an accidental oil spill into open water, the numbers of oil eating microbes increase. Oil eating bacteria include *Alcanivorax borkumensis*. As the oil is fragmented into droplets by wave action, *Alcanivorax* cluster on the surface of droplets to break down the oil.

including radiation, cold, dehydration, vacuum, and even acid. It has been listed as the world's most radiation-resistant life form by Guinness World Records. It has been genetically modified to make a protein that converts highly toxic ionic mercury to less toxic elemental mercury. It is used to help reduce the toxicity of mercury in radioactive waste. Some of these bioremediating bacteria may even generate useful products. For example, bacteria that degrade sulfur liquor, a waste product of the paper manufacturing industry, produce methane that can be collected and used for fuel.

Conservation

Wildlife conservationists have embraced information that can be generated using biotechnological techniques such as **DNA profiling**. In conservation biology, it is important to understand as much as possible about the population being studied. Endangered animal populations can be genetically mapped using biotechnological techniques. For example, a genetic variability map of African elephants can track the origins of contraband ivory and identify the elephant populations being hit by poachers. The genetic variability map can be used by law enforcement officials to catch poachers and to protect the populations of elephants in greatest need.

Investigations into biodiversity are being taken to another level through a technique called **DNA barcoding**. As the name suggests, DNA barcoding has the goal of making species identification as easy as scanning items in a supermarket. A DNA barcode is a short sequence of an organism's DNA that can be used for identification. Through advances in DNA sequencing, information technology, and bioinformatics, a unique genetic barcode can theoretically be created for each organism (see Figure 1.10). DNA barcoding helps researchers identify new species, provides an easy way to look at genetic diversity in any geographical location, and helps taxonomists understand relationships among species (see Chapter 6).

Biodefense

One goal of government laboratories is to use biotechnology to protect countries from terrorist attacks involving biological weapons. Researchers look for ways to neutralize **pathogens** that could be used in a bioterrorist attack. In addition, biosensors that use antibodies or enzymes are being developed to detect pathogenic bacteria or nerve agents in the air or water supply in real time, allowing a swift response should an attack occur.

Forensics

The criminal justice system has been revolutionized by biotechnology. The use of DNA evidence has become routine in court cases, and yet the first DNA fingerprinting was performed only in 1986 (see Chapter 4). DNA fingerprinting now uses PCR to

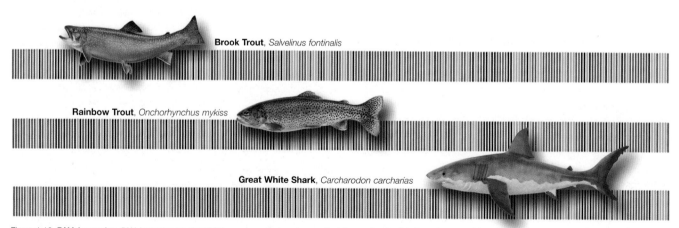

Figure 1.10. **DNA barcodes.** DNA barcodes are short DNA sequences that can be used to tell even closely related species apart. Here the first 243 bases of the barcoding region of brook trout (*Salvelinus fontinalis*) rainbow trout (*Oncorhynchus mykiss*), and great white shark (*Carcharodon carcharias*) are shown. Each color represents a different nucleotide, adenine, thymine, cytosine, or guanine. There are fewer differences between the barcode sequences of the closely related rainbow trout and brook trout than between either trout and the great white shark.

amplify regions of DNA that are different among individuals to create a pattern that is unique to every individual, except identical twins. DNA evidence has also helped to free people wrongly convicted of crimes. The Innocence Project, a non-profit legal clinic in the U.S., has used DNA evidence to overturn more than 350 convictions (see Figure 1.11).

Figure 1.11. **Exonerated inmates released from Georgia prisons**. Harvard Law School, March 25, 2007. From left to right, front row: Clarency Harrison (2004), Robert Clark (2005), and Pete Williams (2007), who collectively spent 57 years in prison. Back row: members of the Georgia Innocence Project: Lisa George, Greg Hampikian, Amy Maxwell, and Cliff Williams. Photo courtesy of Greg Hampikian.

Human Origins

DNA fingerprinting has also helped anthropologists studying human origins (see Figure 1.12) and archaeologists studying early human civilizations. To better understand how humans lived, archaeologists now perform genetic analyses of human refuse found in archaeological digs to identify the plants and animals humans used and ate. Genetic anthropologists are mapping the ancestry of the human race using DNA sequences to determine where different human populations came from and who their closest relatives are. The Genographic Project, initiated by the National Geographic Society, endeavors to create a detailed world map of human migration patterns. The project invites anyone in the world to pay $200 to have his or her genotype determined and uses the money to fund the project to genotype indigenous populations around the world. At the time of writing, nearly 1 million participants across 140 different countries had signed up for the project.

DNA fingerprinting is also in wide use for people interested in their personal ancestral origins. Companies like Ancestry, 23andMe, and MyHeritage offer kits for home use that can return information about the regions of the world a person's ancestors came from (see Personal Genetic Information in Chapter 4).

Nanotechnology

Nanotechnology is the manipulation and use of materials at the atomic and molecular scales to perform functions that cannot be performed by larger particles or materials. When nanotechnology involves biology, it is termed **nanobiotechnology**.

Figure 1.12. **Collecting DNA samples**. Members of the Genographic project collect DNA samples from participants all across the globe including Tubu tribesmen in Gouro, Chad, Africa (pictured). Using DNA sequencing, scientists and anthropologists are able to map historical human migration. Photo credit: David Evans/National Geographic/Getty Images.

Biological molecules, which are all nanoscale in size, are used in nonbiological contexts in many industries. For example, we may one day have biological computers that process and store information using DNA and enzymes instead of silicon chips. Due to its complementary base-pairing ability, DNA is being investigated as a structural material for use in nanotechnology applications. Although DNA usually forms a double helix, it can also form different shapes. This property is being investigated by nanotechnologists who plan to use DNA as a structural scaffold to hold other molecules in place.

Advances in nanotechnology have enabled the development of targeted therapies that can be delivered directly to diseased tissue and greatly reduce side effects. Some drugs, like those used in chemotherapy, have serious side effects such as nausea and hair loss. This is because the drugs treat the whole body and not just the diseased tissue. Rexin-G, a drug that targets pancreatic cancer, has been developed to be delivered on **nanoparticles**, particles that are nanometers in diameter. Rexin-G is a genetically engineered virus designed to seek out and kill cancer cells that have metastasized or escaped the main tumor. This drug was so effective in clinical trials that the FDA, which regulates clinical trials, gave it special status in 2010 to help speed its approval for clinical use. Rexin-G is an example of nanotechnology and biotechnology being combined to create the emerging field of nanobiotechnology.

The unique properties of proteins are also being used on the nanoscale. For example, aquaporin, a protein that naturally regulates the flow of water in and out of cells, is used in artificial membranes to ultrapurify water for the semiconductor industry.

Biotech In The Real World

Bio-Rad: Then and Now

Biotechnology companies often start with an idea. In 1952, a group of students from the University of California, Berkeley were at a bridge game joking about products that were not on the market that should be. One student said, "It's too bad nobody is selling tobacco mosaic virus." At the time tobacco mosaic virus (TMV) was being used as a model system to study viruses. Researchers, including biochemist Alice Schwartz, part of the husband-and-wife team that founded Bio-Rad Laboratories, had to take on the laborious task of producing their own TMV, which took many days to prepare. David Schwartz, who had recently graduated with a degree in chemistry, questioned why no one was manufacturing it. Within a matter of months, he and Alice were making TMV out of a World War II Quonset hut in Berkeley, CA. This was the start of Bio-Rad Laboratories. TMV did not really take off but other products did, including chromatography resins and stable isotopes.

Keeping a start-up company afloat requires wearing many hats. A photographer for Life magazine shot the photo at left in 1955, commenting that David Schwartz "still mows the grass" even though he was the company president. David was responsible for sales and marketing, and Alice developed the products. David was president, CEO, and chairman from the time the company was incorporated until 2003. He remained chairman until he died in 2012 at age 88. Alice Schwartz has been on the board of directors since 1967. Their son, Norman is the president and CEO today. When the company incorporated in 1957, its annual sales were $25,000. In 2017 Bio-Rad revenues exceeded $2.1 billion.

Today, Bio-Rad operates globally in two industry segments: clinical diagnostics and life science. With more than 8,250 employees, Bio-Rad serves over 150,000 customers annually. Bio-Rad's Clinical Diagnostics Group develops products for medical screening and clinical diagnostics, including products for diabetes monitoring, blood virus testing, and genetic disorders testing. The Life Science Group develops laboratory instruments, apparatus, and consumables used for research. These products, which are used to separate, purify, identify, analyze, and amplify biological materials, are based on technologies such as electrophoresis, chromatography, and PCR (the type of equipment and reagents that will be used in this course). The products serve as tools to support life science research in universities, government laboratories, and biotechnology and pharmaceutical companies exploring questions about proteins, genes, and cells, and making discoveries related to healthcare and food safety.

In 1996, an educational arm of the Life Science Group, the Bio-Rad Explorer™ Program, was started to advance biotechnology education in high schools and colleges. The program develops high-quality, engaging, and relevant hands-on laboratory activities and this textbook to teach biotechnology skills to students around the world and spark their interest in science and its global influence.

1.3 The Biotechnology Industry

Biotechnology is a major part of the global economy, and it is expected to reach a worth of over $700 billion by 2025.

Biotechnology companies often start as private companies and then "go public" through an initial public offering (IPO) once they have developed technologies or products that have the potential to be profitable. In the U.S. privately held biotechnology companies outnumber public ones.

A brand-new biotechnology company is commonly referred to as a "start-up." Start-ups are usually dynamic companies with energetic, groundbreaking teams. Employees have the potential to gain experience in many different areas due to the small size of such companies. If the technology offered by a start-up is successful, there may be an opportunity to make money from stock options and bonuses. The downside is the lack of job security and long work hours.

What Is a Biotechnology Company?

Biotechnology first emerged as a distinct industry in the late 1970s, with its first products based on recombinant DNA technology. One of the first biotechnology companies was Genentech, which was founded in 1976. Genentech developed a process to make a drug called Humulin, a recombinant human insulin to treat diabetes, that was produced in bacteria. As mentioned earlier, not all biotechnology companies make medical products or drugs. Some, like Bio-Rad Laboratories, develop and sell instruments, reagents, and other tools for research, others sell agricultural technologies and products, etc. Healthcare, however, is the largest segment of the biotechnology industry.

Medical, or pharmaceutical-focused, biotechnology companies are distinct from traditional pharmaceutical companies. Even though pharmaceutical companies use a great deal of biotechnology in their business, they conduct their business and manufacture and market products differently (see Table 1.2). Many biotechnology companies do not manufacture the drugs they develop; instead, they sell drug candidates to larger biotechnology or pharmaceutical companies with expertise in manufacturing biological drugs. Biotechnology companies are sometimes acquired by other companies to introduce new technology and products into their portfolio. For example, in 2009 Roche acquired Genentech for $46 billion and in 2016 Bayer acquired Monsanto for $66 million.

Examples of large public biotechnology companies are Amgen, Gilead Sciences, and Biogen. Based on 2017 revenues, the world's largest pharmaceutical companies were Johnson & Johnson, Roche, and Pfizer, each with revenues of over $50 billion. Small biotechnology firms are the rule rather than the exception — the 2015 OECD report claimed 72% of the biotechnology firms in the U.S. had 50 or fewer employees.

Table 1.2. **Typical differences between pharmaceutical and medical biotechnology companies.**

	Pharmaceutical	Biotechnology
Size	Medium to large	Start-up to medium
Development process	Traditional	Innovative
Expertise	Chemical	Biological
Focus	Drug discovery and development, manufacturing, and marketing	Drug discovery and development, manufacturing, and marketing
Primary Methods	Chemical synthesis	Biological processes
Financing	Shareholders, private investors, revenues	Shareholders, private investors, revenues

Biotechnology Product Development

Biotechnology companies can produce a wide variety of products, from new crops to industrial enzymes to therapeutics. The details of the product development timeline depend on the type of product being manufactured. A general timeline includes the following major phases (see Figure 1.13):

- **Conceptualization** — marketing research identifies a need for a product and researchers confirm such a product can and should be developed. Companies factor in public need, other products already on the market, novel methods for making products, and overall commercial value. This phase may also include legal work to determine any conflicts and to establish patents

- **Research** — scientists create the initial products; the work done during this phase depends on the type of product being produced. For example, agricultural companies will work to generate a desired trait in a crop whereas biomedical companies will work to identify drug targets and promising drug candidates

- **Development** — begins when a candidate product is in hand and involves refining the product, investigating how best to manufacture it, rigorously detailing its specifications, developing assays to test them, and then testing the safety and efficacy, as required for approval by regulatory agencies

- **Commercialization** — during this stage, the details of marketing, manufacturing, and distribution are decided and the product is prepared for sale

In choosing which products to make, companies must weigh the benefit and purpose of the product against their own financial gain. As mentioned previously, drug therapies require an enormous financial investment, so it is no surprise that companies often show little interest in developing drugs for rare diseases, as not many people would need to purchase them. To address this, the U.S.

government passed the Orphan Drug Act (ODA) of 1983. The ODA provides a fast track through the FDA and extended patent rights, among other benefits, to encourage companies to pursue treatments for these diseases, and it has been successful. In fact, in 2010, drug maker Pfizer even established a division to focus on the development of such drugs.

Funding Biotechnological Advancements

Sources of funding depend on the type of organization where the research is being performed and the individual(s) performing it (see Table 1.3). Established companies use the profits from product sales to fund research and development (R&D). This creates a pipeline in which products that have already been launched pay for the development of new ones. The new products, in turn, should eventually pay back their own development costs (this is called return on investment, or ROI). In 2016, public biotechnology companies in the U.S., Europe, Canada, and Australia spent around $70 billion on R&D.

As mentioned earlier, the company that sells a product may not have performed the R&D for that product. Companies frequently partner with one another or with an academic laboratory to develop technology. The company or academic laboratory that performed the R&D is then paid either by sharing the revenues through royalties when the product sells, or by granting rights to the technology through licensing.

Research in academic, institutional, and government laboratories is usually funded through grants. Researchers write grant proposals that describe the research they plan to conduct and why that research is important. The grant funding body then assesses the applications and decides which applicants are worthy of funding. Grants can range from thousands to millions of dollars and can fund individual researchers, whole laboratories, partnering institutions, or large international projects.

In the U.S., the major source of funding for medical research is the National Institutes of Health (NIH), which is part of the U.S. Department of Health and Human Services. In 2017, the NIH funded around 50,000 grants totaling more than $32.3 billion to research institutions, colleges, universities, small businesses, non-profit organizations, and hospitals. The National Science Foundation (NSF) is also a major source of biotechnology funding. The NSF funds nonmedical science and engineering research, including biotechnology training grants. Other U.S. government agencies, such as the Department of Energy (DOE), the Department of Defense (DOD), and the United States Department of Agriculture (USDA) also fund biotechnology research that impacts their areas of governance. For example, the DOE funds technologies that may result in new energy sources. Most countries have similar governmental funding bodies. Funding can also come from charities, foundations, and businesses that award grants to researchers who meet certain criteria.

Table 1.3. **How different organizations fund biotechnology research.**

Type of Organization	Funding Sources
Medium to large biotechnology companies	Profits from the sales of existing products
Small biotechnology companies and start-ups	Angel investors, venture capital, and government grants
Academic institutions, research institutes, hospitals, government laboratories	Government grants and other funding, foundations, endowments, and charities

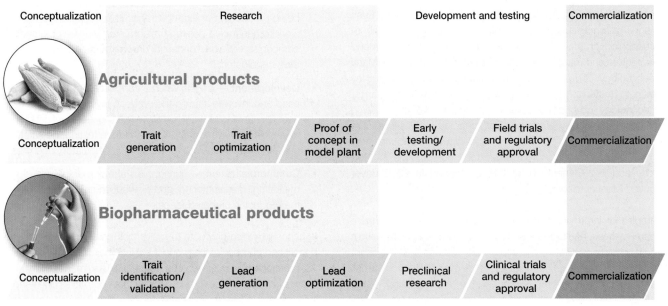

Figure 1.13. **Biotechnology development pipelines.**

Biotechnology companies use applied research to develop beneficial products or processes and rely on financial gain and profits for their continued operation. In contrast, the goal of research at public and academic institutions is to advance human knowledge that leads to improvements in society; such research usually takes a more long-term view and allows for curiosity-driven basic research. Basic research acts as a foundation for applied research; therefore, the goals of these two types of research complement each other and help drive biotechnological advances.

Patents

Product research is an expensive endeavor. If the goal of the research is to make money by commercializing its product, new inventions are patented to ensure a return on investment. A patentable invention can be anything that is new, nonobvious, and useful, such as a novel drug or a new process to manufacture an existing drug. A patent defines a set time period (usually 20 years) during which everyone but the inventors (or patent holders) is prevented from using the invention. This allows the inventors to make back the money they invested and earn a profit on their original work or idea. In exchange for these rights, the invention must be disclosed and shared with the scientific community.

Before 2011, a patent in the U.S. was awarded to whomever was the first to invent a patentable technology. If information about that technology leaked to a competitor before filing for the patent, a company could still receive the patent if they could prove that they had invented it before the competitor. In 2011, the America Invents Act changed the assignment of rights for a patent from "first to invent" to "first to file." Now, if a patentable technology leaks to a competitor and that competitor files for the patent first, they will receive the patent regardless of who actually invented the technology. With the cost of developing and commercializing products, it is more essential than ever to maintain confidentiality of company data.

If others wish to use a technology that is under patent, they negotiate terms with the owner of the patent for permission to use or license the technology. For example, Cetus Corporation, Kary Mullis's employer at the time he conceived PCR, patented the process in 1985. As a result, all manufacturers of thermal cyclers had to pay royalties to Cetus Corporation (the patent owner) for each instrument sold until the patent expired in March 2005. Companies usually seek patents to protect their investment, although many academic researchers and institutions have also been awarded patents. Patents are awarded by a national governmental agency. In the U.S., the U.S. Patent and Trademark Office, part of the U.S. Department of Commerce, is responsible for awarding patents.

1.4 Governmental Regulation of Biotechnology

Biotechnology is powerful and so is highly regulated by many governmental agencies to ensure that research is performed safely, that products are safe and effective, and that neither the products

nor their manufacturing processes harm the environment. This section concentrates on U.S. governmental agencies, but other countries have similar regulatory processes and agencies.

In the U.S., the main agencies governing biotechnology are the following:

- Food and Drug Administration (FDA) — ensures that food and beverages are safe for human consumption and that therapeutic drugs and devices are safe and effective

- Environmental Protection Agency (EPA) — protects human health and the environment

- United States Department of Agriculture (USDA) — ensures that agriculture and livestock are protected

The Occupational Safety and Health Administration (OSHA) also regulates the safety of workers in the workplace. Each agency regulates the aspects of biotechnology that fall into its area of responsibility, so a technology is often regulated by multiple agencies (see Figure 1.14).

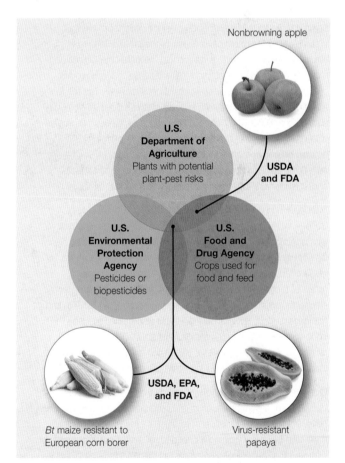

Figure 1.14. **U.S. government agencies overseeing biotechnology.** Each agency regulates the aspects of biotechnology that fall into its area of responsibility, so a technology is often regulated by multiple agencies.

Careers In Biotech

Timothy Balmorez
Chemist, Bio-Rad Laboratories

Timothy Balmorez is a production chemist at Bio-Rad Laboratories in Hercules, California. In one aspect of his job, he uses techniques such as chromatography, electrophoresis, and spectrophotometry to manufacture and purify stained proteins, antibodies, and other

biochemical reagents that scientists use in their research. In another aspect, he incorporates his interests in technology and psychology by also working with Bio-Rad's digital marketing group to develop and manage the search engine marketing (SEM) campaigns that help customers find Bio-Rad products online.

Born and raised in the San Francisco area, Timothy remembers being interested in his surroundings since childhood. He was curious about the chemical makeup of the smoke coming from the refineries around where he lived, in the different cultures and religions of his classmates, and in the disease that robbed his grandmother of her memory — Alzheimer's disease. The toll that Alzheimer's took on her and his family motivated him to someday study the disease. The community in which he grew up struggled with poverty, and this motivated him to always be mindful of helping others who are less fortunate. To that end, he volunteers as a Bio-Rad Science Ambassador, traveling to local schools and science fairs to demonstrate Bio-Rad's educational kits.

After earning his high school diploma through Contra Costa College in 2009, Timothy went on to study Biological Chemistry, with an emphasis in drug development, at the University of California, Berkeley. While in college, he worked as a student researcher, where he was able to learn, apply, and master a range of techniques, including western blotting, chromatography, electrophoresis, ELISA, PCR, protein purification, pH testing, and nuclear magnetic resonance (NMR). He also helped develop the code for simulations of the structure of amyloid-beta, a protein related to Alzheimer's disease. He earned his B.S. in 2013 and started working for Bio-Rad later that same year.

Over the next decade, Timothy projects that many companies will outsource or automate much of the production chemistry work they have. The best way to respond to these changes is to anticipate them and gain new training as needed, or even branch out and explore other professional interests, as he has. He hopes that more companies will follow Bio-Rad's example with the Science Ambassador program and engage more with their communities. His advice to those starting out in biotechnology is to stick to your principles, keep your head up, and never stop fighting to make the world a better place.

Regulation of Genetically Modified Organisms

Multiple U.S. agencies regulate the use and production of genetically engineered, or "modified," organisms (GMOs), depending on where the GMOs are used or developed. The use of **genetically modified organisms** in research laboratories is regulated under guidelines from the NIH. If the GMO becomes a commercial product or part of a commercial process, it falls under the guidance of a different governmental agency (see Figure 1.14). Foods derived from genetically engineered plants must meet the same safety, labeling, and other regulatory requirements that apply to all foods regulated by the FDA.

On July 1, 2016, a law requiring labeling of foods containing greater than 75% genetically engineered ingredients went into effect in Vermont. Because it is expensive to change labels, many major food manufacturers elected to change their labels across the U.S. with the words "produced with genetic engineering" or "partially produced with genetic engineering" on many of their packaged foods (see Figure 1.15).

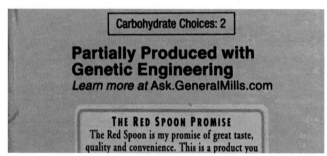

Figure 1.15. **Food labels on some foods in the U.S. indicating ingredients are genetically engineered**.

The USDA regulates GMOs in the food supply. On July 29, 2016, the National Bioengineered Food Disclosure Law was signed, requiring the USDA to establish guidelines for labeling foods containing GMOs by July 29, 2018. This law overrides the Vermont law, and food producers will be allowed 1–3 years to make their disclosures. The USDA will have to establish a threshold amount below which labeling would not be required. In Japan, if genetically modified ingredients exceed 5% of all ingredients, the food must be labeled as genetically modified (in Europe, this limit is 0.9%).

In the U.S. many companies choose to label the lack of GMOs, using the label as a marketing tool. The Non-GMO Project is a not-for-profit agency that provides third-party verification and labeling for foods and products not containing GMOs. Many companies seek their certification even for foods for which there is no commercially available counterpart.

Field testing of genetically modified plants (food crops and nonfood crops, such as cotton) is mainly regulated by the Animal and Plant Health Inspection Service (APHIS), an agency of the USDA. APHIS is mainly concerned about the effect GMOs may have on agriculture, such as the evolution of pesticide-resistant pests, or herbicide-resistant weeds. The EPA is involved in regulating genetically modified crops and is concerned about the effects that GMO cultivation may have on the environment, such as causing harm to wildlife that eats the genetically modified crop. The EPA is also concerned with the impact GMOs used in industry and commercial processes have on the environment. Under the Toxic Substances Control Act Biotechnology Program, GMOs regulated include those used to produce industrial enzymes and other chemicals, and those that break down into chemical pollutants in the environment. In addition, the EPA regulates microorganisms used in agricultural practices (for example, biofertilizers), biosensors, and biofuels. The EPA is concerned with the impact that the release of commercialized GMOs has on the environment and whether those GMOs need to be regulated to protect the environment.

Regulation of Products in Health Care

The FDA regulates therapeutic drugs, medical devices, and diagnostic tools. It is the responsibility of the manufacturer to prove the safety and effectiveness of its products to the satisfaction of the FDA, and the approval process can take many years. All new therapeutic drugs distributed in the United States must receive approval from the FDA, through a step-wise process of approvals.

The FDA also regulates medical devices such as glucose meters, artificial joints, diagnostic tests, and tools. The manufacturer must prove the device functions as intended and is safe to use. Diagnostic tests and tools must be proven to diagnose the disease for which they are intended before approval for sale.

Preclinical Research

During the phase of drug development called preclinical research and testing, various cell culture and animal models (and even computer models) are used to test whether the drugs work as expected and that they are safe. **In vitro** cell cultures or **in vivo** small animal models (for example, mouse or rat) are inexpensive and have fast generation times and so are often used in the initial stages of testing. But because these systems may not provide a complete picture of drug safety or efficacy (effectiveness), larger animals may be used in later stages of testing, if needed. Preclinical testing also helps to determine the concentration at which drugs work best and with the fewest side effects.

Results obtained through animal models may not always translate to humans, and the use of animals can be costly and raise ethical concerns. For these reasons, alternative synthetic models are being developed. For example, "organs-on-chips" (see Figure 1.16) are microfluidic chips whose channels are lined with different human cell types to simulate the activities and responses of organs and organ systems. Emulate, a producer of these organs-on-chips based in Boston, partners with biotechnology and pharmaceutical companies like Johnson & Johnson and Roche to use these chips for preclinical research.

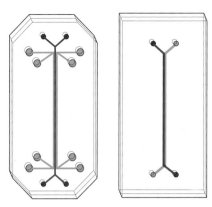

Figure 1.16. **An organ-on-a-chip.** The channels on chips like these (blue, red) can be lined with various human cell types – such as the lung, liver, brain, intestine or kidney, to recreate the natural physiological and mechanical forces that cells experience within the human body. These chips can be used in place of animal models for some preclinical research.

Transgenic animals may also be used to test drugs. Transgenic animals are laboratory animals that have been genetically engineered to express a specific gene or genes or have had the expression of a gene or genes inhibited or deleted. Transgenic animals may express the drug target or may have been specifically designed to model the human disease. Transgenic animals are often developed during drug discovery and are used again during drug development.

Preclinical trials must be performed using **good laboratory practice** (**GLP**), which ensures that the research and the data generated by the company meet an internationally recognized level of quality. GLP, which is a standardized method for conducting research in the laboratory, is discussed later in this chapter, in the Industry Practices section.

Clinical Trials

Once a new drug passes preclinical trials it can be tested in human **clinical trials**. The first stages of clinical trials test how safe the drug is in a small, often healthy population. Later trials test how effective the drug is against the disease in actual patients.

The design of clinical trials is very important and depends on the severity of the target disease and on how common it is. Clinical trials are usually conducted by splitting patients randomly into two groups, sometimes called "arms." One group receives the drug, while the other receives a placebo (often referred to as a sugar pill) that has been designed to visually resemble the actual drug. Usually neither the physicians nor the patients know which treatment patients have received — this is referred to as a double-blind study. Double-blind studies reduce the possibility of the physicians biasing the results.

Clinical trials are conducted in at least three phases. The results of each phase of a clinical trial are assessed by the FDA before the drug is approved to move to the next phase. If a drug passes all three phases of a clinical trial, the company applies to the FDA for approval to sell and market the drug (see Figure 1.17).

- In phase 1, a small group of often healthy individuals, usually fewer than 100, is given the drug to test for safety and dosage levels, as well as to determine how the body metabolizes the drug (for example, how long the drug remains active in the bloodstream)

- In phase 2, the trial is expanded to 100 to 300 participants to investigate whether the drug helps people in the specific patient population who are suffering from the disease

- In phase 3, a large trial is conducted in the specific patient population, with hundreds to thousands of patients monitored for the drug's effectiveness and side effects

In 2018, there were 2,802 phase 3 clinical trials being conducted in the U.S., the majority of which were for cancer therapies. The annual number of registered clinical trials worldwide increased from 23,384 in 2013 to 76,460 in 2018.

Regulation of Animal Research

Animal testing is a vital part of drug development and is valuable to life science research. Animal models allow scientists to understand how a drug, gene, or protein functions in a whole organism, something that is not possible using cell culture or computer-modeling techniques.

In the U.S., research institutions are required by law to establish an Institutional Animal Care and Use Committee (IACUC) to oversee their work with animals. IACUCs require researchers to justify their need for animals, select the most appropriate species, ensure the animals are treated humanely, and study the fewest number of animals possible to answer a specific question. As mentioned earlier (see Preclinical Research), different animal models serve different purposes and benefits in research and testing. Small animal models, such as mouse and rat, tend to be more cost-effective but may not provide the best representation of the human body. For this purpose, larger animal models, such as chimpanzees, may be more desirable, but there are more regulations and rules surrounding their use.

The USDA, through APHIS, has set forth federal regulations governing the care and use of animals in biomedical research in the Animal Welfare Act (AWA). The AWA sets standards of care for research animals regarding their housing, feeding, cleanliness, ventilation, and medical needs. It also requires the use of drugs such as anesthesia during and after potentially painful procedures. The AWA currently does not cover rats and mice, which comprise more than 90% of all animals used in laboratory research.

The U.S. Public Health Service (PHS) Act covers all laboratory animals used in institutions that receive research funds from the NIH, the FDA, or the Centers for Disease Control and Prevention (CDC). These researchers must adhere to the standards set out in the "Guide for the Care and Use of Laboratory Animals," authored by the National Research Council of the National Academies.

Regulation of Human Research

For pharmaceutical and therapeutic research, human subjects are often used in testing, and testing in human clinical trials is essential for any final approval. Generally, testing in humans does not occur until the therapy has been demonstrated to be safe in animal models. The International Conference on Harmonisation provides global guidelines for the development of pharmaceuticals. It defines **good clinical practice (GCP)**, a quality standard that governments can use to establish their own regulations for clinical trials involving human subjects.

Many countries use national, regional, or local Institutional Review Boards (IRBs) to make sure research involving human subjects is both lawful and ethical. IRBs review the purpose and methods of the proposed research to ensure that they are ethical. They may also conduct some form of risk-benefit analysis to determine whether or not research should be conducted. IRBs ensure that appropriate steps are taken to protect the rights and welfare of humans participating as subjects in a research study, including the following:

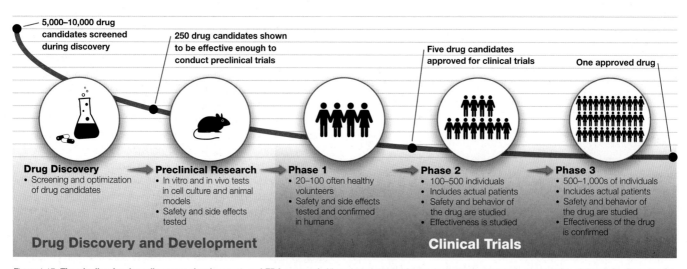

Figure 1.17. **The pipeline for drug discovery, development, and FDA approval.** All products intended for use in diagnosis, treatment, or prevention of disease must go through this regulated process in order to be approved by the FDA for use in humans. The process can take 10–15 years and costs over $2.5 billion for each approved drug. This cost includes the research into all the drugs that were rejected. For every one drug that is approved by the FDA, up to 10,000 potential drug candidates are screened during the discovery phase.

- The research methods and purpose should protect the rights, safety, and well-being of all trial subjects, paying special attention to vulnerable subjects, such as pregnant women, children, prisoners, the elderly, or persons with diminished comprehension

- Research subjects should be provided full disclosure of research purpose and methods and of possible risks

- Researchers must provide proposed subject recruitment plans (including all advertising materials), obtain written consent from the subjects, and ensure patient confidentiality

- The method, type, and amount of compensation to be given to subjects must show no coercion or undue influence on the trial subjects

In the U.S., IRBs are required for all research that receives federal support. They are regulated by the Office for Human Research Protections (OHRP) within the Department of Health and Human Services (HHS).

Regulation of Waste Disposal

Biotechnology generates different types of waste, including nonhazardous liquid and solid waste, biohazardous waste, and toxic waste. Agencies such as the EPA regulate how companies conduct their manufacturing processes and how they generate and dispose of their waste. Each country, state, and institution has specific guidelines that should be followed for waste disposal. These regulations protect the environment from pollutants.

Biohazardous waste includes any type of medical waste, such as tissue samples, blood, cell and microbial cultures, stocks of viruses, and sharps. Sharps are discarded medical articles, such as hypodermic needles, that may cause puncture wounds. Most biohazardous waste is treated to kill microorganisms or infectious agents by steam sterilization (autoclaving), gas sterilization, or chemical disinfection such as submersion in bleach. Once treated, the waste can be disposed of as nonhazardous. Sharps and other waste that cannot be disposed of by these methods are transported to specialized facilities for appropriate disposal.

Toxic waste can be poisonous, radioactive, flammable, explosive, corrosive, carcinogenic (causing cancer), mutagenic (damaging DNA), teratogenic (causing defects in the unborn), or bioaccumulative (accumulating in the bodies of plants and animals and thus in the food chain and in the human body). In the laboratory, most chemicals and reagents are labeled to provide information on their health and environmental effects. Harmful laboratory chemicals that have special disposal requirements include some DNA and protein stains, acids, bases, and heavy metals (for example, mercury in thermometers). The use of radioactive materials in laboratory research is on the decline; however, radioactivity is still used in some biotechnology research and strict guidelines regulate its use and disposal.

Toxic waste may be sealed and disposed of by burying it in a hazardous waste landfill. More often, toxic waste is incinerated (burned) to destroy the toxic chemicals. Due to the high cost of disposal, companies are working to reduce the toxic waste they produce. Some industrial plants save some money by repurposing the heat generated by incineration to create steam for power. Biotechnology is also used as a tool to reduce, eliminate, or remediate waste (see the Pollution Monitoring and Waste Management section earlier in this chapter).

Regulation of Nanotechnology

Nanotechnology and nanobiotechnology are emerging fields and there are currently no specific regulations guiding the use or disposal of the nanoparticles they produce. Concerns have been raised that nanoparticles may be health risks and may pollute the environment. In 2006, the FDA formed a Nanotechnology Task Force to determine regulatory approaches that encourage the continued development of innovative, safe, and effective FDA-regulated products that use nanotechnology materials. Currently the FDA does not consider products containing nanomaterials as intrinsically either benign or harmful. Therefore, it regulates nanotechnology products using existing rules for each type of product.

The EPA requires recordkeeping on and one-time reporting of existing exposure and health and safety information on nanoparticles under the Toxic Substances Control Act section 8(a). This rule requires companies that manufacture, import, or process certain chemical substances already in commercial use as nanoscale materials to notify the EPA as to the substance's chemical identity and its method and volume of manufacture, and to provide the agency with use and exposure information and health and safety data for the substance. The U.S. National Institute for Occupational Safety and Health is investigating potential health concerns caused by nanoparticles.

Workplace Safety Regulation

Working in a laboratory or biotechnology manufacturing plant can be hazardous, and the safety of workers is regulated in the U.S. by the Occupational Health and Safety Administration (OSHA), part of the U.S. Department of Labor. OSHA encourages employers to reduce workplace hazards and develops mandatory job safety and health standards, which are enforced by worksite inspections. These standards include a requirement that employees wear appropriate personal protective equipment (PPE). In the laboratory, PPE includes a laboratory coat, safety glasses, and gloves. Specific dangers in a laboratory include hazardous chemicals and biohazards such as microbial cultures and bloodborne pathogens. OSHA requires that workers be trained in the safe use of hazardous materials and in general laboratory safety.

Businesses are also required to keep a log of workplace injuries and illnesses and to show improvements in their safety record. All employees in the U.S. have the right to a safe workplace under the OSHA Act. The purpose of this law is to reduce workplace hazards and implement safety and health programs for employers and their employees. All employees are protected under the law from discrimination that might occur as a result of exercising the right to a safe workplace.

How To...

Write an SOP

Anyschool University Standard Operating Procedure	
SOP. 001	**Laboratory Notebook Entries**
Department: **Biotechnology Dept**	**Page 1 of 2**
Version: **1**	Date Created: **23/May/2018**
Author: **John Smith**	Authorization: **Dr. Summer**

Purpose
This SOP describes how to enter laboratory activities into a laboratory notebook.

Scope
This SOP applies to all students and course instructors in the Biotechnology program at Anyschool University. It is the responsibility of the students to follo
the procedures described in the SOP. It is the responsibility of the course instructors to ensure that students are complying with the SOP and to provid
adequate training to ensure compliance.

Additional Documentation
Laboratory notebooks will be permanently bound with sequential numbered pages. Each page will have a space for a title. Each page will have "To" an
"From" page indicators to link experiments that continue on multiple pages. Each page will have a space for a signature of the author and a witness an
the date the page was signed.

Definitions
SOP: Standard Operating Procedure
Laboratory Notebook: Any permanently bound book used to record laboratory experiments

Procedure
1. General Guidelines
    a. Make all laboratory notebook entries using a black or blue pen.
    b. Page linking identifiers: When an entry continues onto an additional page, write the page number on which the entry is continued in the To Page

Figure 1.18. **Example of an SOP header**.

A standard operating procedure (SOP) is a document that describes how to complete a common task. SOPs are necessary to ensure workers comply with company policy and industry standards. SOPs provide detailed instructions so that a task can be correctly and consistently performed by anyone at any time. SOPs are usually written on a template specific to the company or organization. An SOP should be written clearly so that it can be understood and followed by anyone expected to perform the procedure. Once drafted, an SOP is reviewed and approved by a department manager.

The header of the SOP must be present on every page of the SOP and usually contains the SOP title and document number, the author's name and department, the date of creation and approval, and page numbers (See Figure 1.18). The document number allows the SOP to be tracked through the company's or organization's quality system.

An SOP also has defined sections that must be addressed by the author. Depending on the organization and the regulatory and quality control needs, these categories may be compacted or expanded. If the sections do not apply, "N/A" or "none" is written under the heading. Common sections are defined here:

- **Scope:** who the SOP applies to, the responsibilities of each party, and what standards must be met. Scope can also include a brief description of the objective of the SOP

- **Purpose:** the subject of the SOP, any necessary background information, and what the SOP is supposed to achieve

- **Responsibility:** a definition of all functional roles involved

- **Definitions:** a list of the acronyms, abbreviations, or specialized terms (each with definitions) that are used in the SOP

- **Additional Documentation:** a list of related documents or attachments

- **Procedure:** the actual procedure written as a numbered list of step-by-step instructions, often including equipment and supplies

- **Revision History:** the SOP may be a revision of a previous version. A revision history is included at the beginning or end of the document. The revision history should state the version number, the name of the reviser, the date of the revision, and a description of changes from the previous version. Revision history is part of the quality record of the SOP

- **Appendices:** may include data templates, diagrams, and supporting materials

- **Approvals:** a list of reviewers, including name, functional role, signature, and date of approval

1.5 Industry Practices

In addition to ensuring their business practices meet government regulations, companies voluntarily follow industry standards. Industry standards are rules established by international organizations, such as the Organisation for Economic Co-operation and Development (OECD) or the International Organization for Standardization (ISO), to ensure products are researched, developed, and manufactured correctly and consistently. These organizations certify that business processes, practices, and products adhere to a specific set of standards called a quality system or quality management system. Each company chooses the standards to follow (the ones that apply to its products), and the standards organization verifies that the standards are being met.

The **quality assurance** (**QA**) department within a company ensures that the company complies with its own quality system and any applicable government regulations. Quality systems require that records be maintained on the operations happening in the company. For example, a laboratory notebook is a record of the research. Companies also keep records on raw materials they receive and any testing performed on both these materials and finished products. One way companies ensure consistency in their practices is by determining how each task or process should be performed and documenting instructions in a **standard operating procedure** (**SOP**). SOPs must be followed by anyone in the company who performs that task. The task may be a laboratory procedure or a method to calibrate laboratory equipment, but it may also be a general business practice related to other departments, such as customer service or accounting (see How To… Write an SOP).

Industry standards harmonize business practices around the world, making communication and business dealings run more smoothly. Conformance to standards guarantees the company is working under a specific set of rules, which reduces risks when working with that company. Following these standards is also good for the company since they ensure consistent processes leading to reproducible and anticipated outcomes. Governmental organizations may also require that preclinical trials be conducted using GLP and the manufacture of some products be conducted following **good manufacturing practice** (**GMP**). Good manufacturing practice is sometimes referred to as current good manufacturing practice (cGMP) because it changes over time.

Good Laboratory Practice

GLP is a quality system, a set of regulated practices used to collect safety data on a drug or product being developed. As such, GLP is a critical component of product testing and approval. In nonclinical health and environmental safety studies, for example, scientists involved in GLP would investigate factors such as biodistribution (where the product goes within the body), how long it remains in the body, side effects, breakdown products, and whether it is passed on to progeny. These data are then evaluated by the FDA (or other regulatory agencies outside the U.S.) to determine whether the product is safe to use in human clinical trials, which have stringent GCP requirements.

GLP specifies how studies must be performed and requires that a QA program be in place to ensure compliance. GLP is extremely detailed and includes the following requirements:

- The responsibilities of all individuals involved in the study — from management to laboratory workers — must be stated

- The qualifications and training required for personnel to conduct the study must be recorded

- SOPs must state how samples, materials, and controls will be received, labeled, and stored; how apparatus will be maintained and calibrated; and how computer systems will be validated and backed up

- A detailed plan must be written prior to the start of the study; it must outline the purpose of the study, detail test methods, and describe how the results will be reported (laboratory notebooks are an integral part of how results are reported)

- The results and data generated by the study must be properly stored and archived in case follow-up experiments are required

Good Manufacturing Practice

GMP is a quality system with a set of standardized practices that ensure steps are taken to produce consistently safe and effective products. It primarily focuses on the manufacturing, processing, testing, packaging, and storage of drugs, medical devices, some food items, and blood products. GMP is designed to minimize errors thus protecting the customer from purchasing a product that may be ineffective or even dangerous. As with GLP, training, recordkeeping, archiving, data monitoring, SOPs, and controlled documents are important in GMP. Environmental control to minimize contamination, equipment calibration, process validation, and complaint handling are also important components of GMP. GMP is enforced by the FDA in the U.S. and by similar national regulatory agencies in European countries. Failure to comply with GMP regulations can result in very serious consequences including recall, fines, or even jail time.

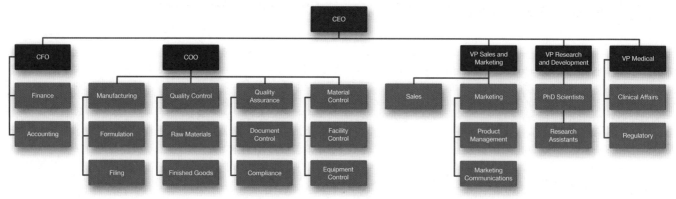

Figure 1.19. **An example of the structure of a biotechnology company**.

1.6 Biotechnology Company Structure

The structure of a biotechnology company depends largely on the types of research it conducts, the products it makes, and the number of people it employs (see Figure 1.19). The departments have distinct functions but must work together to make, test, and review products following appropriate requirements. Having separate departments for each function also reduces the risks of conflicts of interest. Imagine a technician who has spent significant time purifying a protein product and is anxious to move on to the next task. When **quality control (QC)** discovers that the product does not meet the purity specification limit by a small margin, QA must reject it for sale while the technician may have been tempted to call it "good enough." This separation of responsibilities creates a system of checks and balances to ensure only products that meet their specifications are released for sale.

Most biotechnology companies include the following departments and functions:

- Research and Development (R&D) — develops new products and supports existing products through product improvement, documentation, and new applications

- Medical or Clinical — for companies that produce drugs or medical devices, this department oversees the clinical, medical, and regulatory affairs related to its products, including clinical trials, ensuring that all procedures required for regulatory compliance are met

- Sales and Marketing — conducts market research, promotes and sells products, and develops and supports the customer base

- Legal Counsel — ensures the company breaks no laws, advises regarding business deals, and handles SEC filings, patent filings, etc.

- Manufacturing and Distribution — responsible for the manufacturing of products, their packaging, and shipping to customers. Manufacturing may involve a variety of manufacturing techniques such as chromatography, fermentation, formulation of bulk liquids or powders, etc. This group uses GMP to ensure products meet specifications

- Quality Control (QC) — inspects and tests materials at all critical stages of the manufacturing process from raw materials to finished product. QC follows test records to record data from assay measurements for each product specification

- Quality Assurance (QA) — oversees regulatory compliance and the company quality system including document control, internal and external audits, and record review. QA reviews the records and data from QC testing and other requirements prior to formally releasing materials to the next stage of manufacturing or products to sale

- Customer and Technical support — provide help to customers with their accounts, orders, and technical questions about products

- Finance and Accounting — regulate the flow of money into and out of the company; collect money from sales, pay workers, vendors, and suppliers, and balance the books

In addition to these departments, other support functions can be filled either internally or through use of contractors and vendors. These include facilities management, information technology (IT) support, and public relations. All these job functions must be coordinated in ways that make sense for the company, so they may be combined or separated as needed. For example, some companies combine manufacturing, distribution, QA, and QC into a single Operations department. Others may separate sales from marketing.

1.7 Careers in Biotechology

Working in biotechnology means joining an industry that is pushing the frontiers of technology with a mission to benefit humanity. Biotechnology is used in many industries; therefore, if you have an interest in biotechnology, there are many different careers open to you. Some careers involve working in the laboratory; many others do not (see Figure 1.20). The career stories of a variety of people are told throughout this book to provide an idea of the diversity of careers available to you.

Figure 1.20. **Scientists working in a laboratory**.

Jobs in biotechnology are also increasing. The U.S. Bureau of Labor predicts that the number of jobs for people with biotechnology or biology skills will increase by 10% between 2016 and 2026. In 2014 the biosciences subsector employed 1.7 million people in the U.S. The options open to you depend on the level of education you achieve. In 2016, the median salary for biosciences subsector jobs in the U.S. was nearly $95,000 per year, 85% greater than the private sector average.

Careers in the Laboratory

Working in a laboratory can be fun and exciting and could involve making novel scientific discoveries that help humanity. Jobs in laboratories are available in the biotechnology and pharmaceutical industries, academia, hospitals, and government institutions. Many types of jobs are available in different types of labs, including positions in R&D, clinical, quality control, and manufacturing. The titles of positions in different laboratories are not consistent, as discussed below. It is best to look at job descriptions and required qualifications, in addition to the job title, to determine whether an advertised job may be of interest. All laboratory jobs require great attention to detail and hands-on technical skills to ensure that procedures are performed properly. Good communication skills are also required to relay results and ideas.

Laboratory Technician

Depending on the type of laboratory, laboratory technicians are usually required to have an associate's degree and are often required to have a bachelor's degree in a discipline related to the work conducted in the laboratory. Occasionally, the job may require only a high school diploma and include training toward a two-year degree. A laboratory technician works under the supervision of a scientist and carries out the hands-on work, such as setting up assays, using instrumentation, and recording data.

Quality Assurance Technician

Quality assurance technicians manage the quality system and regulatory compliance procedures to prevent mistakes and ensure products are produced in a manner consistent with specifications and regulations. QA technicians are usually required to have an associate's or bachelor's degree. Possible job duties are as follows: check equipment to make sure it is installed, maintained, and operating properly, inspect storage areas to make sure that conditions will maintain the quality of raw materials and products, issue copies of controlled documents, and review completed documents to verify they were properly followed and completed by employees.

Research Associate

A research associate requires a bachelor's or master's degree in a science. Similar to a technician, a research associate carries out hands-on work in the laboratory under the supervision of a scientist, but with more responsibility for experimental design and data analysis. Research associates may also be expected to produce reports on their data and present those data to senior team members. The type of work performed by a research associate depends on the type of laboratory they work in.

Research and Development Scientist

An R&D scientist is usually required to have a doctoral degree, although a master's degree with experience in the field can be sufficient. In the field of biotechnology, scientists study how biological processes work, develop methods to investigate biological processes, and use biological knowledge to develop products and processes for the biotechnology industry. R&D scientists are responsible for designing experiments, analyzing data, and drawing conclusions. They also perform hands-on laboratory work and usually have a great deal of technical skill in their area of specialization. Scientists also train and supervise technicians and research associates.

Clinical Scientist

A clinical scientist is required to have a bachelor's degree, a graduate degree, or a specialized certification. A clinical scientist performs all levels of work in biomedical research, from analyzing blood samples at a hematology clinic to designing and overseeing clinical drug trials for a pharmaceutical company. Clinical scientists are responsible for proper handling of medical samples and ensuring that the appropriate testing is performed. They also train and supervise clinical technologists and technicians.

Engineer

An engineer is required to have a bachelor's degree, although some positions require a graduate degree. Biological engineers (also called biomedical engineers and bioengineers) use science and math to solve biological problems and have formal training in both engineering and in life sciences. A biological engineer can work in many different fields to develop medical devices or biological instrumentation. The biotechnology industry employs biological, agricultural, chemical, mechanical, electrical, and computer engineers. Engineers working in the biotechnology industry have many options, such as solving agricultural problems, developing instrumentation and machines, improving manufacturing processes, and developing software to analyze data.

Non-Laboratory Careers

Most jobs in biotechnology companies are not in the laboratory. Common in many companies are positions for administrators, sales representatives, marketing specialists, public relations specialists, technical and customer support representatives, lawyers, project managers, biostatisticians, accountants, buyers, logistics experts, quality assurance specialists, technical writers, and graphic designers.

Many of these positions are filled by people with experience in the biological sciences. For example, a marketing specialist may have a bachelor's degree in biology and a master's degree in business administration (MBA). Having a background in biology is an advantage when applying for a non-laboratory position in the biotechnology industry, as candidates with this knowledge can understand the company's products and communicate clearly with company scientists and customers.

Biotechnology affects many industries, and there are jobs using biotechnological advances in many nonbiotechnology industries. Biotechnology teachers are needed to train the new workforce as more jobs are created in biotechnology. Writers, editors, and graphic designers are needed to communicate biotechnological developments in the media and other materials such as textbooks. Bioethicists are needed to ensure that advances in biotechnology do not adversely affect society or the environment. Physicians and others in the medical field are intimately involved in communicating biotechnology-based medical advances to their patients.

Biotechnology is changing the way we live our lives and there is an opportunity for you to be involved in these changes in your chosen career or as an informed member of the public. Use the skills and information you gain in this course to decide what role you could play in this exciting future.

Chapter 1 Essay Questions

1. Perform research into the biotechnology industry in your area. Write a report on local biotechnology companies including a discussion on their research and/or products.

2. Search for three to five different job openings in the biotechnology industry in your area. Compare and contrast the required skills and qualifications, and the expected job duties between the different opportunities.

Additional Resources

Recent reports on the biotechnology industry:

The Value of Bioscience Innovation in Growing Jobs and Improving Quality of Life (2016). TEConomy/BIO. bio.org/jobs2016, Accessed April 9, 2018.

Ernst & Young (2017). Biotechnology Report 2017: Beyond Borders — Staying the course. ey.com/Publication/vwLUAssets/ey-biotechnology-report-2017-beyond-borders-staying-the-course/$FILE/ey-biotechnology-report-2017-beyond-borders-staying-the-course.pdf, Accessed Apr 9, 2018.

Information on laboratory safety and regulations:

Biosafety in Microbiological and Biomedical Laboratories Manual (2009) U.S. Department of Health and Human Services (5th edition). cdc.gov/biosafety/publications/bmbl5/BMBL.pdf, Accessed Apr 6, 2018.

NIH Guidelines for Research Involving Recombinant DNA Molecules (2016). U.S. Department of Health and Human Services. osp.od.nih.gov/wp-content/uploads/2013/06/NIH _Guidelines.pdf, Accessed Apr 6, 2018.

Information on careers:

Occupational Outlook Handbook from the Bureau of Labor. bls.gov/oco/, Accessed Apr 3, 2018.

Reaser A (2002). Jobs in biotechnology: applying old sciences to new discoveries. Occupational Outlook Quarterly Fall 26–35. bls.gov/careeroutlook/2002/fall/art03.pdf, Accessed Apr 3, 2018.

Access Excellence Career Center. accessexcellence.org/RC/CC/, Accessed Apr 3, 2018.

Information on clinical trials:

NIH, U.S. National Library of Medicine, Clinical trials database: ClinicalTrials.gov, Accessed Jun 13, 2018.

CenterWatch clinical trials information and database: CenterWatch.com, Accessed Jun 13, 2018.

Information on Nanotechnology Regulation:

Environmental Protection Agency Control of Nanoscale Materials under the Toxic Substances Control Act. epa.gov/reviewing-new-chemicals-under-toxic-substances-control-act-tsca/control-nanoscale-materials-under, Accessed Apr 5, 2018.

Laboratory Skills

Summary

Working in the laboratory requires skills and behaviors that enable a scientist or technician to operate in a safe environment and generate reproducible data. Biological laboratories are classified by the types of biohazards scientists may face, including microorganisms. Each person must wear appropriate personal protective equipment (PPE) and be properly trained on how to deal with hazards to reduce potential injury. Laboratory workers are expected to know the proper names of laboratory and safety equipment and how to use it correctly. An important basic skill is the ability to make solutions and buffers accurately and reproducibly. Individuals must understand how to perform calculations to determine the quantities of chemicals needed when preparing a solution and how to scale calculations to make larger or smaller volumes of solutions. In any laboratory environment, the laboratory notebook is as important as any piece of equipment. Learning how to record information accurately and with detail is one of the most important skills that one can master. This chapter covers these basic skills and provides practice with common laboratory equipment and solution making.

2.1 Laboratory Safety

Operating Safely in a Laboratory

Each person in a laboratory is responsible for ensuring that everyone is operating safely and that accidents do not occur. The typical hazards encountered in a laboratory include chemicals, biohazards, electrical hazards, and fire hazards. As discussed in Chapter 1, safety in the workplace is regulated by OSHA, which has specific recommendations to ensure the safety of laboratory personnel.

Safety Training

Any good laboratory safety program begins with training and includes ongoing training for personnel. When starting any new laboratory experience, it is vital that a general safety briefing is conducted. This should include training personnel in safe handling procedures for chemicals, biohazards, and waste. As new work requires new equipment and techniques, it is imperative that training be provided and updated as conditions or hazards change. A crucial part of safety training includes deliberate identification and awareness of all lab safety equipment. Personnel should know the locations and usage instructions for first aid stations, eye washes, and safety showers. Upon completion of training, it is important to test for mastery before allowing personnel to work in the laboratory. Periodic updates and testing are required to ensure safe operation.

Personal Protective Equipment

Personal protective equipment (PPE) protects the user against specific hazards and should be worn at all times (see Figure 2.1). Safety glasses are important pieces of PPE and should be selected based upon the specific risks anticipated. Safety glasses with side protection are appropriate if there is a low risk of splash hazard. If there is a higher risk of splash hazard, as in a chemistry laboratory, splash goggles should be worn. If shortwave ultraviolet light is used, a full-face mask made with material that has UV protection should be worn. Besides eye protection, it is important to limit exposure of skin to microorganisms and chemicals. Gloves that protect against exposure to any potential hazard are critical.

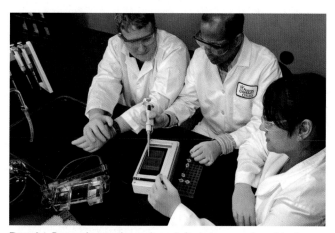

Figure 2.1. **Personal protective equipment.** Students are wearing safety glasses, nitrile gloves, and laboratory coats.

When selecting the type of gloves, keep in mind the hazard identified and any potential allergies to the gloves. Most gloves are made of latex, nitrile, or vinyl. Latex provides a simple barrier, while other materials are selected based upon the time it takes for certain chemicals to penetrate the material. Heat- or cold-resistant gloves are required when extreme temperatures are encountered. Wearing a laboratory coat protects clothing. Feet must also be protected, and open-toed shoes are not permitted in a laboratory.

Chemical Safety

A laboratory contains many different chemicals. Liquid, solid, and gaseous chemicals pose different types of hazards by being acidic, basic, flammable, toxic, or explosive. Each chemical in a laboratory has a **safety data sheet (SDS)** that contains information on storage, hazards, first aid, reactivity, physical properties, and disposal. SDSs should be easily accessible, and training on a particular SDS should take place before working with a chemical for the first time. Each laboratory is required to have a chemical hygiene plan that includes training employees on proper use of chemicals.

When performing experiments that generate hazardous gases or when using flammable or explosive chemicals, fume hoods should be used. These prevent exposure to chemical fumes. Most fume hoods have a movable sash that can be raised to allow access to the hood. The sash should not be raised higher than the level indicated on the outside of the hood and nothing should block the airflow of the hood or fumes may escape from the hood and be inhaled.

Biological Safety

OSHA, the Centers for Disease Control and Prevention (CDC), and the National Institutes of Health (NIH) have guidelines and regulations for the special handling of biological organisms and molecules (see Table 2.1). Biological laboratories are classified by the types of biohazards and risks of exposure that scientists may face. Biohazards include microorganisms, bloodborne pathogens, which are microorganisms in blood that may cause disease, and recombinant DNA molecules that can carry a risk of gene transfer to humans. The classifications are called **biosafety levels (BSL)** and range from BSL-1 to BSL-4 (see Table 2.1). Most high school laboratories are BSL-1. The microorganisms or DNA studied in a BSL-1 laboratory are not known to cause disease in healthy people, for example *E. coli* HB101 bacteria, and can be handled on the open bench.

Most community colleges have BSL-1 laboratories, but some may also have BSL-2 laboratories. Biotechnology and pharmaceutical companies, national laboratories, research universities, and hospitals usually have BSL-2 laboratories. A BSL-2 laboratory has many more regulations than a BSL-1 laboratory since the agents used in a BSL-2 laboratory are of moderate risk to personnel and the environment. Most tissue culture work requires a BSL-2 laboratory. BSL-3 and BSL-4 laboratories handle pathogens that can harm or even kill people. For example, the Ebola virus that is fatal in 90% of infections is handled only in BSL-4 laboratories. BSL-3 and BSL-4 laboratories are rare and require very extensive training for access due to the high level of risk involved. Most BSL-3 and BSL-4 laboratories are found in government facilities.

Table 2.1. **Biosafety level descriptions and precautions.** Each biosafety level includes specific requirements and precautions when working with microorganisms. The Centers for Disease Control and Prevention maintain very detailed information about the requirements for different BSL laboratories.

Description	Personal Protective Equipment	Specific Precautions
BSL-1: Required when working with agents not typically associated with human disease. Minimal health hazard. **Example:** Non-pathogenic strain of *E. coli* such as DH5α, *Saccharomyces cerevisiae* (baker's yeast)	• Protective laboratory coat, gown, or uniform • Eyewear required when performing procedures that might result in splashing of microorganisms or other hazardous materials • Gloves must be worn to protect hands from exposure to hazardous materials	• Do not wash or reuse gloves • Follow handwashing procedure before leaving the laboratory • Eating, drinking, and applying cosmetics and/or contact lenses are prohibited
BSL-2: Required when working with agents that pose moderate health hazards including many bloodborne pathogens. **Example:** Hepatitis B & C viruses, human immunodeficiency virus (HIV), *Candida albicans*	**BSL-1 Requirements plus** • Biological safety cabinet (BSC) for open handling of agents that may splash or produce aerosols of infectious materials • Eye and face protection (goggles, face mask, face shield) required when handling microorganisms outside the BSC and/or if contact lenses are worn • Eye, face, and respiratory protection must be used in rooms with contaminated animals	**BSL-1 Requirements plus** • Access is limited to personnel with appropriate training • Clearly labeled hazard signs posted near every entrance • Careful handling of needles and sharps; disposable needles and syringes must be disposed of in a puncture-resistant container
BSL-3: Required when working with agents that pose significant health or lethal risk when exposed through inhalation. **Example:** Yellow fever virus, Zika virus, West Nile virus	**BSL-2 Requirements plus** • BSCs for open manipulations of all agents	**BSL-2 Requirements plus** • Access is controlled • Laboratory separated from areas with unrestricted traffic with two self-closing laboratory doors and locks • Space around doors and openings must be sealable • Negative airflow into laboratory • Vacuum lines protected with HEPA filters • Exhaust air not recirculated
BSL-4: Required when risk from aerosol exposure results in untreatable life-threatening disease and death or with unknown disease severity or transmission **Example:** Ebola virus, viral hemorrhagic fever	**BSL-3 Requirements plus** • Full body positive pressure protective suits before entering laboratory • Shower upon exiting laboratory • All material decontaminated upon exiting laboratory	**BSL-3 Requirements plus** • Separate building or isolated area • Supply and exhaust, vacuum, and decontamination systems dedicated to BSL-4 laboratory • Rooms arranged for passage through an inner dirty changing area, shower, and outer clean change room upon exiting

General Laboratory Safety

Biological or chemical hazards are not the only kind encountered in a laboratory. Accidents may occur. Spills need to be cleaned up quickly to prevent someone from slipping. Broken glass should be immediately swept up and placed in a broken glass container. All items stored on laboratory shelves should be secured so that they do not pose a falling hazard. Electrical cords should not pose a tripping hazard or obstruct any laboratory work in progress. In summary, work tidily and clean up any accidents. The Globally Harmonized System of Classification and Labeling of Chemicals (GHS) includes standards for the classification of environmental, health, and physical hazards as well as the information to be included on labels of hazardous chemicals and safety data sheets (see Figure 2.2).

Waste Disposal

Many activities generate compounds or microorganisms that need to be disposed of safely. Chemical or toxic waste and biohazardous waste should be disposed of separately. Microorganisms should be disposed of in a clearly labeled biohazard container and destroyed by autoclaving. If an autoclave is not available, most microorganisms used in a teaching laboratory can be killed by soaking in 10% bleach for 10 minutes. Many chemicals are considered hazardous waste and must be disposed of according to local regulations.

Oxidizing   Health hazard   Environmental hazard

Compressed gas   Flammable   Explosive

Toxic   Corrosive   General irritant, transient/acute toxicity

Figure 2.2. **Globally Harmonized System hazard symbols** communicate the potential hazards of chemicals substances. Eight of the symbols are required to be used when applicable. The symbol representing "Environmental Hazard" is optional.

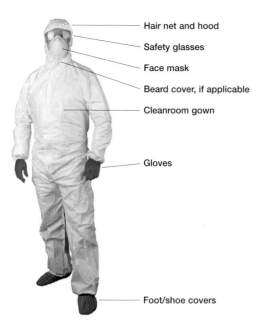

Figure 2.3. **Typical cleanroom attire.** Personnel working in a cleanroom must wear appropriate attire based on the requirements of the work being done.

Hair net and hood
Safety glasses
Face mask
Beard cover, if applicable
Cleanroom gown
Gloves
Foot/shoe covers

Cleanrooms

In addition to ensuring the safety of laboratory workers, laboratory protocols must ensure that products are safe to use for the end user. It is extremely important that steps are taken to prevent the contamination of samples and products throughout production. In fact, this is an essential component of demonstrating adequate control of the production process for U.S. Food and Drug Administration (FDA) approval of a drug or therapy.

Cleanrooms are used in the biotechnology and pharmaceutical industries to control the exposure of raw materials and products to contaminants. Cleanrooms are designed to reduce the presence of particulates and microorganisms in the air and control environmental factors such as temperature, humidity, and pressure during product manufacturing.

Cleanrooms can have different ratings based on the allowable level of contaminants within. The design and protocol for using the cleanroom varies accordingly. In general, personnel enter and leave a cleanroom through an airtight space with two interlocking doors, called an airlock, sometimes including a showering step. Personnel are required to wear protective clothing, including a hood, face mask, gloves, boots, and coveralls or a cleanroom gown (see Figure 2.3). One of the most important components of a cleanroom is a High Efficiency Particulate Air (HEPA) filter. All air that is delivered into a cleanroom must pass through a HEPA filter. Some cleanrooms have heating, ventilation, and air conditioning (HVAC) systems that keep the humidity low, which requires ionizing equipment to control problems with electrostatic discharge. Cleanrooms also have rounded, seamless edges and corners on both the floor and ceiling for easy cleaning.

The environment within a cleanroom, including the air and surfaces, must be continuously monitored for contamination. Methods for detecting particles and microbes may include exposing the cleanroom air to a sterile petri dish, doing swab testing of surfaces, or laser particle detection systems.

Chapter 2

Bioethics
Waste Disposal

Most biotechnology facilities — including your classroom laboratory — manage and regulate hazardous waste from the moment it is generated until it reaches its destination for final disposal or treatment. Everyone is responsible for understanding the hazards associated with the materials they use and generate and for how to store and dispose of waste safely. The consequences for not doing so can be severe.

On June 28, 1988, three chemical engineers, Carl Gepp, William Dee, and Robert Lentz (the "Aberdeen Three"), were criminally indicted for improperly storing, treating, and disposing of hazardous waste at the Aberdeen Proving Ground. A U.S. Army facility in Maryland, Aberdeen Proving Ground has been a site for the manufacture and testing of chemical weapons since 1918. The Aberdeen Three were all experts in chemical weapons, and all were managers at Aberdeen when the facility was inspected and found to have many chemicals that had been misplaced and were unlabeled or poorly contained. The inspection also revealed that a part of the roof of the facility had collapsed weeks before, smashing storage containers and spilling chemicals, which had not been cleaned up. Although the Aberdeen Three did not mishandle the chemicals themselves, as managers they had permitted the improper handling of the chemicals, did nothing to fix the situation, and had not even reported any of the incidents to their superiors. Each defendant was charged and initially faced up to 15 years in prison and up to $750,000 in fines, though they were finally sentenced to only three years of probation and 1,000 hours of community service (United States v. Dee, 912 F.2d 741 4th Cir. 1990).

Your institution should have clear guidelines on waste disposal, and you should be trained to properly dispose of the waste generated by your experiments. Whenever you work with potentially hazardous substances, consider your personal safety as well as the impacts of waste disposal on your work environment, your community, and the environment as a whole.

2.2 Laboratory Notebooks

Documenting work in a laboratory notebook or any controlled document is an important skill to master and can be mandatory in some laboratories. The laboratory notebook is used to document the experiment being performed so that the work can be reproduced by following the information entered in the book. New researchers often make the mistake of writing insufficient information in their laboratory notebooks. The phrase "commit nothing to memory, write everything down," is a good rule. A laboratory notebook should record the purpose of the experiment, methods used, results observed, and conclusions drawn at the end of the experiment. In addition, any novel ideas regarding the work should be recorded. Although it is no longer possible to establish intellectual property rights for inventions based on the record of a laboratory notebook, it is still crucial that experiments and ideas are properly recorded. Transferring new inventions from the research lab to manufacturing is not possible if the procedures and observations used are not well documented.

Laboratory Notebook Structure

A laboratory notebook should be bound with pages that are permanently numbered to ensure that the notebook cannot be modified or have pages added or removed. There should be places to link experiments that occur on nonconsecutive pages. Usually at the top and bottom of each page, there are spaces for "from page___" and "to page___" entries (see Figure 2.4). At the top of the page, there should be a space for the title so that the experiment being conducted can be identified. Often there are additional fields for entering information, such as a project or book number, on the page. At the bottom of each page, there should be a space for the researcher and a witness to sign and date.

Components of a Laboratory Notebook Entry

All notebook entries should be able to be read by anyone picking up the book without asking any questions. Therefore, the entries should be neat, legible, and complete (see Figure 2.5). Choose a black pen that does not bleed when wet or put too much ink on the page.

- **Descriptive Title** — clearly identifies the experiment or protocol being conducted

- **Hypotheses** — stated only if appropriate; for example, a hypothesis is not needed when reagents or solutions are being prepared

- **Materials** — should record all specialized equipment, reagents, the model number of equipment, and, if appropriate, the serial numbers of equipment

- **Procedures or Methods** — should document each action completed, including specific quantities and concentrations of all reagents, reaction times, and any written calculations used to deduce quantities

- **Results** — should have detailed observations, sketches and graphs labeled as a figure with a figure number and a caption, and tables clearly labeled with a table number and a caption

- **Conclusion** — should discuss how results relate to experiment purpose or hypothesis, discuss ideas for future investigations, and summarize any reading and references that relate to the experiment at hand, especially for protocols taken directly from another source

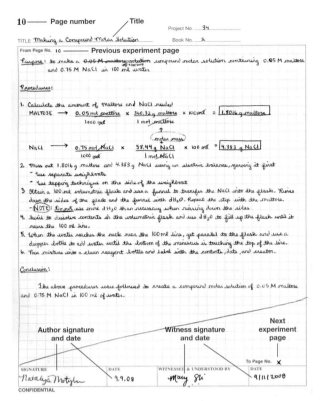

Figure 2.4. Laboratory notebook page. The places for title, signatures, page directions, and page numbers are indicated. Note also that the notebook entry has a purpose, procedure, and conclusion. Extra space and errors are crossed out with a single line, and calculations are shown.

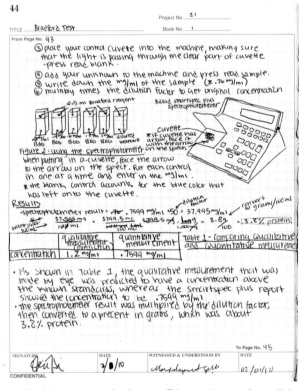

Figure 2.5. Laboratory notebook page. This page is part of a multiple-page experiment. The procedural steps that are continued from the previous page are numbered. The sketch and table are labeled with titles. The full calculation is shown. The results are presented in a table.

Sometimes it is necessary to stop one experiment and start another. The laboratory notebook should be written in the order of the work being performed. No space should be left in the book to fill in later. As the documentation of an experiment continues from page to page in the laboratory notebook, it is important to record "to" and "from" page numbers. Any blank space or pages should have a line drawn diagonally across them. Any mistakes that are made should be crossed out with one line. The words crossed out should still be easy to read. If large sections are to be crossed out, just put one diagonal line across the section. Any spaces, mistakes, or sections that are crossed out should always be initialed and dated.

2.3 Laboratory Equipment

There are many types of instruments and apparatus in a laboratory. It is important to learn the names of the equipment and to understand how it is used so that protocols can be performed. For experiments to be replicated, reagents must be accurately measured and reaction conditions duplicated. Many pieces of equipment in a laboratory are designed to measure variables such as volume, mass, temperature, and pH. The choice of equipment depends on the task to be performed and the accuracy required for achieving the task. Different pieces of equipment have different levels of accuracy. For example, a beaker should be used to hold liquids, but the measurement marks on the beaker should be used only as a rough indication of the volume. The quality of the laboratory equipment is judged by its accuracy, precision, and durability. Laboratory glassware is generally of higher quality and accuracy than plasticware.

For some types of laboratory equipment it is helpful to maintain a usage log. A usage log is a document in which users record information about the frequency and duration of equipment use. These logs are helpful not only to estimate the remaining useful lifetime of equipment parts, but also to inform potential users about when equipment will be available. For instance, a UV-enabled spectrophotometer usage log might include entries for "duration (minutes) of UV lamp use" to keep a running tally of the cumulative running hours for the UV lamp. This can help to estimate when the lamp is approaching time for replacement. An autoclave usage log might include entries for start time, type of sterilization cycle used, and cycle duration so that the next user can estimate when the autoclave will be available.

Measuring Volumes
Volume is measured in liters (L) or fractions of liters. A milliliter (ml) is one thousandth of a liter and a microliter (μl) is one millionth of a liter. Biotechnology laboratories commonly measure volumes ranging from microliters to liters, and a variety of equipment is needed to accurately measure volumes in each range.

Who Invented Claritin?
How Laboratory Notebooks Played Their Part

The rights to a method for making the antihistamine loratadine (the active ingredient in Claritin) from two precursor chemicals came down to one Spanish company, Medichem, using laboratory notebooks to try to prove that it had invented the method before another Spanish company, Rolabo (MediChem S.A. v. Rolabo S.L., 437 F.3d 1157 Fed. Cir. 2006). Medichem's method required that the reaction be carried out in the presence of a tertiary amine, while Rolabo's method permitted, by not excluding or requiring, the presence of a tertiary amine. Thus, the Medichem invention was a subset of Rolabo's invention. Since Rolabo had filed its patent first, it was up to Medichem to sue Rolabo and show evidence that Medichem had actually performed the patented method before Rolabo; in legal terms, Medichem had to prove that it had been the first to reduce the invention to practice.

The case was heard by several courts, and initially one lower court found in Medichem's favor. However, the ruling was reversed by two upper courts, including the U.S. Federal Circuit, which questioned Medichem's evidence that it had actually performed the patented method before Rolabo. Medichem's evidence was laboratory notebooks of the co-inventors of the process. The notebook contained the results of the reaction but not the methods. Two other laboratory notebooks that may have held important information to support the case were unsigned; therefore, they could not stand up to legal scrutiny and Medichem lost the case. Claritin sales in the U.S. in 2017 were approximately $238.1 million. Imagine if it was your laboratory notebook that had lost such a valuable patent!

The role of lab notebooks in patent cases has changed since the United States converted from a "first to invent" to a "first to file" system of priority under the America Invents Act (AIA) in March 2013. Under the AIA, the first inventor to file a patent application in the United States Patent Office would win a patent dispute. This change decreased the importance of lab notebooks in determining who was first to invent. However, lab notebooks are still important in determining who the inventors on a patent are and helping an attorney obtain a patent application.

Careers In Biotech

Joshua Moore

Bio-Containment Protocol Support Supervisor
U.S. Army Medical Research Institute
of Infectious Diseases (USAMRIID)

Joshua Moore started working in the biotechnology industry straight out of high school. Within six months of graduating in 2000 from Frederick High School in Frederick, Maryland, he was offered a position at the National Institutes of Health (NIH) as an entry-level animal caretaker. Six months after that, he was promoted to assistant supervisor. He worked in an animal holding facility, taking care of the

Photo courtesy of Joshua Moore

animals and facilities. He also wrote SOPs, ensured all procedures complied with the facility's SOPs and guidelines, and trained others in the proper care and handling of the animals. After two years, he transferred his skills to a private research company, BIOQUAL, Inc., in Rockville, Maryland, where he managed their animal facility. Among other duties, Joshua ensured the animals were cared for in accordance with the Animal Welfare Regulations, U.S. Public Health Policy on Humane Care and Use of Laboratory Animals, and the Guide for Care and Use of Laboratory Animals.

In 2004, Joshua moved to a laboratory technician position at USAMRIID, an Army laboratory within the Department of Defense, which develops vaccines, drugs, and diagnostic tools to help defend against biological threat agents. There, he cared for the lab animals and assisted in experiments, adhering to SOPs and government regulations. He also gained experience in bio-containment, which is the physical containment of pathogens such as bacteria, viruses, and natural toxins to prevent accidental infection of workers or the surrounding community. Within two years, he was promoted to Bio-Containment Supervisor, supervising all bio-containment technicians and ensuring all lab animals are cared for and maintained in accordance with federal and state rules and regulations, SOPs, and good laboratory practice (GLP) requirements.

Joshua became interested in biotechnology because it is possible to grow and advance in the field quickly. He enjoys the technical challenges and high standards associated with the work he does, and he sees the field of bio-containment becoming more advanced over the next ten years. He believes medical biological defense will be at the forefront of research and development.

Joshua's advice to people starting out in biotechnology is to set a few short- and long-term goals and focus on them first. Once those goals have been achieved, the sense of accomplishment and the experience gained will promote further career growth.

Graduated Cylinders

To measure a moderate volume of a liquid, a graduated cylinder is the tool of choice (see Figure 2.6). The capacity of graduated cylinders ranges from 5 ml to 5 L. In general, the larger the cylinder, the less accurate the volume measurement; this is due to the large diameter of the **meniscus**, which is more sensitive to changes in atmospheric pressure. The meniscus is the dip in the surface of liquid in a tube. Volume is measured by reading the bottom of the meniscus at eye level. The rule of thumb is to use the smallest available cylinder of sufficient capacity for your measurements.

Figure 2.6. **Graduated cylinders.**   Figure 2.7. **Volumetric flasks.**

Volumetric Flasks

A volumetric flask is used to accurately prepare a standard volume of solution. Volumetric flasks are made of glass and have a very narrow neck that has an etched circle marking the fixed volume, which ensures an accurate volume measurement (see Figure 2.7). A volumetric flask can measure within 0.04% of the target volume, while a graduated cylinder measures within 1% of the target volume. The most common capacities for volumetric flasks range from 10 ml to 2 L.

Vacuum-Assisted Pipets

To measure volumes ranging from 1 to 25 ml, the equipment of choice is a vacuum-assisted pipet. Liquid is drawn into the pipet by creating a vacuum with a pump. (The pump can be as simple as a rubber bulb or as complex as a motorized pipet pump.) There are various types of vacuum-assisted pipets with different degrees of accuracy. As with graduated cylinders, the volume of liquids in pipets should be read by measuring the bottom of the meniscus. Many vacuum-assisted pipets are packaged with a cotton plug inside the top of each pipet to prevent the contents from being drawn into the pipet pump.

Pasteur pipets are usually made of glass (see Figure 2.8). Since they do not have graduation marks, Pasteur pipets are used for transferring liquid from one receptacle to another rather than for measuring liquid.

Disposable plastic transfer pipets, also called transfer or bulb pipets, have an integrated bulb and are similar to Pasteur pipets. Although transfer pipets are not accurate, they can be graduated for measuring volumes (see Figure 2.9). Transfer pipets are often available in individual, sterilized packages for microbiological work.

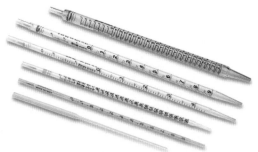

Figure 2.8. **Serological and Pasteur pipets.** From top to bottom, five plastic plugged serological pipets, 25, 10, 5, 2, and 1 ml, and a glass Pasteur pipet.

Volumetric pipets are the most accurate types of vacuum-assisted pipet. These pipets measure a single specific volume very accurately; for example, a 10 ml Class A volumetric pipet is accurate to within 0.02 ml of the volume being measured. Volumetric pipets are made of glass and have a recognizable shape, with a reservoir in the middle. Most volumetric pipets are accurate when their contents are delivered by gravity, and a small amount of liquid that remains in the pipet should not be blown out.

Graduated pipets can be made of glass or plastic. These pipets are cylindrical in shape and are similar to a graduated cylinder with fine graduations (see Figure 2.8). Graduated pipets are not as accurate as volumetric pipets; for example, a 10 ml disposable serological pipet measures to within 0.15 ml. However, unlike a volumetric pipet that measures a fixed volume, a graduated pipet can measure any volume in its range. The end of the graduated pipet tapers to a narrow tip to deliver the liquid as a fine stream or drops. There are two types of graduated pipets: Mohr pipets and serological pipets. Mohr pipets have a dead volume that should not be emptied out beyond the zero mark. Serological pipets are designed to deliver the entire contents; therefore, the last drops from a serological pipet should be blown out with the pipet pump (see Figure 2.10).

Figure 2.9. **1 ml graduated disposable plastic transfer pipets.**

Pipet pumps fill and deliver the contents of a vacuum-assisted pipet. The simplest pump is a bulb, such as an eye dropper, attached to the end of the pipet. Air or liquid is pushed out of the tube when the bulb is squeezed. The change in air pressure sucks air or liquid back in when the bulb is released. The concept is the same for all other pipet pumps. Tri-valve pipet fillers use a large bulb with three pinch valves for filling, delivery, and exhaustion of suction (see Figure 2.10). A manual pipet pump has a plunger that is controlled by a wheel turned by the operator's thumb (see Figure 2.10). The most commonly used pipet pump in research laboratories is a motorized pipet pump. These pumps have a pistol-style grip with two buttons: one button applies suction, while the other blows out to release the liquid (see Figure 2.10).

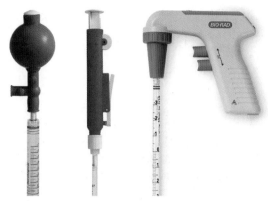

Figure 2.10. **Pipet pumps.** From left to right, a tri-valve pipet filler, a manual pipet pump, and a motorized pipet pump.

Micropipets

Micropipets are used to measure volumes of 1 ml or less (see Figure 2.11). Some micropipets are designed to measure fixed volumes, while others are adjustable and accommodate a variety of volume ranges. The most common volume ranges are 0.5–10 µl, 2–20 µl, 20–200 µl, and 100–1,000 µl. The accuracy of a micropipet varies depending on its size and the volume being measured, but it is usually 0.5–4%. A pipet is least accurate when it is used to measure a volume that is at the bottom of its range; thus, a 20–200 µl pipet is most accurate when measuring 200 µl and least accurate when measuring 20 µl. Therefore, when choosing between two pipets that are capable of delivering the same desired volume, select the pipet whose middle to upper range is closest to the desired volume.

Micropipets require disposable tips. To prevent contamination, a fresh tip should be used for every new reagent. A standard mechanical micropipet has a plunger that has at least two stops (see How to... Use an Adjustable-Volume Micropipet). Many other formats and types of micropipets are available, such as multiple channel, repeater, and electronic micropipets.

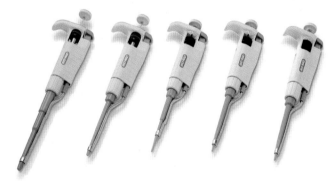

Figure 2.11. **Adjustable-volume micropipets.**

Burettes

A burette is a graduated long glass tube that is similar to a graduated pipet; however, it is filled by pouring a solution into the top and emptied via a stopcock at the base (see Figure 2.15). A burette is usually held upright with a stand and clamp. This glass

How To...

Use an Adjustable-Volume Micropipet

Setting Micropipet Volume

The volume of a micropipet is changed by twisting either a ridged cylinder on the micropipet handle or the top of the plunger, depending on the style of micropipet. When the volume is changed, the new volume is displayed on the readout dial. Figure 2.12 demonstrates how to read the volume on three common micropipet sizes.

Figure 2.12. **Dial readout from adjustable-volume micropipets.** From left to right, 2–20 µl, 20–200 µl, and 100–1,000 µl micropipets reading 15.16 µl, 127.4 µl, and 758 µl, respectively.

Installing a Pipet Tip

A micropipet is always used with a pipet tip. The end of the micropipet is inserted into the open end of the tip and tapped gently while the tip is in the box. This method ensures that the end of the tip remains sterile.

Loading and Dispensing Sample

The standard method for loading and dispensing sample is called forward pipetting. Once the volume is set and a pipet tip is installed, depress the micropipet plunger. The micropipet stops naturally as the target volume is reached—this is called the first, or soft, stop (see Figure 2.13). Place the tip into the solution, and gently release the plunger. Once the set volume is drawn into the tip, move the pipet to the desired location and depress the plunger to deliver the contents. Depressing the plunger further to the second, or hard, stop blows out any residual contents. Remove the tip from the solution before releasing the plunger to avoid sucking the solution back into the pipet.

Always watch the solution going into and out of the pipet tip so that you will notice if the tip is loose or blocked. Never use the hard stop except to blow out contents when using forward pipetting technique.

Figure 2.13. **Plunger positions of micropipets.** From left to right, plunger at rest, at soft stop (first stop), and at hard stop (second stop).

Ejecting a Pipet Tip

Once the liquid has been dispensed, eject the tip into a waste container. Tips are ejected by pressing the tip ejector button.

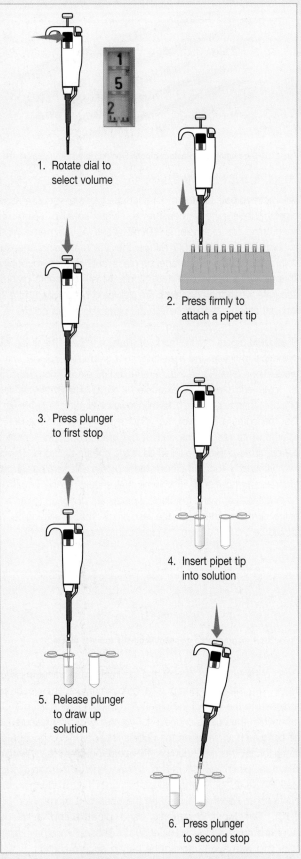

1. Rotate dial to select volume

2. Press firmly to attach a pipet tip

3. Press plunger to first stop

4. Insert pipet tip into solution

5. Release plunger to draw up solution

6. Press plunger to second stop

Figure 2.14. **Steps for using an adjustable-volume micropipet.**

tube is commonly used during titration when the volume of liquid to be dispensed is variable and needs to be measured after dispensing. A measurement is taken before and after dispensing the liquid, and the volume dispensed is then calculated.

$$V_{Final} - V_{Initial} = V_{Delivered}, \text{ where } V = \text{volume}$$

Burettes are usually used to accurately measure volumes between 10–100 ml; for example, a 25 ml burette can have an accuracy of ±0.06 ml.

Figure 2.15. **Burette.**

Liquid Containers

Erlenmeyer Flasks

An Erlenmeyer flask has slanted sides and a narrow opening that facilitate swirling and mixing of the contents without spilling (see Figure 2.16). Erlenmeyer flasks are typically not used for measuring, although the graduations on the flasks provide a rough guide for measuring liquids within approximately 5% of the target volume. Erlenmeyer flasks with the capacity to handle from 25 ml to 4 L are commonly found in a laboratory. Erlenmeyer flasks with fire-polished tops are primarily used for mixing and moving solutions from one location to another; these flasks can be used for short-term storage when covered with Parafilm or another impermeable cover. Erlenmeyer flasks with a culture top or screw cap top are used to grow liquid cultures of microorganisms. Such flasks may have baffles added to the bottom to aid in the agitation and oxygenation of the liquid medium as it is swirled by the shaking incubator. Erlenmeyer flasks may also have ground glass stoppers that enable the contents to be stored temporarily.

Figure 2.16. **Erlenmeyer flask.**   Figure 2.17. **Reagent bottles.**

Reagent Bottles

Reagent bottles are primarily used to store solutions and are available in many shapes and sizes (see Figure 2.17). Reagent bottles can be made of plastic or glass; the type of plastic or glass used depends on how the bottles are to be used. For example, 1 M NaCl can be stored at room temperature in a bottle made of any type of plastic or glass. However, some chemicals, such as strong acids, may react with specific types of plastic and are almost always stored in glass containers. It is also important to note that

some chemicals, such as strong bases, may present a hazard if stored in glass containers. A heat-resistant bottle such as Pyrex glassware should be used if the bottle is to be heated or autoclaved, for example to make an agarose gel, as used in DNA electrophoresis (see Chapter 4). Glass reagent bottles that are made by Wheaton (and commonly referred to as Wheaton bottles) are generally appropriate for storage of solutions; however, these bottles are not constructed of heat-resistant glass and should not be heated or autoclaved. In summary, it is important to confirm that bottles are made of the appropriate material for the task at hand.

Measuring Mass

Mass is measured in grams or in multiples or fractions thereof. A kilogram (kg) is one thousand (10^3) grams, a milligram (mg) is one thousandth (10^{-3}) of a gram, and a microgram (µg) is one millionth (10^{-6}) of a gram. Biotechnology procedures require mass measurements ranging from kilograms to micrograms and use balances to measure these amounts. Biotechnology calculations may even calculate mass to the nanogram (ng) (10^{-9}) or picogram (pg) (10^{-12}) level, but it is not common to actually measure these small quantities on a balance.

Figure 2.18. **Electronic balance.** Photo courtesy of OHAUS Corporation.

Balances

An electronic balance is used to measure mass accurately (see Figure 2.18). Different types of electronic balances address different purposes and needs for accuracy. In general, the more places a balance can read to the right of the decimal point, the more expensive it is. A balance that measures to the nearest gram is less expensive than one that measures to the hundredth of a gram. Balances also have a maximum capacity; the larger the capacity, the more expensive the balance. Never exceed the capacity of a balance to avoid irreparably damaging it. The most common types of balances measure to the hundredth of a gram but laboratories frequently have balances that measure to the thousandth or ten thousandth of a gram. Balances that measure to the thousandth of a gram and below are referred to as analytical balances. These balances generally have a draft shield that prevents the movement of air in the room from affecting the mass display. Analytical balances may also be placed on a granite block to minimize vibrations from the ground.

Mechanical balances are commonly referred to as scales. They physically balance material against an opposing mass to accurately weigh to the tenth of a gram. Although mechanical balances are not as common as electronic balances, many laboratories still use mechanical balances for repetitive tasks that do not require a high degree of precision. Mechanical balances are provided in various formats; for example, the triple beam has sliding mass readings for 100, 10, and 1 g weights (see Figure 2.19). Other mechanical balances use a combination of sliding mass readings and a dial.

Figure 2.19. **Triple beam balance.** Photo courtesy of OHAUS Corporation.

All balances should be used with weigh boats (see Figure 2.20) or weighing paper. Weigh boats and weighing paper are designed so that they do not interact with the material being weighed. Weigh boats and weighing paper are also hydrophobic to enable easy removal of the contents when rinsed with water. Ordinary paper should never be used because some of the material being weighed becomes trapped in the paper and does not make it into the solution. Weigh boats and weighing paper come in various sizes and should be chosen to easily accommodate the amount of solid being weighed. If the weigh boat or weighing paper is too small, the material could spill out onto the balance and contaminate the surrounding area.

To weigh a solid chemical from a stock bottle, gently shake the bottle over a weigh boat or weighing paper. Alternatively, use a clean spatula to transfer the chemical to the weigh boat or weighing paper. To prevent contamination of the remaining stock powder, use a clean spatula. Once removed, the excess chemical should not be returned to the stock bottle.

Figure 2.20. **Weigh boats.**

Measuring Temperature

Temperature in a molecular biology laboratory is measured in degrees Celsius (°C). Many biology experiments are conducted at 37°C, which is normal human body temperature, or at room temperature, which is usually around 22°C. If a reagent is temperature sensitive, it may be stored in a refrigerator at 4°C or in a freezer at –20°C. If a reagent needs to be stored at even lower temperatures, use an ultralow freezer at –80°C or a liquid nitrogen tank at –196°C.

Thermometers

Thermometers range widely in purpose, function, and price. The most common laboratory thermometer is the alcohol thermometer, which contains a colored alcohol solution that expands and contracts in response to changes in temperature and moves up and down a graduated scale (see Figure 2.21). In the past, many thermometers used mercury, but such thermometers are now rare due to the serious safety hazard caused by the exposed mercury if the thermometers are broken. Some laboratory instruments, such as pH meters, contain digital thermometers that provide an electronic readout of the temperature. Infrared thermometers measure and display temperatures when the triggering mechanism is pressed while the thermometer is aimed at the item being measured. Regardless of how a thermometer is used, it needs to be kept clean. Note that most thermometers cannot be autoclaved.

Figure 2.21. **Alcohol thermometers.**

Measuring pH

pH is a measure of the acidity, or the amount of hydrogen ion (H^+) in a solution. The pH of a solution is commonly measured and adjusted using a pH meter. For the purposes of a biology laboratory, pH is the negative log of the H^+ concentration expressed as moles per liter. For example, pure water contains 1×10^{-7} mol/L of H^+ ions and has a pH of 7. The pH of a solution is affected by temperature; therefore, the pH specification is generally given hand in hand with a temperature specification (for example, pH 6.9 at 24.5°C, as in Figure 2.22).

pH Meters

pH meters can be small and handheld (see Figure 2.22) or stationary with arms to hold the electrodes apart from the base unit (see Figure 2.23). The small handheld models are useful and portable but they run on battery power and can be less accurate than stationary models. A pH meter uses two electrodes: a reference electrode and an electrode that is sensitive to H^+ concentration. These two electrodes are frequently housed in a single unit as a combination electrode. The electrodes of a pH meter need to be stored properly—either capped or in a storage solution—to prevent them from drying out. A pH meter is actually a voltmeter that reads the voltages of the reference and measuring electrodes, and then converts the measurements into a pH reading. Prior to use, a pH meter should be calibrated using color-coded buffers of fixed pH (pH 4.0, 7.0, and 10.0 are standard). There is a linear relationship between the voltage measured and the pH of the solution. Thus, two buffers are used for calibration and should bracket the desired pH. For example, if pH 5 is desired, buffers of pH 4.0 and 7.0 should be used. On the other hand, if pH 8 is desired, buffers with pH 7.0 and 10.0 should be used. A pH meter reads the voltages of the calibration buffers, calculates a slope, and uses the slope to calculate the pH of the solution being measured. A modern pH meter guides the user through the calibration steps. First, the meter asks which range to calibrate. It then requests that the electrode be placed in one standard buffer until the readout is steady, then that the electrode be placed in a second buffer until the reading is steady. The meter indicates that it is calibrated and ready for use.

To measure pH, the electrodes should be rinsed with distilled water (dH$_2$O) and then placed in the solution to be measured. The initial reading will fluctuate, so the pH should only be read once the reading is steady. When the reading is completed, the electrodes should be rinsed again and then capped or put into storage solution.

 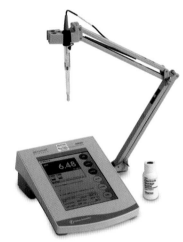

Figure 2.22. **Handheld pH meter.**   Figure 2.23. **Stationary pH meter.**

Calibration

All measuring instruments must be calibrated to ensure they are measuring at the expected accuracy and precision (how consistent the measurements are). The method of calibration is different for each type of instrument. **Calibration** can sometimes be performed by laboratory personnel, but it often requires a specialized technician. Some instruments need to be calibrated more frequently than others; for example, a pH meter should be calibrated each day it is used, while a micropipet or balance should be calibrated at least once a year unless it is out of calibration.

Although specialized technicians are available to calibrate micropipets and balances, most micropipets and balances can be calibrated by laboratory personnel using the instructions provided in the instrument's instruction manual. Micropipet calibration is usually based on comparing the setting of the pipet with the mass of water dispensed by the pipet, since 1 ml of water weighs 1 g. Balances are calibrated based on standard weights. If a laboratory is regulated by industry standards it is required to keep equipment calibration records, which typically include information regarding equipment serial numbers, calibration test results, and next scheduled calibration dates.

Washing Glassware

All glassware must be clean before use. In most laboratories, each individual is responsible for washing and sterilizing his/her own glassware. In larger facilities, a central service may wash and sterilize the glassware. The basic steps to wash glassware are as follows:

Step 1   Add a small amount of laboratory soap, such as Alconox, to the bottle and use a bottle brush (see Figure 2.24) to loosen any material adhering to the walls of the bottle. Make sure all surfaces of the bottle are scrubbed.

Step 2   Rinse with tap water until no more suds are evident. At least three rinses are required.

Step 3   Rinse two times with distilled water (dH$_2$O).

Step 4   Air dry on a rack or in an oven.

Occasionally, more stringent washing procedures are required to clean bottles that are used to store certain chemicals. The washing procedures can involve hazardous chemicals, such as strong acids, bases, or organic solvents.

Figure 2.24. **Bottle brush.**

Disinfecting and Sterilizing

Disinfection is the removal of most microorganisms, while sterilization destroys all microorganisms. Maintaining cleanliness and sterility is crucial when working in a laboratory. Even though a container has been cleaned properly, it may also need to be sterilized. Autoclaving is the primary method used to sterilize media and glassware. Alternative methods can be used to sterilize media and containers if the container cannot be autoclaved due to its constitutive material or if there is no access to an autoclave.

Autoclaving

An **autoclave** is a piece of equipment (see Figure 2.25) that sterilizes liquids, containers, and instruments by increasing the pressure such that the boiling point of water is raised beyond 100°C, which kills most microorganisms. The same phenomenon is used by pressure cookers to cook food more quickly than in conventional pots. Pressure cookers are commonly utilized in many laboratories that cannot afford an autoclave. When using an autoclave, the standard conditions used to kill bacteria and their spores are 121°C and 15–20 pounds/square inch (PSI); however, conditions can vary depending on the materials being sterilized. Autoclaves use a dry cycle for empty containers and instruments and a liquid cycle for solutions. The time needed for sterilization depends on the amount of material in the autoclave but is usually between 15 minutes and 1 hour. For example, 1 L of solution would be autoclaved for 15–20 minutes on the liquid cycle.

It is extremely important to follow the manufacturer's instructions when using an autoclave. When operating the autoclave one must verify that the autoclave is working. Most autoclave bags have an indicator that changes from a light color to a dark color if the sterilization was successful. Autoclave tape is used for items that do not have a built-in indicator. Autoclave tape looks very similar to masking tape but it has white diagonal lines on it that turn black upon sterilization.

Figure 2.25. **Benchtop autoclave.**

Disinfecting Surfaces with Bleach

Basic surface disinfection can be accomplished with a 10% bleach solution. After washing with laboratory soap, the container should be filled to the top with 10% bleach and allowed to sit for 15 minutes. If a cap is present, the cap should also be soaked in the bleach solution. After soaking, the container should be rinsed in sterile dH_2O and covered to prevent recontamination. Bleach also destroys DNA that can remain intact even after autoclaving and treatment with alcohol. Therefore, bleach is useful if contamination with DNA is a concern.

Disinfecting Surfaces with Alcohol

Surface disinfection can be accomplished with 70% isopropyl or ethyl alcohols. Alcohol is usually used for disinfecting exposed surfaces rather than the inside of containers intended for storing solutions. Unlike bleach, alcohol does not destroy DNA and some bacterial spores are unaffected by alcohol.

Disinfecting Using a Microwave Oven

A microwave oven can disinfect containers if they are wet and microwaved on a high setting for 5 minutes. The container must be heat resistant and microwave safe.

Disinfecting with Germicidal UV Light

Available in most standard biological safety cabinets, shortwave ultraviolet (UV) light kills organisms by destroying their DNA. Germicidal UV light disinfects material that cannot be heated or exposed to the other methods described. Because UV light is also useful for preventing potential DNA contamination, it is commonly found in workstation cabinets used to set up polymerase chain reactions (PCR).

Sterilizing Solutions

Solutions used for cell culturing and those that need to last a long time must be sterilized. Many solutions can be sterilized using an autoclave on the liquid cycle. However, solutions that are heat sensitive can be sterilized by filtering them through a 0.22 μm membrane that is impermeable to any contaminants with a diameter >0.22 μm, such as bacteria. Disposable plastic vacuum filter units are available with 0.22 μm membranes that can be attached to a vacuum source in the laboratory. The vacuum source pulls the solutions through the membrane into a sterile container below. Disposable plastic vacuum filter units are available in sizes from 50 ml to 1 L. For smaller volumes, 0.22 μm filters that fit onto a syringe can be used. When sterilizing a small volume of solution, the solution is transferred into the syringe barrel, pushed through the filter disc, and into a sterile container.

Figure 2.26. **Labeled reagent bottle.** The bottle is labeled with the name of the solution, the concentration, the date it was made, and the initials of the person who made the solution.

Labeling

Once a container has been properly cleaned and the contents added, it is imperative that the container be labeled immediately. It is good practice to write the label and place it on the container prior to adding the contents. There are four pieces of information that should be on any filled container: the name of the solution, the concentration, the date it was made, and the name of the person who made the solution (see Figure 2.26). The necessary information should be printed clearly on the label. If the laboratory has a number or a principal investigator, this information should also be included on the label so that anyone who finds the bottle can easily locate its owner. Any safety requirements should also be plainly displayed on the container. Many commercial labels use the National Fire Protection Association (NFPA) hazard identification system. The system uses a diamond-shaped symbol to call out health (blue), fire (red), reactivity (yellow), and other (white) hazards (see Figure 2.27). Hazards are rated on a scale of 1–4 with 4 being the most hazardous. This information is vital to helping emergency personnel determine if chemicals pose a further risk in case of an accident.

Figure 2.27. **National Fire Protection Association (NFPA) hazard identification system.**

2.4 Numerical Data

Scientific research generates data, and much of the information obtained is in the form of numbers, or numerical. Many of the numbers generated are either very large or very small. In addition, the precision of the numerical data depends on the method used to generate them. Scientists use rules and conventions to deal with numerical data and convey the information in the simplest format without having to count multiple zeros.

Significant Figures

It is common for students to report many more digits than are significant. A calculator or instrument may display many more digits than the precision of the instrument, so it is important to understand how the actual digits that convey the limit of precision are determined. The digits that contribute to the precision of the instrument are referred to as **significant figures**, or digits.

When a measurement is taken, the precision of the instrument determines the number of significant figures reported in the result. For example, when an individual uses a ruler to measure a small fish and determines that the fish is 12.5 cm, the result would not be recorded as 12.500 cm since the zeros to the right of the .5 are not significant. Because an ordinary ruler has markings down to the nearest millimeter (mm), one could report only to that level of precision. The easiest way to remember how to determine significant figures is to follow certain rules.

Rule 1: The number of significant figures is the number of digits to the right of the first non-zero number.

For example, 0.0003, 0.002, and 0.01 all have one significant figure. *Exceptions to the rule* are as follows:

- The zeros to the right of a whole number are not significant. For example, 1, 40, 500, 7,000, and 7,000,000 all have one significant figure, while 420, 7,200, and 85,000 all have two significant figures

- If zeros are in between two numbers, they are significant. For example, the numbers 402, 1,020,000, and 1,040 all have three significant figures. **Note:** If a whole number followed by zeros is precise and not an estimate, this can be indicated by placing a decimal point after the number. For example 1,040. has four significant figures

- If the number has a decimal point, the zeros to the right of the decimal point are significant. For example, 12.0000 has 6 significant figures

When performing calculations, it is important to use the correct number of significant figures in the answer. The number of significant figures is based on the least precise measurement. Calculators provide an answer with more digits than are significant. The calculated answer is rounded off to show the correct number of significant figures. The rules for addition and subtraction are different from those used for multiplication and division.

Rule 2: When adding or subtracting numbers, the resulting numbers cannot have any more significant figures to the right of the decimal point than any value used in the calculation.

Example:

21.3 g − 0.232 g = 21.068 g is rounded to 21.1 g.

The value 21.3 is the least precise measurement and has one significant figure to the right of the decimal.

Rule 3: When multiplying or dividing numbers, the answer cannot have any more significant figures than the value in the calculation with the fewest significant figures.

Example:

$$\frac{30.5 \ \mu l \times 2.223 \ mM}{5.00 \ mM} = 13.5603 \ \mu l \text{ is rounded to } 13.6 \ \mu l.$$

The values 30.5 and 5.00 have the fewest number of significant figures (3) than the other values in the calculation.

Scientific Notation

To make numbers easier to read and to communicate the level of precision at which numbers were recorded, scientific notation has been developed. The basis of **scientific notation** is to express a very large or very small number in a shorthand form that uses a number multiplied by 10 raised to a positive or negative number, for very large or very small numbers, respectively. For the value 10^4, 10 is the base and 4 is the exponent; this means that 10 is multiplied by itself 4 times (10 x 10 x 10 x 10). For numbers less than 1, a negative exponent, or reciprocal, of the base 10 is used. For example, the value 10^{-5} means 1/10 is multiplied by itself 5 times (1/10 x 1/10 x 1/10 x 1/10 x 1/10).

To write **1,200,000** in scientific notation:

1. Move the decimal point one digit to the right of the first non-zero number starting from the left. Delete any nonsignificant zeros to the right of the decimal point. In this example, that would leave the number as **1.2**.

2. Place a multiplication sign and the number 10 to the right of that number to give **1.2 x 10**.

3. Use an exponent to indicate the number of places the decimal point was moved. For numbers greater than 10 where the decimal has moved to the left, the exponent will be positive. For numbers less than 1 where the decimal has moved to the right, the exponent will be negative. In this example, the resulting number would be **1.2 x 10⁶**.

To write **1,050,000** in scientific notation:

1. Move the decimal point, keeping the significant digits. The resulting number will be **1.05**.

2. Add the base and exponents to indicate the number of places the decimal point has been moved. The resulting number will be **1.05 x 10⁶**.

To write **0.0000102** in scientific notation:

1. Move the decimal point, keeping the significant digits. The resulting number will be **1.02**.

2. Add the base and exponents to indicate the number of places the decimal point has been moved. The resulting number will be **1.02 x 10⁻⁵**.

Units of Measure

Mass, volume, temperature, and voltage are all measurements that are given in numerical values, followed by the unit of measure (for example, 5 liters and 7.36 grams). Biology is a global science; thus, it is rare for imperial units such as gallons and ounces to be used in experimental procedures. It is vital that all protocols and calculations include the units of measure and that the correct metric prefixes are used (see Table 2.2). Errors in units of measure often lead to experimental failure in the laboratory.

Metric Prefixes

Scientists use metric prefixes to deal with very large or very small quantities. The prefix of the unit of measure is adjusted depending on the size of the number. For example, instead of stating the weight of a compound as 0.0000102 g or 1.02×10^{-5} g, the grams would be converted to micrograms (µg) and stated as 10.2 µg. Metric prefixes are Latin terms used to describe the relationship of the number to the basic unit of measure; for example, "milli" is one thousandth, or 10^{-3}, while "kilo" is one thousand times, or 10^{3}. Each of these prefixes is designated by a specific symbol; for example, milli is designated by m and kilo is designated by k (see Table 2.2). The easiest way to convert from one metric prefix to another is to move the decimal point to the left or to the right the same number of places as the change in the exponent of the base 10.

To convert 0.0000102 grams (g) into milligrams and micrograms, two methods can be used.

Method 1

1 gram = 1,000 (or 10^{3}) milligrams (mg)
1 gram = 1,000,000 (or 10^{6}) micrograms (µg)

To convert 0.0000102 grams to milligrams, the decimal point would be moved 3 places to the right (0.0102 mg). To convert grams to micrograms, the decimal point would move 6 places to the right (10.2 µg).

Alternatively, the unit cancellation method can be used. This method is particularly useful because it also allows conversion of nonmetric units. In this method, a conversion factor is used. The conversion factor always equals 1 and the units of the numerator and the denominator must cancel out.

Method 2

Conversion factors:

$$\frac{1,000 \text{ mg}}{1 \text{ g}} = 1 \qquad \frac{1,000 \text{ µg}}{1 \text{ mg}} = 1$$

Multiply the value by the conversion factor and cancel the units.

$$0.0000102 \text{ g} \times \frac{1,000 \text{ mg}}{1 \text{ g}} = 0.0102 \text{ mg}$$

$$0.0000102 \text{ g} \times \frac{1,000 \text{ mg}}{1 \text{ g}} \times \frac{1,000 \text{ µg}}{1 \text{ mg}} = 10.2 \text{ µg}$$

When learning about unit conversions, try both methods to confirm that they give the same answer.

Table 2.2. **Metric prefixes and symbols.**

Prefix	Symbol	Scientific Notation	Meaning	
kilo	k	10^{3}	1,000	Thousand
deci	d	10^{-1}	0.1	Tenth
centi	c	10^{-2}	0.01	Hundredth
milli	m	10^{-3}	0.001	Thousandth
micro	µ	10^{-6}	0.000001	Millionth
nano	n	10^{-9}	0.000000001	Billionth
pico	p	10^{-12}	0.000000000001	Trillionth

2.5 Preparing Solutions

Scientists make solutions by dissolving solutes into solvents. A solute is the dissolved matter in solution, while a solvent is a liquid that can dissolve other substances. Of the many types of solutions that scientists make, most are aqueous and utilize water as the solvent. Distilled water (dH₂O) or ultrapure water that has been passed through a laboratory purification apparatus, such as a Millipore laboratory water purification system, should be used for making solutions. Deionized water is usually not suitable for laboratory applications. Scientists follow standard methods to make solutions so that the steps can be replicated by others. Concentration is the amount per volume and the concentration of solutes in a solution is described and calculated in many different ways that can potentially be confusing for new scientists. These descriptions include percentages, ratios, proportions, parts, molarity, and normality. The variety of solutions and examples of how to make them are discussed in the following sections.

Percent Solutions

Percent solutions are based on a percentage of the solute in solution using the concept of per 100 parts. Therefore, a 1% solution has 1 part of solute in 100 parts of solution. Percent solutions are expressed either as mass or weight per volume (m/v or w/v) and volume per volume (v/v). A mass or weight per volume measurement describes the amount of a solid compound that is dissolved in the solution. A volume per volume measurement describes the dilution of a liquid solute.

Mass per Volume Percent Solutions

A 5% (m/v) NaCl solution has 5 parts of NaCl for every 100 parts of solution. Thus, to make 100 ml of 5% NaCl (m/v), weigh 5 g of NaCl and dissolve it in approximately 50 ml of water. Then raise the total volume to 100 ml in a graduated cylinder. When making solutions using dry chemicals it is important to start with a volume of solvent that is much less than the final volume. This is to ensure that when the solute is added, the final volume is not overshot. This is called "bringing to volume". When preparing a 5% (m/v) NaCl solution that is not a multiple of 100, a proportion can be used calculate the amount of NaCl that is required.

Example: To make 37 ml of 5% (m/v) NaCl solution, use the following calculation to determine a proportion:

$$\frac{5.0 \text{ g}}{100 \text{ ml}} = \frac{x}{37.0 \text{ ml}}$$

$$x = \frac{5.0 \text{ g}}{100 \text{ ml}} \times 37.0 \text{ ml} = 1.9 \text{ g of NaCl}$$

To prepare 37 ml of 5% NaCl, weigh out 1.9 g of NaCl and dissolve it in approximately 20 ml of water. Then bring the volume up to 37 ml in a graduated cylinder.

Volume per Volume Percent Solutions

A 5% (v/v) 2-propanol solution has 5 parts of 2-propanol for every 100 parts of solution. Thus, to make 100 ml of 5% (v/v) 2-propanol, pipet 5 ml of 100% 2-propanol into a graduated cylinder. Then raise the volume to 100 ml with water.

Example: To make a 5% (v/v) 2-propanol solution whose volume is not a multiple of 100 ml, for example, to make 37 ml of 5% (v/v) 2-propanol, a proportion can be used:

$$\frac{5.0 \text{ ml}}{100 \text{ ml}} = \frac{x}{37 \text{ ml}}$$

$$x = \frac{5.0 \text{ ml}}{100 \text{ ml}} \times 37 \text{ ml} = 1.9 \text{ ml of } 100\% \text{ 2-propanol}$$

To prepare 37 ml of 5% 2-propanol, pipet 1.9 ml of 100% 2-propanol to a graduated cylinder and then bring the volume up to 37 ml with water.

Percentages, Ratios, and Proportions

Percent solutions can also be referred to by a ratio or a proportion. For example, a 1% solution has proportionally 1 part of solute in 100 parts of solution, which can also be stated as a 1:99 ratio of solute to solvent. Other common ways of referring to this ratio are

1 part in 100, 1/100, or a 1:99 ratio. Similarly, a 25% solution can also be stated as a 1/4 or a 1-in-4 solution, which is also a 1:3 ratio since there is 1 part of one solution and 3 parts of another.

Diluting Percent Solutions

To make dilutions of percent solutions, use the formula:

$$C_1 V_1 = C_2 V_2$$

where:
- C_1 = concentration of stock solution
- C_2 = desired concentration of final solution
- V_1 = volume of stock solution
- V_2 = desired volume of final solution

This formula is versatile and can be used to dilute any solution from a more concentrated one. When using this formula it is important to ensure that the units of measure are the same on each side of the equation. If not, the answer can be wrong by an order of magnitude.

Example: The formula can be used to prepare 50 ml of 70% alcohol from a stock solution of 95% alcohol as shown:

$$C_1 V_1 = C_2 V_2$$

Replace the appropriate values for the known values and solve V_1.

$$95\% \text{ alcohol} \times V_1 = 70\% \text{ alcohol} \times 50 \text{ ml}$$

$$V_1 = \frac{70\% \text{ alcohol} \times 50 \text{ ml}}{95\% \text{ alcohol}}$$

$$V_1 = 37 \text{ ml of } 95\% \text{ alcohol}$$

Therefore, to make 50 ml of 70% alcohol, transfer 37 ml of 95% alcohol to a graduated cylinder and add water until the volume reaches 50 ml.

Using Stock Solutions Given in Terms of "x"

Sometimes the concentration of stock solutions is given in terms of "x," where the x refers to the number of times more concentrated the stock solution is compared to the normal working concentration of the solution. A 10x solution is 10 times more concentrated than the working solution. Some common laboratory solutions such as TAE electrophoresis buffer are prepared and stored as a 50x solution, which is then diluted to 1x before use. The formula $C_1 V_1 = C_2 V_2$ calculates the quantity of concentrated solution to make the 1x final solution.

Example: To calculate the volume of 50x TAE needed to make 2 L of 1x TAE electrophoresis buffer, use the formula:

$$C_1V_1 = C_2V_2$$

$$50x\ TAE \times V_1 = 1x\ TAE \times 2\ L$$

$$V_1 = \frac{1x\ TAE \times 2\ L}{50x\ TAE}$$

$$V_1 = 0.04\ L\ or\ 40\ ml\ 50x\ TAE$$

To make 2 L of 1x TAE, measure 40 ml of 50x TAE buffer in a 100 ml graduated cylinder add it to a 2 L graduated cylinder. Add water until the meniscus is at the 2 L mark.

Molar Solutions

The Mole

By definition, a **mole** (mol) of a substance has the same number of molecules as 12 g of carbon and a value of 6.02×10^{23}; this number is also known as Avogadro's constant. The atomic mass of an element in grams is equal to one mole of the atoms of that element. By adding the atomic masses of the atoms in a compound, the molar mass, molecular weight (MW), or formula weight (FW) can be calculated to obtain the mass of the compound that is equal to one mole.

Example: To determine the mass of 1 mole of NaCl:

1. Determine the atomic mass of each atom in the compound by looking at the periodic table.

Na: 22.99
Cl: 35.45

2. Determine the FW by adding the atomic mass of each atom in the molecule. **Note:** If there is more one atom of an element in the molecule, such as with $CaCl_2$, the FW would be the mass of 2 atoms of chlorine added to 1 atom of calcium.

22.99 + 35.45 = 58.44 g/mol
Thus, one mole of NaCl weighs 58.44 g.

Molarity

To make a 1 molar solution, take 1 mole of a compound and dissolve it in solvent to make a 1 L volume. A 1 molar solution has 1 mole of solute per liter of solution. In other words, there are 6.02×10^{23} molecules of that compound in every liter of solution. The symbol for **molarity** is **M**, which stands for moles/liter. Mole and molar are commonly mistaken for one another when solutions are being made. **Mole and M are not the same and should not be used interchangeably.** A mole refers to the specific number of molecules, while molar refers to the concentration. There are multiple ways to calculate the amount of a solute needed to make molar solutions. The first step for all methods is to determine the formula weight of the solute.

Example: To calculate the formula weight of $CaCl_2$ when preparing 1 L of 1 M solution:

1. Determine the atomic mass of each atom involved in the compound by looking at the periodic table.

Ca: 40.08
Cl: 35.45 x 2

2. Determine the FW by adding up the atomic mass of each atom.

FW of $CaCl_2$ = 40.08 + 35.45 + 35.45 = 110.98 g/mol

So, 110.98 g of $CaCl_2$ is equivalent to 1 mol of $CaCl_2$ and would yield a 1 M solution when dissolved in water to make 1 L.

Making Molar Solutions

To make 1 M $CaCl_2$, use a powder funnel to transfer 110.98 g of $CaCl_2$ crystals to a 1 L volumetric flask. Use a squirt bottle filled with water to rinse any remaining $CaCl_2$ crystals into the flask, then add approximately 500 ml of water and dissolve by swirling. Once the crystals are dissolved add the rest of the water being careful that the meniscus just touches the mark on the neck of the flask. This would yield 1 M $CaCl_2$ because 110.98 g of $CaCl_2$ is dissolved to make 1 L of solution. To calculate the mass of solute required for 1 L of a 1 M solution is simple, but making varying volumes of solutions of varying molarities requires additional calculation. There are many ways to calculate molarities; the proportional method and the unit cancellation method are discussed next.

Proportional Method

To use the proportional method, first convert the molarity into moles/1,000 ml, then move the decimal over one place on both the numerator and denominator.

Example: if 110.98 g of $CaCl_2$ in 1 L is a 1 M solution, then 11.1 g $CaCl_2$ in 100 ml also is a 1 M solution.

$$\frac{110.98\ g\ CaCl_2}{1\ L} = \frac{110.98\ g\ CaCl_2}{1,000\ ml} = \frac{11.1\ g\ CaCl_2}{100\ ml}$$

To make 200 ml of 1 M $CaCl_2$, multiply 11.1 g of $CaCl_2$ by 2 to get the quantity of $CaCl_2$ needed. To make 50 ml of 1 M $CaCl_2$, divide 11.1 g of $CaCl_2$ by 2.

Unit Cancellation Method

The unit cancellation method is appropriate when the proportion method is not useful for determining the amount of a solute needed to prepare a molar solution. Many permutations of the unit cancellation method exist; only one is given here.

Example: To make 100 ml of 0.35 M $CaCl_2$, the following formula determines the amount of $CaCl_2$ required:

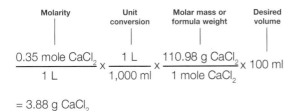

Molarity   Unit conversion   Molar mass or formula weight   Desired volume

$$\frac{0.35 \text{ mole } CaCl_2}{1 \text{ L}} \times \frac{1 \text{ L}}{1,000 \text{ ml}} \times \frac{110.98 \text{ g } CaCl_2}{1 \text{ mole } CaCl_2} \times 100 \text{ ml}$$

$= 3.88 \text{ g } CaCl_2$

The formula used for the unit cancellation method is universal and the unit conversion does not need to be updated. If the volume is already in liters, unit conversion is not necessary when calculating the amount of a compound needed to make liters of solutions.

Example: When making 5 L of 0.22 M NaCl, the formula is adjusted as shown:

Molarity   Molar mass or formula weight   Desired volume

$$\frac{0.22 \text{ mole NaCl}}{1 \text{ L}} \times \frac{58.44 \text{ g NaCl}}{1 \text{ mole NaCl}} \times 5 \text{ L} = 64.28 \text{ g NaCl}$$

Making Compound Molar Solutions

Molar solutions can be composed of multiple compounds, with each compound having a different molarity.

Example: The amount of salts needed to make 100 ml of a solution that contains 0.25 M NaCl and 0.05 M $CaCl_2$ would be calculated as shown:

$$\frac{0.25 \text{ mole NaCl}}{1 \text{ L}} \times \frac{1 \text{ L}}{1,000 \text{ ml}} \times \frac{58.44 \text{ g NaCl}}{1 \text{ mole NaCl}} \times 100 \text{ ml}$$

$= 1.46 \text{ g NaCl}$

$$\frac{0.05 \text{ mole } CaCl_2}{1 \text{ L}} \times \frac{1 \text{ L}}{1,000 \text{ ml}} \times \frac{110.98 \text{ g } CaCl_2}{1 \text{ mole } CaCl_2} \times 100 \text{ ml}$$

$= 0.55 \text{ g } CaCl_2$

To make a 100 ml solution of 0.25 M NaCl and 0.05 M $CaCl_2$, calculate the required mass for each compound separately, then add each compound to a 100 ml volumetric flask. Fill the flask halfway with water, swirl to dissolve the compounds, and then add more water to the 100 ml mark.

Diluting Molar Solutions

Many molecular biology protocols involve diluting stock solutions. The $M_1V_1 = M_2V_2$ formula, which is similar to the one used for diluting percentage solutions, is appropriate when diluting molar solutions.

M_2 = desired molarity of final solution

M_1 = molarity of stock solution

$M_1V_1 = M_2V_2$

V_2 = desired volume of final solution

V_1 = volume of stock solution

Example: To make 50 ml of 0.32 M NaCl from 0.8 M NaCl, the $M_1V_1 = M_2V_2$ formula can be used to calculate the amount of 0.8 M NaCl required:

$M_1V_1 = M_2V_2$

$0.8 \text{ M NaCl} \times V_1 = 0.32 \text{ M NaCl} \times 50 \text{ ml}$

$$V_1 = \frac{0.32 \text{ M NaCl} \times 50 \text{ ml}}{0.8 \text{ M NaCl}}$$

$V_1 = 20 \text{ ml of } 0.8 \text{ M NaCl}$

To make 50 ml of 0.32 M NaCl, add 20 ml of 0.8 M NaCl to a graduated cylinder and then add water until the meniscus reaches 50 ml.

Preparing Small Volume Dilutions

Dilutions can be made using a micropipet if the volumes desired are very small. In many instances, the total volume needed does not exceed 100 µl. The $C_1V_1 = C_2V_2$ or $M_1V_1 = M_2V_2$ formula applies for calculating dilutions, but the volumes are in microliters and require the use of a micropipet.

Example: To make 100 µl of 50 mM $CaCl_2$ from 0.5 M $CaCl_2$, the $M_1V_1 = M_2V_2$ formula still applies:

$0.5 \text{ M } CaCl_2 \times V_1 = 0.05 \text{ M } CaCl_2 \times 100 \text{ µl}$

$$V_1 = \frac{0.05 \text{ M } CaCl_2 \times 100 \text{ µl}}{0.5 \text{ M } CaCl_2}$$

$V_1 = 10 \text{ µl}$

To make 100 µl of 50 mM $CaCl_2$, pipet 10 µl of 0.5 M $CaCl_2$ into a microcentrifuge tube and add 90 µl of water to the tube.

Normality

The **normality (N)** of a solution is the molarity multiplied by the number of equivalents per mole. In biology, normality usually refers to the number of moles per liter of hydrogen or hydroxide ions donated by acids or bases, respectively. Since HCl only donates one H^+ ion per molecule, molarity is equal to normality; therefore, 1 M HCl is 1 N HCl. For sulfuric acid (H_2SO_4), the acid donates two H^+ ions to the solution, so a 1 M H_2SO_4 solution is 2 N.

Chapter 2 Essay Questions

1. Perform the calculations, and describe the method used to prepare 5 L of 50x Tris/acetic acid/EDTA (TAE) buffer using a solid form of Tris and EDTA and a liquid form of acetic acid. The recipe for TAE can be found on the Internet.

2. Describe how you would produce 10 bottles of 750 ml of 1x TAE from a stock of 50x TAE. Include any necessary calculations and steps that should be taken.

3. Write a standard operating procedure (SOP) for the scalable production of 500 ml to 5 L of sterile TE (Tris/EDTA) from Activity 2.D, to be used by technicians in a manufacturing facility. State any assumptions you have made about the facility, including specifications on acceptable margins of error.

Additional Resources

For more biotechnology calculation reference materials and additional problem sets, refer to:

Stephenson F H (2016). Calculations in Molecular Biology and Biotechnology, Third Edition (San Diego: Academic Press).

Seidman L A (2008). Basic Laboratory Calculations for Biotechnology (San Francisco: Pearson Benjamin Cummings).

For additional information on laboratory safety and biotechnology equipment and its use in the laboratory, refer to:

Seidman L A and Moore C J (2009). Basic Laboratory Methods for Biotechnology: Textbook and Laboratory Reference (San Francisco: Pearson Benjamin Cummings).

Activity 2.A DNA Extraction and Precipitation

Overview

This is a very simple and fun laboratory activity in which you will extract DNA from your own cheek cells and precipitate it using salt and alcohol. Once the DNA becomes visible, you will make a necklace with it. This activity gives you a chance to practice writing up experiments in your laboratory notebook.

Deoxyribonucleic acid (DNA) is a molecule that is present in all living things, including bacteria, plants, and animals. DNA carries genetic information that is inherited, or passed down from parents to offspring. It is responsible for determining a person's hair, eye, and skin color, facial features, complexion, height, blood type, and just about everything else that makes an individual unique.

Figure 2.28. **DNA double helix.**

DNA contains four chemical units, referred to by the first letters in their names: A (adenine), G (guanine), T (thymine), and C (cytosine) (See Figure 2.28). These four DNA "letters" make up a code for genetic information. The letters of the DNA code are similar to the letters of our alphabet. The 26 letters in our English alphabet spell words, which can be arranged in infinite ways to create messages and information. Similarly, the four chemical letters of DNA are organized to make messages, called genes, that can be understood by cells. Genes contain the information needed to make proteins, which are responsible for almost all of your body's structures and functions. A gene is like a recipe, since it contains all the information needed to make a protein.

To study DNA, it must first be extracted from cells. To get the DNA out of cells, the cells must be ruptured, or lysed. This is accomplished using a detergent such as sodium dodecyl sulfate (SDS). The proteins that are bound to DNA are digested using a protease that has been added to the lysis buffer. The protease also digest DNases, enzymes that would break down the DNA being harvested. Salt is added to the cell extract to enable the negatively charged DNA polymers to clump together and precipitate when alcohol is added. DNA extraction will be discussed in detail in later chapters.

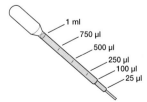

Figure 2.29. **1 ml graduated transfer pipet indicating volume markings.**

Tips and Notes

- A 1 ml graduated transfer pipet has two markings labeled with 0.5 ml, which is 500 µl, and 1 ml. The other markings denote 750 µl, 250 µl, 100 µl, and 25 µl, as shown in Figure 2.29

Safety Reminders: Alcohol is flammable; therefore, take precautions to reduce any possibility of ignition. Always handle your own saliva sample and not anyone else's. All laboratory safety protocols should be followed and PPE must be worn for this activity. Read and be familiar with all SDSs for this activity.

Objectives

- Extract and precipitate cheek cell DNA
- Make a necklace containing DNA
- Record a procedure in a laboratory notebook

Skills to Master

Refer to Laboratory Skills Assessment Rubric (Appendix E) and Laboratory Notebook Rubric (Appendix F) for more details.
- Record laboratory notebook entries
- Follow laboratory protocols
- Select and wear proper PPE
- Extract DNA from cells
- Precipitate DNA
- Maintain a laboratory notebook

Student Workstation Materials

Items	Quantity
Water bath at 50°C with test tube rack or beakers (shared)	1
Transfer pipets	6
Test tube rack or beaker	1
Waste container	1
Materials for making necklaces	4
Laboratory marking pen	1
15 ml tubes with 3 ml of water	4
Lysis buffer	10 ml
Protease and salt solution	1.25 ml
Alcohol on ice	45 ml

Prelab Focus Questions

1. Name six components that should be present in a laboratory notebook entry.
2. Why are sketches important? Should they be labeled?
3. Where is genomic DNA found in a eukaryotic cell? Is DNA found anywhere else in a cell?
4. What proteins might be associated with DNA in a cell?
5. Meat tenderizer is often used to tenderize tough pieces of meat such as steak. Knowing that steak is made of protein-rich muscle tissue from cows, can you think of an explanation for how meat tenderizer works?
6. Discuss one purpose of DNA extraction and precipitation in research or industry.

Activity 2.A DNA Extraction and Precipitation

Protocol

1. Label a 15 ml tube containing 3 ml of water with your initials. (Each student in the team should have a tube.)

2. Gently chew the insides of both cheeks for 30 sec.

3. Take the water from the 15 ml tube into your mouth and swish the water around in your mouth vigorously for 30 seconds.

4. Carefully transfer all of the water back into the 15 ml tube.

Lysis buffer

5. Using a clean transfer pipet, add 2 ml of lysis buffer to the tube containing your cells.

6. Place the cap on the tube and gently invert the tube 5 times to lyse your cells. Do not shake the tube since this could fragment your DNA. Record any observations regarding changes in the cell lysate.

Add 5 drops of protease + salt solution

Protease and salt solution

7. Add 5 drops of the protease and salt solution to the 15 ml tube containing your cell extract. Cap the tube and gently invert it again 5 times to mix.

8. Place the tube containing your cell extract in the 50°C water bath for 10 min. to allow the protease to work.

Alcohol

9. Hold the tube containing the cell extract at a 45° angle and slowly add approximately 10 ml of cold alcohol to the inside wall of the tube. It will take repeated additions with the transfer pipet to add 10 ml of alcohol. Screw the cap back on the tube.

10. Place the tube upright and leave it undisturbed at room temperature for 5 min. Record any observations.

5 min

11. After 5 min, look again at the contents of the tube, especially in the area where the alcohol and cell extract layers meet. Record your observations. Compare your sample with those of your classmates.

12. While the tube is tightly capped, mix the contents by slowly inverting the tube 5 times. Look for any stringy or white material. This is your DNA!

13. Using a transfer pipet, carefully transfer the precipitated DNA along with approximately 750 µl of the alcohol solution into the vial. Then seal the vial to complete making the necklace.

Self Assessment

Assess whether you have mastered the skills of this activity using the Laboratory Skills Assessment Rubric in Appendix E.
Assess your experimental write up in your laboratory notebook using the Laboratory Notebook Rubric in Appendix F.

Postlab Focus Questions

1. Why is there a detergent in the lysis buffer?
2. Why is protease added to the cell lysate?
3. Why is sodium chloride (salt) added to the cell lysate?
4. Why is alcohol added to the cell lysate?

Activity 2.B Pipetting

Overview

Scientists routinely measure various volumes of solutions (see Figure 2.30). The equipment used depends on the volume being measured and the accuracy required. For example, a 100–1,000 µl adjustable-volume micropipet can measure 1 ml of liquid accurately, while a transfer pipet can measure the same volume but with much less accuracy. In these activities, you will be measuring and transferring various volumes of liquid with the proper equipment.

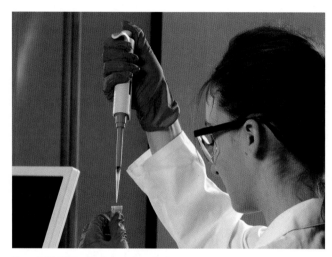

Figure 2.30. **Student using a micropipet.**

Tips and Notes

- When using a transfer pipet to withdraw liquid press the bulb prior to inserting the pipet into the liquid. Once the liquid has been delivered into the appropriate container, keep the bulb pressed until the pipet tip has been removed from the liquid

- Review the How To... Use an Adjustable-Volume Micropipet section on page 36 prior to performing the activity

- Remember to record the correct number of significant figures from your measurements

Safety Reminders: All laboratory safety protocols should be followed, and PPE must be worn for this activity. There are no particular safety issues with this activity. Colored water may stain clothing; therefore, it is advisable that laboratory coats be worn.

Objectives

- Accurately measure volumes using a serological pipet
- Accurately measure volumes using adjustable-volume micropipets
- Accurately measure volumes using a 1 ml graduated transfer pipet
- Calculate percent error of single measurements
- Calculate mean and standard deviation among measurements

Skills to Master

Refer to Laboratory Skills Assessment Rubric (Appendix E) and Laboratory Notebook Rubric (Appendix F) for more details.
- Record laboratory notebook entries
- Follow laboratory protocols (Activity 2.A)
- Use a serological pipet with a pipet pump or filler
- Use a transfer pipet
- Select an appropriate size of micropipet
- Use an adjustable-volume micropipet
- Calculate percent error
- Calculate mean and standard deviation
- Maintain a laboratory notebook (Activity 2.A)

Student Workstation Materials

Items	Quantity
Balance (shared)	1
Pipet pump or filler	1
5 ml serological pipet	1
2–20 µl adjustable-volume micropipet and tips	1
20–200 µl adjustable-volume micropipet and tips	1
100–1,000 µl adjustable-volume micropipet and tips	1
1 ml graduated transfer pipette	1
Paper cup or beaker	1
10–20 ml test tubes	5
Test tube rack	1
Microcentrifuge tube rack	1
Microcentrifuge tubes	3
Precut waxed paper or Parafilm	1–4 pieces
Laboratory marking pen	1
Colored water	50 ml
Waste container	1

Prelab Focus Questions

1. What is the mass of 1 ml of water? What is the mass of 250 µl of water?
2. When is it acceptable to use a sterile graduated disposable transfer pipet instead of a micropipet with a nonsterile tip that measures the same volume?
3. What is the purpose of the cotton plug at the top of a serological pipet?
4. Why do you need to change the tip on a micropipet when transferring different reagents?

Activity 2.B Pipetting

Part 1: Using a Serological Pipet and Pipet Pump

1. Label test tubes as **empty**, **2 ml**, **3 ml**, **4 ml**, and **5 ml**.

2. Determine the mass of a beaker containing one empty test tube: zero the balance and place a beaker containing one empty test tube on the balance. Record the mass in your laboratory notebook.

3. Insert the top of a 5 ml serological pipet (the flat end that has the cotton plug) into the pipet pump. If the pipet is sealed in a plastic wrap, open the wrap from the end with the cotton plug, not from the narrow end with the tip.

4. Place the pipet tip into the beaker of colored water and apply suction to the pipet pump to draw in water until the meniscus just touches the 2 ml mark.

5. Move the pipet to the test tube labeled **2 ml** and release the contents of the pipet into the tube.

6. Repeat steps 4–5 to deliver 3, 4, and 5 ml of colored water into appropriately labeled test tubes.

7. Using the same beaker you weighed in step 2, weigh each test tube containing the 2 ml, 3 ml, 4 ml, and 5 ml of water. Record your results. Calculate the mass of the pipetted solution by subtracting the mass of the beaker with the empty test tube from the mass of the beaker with the filled test tube.

8. Repeat steps 3–7 two more times. Do not pour out any solution. Each new liquid measurement should be added to the liquid already in the test tube. The previous mass measurement of the test tube should be subtracted from the new mass of the test tube. This will give the mass of the liquid newly added to the test tube.

9. Record the results in a table in your laboratory notebook, making sure to include the volume pipetted, the total mass of the beaker plus the test tube, and the final mass of the solution pipetted.

Part 2: Using a 2–20 μl Micropipet

2–20 μl
micropipet

1. Set a 2–20 μl micropipet to deliver 2 μl and insert a clean pipet tip on to the end of the pipet.

2. Press down on the pipet plunger until it stops at the first stop.

3. Lower the tip into the colored water and slowly release the plunger to draw 2 μl of water into the tip.

4. Move the micropipet to the small piece of waxed paper or Parafilm and deliver the 2 μl droplet by pressing down on the plunger until it stops at the first stop.

5. Press the plunger to the second stop to blow out the remaining contents.

6. Continue adding 2 μl droplets to the waxed paper or Parafilm and write your initials with the droplets. Is each droplet the same size? If not, wipe off and repeat. When each droplet is the same size, show your instructor.

Activity 2.B Pipetting

20–200 µl micropipet
2 x 200 µl
1 x 100 µl

p200

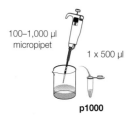

100–1,000 µl micropipet
1 x 500 µl

p1000

500 µl

TP

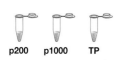

p200   p1000   TP

7.   Wipe off the waxed paper or Parafilm and deliver a fresh 2 µl droplet.

8.   Set the pipet to 4 µl and deliver a 4 µl droplet next to the 2 µl droplet. Then deliver 6, 8, and 10 µl droplets. When comparing the various droplets, do they gradually increase in size? If not, repeat until they do. Show these drops to your instructor.

Part 3: Using 20–200 µl and 100–1,000 µl Micropipets and Transfer Pipets

1.   Label 3 microcentrifuge tubes **p200**, **p1000**, and **TP**. Weigh each microcentrifuge tube and record the mass in your laboratory notebook.

2.   Set a 20–200 µl micropipet to 200 µl. Insert a clean pipet tip on to the end of the micropipet.

3.   Draw in 200 µl of colored water and transfer the liquid to the tube labeled **p200**. Add another 200 µl of colored water to the tube.

4.   Set the micropipet to deliver 100 µl and transfer 100 µl of colored water to the tube labeled **p200**. There should now be a total of 500 µl of colored water in the tube.

5.   Set a 100–1,000 µl micropipet to deliver 500 µl. Insert a clean pipet tip on to the end of the pipet. Transfer 500 µl of colored water to the microcentrifuge tube labeled **p1000**.

6.   Press the bulb of a 1 ml graduated transfer pipet and lower the pipet into the colored water. Gently release the pipet bulb until the liquid passes the 0.5 ml mark on the pipet. Then gently press the bulb down until the meniscus touches the 0.5 ml mark to measure 500 µl. Without releasing the bulb, lift the transfer pipet out of the liquid.

7.   Move the transfer pipet to the microcentrifuge tube labeled **TP** and press the bulb to deliver the 500 µl of colored water.

8.   Compare the volume of the colored water in each of the three 1.5 ml microcentrifuge tubes; the tubes should each contain 500 µl of water.

9.   Weigh all three microcentrifuge tubes and record their mass in your laboratory notebook.

10. Calculate the mass of the liquid in each microcentrifuge tube by subtracting the mass of the empty tube. Remember to use the correct number of significant figures.

Activity 2.B Pipetting

Results Analysis

Part 1: Generate a table to record your results (see Table 2.3). Input the mass of the beaker with the empty test tube and with each filled test tube. Then calculate the following:

• The mass of the newly added liquid in each tube

• The difference between the actual volume pipetted and the expected/target mass of the pipetted liquid (2.00, 3.00, 4.00, and 5.00 g, for volumes 2, 3, 4, and 5 ml, respectively). Remember to use the correct number of significant figures

• The percent error, which is obtained by taking the difference between the actual mass and the expected mass of the pipetted liquid and dividing it by the expected mass of the measured liquid. This is multiplied by 100 to give a percentage. For example, if the value for the 2 ml measurement was 2.05 g, the difference between the actual mass and the expected mass is 0.05 g. The percent error is 0.05 g/2.00 g x 100 which is 2.5%

Determine whether the measurements meet the specifications provided by your instructor. If the specifications are not met, repeat the task and record each result. See Table 2.3 for an example of where, if the specification is +/–0.2 g, the measurement fails the specification. The failure means the task should be repeated.

Table 2.3. **Example of results from Part 1**.

Sample Name	Expected Mass of Measured Liquid	Mass of Beaker and Empty Test Tube	Mass of Beaker and Test Tube with Measured Liquid	Actual Mass of Measured Liquid	Difference Between Actual and Expected	Percent Error	Pass or Fail Specification
2 ml, #1	2.00 g	51.45 g	53.68 g	2.23 g	0.23 g	11.5%	Fail

Once measurements have been recorded, calculate the average (mean), and the standard deviation of the actual mass of the newly added liquid for the three separate measurements of each volume. (See Chapter 9 for a refresher on statistical analysis.) Record these results in a new table. How consistent were your measurements?

Part 2: Sketch the pipetted droplets to provide an indication of droplet size and uniformity. Make sure to sketch all attempts so that you can demonstrate improvement and final mastery of the technique.

Part 3: Record your results in a table similar to the example shown in part 1. Comment on the limit of accuracy of the balance used for weighing compared to the accuracy of the pipet.

Self Assessment

Assess whether you have mastered the skills of this activity using the Laboratory Skills Assessment Rubric in Appendix E.
Assess your experimental writeup in your laboratory notebook using the Laboratory Notebook Rubric in Appendix F.

Postlab Focus Questions

1. What could lead to inaccurate volumes being delivered with a serological pipet?
2. What could lead to inaccurate volumes being delivered with a micropipet?
3. How could you determine the accuracy of a pipet?
4. When measuring 20 µl of liquid, should you use a micropipet with a 2–20 µl range or one with a 20–200 µl range? Why?
5. Which type of pipet should you use to most accurately measure 750 µl of liquid?

Activity 2.C Kool-Aid Column Chromatography

Overview

Preparing solutions is an important skill to master in the laboratory. This activity teaches you how to make volume per volume (v/v) percent solutions. You will make solutions containing different percentages of alcohol. These solutions will be subsequently used to separate the colors of Kool-Aid, a common drink mix (see Figure 2.31), on a chromatography column by using differences in the polarity of the pigments, the chromatography resin, and the alcohol solutions.

Although the pigments of Kool-Aid drink mix have different polarities due to their chemical composition, they are all nonpolar. Alcohol is nonpolar, and as the percentage of alcohol in a solution increases, the polarity of the alcohol solution decreases. When dissolved in a polar environment, such as water, the nonpolar pigments of Kool-Aid mix bind to the nonpolar resin of a Sep-Pak C18 cartridge. Solutions of gradually increasing percentages of alcohol are passed over the cartridge. When the polarity of one of the pigments matches the polarity of the alcohol solution running past, the pigment is released from the resin in the column, dissolved in the alcohol solution, and passed out of the cartridge. The other pigments remain in the cartridge until a solution of alcohol with a matching polarity releases them from the resin.

Note: This activity is based on Bidlingmeyer BA and Warren FV (1984). An inexpensive experiment for the introduction of liquid column chromatography. J Chem Educ, 61, 716–720.

Figure 2.31. **Packet of Kool-Aid drink mix.**

Safety Reminders: Remember that alcohol is flammable; therefore, take precautions to reduce any possibility of ignition. All laboratory safety protocols should be followed, and PPE must be worn for this activity. Read and be familiar with all SDSs for this activity.

Research Question

Can pigments be separated by using differences in their chemical properties?

Objectives

- Make 5% and 25% 2-propanol solutions
- Separate Kool-Aid pigments using column chromatography
- Write up the experiment in your laboratory notebook

Skills to Master

Refer to Laboratory Skills Assessment Rubric (Appendix E) and Laboratory Notebook Rubric (Appendix F) for more details.

- Record laboratory notebook entries
- Perform calculations using the $C_1V_1 = C_2V_2$ equation
- Follow laboratory protocols (Activity 2.A)
- Wash glassware properly
- Select and use appropriate equipment to measure volumes
- Make percent solutions
- Write a label for a reagent bottle
- Use a Sep-Pak C18 cartridge and syringe
- Use graduated cylinders, serological pipets, and micropipets (Activity 2.B)
- Maintain a laboratory notebook (Activity 2.A)
- Interpret results and draw conclusions from data

Student Workstation Materials

Items	Quantity
50 ml graduated cylinder	1
100 ml graduated cylinder	1
Pipet pump or filler	1
10 ml serological pipet	1
10 ml syringe	1
250 ml beaker for liquid waste	1
50 or 100 ml empty reagent bottles	2
Funnel	1
20 ml test tubes	10
Test tube rack	1
Laboratory marking pen	1
Bottle brushes and laboratory soap (shared)	1
Distilled water (dH$_2$O)	250 ml
Concentrated 2-propanol	100 ml
Grape-flavored Kool-Aid drink mix	10 ml
Sep-Pak classic C18 cartridge	1

Prelab Focus Questions and Calculations

1. What is the abbreviation used to describe a percent solution that has a solid added to a liquid?
2. What is the abbreviation used to describe a percent solution where one liquid is added to another?
3. What is the formula used to calculate the quantities needed when making percent solutions?
4. Using the formula $C_1V_1 = C_2V_2$, calculate the volume of 91% 2-propanol needed to make 100 ml of 5% 2-propanol.
5. Calculate the volume of 91% 2-propanol needed to make 100 ml of 25% 2-propanol.

Activity 2.C Kool-Aid Column Chromatography

Protocol

Part 1: Making 5% and 25% 2-propanol Solutions

1. Verify the concentration of 2-propanol is the same as the one used in your calculations and have your instructor check calculations from the prelab focus questions.

2. Wash all test tubes, graduated cylinders, and reagent bottles with laboratory soap and tap water using a brush. Rinse three times with tap water and then two times with dH$_2$O.

3. Label a 100 ml reagent bottle with the name of the solution, concentration, date, and name of the person who prepared the solution. Remember to include the hazards; in this case, label the solution as flammable.

4. Using a serological pipet, or a 50 ml graduated cylinder for a larger volume, measure the calculated volume of 91% 2-propanol needed to make a 5% alcohol solution. Add the 2-propanol to a 100 ml graduated cylinder.

5. Carefully add dH$_2$O to the 2-propanol until the meniscus just touches the 100 ml mark. The alcohol in the graduated cylinder is now at the desired percentage.

6. Using a funnel, add the 5% alcohol solution to the reagent bottle. This is 5% 2-propanol. Store at room temperature.

7. Repeat steps 3–6 to make a 25% alcohol solution.

Part 2: Performing Column Chromatography

1. Place 10 clean test tubes in a test tube rack. Label the test tubes 1 through 10; these test tubes will be used to collect the fractions from the Sep-Pak C18 cartridge as solutions are pushed through. You may need a few additional test tubes depending on your results. Label an empty beaker as waste.

2. Remove the plunger from a 10 ml syringe and attach the long end of the Sep-Pak C18 cartridge to the syringe barrel.

3. Prewet the cartridge by pouring 10 ml of concentrated 2-propanol into the syringe. Attach the plunger and gently push the concentrated alcohol through the cartridge into the waste beaker. (Make sure to push the solution through slowly so that it exits the syringe in fast droplets rather than in a continuous stream of liquid.)

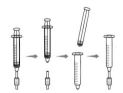

4. Remove the cartridge from the syringe, pull the plunger out, and then reattach the cartridge. (Always perform these steps in this specified order so that the solution flows through the cartridge only in one direction.)

Activity 2.C Kool-Aid Column Chromatography

5. Add 10 ml of dH_2O to the syringe. Attach the plunger and gently push the solution through as described in step 3. Discard the flow-through into the waste beaker.

6. To prepare the syringe, remove the cartridge and pull out the plunger as described in step 4.

7. Add 5 ml of the Kool-Aid drink to the syringe. Do not discard the flow-through. Collect the flow-through fraction in test tube #1 and save it for analysis.

   Note: For this step and the following steps, describe what happens to the cartridge and the flow-through, and record the information in your laboratory notebook. See the results analysis section below for more detail.

8. Prepare the syringe as described in step 4.

9. Add 10 ml of 5% 2-propanol to the barrel. Insert the plunger to slowly push the alcohol through. Collect the flow-through in test tube #2.

10. Repeat steps 8 and 9 to push another 10 ml of 5% 2-propanol and collect the flow-through in the next test tube. Continue to repeat these steps until the flow-through is clear. Record the tube that each solution went into. Each tube is called a fraction.

11. Once the flow-through is clear, push 10 ml of 25% 2-propanol through the column and collect the flow-through in the next numbered test tube. Repeat this step until the flow-through is clear.

12. Finally, push 10 ml of concentrated 2-propanol through the cartridge. Collect the flow-through in the next test tube.

Results Analysis

Review the observations you recorded in your laboratory notebook. Be sure that you have described the following for each fraction collected:

- Color of the flow-through collected in the test tubes

- Colors of the pigments in the Sep-Pak C18 cartridge

  In addition, sketch the appearance of the pigments in the cartridge, making sure to orient your sketch with respect to the direction of flow. Remember to provide a caption for each figure and to label each figure appropriately.

Postlab Focus Questions

1. At what alcohol concentration did the red pigment come off the column? What about the blue pigment?
2. Can pigments be separated by using differences in their chemical properties? What evidence do you have to support this?
3. Which pigment is more polar? Why? What evidence do you have to support this?
4. What can be concluded about the relative polarities of two pigments that require two different concentrations of alcohol to elute off the Sep-Pak C18 cartridge?
5. Name two chemical properties of molecules in solution that scientists can use as a basis for separation.

Self Assessment

Assess whether you have mastered the skills of this activity using the Laboratory Skills Assessment Rubric in Appendix E.
Assess your experimental writeup in your laboratory notebook using the Laboratory Notebook Rubric in Appendix F.

Activity 2.D Making Solutions

Overview

Learning to use graduated cylinders, serological pipets, volumetric flasks, and electronic balances to make solutions builds fundamental skills that are used again and again in any laboratory career. Percent solutions are based on either mass per volume or volume per volume ratios. In Activity 2.C, dilutions of 91% 2-propanol were prepared using the $C_1V_1 = C_2V_2$ formula to calculate the amount of stock solution needed to make the desired concentration. The same formula can be used when the concentration of a stock solution is given as "x" or as the number of times more concentrated the stock solution is than the working 1x concentration. Two electrophoresis buffers that will be used for activities in the following chapters are Tris/acetic acid/EDTA (TAE) and Tris/glycine/SDS (TGS). TAE is commonly provided as a 50x concentrate, while TGS is provided as a 10x solution. In this activity, 50x TAE and 10x TGS will be diluted to a 1x working concentration.

Solutions based on molarities require different calculations. Remember that molarity refers to the concentration of the molecules in a given amount of solvent, while mole refers to the actual number of molecules present. In this activity, you will make molar solutions. $CaCl_2$ can be used in bacterial transformations in Chapter 5.

Tips and Notes

- Refer to the background section Preparing Solutions for help with calculations needed to make molar solutions and dilutions of solutions that are given in terms of "x"

- When measuring a foaming solution, read the volume at the bottom of the foam. The bubbles add very little to the final volume

- When adjusting pH, remember to add the acid or base slowly. When the solution is close to the desired pH, ensure that the acid or base is well mixed with the solution and that the pH reading is steady before adding more acid or base

- EDTA solution is difficult to make. Unlike most compounds, EDTA does not dissolve until its pH is around 8.0. When preparing EDTA solution, adjust the pH to 8.0 while stirring or swirling to dissolve the crystals. The pH drops as more EDTA gets dissolved, so the pH should be adjusted back to 8.0 and stirred until all the EDTA has dissolved

Safety Reminders: Strong acids and bases can cause severe burns; therefore, extra precaution should be taken when working with chemicals. All laboratory safety protocols should be followed and PPE must be worn for this activity. Read and be familiar with all SDSs for this activity.

Objectives

- Make 500 ml of 1x TAE buffer
- Make 500 ml of 1x TGS buffer
- Make 50 ml of 1 M NaCl from solid NaCl
- Make 50 ml of 50 mM NaCl from a stock solution
- Make 500 ml of 0.5 M $CaCl_2$ from solid $CaCl_2$
- Make 10 ml of TE (10 mM Tris, 1 mM EDTA, pH 7.6) from stock solutions of Tris and EDTA
- Write up the procedures in a laboratory notebook

Skills to Master

Refer to Laboratory Skills Assessment Rubric (Appendix E) and Laboratory Notebook (Appendix F) for more details.

- Record laboratory notebook entries
- Perform calculations using the $C_1V_1 = C_2V_2$ equation (Activity 2.C)
- Calculate molarity
- Follow laboratory protocols (Activity 2.A)
- Use balances, graduated cylinders, and pipets
- Make dilutions from concentrated solutions
- Make molar solutions using a volumetric flask
- Make compound molar solutions
- Use and calibrate a pH meter
- Wash glassware properly (Activity 2.C)
- Write a label for a reagent bottle (Activity 2.C)
- Maintain a laboratory notebook (Activity 2.A)

Activity 2.D Making Solutions

Student Workstation Materials

Items	*Quantity*
Autoclave and autoclave tape (optional) (shared)	1
Balance (shared)	1
Weigh boats	4
pH meter and calibration buffers (shared)	1
Magnetic stir plate and stir bar (optional)	1
100 ml volumetric flask	1
100 ml graduated cylinder	1
500 ml graduated cylinder	1
Pipet pump or filler	1
10 ml serological pipet	1
5 ml serological pipets	2
Transfer pipets	3
100 ml beaker	1
100 ml empty reagent bottles	6
500 ml reagent bottles	2
Funnel	1
100 ml filter sterilization units and a vacuum source (optional)	2
Laboratory marking pen	1
Bottle brushes and laboratory soap (shared)	1
Distilled water (dH$_2$O)	1.5 L
Squirt bottle with distilled water (dH$_2$O)	1
50x Tris/acetic acid/EDTA buffer (TAE)	15 ml
10x Tris/glycine/SDS buffer (TGS)	60 ml
NaOH*	>20 ml
HCl*	>20 ml
NaCl	10 g
CaCl$_2$	10 g
Tris	5 g
EDTA disodium dihydrate	15 g

* NaOH is very caustic and can cause burns. HCl is very acidic and causes burns. When working with acids and bases, be very careful and follow all safety precautions. For this activity, NaOH needs to be 4 M or above and HCl needs to be 1 M or above.

Prelab Focus Questions and Calculations

Part 1

1. Using the correct formula, calculate the amount of 50x TAE buffer needed to make 500 ml of 1x TAE buffer.

2. Describe how you would make 500 ml of 1x TAE buffer from a 50x TAE stock solution.

3. Using the correct formula, calculate the amount of 10x TGS buffer needed to make 500 ml of 1x TGS buffer.

4. Describe how you would make 500 ml of 1x TGS buffer from a 10x TGS stock solution.

Part 2

5. Using the correct formula, calculate the amount of NaCl (FW 58.44 g/mol) needed to prepare 100 ml of 1 M NaCl.

6. Describe how you would make 100 ml of 1 M NaCl.

7. Using the correct formula, calculate the amount of 1 M NaCl needed to make 100 ml of 50 mM NaCl.

8. Describe how you would make 100 ml of 50 mM NaCl from a 1 M NaCl stock solution.

9. Using the correct formula, calculate the amount of anhydrous CaCl$_2$ (FW 110.98 g/mol) needed to make 100 ml of 0.5 M CaCl$_2$.

10. Describe how you would make 100 ml of 0.5 M CaCl$_2$.

Part 3

11. How much EDTA disodium dihydrate (FW 372.2 g/mol) would you need to make 50 ml of 0.5 M EDTA?

12. How much Tris (FW 121.14 g/mol) would you need to make 50 ml of 0.1 M Tris?

13. Calculate the amount of each stock solution of 0.5 M EDTA and 0.1 M Tris needed to make a 50 ml solution containing 10 mM Tris and 1 mM EDTA.

14. Which two of three common pH meter calibration buffers (pH 4.0, 7.0, and 10.0) should be used to calibrate the pH meter if the pH values of the solutions in question are between pH 7.6 and 8.0?

15. If the pH of a solution is 5, would hydrochloric acid (HCl) or sodium hydroxide (NaOH) be used to adjust the pH to 9?

16. If the pH of a solution is 10, would HCl or NaOH be used to adjust the pH to 7?

Activity 2.D Making Solutions

Protocol

Part 1: Preparing Solutions with x Units
Making 500 ml of 1x TAE and 500 ml of 1x TGS Solutions

1. Ensure all the calculations to make all the solutions have been performed in the prelab focus questions.

2. Wash all glassware and plasticware using the proper procedure (See Activity 2.C, step 2).

3. Prepare to make 500 ml of 1x TAE by labeling a 500 ml reagent bottle with the concentration, substance, date, and name of the person preparing the solution.

4. Measure the previously calculated volume of 50x TAE buffer using a graduated cylinder or serological pipet and add it to a 500 ml graduated cylinder. Remember to measure where the meniscus touches the calculated volume.

5. Add dH₂O to the 500 ml graduated cylinder until the meniscus touches the 500 ml mark.

6. Using a funnel, add the contents of the graduated cylinder to the reagent bottle. This is 1x TAE.

7. Make 500 ml of 1x TGS solution by following steps 2–6 using 10x TGS buffer instead of 50x TAE buffer.

   Note: Be sure to use the correct volume of 10x TGS.

8. Store TAE and TGS at room temperature.

Part 2: Making Molar Solutions
Making 100 ml of 1 M NaCl Solution

1. Wash all glassware and plasticware using the proper procedure.

2. Prepare to make 100 ml of 1 M NaCl by labeling a 100 ml reagent bottle.

3. Use a weigh boat and balance to weigh out the previously calculated amount of NaCl. Remember to zero (tare) the balance after placing the weigh boat on the pan.

4. Place a small powder funnel into the top of a 100 ml volumetric flask. Transfer the NaCl into the flask. Use a squirt bottle of dH₂O to rinse the remaining NaCl crystals from the weigh boat into the funnel and to rinse the contents of the funnel into the flask.

5. Fill the flask approximately half full with dH₂O. Swirl to dissolve the NaCl.

6. Once all the NaCl crystals are dissolved, carefully add dH₂O to the flask until the meniscus touches the etched mark on the neck of the flask. (A squirt bottle or transfer pipet makes it easier to add water drop by drop when near the etched mark.)

7. Using a funnel, add the contents of the volumetric flask to the reagent bottle. This is 1 M NaCl. Store at room temperature.

Activity 2.D Making Solutions

Making 100 ml of 50 mM NaCl Solution from 1 M NaCl Stock Solution

8. Wash all glassware and plasticware.

9. Prepare to make 100 ml of 50 mM NaCl by labeling a 100 ml reagent bottle.

10. Measure the previously calculated volume of 1 M NaCl solution using a graduated cylinder or a serological pipet and add it to a 100 ml volumetric flask.

11. Add dH_2O until the meniscus touches the mark on the neck of the flask.

12. Using a funnel, add the contents of the volumetric flask to the reagent bottle. This is 50 mM NaCl. Store at room temperature.

Making 100 ml of 0.5 M CaCl$_2$ Solution

13. Wash all glassware and plasticware.

14. Prepare to make 100 ml of 0.5 M $CaCl_2$ by labeling a 100 ml reagent bottle.

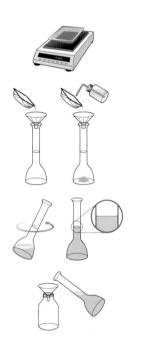

15. Use a weigh boat and balance to weigh the previously calculated amount of $CaCl_2$. Remember to zero (tare) the balance after placing the weigh boat on the pan.

16. Place a small funnel into the top of a 100 ml volumetric flask and transfer the $CaCl_2$ crystals into the flask. Use a squirt bottle of dH_2O to rinse the remaining $CaCl_2$ from the weigh boat into the funnel. Rinse the contents of the funnel into the flask.

17. Fill the flask approximately half full with dH_2O. Swirl to dissolve the crystals.

18. Add dH_2O to the flask until the meniscus touches the etched mark on the neck of the flask.

19. Using a funnel, add the contents of the volumetric flask to the reagent bottle. This is 0.5 M $CaCl_2$. Store at room temperature.

20. (Optional) Sterilize the solution using an autoclave (see Activity 3.A). Alternatively, filter-sterilize the solution.

Part 3: Making Compound Molar Solutions of Specific pH
Making 50 ml TE Solution (10 mM Tris, 1 mM EDTA, pH 7.6)

Note: To make a TE solution that contains 10 mM Tris and 1 mM EDTA, stock solutions of Tris and EDTA must be made at a specific pH. 50 ml of 0.1 M Tris, pH 7.6, and 50 ml of 0.5 M EDTA, pH 8.0, will be made.

1. Wash all glassware and plasticware.

2. Prepare to make 50 ml of 0.1 M Tris, pH 7.6, by labeling a 100 ml reagent bottle.

3. Weigh out the previously calculated amount of Tris needed to make 50 ml of a 0.1 M solution.

Activity 2.D Making Solutions

4. Transfer the solid Tris to a 100 ml beaker containing 30 ml of dH_2O. Dissolve the Tris using a magnetic stir plate with a stir bar or by swirling.

5. Calibrate the pH meter if necessary.

6. When the Tris is dissolved, read the pH of the solution. Decide whether acid or base is required to adjust the pH to 7.6.

7. Adjust the pH of the Tris solution by adding HCl or NaOH to the beaker dropwise with a transfer pipet mixing until the pH is 7.6. Then transfer the solution to a graduated cylinder, and add dH_2O to bring the volume to 50 ml.

8. Transfer the Tris solution to a labeled reagent bottle. This is 0.1 M Tris, pH 7.6. Store it at room temperature.

9. Wash all glassware and plasticware.

10. Prepare to make 50 ml of 0.5 M EDTA, pH 8.0, by labeling a 100 ml reagent bottle.

11. Weigh out the calculated amount of EDTA needed to make 50 ml of a 0.5 M solution.

12. Add the EDTA crystals to a 100 ml beaker containing 30 ml of dH_2O. Mix using a magnetic stir plate and stir bar or by swirling; the EDTA does not dissolve at this stage (see Tips and Notes on page 57).

13. Read the pH of the solution.

14. Adjust the pH to 8.0 by adding HCl or NaOH to the beaker dropwise with a transfer pipet while mixing the solution. The pH will fall as the EDTA continues to dissolve so readjust the pH by mixing until all the EDTA dissolves and the pH is steady at 8.0. Then add dH_2O to bring the volume to 50 ml in a graduated cylinder.

15. Transfer the solution to a labeled reagent bottle. This is 0.5 M EDTA, pH 8.0. Store it at room temperature.

16. Prepare to make 50 ml of TE (10 mM Tris, 1 mM EDTA, pH 7.6) by labeling a 100 ml reagent bottle.

17. Add the calculated volumes of 0.1 M Tris, pH 7.6, and 0.5 M EDTA, pH 8.0, into a graduated cylinder and bring up the volume to 50 ml with dH_2O.

18. Transfer the solution to a labeled reagent bottle. This is TE. Store it at room temperature.

19. (Optional) Sterilize the solution using an autoclave (see Activity 3.A). Alternatively, filter-sterilize the solution.

20. Wash all glassware and plasticware.

Self Assessment

Assess whether you have mastered the skills of this activity using the Laboratory Skills Assessment Rubric in Appendix E.
Assess your experimental writeup in your laboratory notebook using the Laboratory Notebook Rubric in Appendix F.

Postlab Focus Questions

1. State two types of solutions for which you would use the $C_1V_1 = C_2V_2$ formula. State one type of solution for which you would use the $M_1V_1 = M_2V_2$ formula.
2. When would you use a volumetric flask to make solutions instead of a graduated cylinder and vice versa?

Activity 2.E Titration

Overview

Titration is a term used to describe the technique of varying the amount of a substance to produce a specific effect. In medicine, the dosage of a drug is adjusted, or titrated, in a patient until a specific effect is achieved. In chemistry, titration can be used to measure the concentration of a solution. A solution with an unknown concentration is titrated until it reacts with a reactant of known concentration to produce a measurable change, such as a color change. In biotechnology laboratories, titration can measure the concentration of products of an enzyme reaction. This activity gives students the opportunity to practice titration skills using a burette. In this activity, the concentration of sodium hydroxide (NaOH) is determined by reacting it with a known concentration of hydrochloric acid (HCl). The progression of the reaction is determined by using phenolphthalein, a pH indicator that is colorless in an acidic environment and pink in a basic environment. Thus, when the solution turns pink, the NaOH has neutralized the acid; this point is referred to as the end point of the titration.

$$NaOH + HCl \rightarrow NaCl + H_2O$$

According to the chemical reaction equation above, one molecule of NaOH is required to neutralize one molecule of HCl. Thus, upon neutralization, the number of molecules of HCl equals the number of molecules of NaOH, allowing the concentration of NaOH to be calculated using the $M_1V_1 = M_2V_2$ formula.

Tips and Notes

- Patience is crucial for this activity. Make sure to add the NaOH solution dropwise when the end point is close. If the end point is overshot by even a drop, the titration should be repeated

Safety Reminders: Strong acids and bases can cause severe burns; therefore, extra precaution should be taken when working with these chemicals. All laboratory safety protocols should be followed and PPE must be worn for this activity. Read and be familiar with all SDSs for this activity.

Objectives

- Titrate HCl with NaOH
- Determine the concentration of NaOH in a solution

Skills to Master

Refer to Laboratory Skills Assessment Rubric (Appendix E) and Laboratory Notebook Rubric (Appendix F) for more details.

- Record laboratory notebook entries
- Follow laboratory protocols (Activity 2.A)
- Titrate a solution
- Use a burette
- Use graduated cylinders and pipets (Activity 2.B)
- Calculate unknown molarity
- Calculate mean and standard deviation (Activity 2.B)
- Maintain a laboratory notebook (Activity 2.A)

Student Workstation Materials

Items	Quantity
Magnetic stir plate and stir bar (optional)	1
Stand and clamp	1
Flask	1
Pipet pump or filler (optional)	1
10 ml serological pipet or 25 ml graduated cylinder	1
25 or 50 ml burette	1
Transfer pipet	1
Funnel (optional)	1
0.1 M HCl	50 ml
NaOH solution (unknown concentration)	50 ml
1% phenolphthalein in ethanol	0.5 ml
Laboratory marking pen	1
Waste container for liquid waste	1

Prelab Focus Questions

1. What is the end point of a titration and how is it indicated?
2. If 12.4 ml of NaOH solution neutralizes 10 ml of 0.1 M HCl, what is the concentration of NaOH?

Activity 2.E Titration

Protocol

1. Set up the titration apparatus. Clamp the burette into the stand so that it is vertical with the stopcock at the bottom. Close the stopcock.

2. Using a funnel or a pipet, fill the burette with the NaOH solution of unknown concentration. Read and record the volume of NaOH in the burette by reading the bottom of the meniscus.

3. Measure 10 ml of 0.1 M HCl and pour it into a flask.

4. Add 2 drops of phenolphthalein to the HCl solution using a transfer pipet. Add a stir bar to the flask and place the flask on a stir plate directly under the burette. If no stir plate is available, swirl the solution to mix. Place a white background, such a piece of paper, under the flask.

5. Use the stopcock to slowly add the NaOH solution to the HCl solution in the flask. The HCl solution should initially turn pink, which then disappears upon stirring or swirling. Continue adding the NaOH solution until the pink color takes longer to fade. Then add the NaOH solution in single drops, making sure that each drop is thoroughly mixed before adding the next.

6. Once the HCl solution stays pink color, the acid has been neutralized. Read the volume of NaOH on the burette and calculate the amount of NaOH solution required to neutralize the acid.

7. Repeat steps 1–6 two more times and then calculate the mean (average) volume of the NaOH solution required to neutralize the HCl solution.

8. Calculate the concentration of NaOH.

Postlab Focus Questions

1. What is the concentration of NaOH?
2. Why was the experiment repeated three times? Calculate your mean and standard deviation from the volumes of NaOH. (See Chapter 9 for a refresher on statistical analysis.) How accurate were your measurements?
3. In your experiment, did you overshoot the end point? Considering the known factors in this experiment and the fact that the color change is reversible, how could this method be rescued after the end point has been overshot?
4. Name two places in the experiment where errors could have been introduced, thereby changing the calculated molarity of the NaOH solution.

Self Assessment
Assess whether you have mastered the skills of this activity using the Laboratory Skills Assessment Rubric in Appendix E.
Assess your experimental writeup in your laboratory notebook using the Laboratory Notebook Rubric in Appendix F.

Activity 2.F Writing a Standard Operating Procedure

Overview

Standard operating procedures (SOPs) are necessary in biotechnology laboratories to ensure that all members of the laboratory conduct the same task using the same protocol. For example, there are SOPs for calibrating instruments and for making common solutions. In this activity, you will write an SOP for how to make an entry in a laboratory notebook. The section in this chapter on laboratory notebooks describes how to use them and discusses the different sections and requirements of a good laboratory notebook. Use this information to write an SOP that you will then follow for the laboratory activities in this class. For additional information, refer to How To...Write an SOP in Chapter 1.

Objectives

- Write an SOP entitled **SOP.001. Laboratory Notebook Entries**

Skills to Master

Refer to Laboratory Skills Assessment Rubric (Appendix E) and Laboratory Notebook Rubric (Appendix F) for more details.
- Write an SOP
- Maintain a laboratory notebook (Activity 2.A)

Student Workstation Materials

Items	Quantity
Computers	1–4
Printer (shared)	1

Prelab Focus Questions

1. What does SOP stand for?
2. State two reasons SOPs are necessary.

Activity 2.F Writing a Standard Operating Procedure

Protocol

1. Generate the header for the SOP (see Figure 1.18 for an example). The header should include the name of the institution (your school or college), the SOP number and title (in this case, SOP.001 and Laboratory Notebook Entries, respectively), the author, the person who authorizes the SOP (in this case, the name of your instructor), and the effective date of the SOP.

2. Add a heading entitled Purpose. Write a short statement describing the purpose of the SOP.

3. Add a heading entitled Scope. Write a statement that describes the personnel covered by the SOP. (Since the SOP covers all students in the program, make sure to also include the name of the program or course.) State the responsibilities of students and course instructor(s) regarding the SOP. The responsibilities of students and instructor(s) will be different. The statement on the responsibilities should start with the phrase "It is the responsibility of the students to ..." or "It is the responsibility of the course instructor(s) to ..."

4. Add a heading entitled Additional Documentation. Record the type of laboratory notebook used in the course. State any specific requirements for the book and sections that should be present in a laboratory notebook (see the Laboratory Notebook Structure section on page 32 for guidance).

5. Add a heading entitled Procedure. Using the Components of a Laboratory Notebook Entry section on page 32 as a guide, write step-by-step instructions on how to document an experiment in a laboratory notebook. The Procedure section should be split into major steps that are sequentially numbered starting from 1; each major step can then be split into sub-steps starting from 1.1. If necessary, add any additional requirements from the course instructor. The steps should be brief and clear and should avoid using the word "you."

6. Add a heading entitled Definitions. In this section, define any acronyms, abbreviations, or new terms used in the SOP. This section should include terms such as SOP, laboratory notebook, and any other terms that may need to be defined to an individual using this SOP.

7. Add a heading entitled Revision History. In this section add a table with the headings: Version, Date of Revision, Author of Revision, and Description of Changes. Complete the table with the appropriate information. Since this SOP document is going to be the first version, the description of changes should be "new document."

8. Compare your draft SOP to those of your classmates and see how your SOP can be improved. If desired, combine the draft SOPs from the entire class to create a single document that will be followed by everyone throughout the course.

When performing the next laboratory activity, use your SOP or the SOP generated by the class to test whether it contains all the directions necessary to make a laboratory notebook entry. As the course progresses, consider revising the SOP to make it even clearer and more thorough as you gain experience in making notebook entries.

Self Assessment

Assess whether you have mastered the skills of this activity using the Laboratory Skills Assessment Rubric in Appendix E.

Postlab Focus Questions

1. Name three essential components of a laboratory notebook entry.
2. Name four headings required in an SOP.
3. Why is a revision history necessary in an SOP?

Chapter 2 Extension **Activities**

1. Read and review SDSs for chemicals (for example, sodium chloride, sodium hydroxide, and phenol).

2. Find various materials, such as leaves, fruit juice, and candy, containing pigments that could be separated using alcohol percentages and Sep-Pak C18 cartridges.

3. Compare Sep-Pak C18 chromatography with other forms of column chromatography, such as hydrophobic interaction and size exclusion chromatography.

4. Practice making additional solutions, such as 50x TAE and 10x TGS, from individual component chemicals.

Microbiology and Cell Culture

Summary

Cells are the basic unit of life. Prokaryotes are single-celled organisms, while eukaryotes can be single-celled or multicellular organisms. A little more than 150 years ago, scientists discovered that infectious diseases were mainly caused by microorganisms. Since that discovery, humans have been looking for ways to combat disease-causing microorganisms. Microorganisms have also been used for millennia to serve people's needs. For example, lactobacilli and yeast are used to make yogurt and bread respectively. In modern biotechnology, eukaryotic cells and microorganisms are used in cell culture systems as microscopic factories. Microorganisms are also often used to produce proteins that are used as drugs in health care or enzymes for industrial processes. The manipulation of cells in biotechnology is not without controversy; many people question the creation of genetically engineered organisms and the use of human embryonic stem cells. The activities in this chapter introduce students to the basic skills needed to study and culture cells in the laboratory.

3.1 Microbiology and Cell Biology

Cell biology is the study of cells. Microorganisms are comprised of single cells or small clusters of cells; in contrast, human tissues are made of millions of cells working in concert. Cells are either prokaryotic or eukaryotic. Eukaryotic cells have a nucleus and other membrane-bound organelles. Prokaryotic cells do not have a nucleus or membrane-bound organelles and are much smaller than eukaryotic cells (see Figure 3.1).

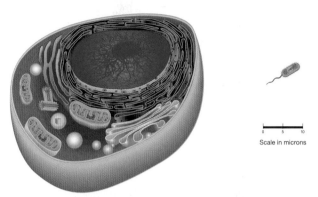

Scale in microns

Figure 3.1. **Size comparison between eukaryotic and prokaryotic cells**. Eukaryotic cells (left) are much larger than prokaryotic cells (right).

Part of cell biology, **microbiology** is the study of microorganisms and their effects on other living organisms. The terms microbe and **microorganism** refer to organisms that must be viewed at the microscopic level such as bacteria, yeast, algae, protozoa, fungi, and viruses. (**Note:** There is no consensus on whether viruses are living organisms; however, since viruses can be viewed with a microscope, they are often included under the umbrella of microbiology.) The majority of naturally occurring microbes are harmless, and many are used in biotechnology to benefit mankind. However, harmful microbes cause many types of disease affecting people, animals, and plants.

An understanding of cell biology and microbiology is necessary in biotechnology because cells do much of the work of biotechnology, such as making **recombinant DNA** and proteins, and because cells are the targets of many biotechnological products. For example, drugs are engineered to target specific proteins on the plasma membranes of human cells.

Three Domains of Life

Life on earth is organized into three domains: Archaea, Bacteria, and Eukarya (see Figure 3.2). Archaea and Bacteria are prokaryotes. Eukarya include yeast, algae, and humans. There are many more prokaryote species than eukaryote species on earth, and there are more Bacteria than Archaea.

Archaea were not recognized as a separate domain until the 1970s. Currently, there are no known human diseases caused by Archaeans. Archaea are found in most habitats on earth, including the human digestive system. Many Archaea live in very harsh

environments, such as deep ocean volcanic vents or salt lakes, and are called extremophiles. Archaea are being used as sources of enzymes used for harsh manufacturing or experimental conditions (for example, enzymes used to process food at high temperatures or for chemotherapeutics and antiviral compounds). Bacteria and eukaryotic cells are described in detail in this chapter.

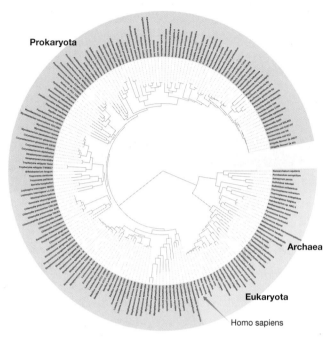

Figure 3.2. **The three domains of life**. This is a phylogenetic tree based on the genetic similarities between organisms that have had their genomes sequenced. Three domains can be seen: Archaea (in green), Bacteria (in blue), and Eukarya (in pink). The position of humans (*Homo sapiens*) is shown. This image was generated using Interactive Tree Of Life (iTOL), an online phylogenetic tree viewer. iTOL (2016) itol.embl.de. Accessed October 17, 2017.

Microorganisms and History

Throughout history, humans have been afflicted by diseases. The impact of diseases such as malaria, anthrax, and plague has changed the course of history. Analyses of the remains of King Tutankhamun in Egypt (see Figure 3.3) indicate that malaria, which is caused by a single-celled eukaryotic protozoan from the genus *Plasmodium*, contributed to the king's death more than 3,000 years ago. Malaria also had a role in the decline of the Roman Empire 1,500 years ago. Anthrax is caused by the *Bacillus anthracis* bacterium, whose spores are common in soil. Anthrax is thought to be the source of the fifth plague of Egypt, mentioned in the Book of Exodus in the Bible, that devastated Egyptian livestock. Plague, which is caused by the bacterium *Yersinia pestis*, is believed to be the source of the Black Death that killed millions of people during the Middle Ages and resulted in the loss of 30–60% of Europe's population.

Almost half of the world's population is at risk for malaria. In 2015, there were approximately 212 million malaria cases and around 429,000 deaths, most of which occurred among children in sub-Saharan Africa. A major effort is under way by the World Health Organization (WHO) to reduce the number of people killed by malaria. Plague infects thousands of people worldwide each year, but advances in antibiotics and other treatments have dramatically reduced fatalities. Anthrax rarely infects humans (see Figure 3.4) but remains a threat to society because it could be used as a biological weapon in terrorist attacks. For example, in 2001 anthrax spores mailed to numerous news offices and two U.S. senators killed five people.

Figure 3.3. **Death mask of King Tutankhamun**. Research published in 2010 by a team of Egyptologists and geneticists led by Zahi Hawass found genetic evidence of malaria in King Tutankhamun's mummy, suggesting it contributed to his early death.

In the 19th century, medical researchers started to understand the causes of diseases, their transmission, and ways to combat them. Anthrax infection was frequently used as a model system by these early scientists. **Model systems** are used by scientists to investigate something new in an organism or biological system they already know a lot about. Model systems are useful for gaining knowledge that can be applied to other systems. In this case, 19th century scientists applied the knowledge they gained from studying anthrax in livestock to diseases that infect humans. In the 1850s and '60s, French physician Casimir Davaine and dermatologist Pierre Rayer discovered *B. anthracis* and were the first to show it caused anthrax. Davaine demonstrated that anthrax could be directly transmitted from one cow to another. In the 1870s, the German physician Robert Koch further investigated anthrax and established that the true cause of the disease was anthrax endospores that persisted in the soil. In the 1880s, French chemist Louis Pasteur used weakened forms of anthrax to immunize sheep. Pasteur used what he learned from anthrax to develop a vaccine against rabies.

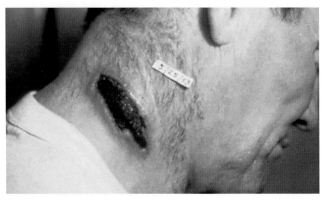

Figure 3.4. **Cutaneous anthrax lesion on the neck**. Courtesy of Centers for Disease Control and Prevention.

Koch and Pasteur are recognized as two founders of microbiological research. Koch used the knowledge he gained from his research into anthrax (and also tuberculosis, or TB) to develop a series of postulates to establish whether a microorganism is the causative agent in a disease.

Koch's postulates are:

1. The microorganism must be found in abundance in all organisms suffering from the disease but should not be found in healthy animals.
2. The microorganism must be isolated from a diseased organism and grown in pure culture.
3. The cultured microorganism should cause disease when introduced into a healthy organism.
4. The microorganism must be re-isolated from the inoculated, diseased experimental host and identified as being identical to the original specific causative agent.

Koch used these postulates to discover the organisms responsible for diphtheria, pneumonia, meningitis, leprosy, plague, and syphilis, to name a few. Koch was awarded the Nobel Prize in 1905 for the identification of the bacterium *Mycobacterium tuberculosis* as the cause of TB. At the time, TB caused one in seven deaths. Koch's postulates apply to many infectious diseases, but as Koch himself discovered, some diseases—such as cholera, which can be carried by asymptomatic hosts—do not meet all the criteria.

Antibiotics are the main line of defense against bacterial infections. The first antibiotic was discovered in 1928 by Alexander Fleming, when he came back to his laboratory after vacation. He observed a mold growing on one of his plates containing staphylococci. He saw a clear zone around the mold, although bacteria were still growing farther away from the mold. This "zone of inhibition" was caused by penicillin produced by the mold. Large-scale production and use of antibiotics began in the 1940s to treat soldiers injured in World War II.

As humankind develops treatments, microbes evolve to resist treatment. A pathogenic bacterium that has recently evolved is methicillin-resistant *Staphylococcus aureus,* or MRSA. MRSA is a strain of *S. aureus* that has developed resistance to most antibiotics and is extremely difficult to treat. *S. aureus* is a common inhabitant on the human skin and causes pimples, skin infections, wound infections, and toxic shock syndrome. Methicillin, an antibiotic related to penicillin, was first used in English hospitals in 1959. Two years later, methicillin-resistant *S. aureus* was detected in hospitals. In 2005, MRSA caused almost 100,000 infections and 18,000 deaths in the U.S. alone. New therapies are being developed to combat MRSA, including new antibiotics, viruses that attack bacteria (phage therapy), and maggot therapy to help clean out wounds. As a result, although MRSA remains a significant public health problem in the United States, rates of hospital-onset severe MRSA infections are declining.

Work is being done to develop new antibiotics and strengthen the antimicrobial pipeline. Also, resources are being provided to support proper use of antibiotics in developing countries in order to slow the spread and development of antibiotic resistance.

3.2 Bacteria

Bacteria are much less complex than eukaryotic cells. The inside of the cell is referred to as the cytoplasm and contains 70S ribosomes used in protein synthesis during translation. Most bacteria contain a single loop of genomic DNA (gDNA) in a centralized area called a nucleoid and sometimes have small extra loops of DNA called plasmids, which are discussed in Chapter 5 and are important tools in genetic engineering.

Bacteria are enclosed by a plasma membrane and a **peptidoglycan** cell wall (see Figure 3.5). The cell wall is composed of two alternating sugars, N-acetylmuramic acid (NAM) and N-acetylglucosamine (NAG), in a polymer that is crosslinked with small peptides. Penicillin-based antibiotics prevent bacterial growth by disrupting the formation of the bacterial cell wall by inhibiting peptide crosslinking, which makes the cell wall weak. Bacteria are classified into two groups based on the thickness of their peptidoglycan cell wall (see Figure 3.6). Thick cell walls enable bacteria to absorb a microbiological stain called **Gram stain**, while thin walls cannot retain the stain. Bacteria are classified as **gram-positive** if they take up the stain or **gram-negative** if they do not. Bacteria classified as *Mycoplasma*, however, do not take up Gram stain because they lack a cell wall and have only a thick cell membrane. Bacteria may also be enclosed in a protective polysaccharide capsule that lies outside the cell wall.

Small hair-like projections called pili are often present on the outside of bacteria and help with cell-cell contact and adhesion. Many bacteria also have flagella that enable them to move and swim in aquatic environments.

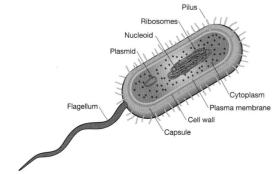

Figure 3.5. Bacterial cell. Bacteria have a plasma membrane, cell wall, nucleoid, and ribosomes. Many also have pili and flagella for locomotion, and some have plasmids.

Names and Shapes of Bacteria

Bacteria are named and classified by their shape. There are three major shapes of bacteria: coccus, bacillus, and spirillum. Cocci bacteria are spherical in shape and look like small balls under a microscope. Bacilli bacteria are oval or rod shaped and look like hot dogs. Spirilli bacteria are spiral-shaped (see Figure 3.7). The way bacteria arrange themselves when growing is also used in their

names. For example, the prefix "strepto" comes from the Greek word *streptos*, which means twisted chain. When used in combination with the shape of a bacterium; for example, streptococcus, the name describes what the bacterium looks like. The term "staphylo" in staphylococcus comes from the Greek word *staphule*, which means a bunch of grapes.

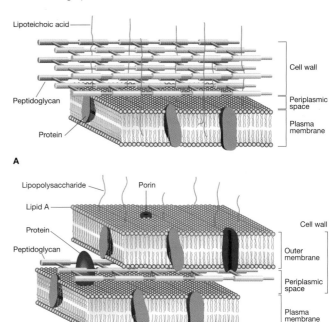

Figure 3.6. Bacterial cell walls. Gram-positive and gram-negative bacteria have cell walls of different thicknesses due to differences in their peptidoglycan content. **A**, gram-positive cell wall; **B**, gram-negative cell wall.

Bacterial Environments

Bacteria have many unique requirements for growth, including oxygen, temperature, and salt levels. Aerobic bacteria prefer high levels of oxygen to grow at their maximum rate. Aerobic bacteria are called obligate aerobes if oxygen is absolutely necessary for their growth or facultative aerobes if they just grow better in the presence of oxygen but it is not required. Anaerobic bacteria prefer to grow in the absence of oxygen. Anaerobic bacteria are called obligate

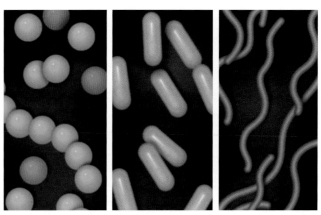

Figure 3.7. Bacterial shapes. Graphical rendering of three common types of bacterial morphology. From left to right: coccus, bacilus, and spirillum bacteria.

anaerobes if the absence of oxygen is necessary for their growth or facultative anaerobes if they just grow better in the absence of oxygen. *S. aureus* is a facultative anaerobe that grows much faster in an anaerobic environment such as puncture wounds.

Bacteria also have ideal growth temperatures and are divided into three major groups: psychrophilic, mesophilic, and thermophilic. Psychrophilic bacteria grow best in cold conditions between −15 and 10°C. Mesophilic bacteria grow best between 15 and 40°C, and thermophilic bacteria grow best between 45 and 80°C. There are extreme thermophiles, also called hyperthermophilic bacteria, which thrive at temperatures above 80°C. The thermophilic bacterium *Thermus aquaticus* was discovered at Yellowstone National Park and was the source of Taq DNA polymerase, which is used in the polymerase chain reaction (PCR). *Escherichia coli* are ubiquitous mesophiles that live in the human colon. Strains of nonpathogenic *E. coli* are commonly used in laboratory research and are grown in incubators at 37°C to mimic the temperature inside the human colon. There are also pathogenic forms of *E. coli* responsible for food poisoning outbreaks.

Salt (sodium chloride) is required for the growth of most bacteria and to maintain the osmotic balance between their environment and the inside of the cell. Bacteria that live in extremely salty conditions are called halophiles (see Figure 3.8).

Figure 3.8. **Salt lake Salar de Tara**. This salt lake in Los Flamencos National Reserve, Atacama, Chile, is close to the border between Bolivia, Chile, and Argentina. Bacteria living in this lake have adapted to extremely high salt conditions and are halophiles.

3.3 Uses of Bacteria in Biotechnology

Food Production

Naturally occurring bacteria are commonly used to make products that benefit humans, including lactic acid fermentation of milk to produce yogurt. *Lactobacillus delbrueckii* subspecies *bulgaricus,* or *L. bulgaricus* (see Figure 3.9) is one of the species of bacteria used to make yogurt. Yogurt is made by first scalding milk to kill bacteria and then inoculating the milk with a very small amount of pure bacterial culture or yogurt. As the bacteria multiply, the acid they produce gives the sour taste and causes milk proteins to precipitate, thickening the yogurt. Lactobacilli bacteria are also used in the production of cheese, sauerkraut, pickles, kimchi, and animal feed such as silage. Naturally occurring bacteria traditionally used in food production are also being genetically modified to make food production more efficient.

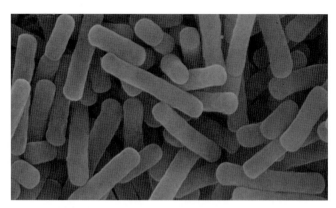

Figure 3.9. **Electron micrograph of *L. bulgaricus***. *L. bulgaricus* are bacilli, or rod shaped bacteria, that are commonly used in making yogurt. Dennis Kunkel/Science Photo Library.

Protein Production

Bacteria have been genetically engineered to manufacture foreign proteins. A protein that is artificially produced in a genetically engineered organism is called a recombinant protein. To produce a recombinant protein in bacteria, the gene for the protein is cloned and transferred into the bacteria. The genetically engineered bacteria are then cultured, which means that they are placed in conditions that enable them to multiply into billions of bacteria. Once a culture has reached the appropriate cell number or density, the bacteria are harvested and lysed (broken open). The recombinant proteins are then purified from the bacteria using chromatography. Uses for recombinant proteins include therapeutic drugs, agriculture, and food production.

Insulin, used to treat diabetes, is manufactured as a recombinant protein. Insulin historically was purified from pig pancreas. However, many patients had reactions to porcine insulin, and it was difficult and expensive to purify. In 1978, Genentech, one of the first biotechnology companies, cloned the human insulin gene and genetically engineered *E. coli* bacteria to produce human insulin protein that could easily be purified and used as a treatment for diabetes. This recombinant insulin was first licensed to Eli Lilly and Company and sold under the name Humulin in 1982. Humulin, which was the first drug made from a recombinant protein, is still the major recombinant human insulin used in the U.S. today. Humulin is now produced in eukaryotic cells. In 2015, 30.3 million people had diabetes in the U.S.; many of these patients are dependent on daily insulin injections. With the projected rise in the incidence of diabetes, recombinant technology will continue to play a significant role in the treatment of this disease.

Bovine somatotropin (BST) is a cow growth hormone that stimulates cows to produce more milk. Recombinant bovine somatotropin (rBST), which is made by expressing cow somatotropin in bacteria, was approved by the U.S. Food and Drug Administration (FDA) in 1994. rBST use in the U.S. has been very controversial. Monsanto, which held the patent for the production of Posilac (the commercial name for rBST), sold the patent to Eli Lilly in 2008 for $300 million.

3.4 Culturing Bacteria in the Laboratory

Growth Media
Bacteria can either be grown on solid **media** that usually contain **agar** as a gelling agent or suspended in liquid media. The choice of growth medium depends on the purpose of the culture or experiment.

Solid Media
Solid media are referred to as agar because the solidifying agent that is derived from marine algae is called agar. Bacteria, yeast, and some plant cultures are grown on solid media in **petri plates**, **slants**, and **deep tubes**.

Petri plates are used to isolate individual colonies of bacteria. The plates provide a large surface area to spread out the culture to yield an isolated clone (see Figure 3.10). Petri plates should be stored upside down to prevent condensation from dripping onto the culture.

Slants, which take up much less storage space than petri plates (see Figure 3.11), are made by pouring the agar into a tube and allowing it to set at an angle, which provides a larger surface area for

inoculating bacteria than just the diameter of the tube. Slants are used to culture bacteria for a short period of time but not to isolate colonies. Bacteria grown on petri plates or slants can be stored for 1–2 weeks at 4°C.

Deep tubes are used to store or study bacteria in **stab cultures**, where bacteria on an inoculation loop or needle are stabbed two or three times into the deep tube of solidified agar (see Figure 3.12). The bacteria are grown for 24 hours and then stored at room temperature for up to 1 year. **Note:** For safe long-term storage of bacteria a glycerol stock that is a liquid culture of bacteria with 10–20% glycerol should be prepared and stored at –80°C.

Stab cultures can also be used to look at motility and gas requirements of bacteria by observing where the bacteria grow (for example, near the surface where there is oxygen or deep in the agar where there is no oxygen) and whether they move through the agar away from the stab.

Solid media come in many formulations, the most generic being nutrient agar that is made from peptone and beef extract. The most common medium used for culturing *E. coli* is **lysogeny broth** (**LB**), which is also known as **Luria-Bertani broth**. LB agar is made with 1.5% agar, 1% tryptone (pancreatic digest of casein) and 0.5% yeast extract. LB also contains salt; the salt concentration depends on the specific formulation. Miller and Lennox formulations of LB have 1% and 0.5% sodium chloride, respectively. There are many other recipes for solid media developed for different purposes.

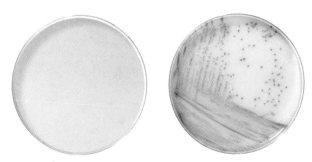

Figure 3.13. **Bio-Rad's MRSA*Select* medium**, selective chromogenic medium for the isolation and direct identification of MRSA. Left, a negative sample showing a few white colonies. Right, a plate positive for MRSA showing multiple pink colonies.

Differential media are formulated with chromogenic (colored) substrates to distinguish among different types of bacteria. For example, eosin methylene blue (EMB) agar makes *E. coli* colonies turn shiny green, lactose-fermenting bacteria turn dark blue/black and non-lactose fermenters, such as *Salmonella enterica*, remain colorless. The composition of some differential media actually selects for specific bacteria. For example, Bio-Rad's MRSA*Select*™ agar is used in clinical microbiology laboratories to screen for MRSA colonization using nasal swabs. This selective medium inhibits the growth of most yeast and bacteria except for MRSA. The agar contains a substrate that is specifically cleaved by MRSA but not by non-methicillin–resistant *S. aureus*. The cleavage of the substrate produces pink colonies and allows for the selection and identification of MRSA (see Figure 3.13).

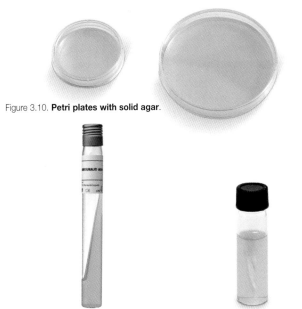

Figure 3.10. **Petri plates with solid agar**.

Figure 3.11. **Slanted tube**. Bio-Rad's Simmons citrate agar medium is designed to differentiate gram-negative bacteria based on their metabolic properties.

Figure 3.12. **Stab culture**. An agar deep tube is inoculated with bacteria to form a stab culture.

Antibiotic Selection

Most genetically engineered cells are made to be resistant to antibiotics, which allows them to be selected from cells that did not incorporate new DNA. Therefore antibiotic selection is the most common bacterial selection method in a molecular biology laboratory. Antibiotics are typically added to the media in which bacteria are grown. Some antibiotics, such as those in the penicillin family, inhibit cell wall formation by disrupting the formation of peptide crosslinks that connect polysaccharide chains in the peptidoglycan in bacterial cell walls (see Figure 3.6).

The **Kirby-Bauer**, or **disk diffusion test**, is used to determine the sensitivity of bacteria to antibiotics and is commonly used in clinical microbiology laboratories (see Figure 3.14). In this test, paper disks impregnated with various antibiotics are placed on an agar plate that has been covered with the bacteria of interest. The zone of inhibition, which is the area surrounding the disks where the bacteria are inhibited from growing, is measured and used to quantify the effectiveness of the various antibiotics against the bacteria.

Figure 3.14. **Kirby-Bauer (disk diffusion) test**. The test is used to determine the effect of various antibiotics on bacteria that are covering the surface of the agar. The larger the zone of inhibition, the greater the antibacterial effect of the compound tested.

Liquid Media

Liquid media are commonly referred to as broth and are used to produce large quantities of bacteria suspended in liquid, usually for protein production or for isolating plasmid DNA. **Liquid cultures** cannot be used for isolation of individual colonies.

Liquid media are often formulated with the same ingredients as solid media but without the agar. For example, LB broth and LB agar have the same formulation but LB agar has agar as an additional component. Another common broth used to culture *E. coli* is Terrific Broth (TB). This medium is used to culture *E. coli* for plasmid purification. Terrific Broth extends the growth phase of recombinant *E. coli* and increases the plasmid or protein yield. As in solid media, antibiotics are used to select for resistant bacteria. The presence of antibiotics in liquid cultures is even more important than on solid media since contaminating microorganisms can more easily overwhelm a liquid culture.

3.5 Microbiological Techniques

Tools

There are some standard tools necessary for microbiological cell culture. An inoculation loop is a platinum or nichrome wire with a small loop at the end, while an inoculation needle is just a straight wire (see Figure 3.15). Both inoculation loops and needles have an aluminum handle for quick heat dissipation. An inoculation loop can also be made of sterile, disposable plastic.

Bunsen burners are used to heat sterilize inoculation loops and the mouths of glass bottles, culture tubes, or culture flasks. Bunsen burners also maintain an upward air current that reduces airborne contamination. Presterilized disposable plastic inoculation loops, which eliminate the requirement for heat sterilization, are also available (see Figure 3.15).

Spreaders spread bacteria evenly over petri plates. They can be disposable and made of sterile plastic or they can be made of a bent glass or metal rod. Small, sterile glass beads can also be used to spread bacteria around the plate.

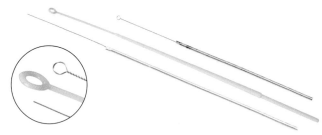

Figure 3.15. **Inoculation loops and needle**.

Incubators maintain a constant temperature; some incubators have a shaking platform that allows liquid cultures to be shaken while being incubated (see Figure 3.16). Some water baths also have a shaking apparatus. Shaking is necessary to oxygenate an aerobic culture for optimal growth. In industrial manufacturing, bacteria are grown on a large scale in specialized vats, called fermenters and bioreactors, under controlled conditions. Fermentors and bioreactors have inlets and outlets to regulate pH, nutrient, and oxygen levels, and are constantly monitored to maintain optimal growth conditions. They can be 5 to 100,000 L in size.

Figure 3.16. **Shaking Incubator**.

How To...

Use Aseptic Technique to
Transfer Bacteria

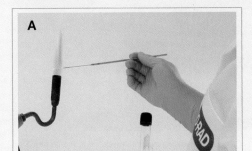

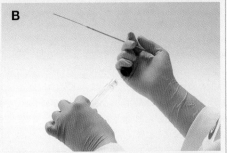

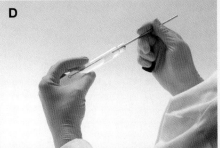

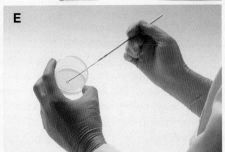

Aseptic technique requires you pay attention to all possible sources of contamination at all times. PPE should always be worn when using aseptic technique.

To transfer bacteria from a liquid culture to a petri plate, sterilize a metal inoculation loop using a Bunsen burner (see Figure 3.17A). (Heat sterilization is not necessary when using presterilized disposable plastic loops.) Pass the loop back and forth across the tip of the inner blue cone in the Bunsen burner flame until the entire length of wire is heated. The inner blue cone is the hottest part of the flame and will heat the wire to red hot. Let the wire cool for 10–20 sec and pick up the culture tube.

Unscrew the cap until it is loose while keeping the sterile loop in your hand (make sure the loop does not touch anything). With the same hand that is holding the loop, use your pinky finger and palm to remove the tube's cap. Keep the cap in your hand—do not set it down (see Figure 3.17B). Using the other hand, pass the mouth of the culture tube through the flame, rotating it while it passes through the flame to ensure that the entire surface is heated (see Figure 3.17C). Do not flame plastic culture tubes.

Place the sterile loop into the culture tube by moving the tube over the loop, rather than moving the loop into the tube. This prevents aerosols from forming if the loop is flicked upon its removal from the culture tube (see Figure 3.17D). Once the loop is coated in culture, ensure that it does not touch anything. Flame the mouth of the tube again and recap it.

Move to the petri plate and lift the lid, holding it over the plate as a shield. Spread the bacteria from the loop back and forth across the surface of the agar. Take care not to dig into the agar—the loop should just skate across the surface (see Figure 3.17E). Replace the lid of the petri plate and then flame the loop before setting it down. When transferring bacteria from the plate to the liquid culture, the loop needs to be sterilized first and then the procedures described above would be performed in reverse.

Figure 3.17. **Sterile transfer of bacteria from a liquid culture to a petri plate using aseptic technique. A**, sterilizing the metal inoculation loop by flaming; **B**, removing the culture tube cap while holding the sterile loop; **C**, sterilizing the mouth of the culture tube by flaming; **D**, placing the sterile loop into the culture tube and coating the loop with culture by moving the tube over the loop; **E**, spreading the bacteria across the surface of the agar.

Aseptic Technique

When performing any cell culture, it is critical to maintain sterility at all times by using a set of specific practices and procedures called **aseptic technique**. Aseptic technique is a method that prevents the introduction of unwanted organisms into an environment. This technique is important in research laboratories and operating rooms in hospitals. In a microbiology laboratory, aseptic technique is used to transfer one bacterial monoculture into a fresh medium without introducing any contaminating microorganisms.

Aseptic technique requires a conscious effort to avoid contact between any nonsterile items and sterile items. This involves keeping hands and all surfaces clean. Always wash your hands thoroughly and wipe down work surfaces with 70% alcohol or antimicrobial cleanser. Change gloves frequently. Take care not to create aerosols when removing gloves to avoid introducing contaminants into your clean work area. Air itself is a source of contamination, so when containers are opened or not in use, always use the cap as a shield to prevent airborne bacteria or fungal spores from settling into the container. For this reason, microbial work is frequently performed next to a Bunsen burner, which causes an upward flow of air due to convection, reducing the chance of airborne contaminants settling into cultures. Similarly, hold the lid over the bottom plate when opening a petri plate or tissue culture dish. Avoid passing items, including arms and hands, over the top of any open sterile container so that contaminants do not fall into the container. Arrange your workspace so that it will not be necessary to reach over sterile containers. Do not put any sterile items, such as bottle caps or pipets, down on the work surface, and be especially careful that pipet tips never touch anything that is nonsterile. Pay constant attention to all the steps taken; a whole day's work can be ruined by one oversight that introduces a contaminant.

When using any cell or microorganism that is known to cause human disease, a **biological safety cabinet (BSC)** must be used to limit the chance of an **aerosol** contaminating a worker. The type of BSC required depends on the potential hazard of the culture being manipulated. In biosafety level-1 (BSL-1) laboratories, a BSC is not necessary because the microorganisms cultured are known not to cause human disease. Human cells should be treated as though they are infected with human pathogens and must always be cultured in a BSC. Tissue culture hoods, which are Class II BSCs, are used to maintain sterility within the hood and reduce the chance of contamination of the culture.

Using Streak Plate Technique to Isolate Single Colonies

Most microbiological work requires that a **cell culture** be started from a single colony. A colony is a single bacterium that has multiplied on a solid medium into millions of clones of itself and looks like a round visible dot on the solid medium (see Figure 3.18). The streak plate method allows isolation of a single colony from a bacterial culture by splitting the plate into quadrants and diluting the bacteria repeatedly as the loop is streaked through each quadrant (see Activity 3.C Part 1B, step 8 for the detailed protocol).

Figure 3.18. **Isolation of single colonies**. An agricultural research laboratory technician isolates single *E. coli* colonies from the stomach contents of cattle using a plate that has been inoculated using the streak plate method. Note the pattern of inoculation has isolated single colonies. Photo courtesy of the Agricultural Research Service, USDA.

Labeling Media

Tubes or petri plates containing media and/or bacterial cultures must always be properly labeled. When pouring a petri plate, the plate should be labeled ahead of time. The label should be on the outside of the bottom plate and written in an arc around the outer edge so that it does not interfere with viewing bacteria on the plate. The label should list the type of agar (including any additives), the date poured, and the initials of the person who poured the plates. Once the plate is inoculated, it should be further labeled with the name of the microorganism, the number in an experimental series, and the date it was inoculated. Deep tube cultures or slants should be labeled on the side of the tubes with the same information described above for plates.

Figure 3.19. **Petri plates** should be labeled with experimental details, initials, and the date.

Quantifying Bacteria
Serial Dilution and Plate Counts

It is often necessary to determine the number of bacteria present in a culture. Inoculating an agar plate with a portion of the culture and counting the colonies is one way to quantify the bacteria in the culture. However, the entire culture cannot be directly inoculated onto the agar plate because even 100 µl of a bacterial culture may contain billions of bacteria that would overcrowd the agar plate. To determine the number of bacteria, the culture must first be diluted. The culture must be diluted at least 1 million times, or 1 ml into 1,000 L, to obtain an agar plate with single colonies. Not only is it unlikely that you have a container that can hold 1,000 L but the cost in broth needed to dilute the bacteria would be prohibitive! Serial dilutions are used instead (see Figure 3.20). A **serial dilution** means diluting the culture several times by the same dilution factor, for example, 10 times, and then diluting that culture another 10 times, and so on. By making six serial dilutions, a 1 million-fold, or a 10^{-6}, dilution is made. To perform a 10-fold serial dilution, 100 μl of culture is added to 900 μl of broth. Then 100 μl from that dilution is added to another 900 μl of broth to make a 100x dilution, and so on, until the original culture has been diluted 1 million times. The culture from each tube is then plated on agar plates. Each individual bacterium spread on the plate will grow into a **colony** and is referred to as a **colony forming unit** (**CFU**). The number of bacteria in the original culture can then be calculated by multiplying the final number of colonies on the plate by the dilution factor. For example, in Figure 3.20, there are 133 colonies on the plate diluted 10^3 times, indicating there were 133 viable bacteria (or CFU) in the 100 μl that was plated. To calculate the number of CFU in the undiluted culture, the number of CFU must be multiplied by the same dilution factor, that is, 133×10^3, or 1.33×10^5 in scientific notation. This is the number of CFU present in 100 μl of the original culture. To convert the answer to CFU per ml rather than CFU per 100 μl, the CFU per 100 μl must be multiplied by a unit conversion to convert μl to ml.

$$\frac{133 \times 10^3}{100 \; \mu l} \times \frac{1,000 \; \mu l}{1 \; ml} = 1.33 \times 10^6 \; CFU/ml$$

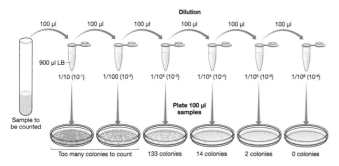

Figure 3.20. Serial dilution and plate counts are used to quantify bacteria. The bacterial culture is diluted tenfold six times so that the final tube has 1 millionth the number of cells as the same volume of the original culture.

Optical Density of Bacteria

A spectrophotometer can also be used to determine the concentration of bacteria in a culture. The value given by a spectrophotometer is referred to as an **optical density** (**OD**), which measures how cloudy the culture is. The cell concentration can be inferred from the OD value; this is usually sufficient since most protocols require bacteria to be grown to a specific optical density but do not require a specific cell number. To determine a cell number, OD readings from the spectrophotometer are used in combination with the results of the serial dilution and the **plate count**. The spectrophotometer is set to read a wavelength at 600 nm. Uninoculated broth is then added to a cuvette and used to "blank" the spectrophotometer. Each serial dilution is added to a cuvette and the OD is read. A standard curve is constructed with the OD plotted against the CFU derived from the plate count (see Figure 3.21). The standard curve can then be reused as a reference without repeating the serial dilution each time. Many modern spectrophotometers have precalculated the cell number versus OD, which can be selected from the menu of the instrument. However, it is always recommended to calibrate the spectrophotometer by doing a serial dilution with each different microorganism used.

Careers In Biotech

Katie Dalpozzo

Laboratory Research Associate,
Genentech, Inc.
South San Francisco, CA

Katie's career in biotechnology began in high school when she attended a hands-on regional program in biotechnology that met every day for an hour before school. She then pursued a BS degree in Ecology and Evolution at the University of California, Santa Cruz (UCSC). During her years at UCSC, she worked in a laboratory at a biotechnology company, where she discovered a knack for performing quality assays and troubleshooting assays when they went wrong. This fueled her desire to continue working in the industry after graduation.

Photo courtesy of Katie Dalpozzo

At Genentech, Katie's work focuses on preclinical trials of drugs targeted toward treating breast and prostate cancers. She is responsible for setting up experiments and for recording and analyzing the data. Every Monday she presents the data she gathered during the previous week to the research team. Katie has come to experience that documenting data by following good laboratory practice (GLP) is critical in preclinical trials. Many hours of training go into becoming familiar with all the regulations that govern development of cancer therapies. One piece of advice from Katie is to record every step in your laboratory notebook while it is being performed and not to wait until the experiment is over.

Katie suggests that anyone interested in a career in biotechnology volunteer in a laboratory, either on campus or at a local company, and then look for internships for recent graduates. She advises that learning all the basic techniques used in a lab (how to pipet, centrifuge, work with tissue culture, make buffers, etc.), will give you knowledge and skills that will serve you well, whether you decide to go to graduate school or straight into working in the industry.

In the next ten years, Katie sees the biotechnology industry pushing aggressively toward personalized medicine. Research generates enormous amounts of data that give clues to what does and does not work for certain types of people, but we do not have tools yet to interpret all these complex biological data. Technology is becoming more and more sophisticated, however, and will soon allow us to pull all of this information together using bioinformatics. The end result will be the ability to select the most effective medicine for each individual patient. For this reason, Katie thinks learning to code in R, a statistical programming language, and understanding bioinformatics will also be helpful to your future.

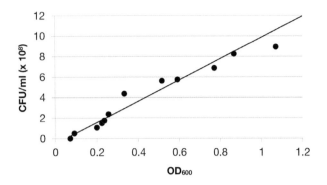

Figure 3.21. **OD$_{600}$ standard curve**. The measured OD$_{600}$ from liquid culture samples is plotted against the colony forming units (CFU) for each sample. There is a linear relationship between the number of colony forming units and the OD$_{600}$. The standard curve can be used to estimate the CFUs for a liquid culture sample with an unknown quantity of microorganisms.

Identification of Bacteria

Identifying microorganisms can be a challenge. Unknown bacteria can be identified using the Gram stain procedure and performing assays that determine the metabolic characteristics of the bacterial cells. **Assays** are procedures carried out to test or measure the activity of a drug or biomolecule in an organism or sample. Assays can be designed to quantify cells, molecules, antibodies, metabolic characteristics, or many other analytes.

The Gram stain was invented by Hans Christian Gram in 1882 to differentiate between two types of pneumonia bacteria. Performing a Gram stain will not only determine the cell wall composition but it will also make the shape of the cells visible at 400x to 1,000x magnification. Bacteria are either gram-positive or gram-negative based on differences in the peptidoglycan content in their cell walls. A gram-positive bacterium has a thick cell wall, while a gram-negative bacterium has a thin cell wall (see Figure 3.6). The Gram stain procedure uses four reagents: crystal violet, Gram's iodine, alcohol, and safranin. Crystal violet binds irreversibly to the thick peptidoglycan of gram-positive bacteria, staining the bacteria dark purple. However, crystal violet binds much less strongly to gram-negative cell walls and is washed away with alcohol. Gram-negative bacteria pick up the safranin and are stained light pink (see Figure 3.22).

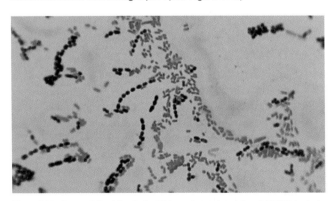

Figure 3.22. **Gram stained bacteria**. Mixture of yogurt and *E. coli* HB101 bacteria. Gram-positive bacteria are purple and cocci (spherical-shaped). Gram-negative bacteria are pink and bacilli (rod-shaped). Image is magnified 1,000x under oil immersion.

Another way to identify bacteria is by assaying their metabolic properties since bacterial strains metabolize some chemicals differently. To assay the metabolic properties of bacteria, they are inoculated on media containing additives that result in a visible difference when metabolized by bacteria. For example, the starch hydrolysis test determines if bacteria produce amylase. When amylase-producing bacteria, such as *Bacillus subtilis* and *Bacillus megaterium*, are grown on medium containing starch, they digest the starch. If the medium is stained with an iodine reagent containing potassium iodide (KI), the agar turns blue if starch is still present in the medium. However, if the starch has been digested, the stain does not bind and the agar remains clear (see Figure 3.23). Commercial assays that test multiple metabolic characteristics simultaneously are available. Depending on the bacterial strain, the agar media change color and the bacteria can be identified.

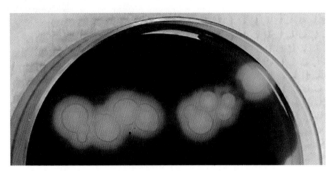

Figure 3.23. **Starch hydrolysis test**. Agar containing starch stained with potassium iodide (KI) turns blue. Amylase from amylase-producing bacteria digests the starch in the agar, resulting in a colony surrounded by a clear circle. If no amylase is present, the agar remains blue due to the presence of the KI-stained starch.

The morphology of bacterial colonies can also identify bacteria (see Figure 3.24). Identifying factors include the color, size, and growth pattern of the colony. Large colonies indicate that the bacteria are also large. A shiny colony indicates that the bacteria have a capsule. A diffuse or spreading margin around the colony indicates that the bacteria are motile. Some bacteria also produce pigments and so colonies are colored.

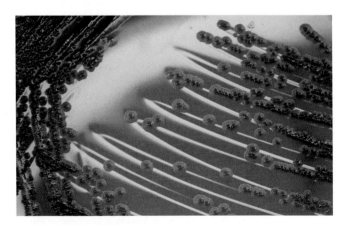

Figure 3.24. **Colony morphology**. The colony morphology displayed by gram-negative *Chromobacterium violaceum* bacteria, which were grown on sheep's blood agar. Note the dark-violet coloration of these bacterial colonies caused by violacein, an antibiotic pigment that is currently being studied for the treatment of colon and other cancers. Photo credit: Pete Seidel of the Centers for Disease Control and Prevention.

3.6 Eukaryotic Cells

Eukaryotes make up the third domain of life. Eukarya includes almost all the organisms we can see, including animals, plants, and fungi. Many microorganisms, such as yeast and protists, are also eukaryotes.

Eukaryotic cells have a much more complex cell structure than bacteria (see Figures 3.25 and 3.26). All eukaryotic cells are surrounded by a plasma membrane (see Figure 3.27) that is composed of phospholipids. The plasma membrane regulates what materials enter and leave the cell. The phospholipids create a hydrophobic zone that prevents water loss and ensures that charged particles enter and leave through transmembrane proteins. Glycoproteins and glycolipids, which are proteins and lipids bound to carbohydrates, respectively, have important roles on the outside of the cell in how it recognizes its environment and other cells. The inside of the eukaryotic cell is referred to as the cytosol or cytoplasm and is filled

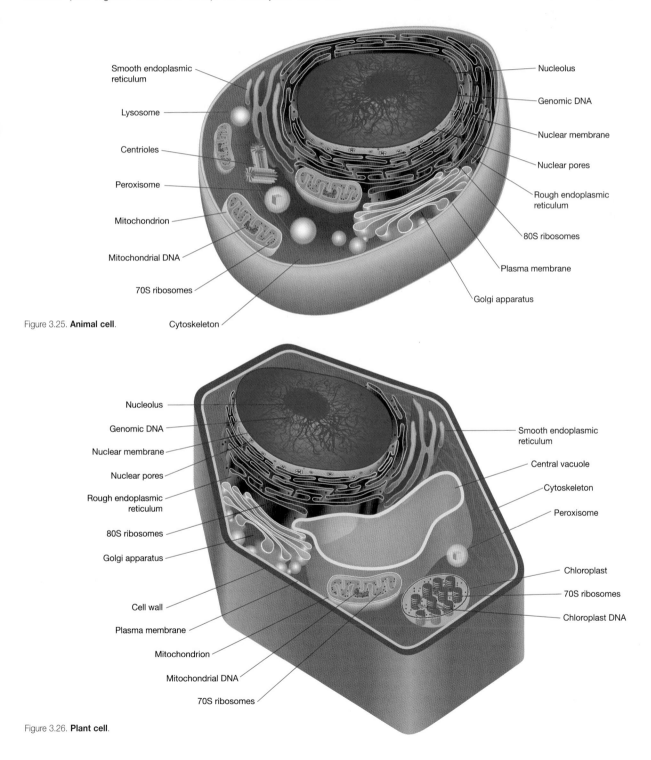

Figure 3.25. **Animal cell**.

Figure 3.26. **Plant cell**.

with organelles, leaving very little empty space. Organelles, most of which are membrane bound, partition cell function into distinct areas. Since the cells of producers (organisms that use the sun to make organic food molecules) are the foundation of the terrestrial

food chain in the ecosystem, they have a number of structures involved in capturing energy from the sun and converting it into energy contained in chemical bonds that can be transferred between organisms via photosynthesis.

Other Eukaryotic Cells

The fungi kingdom includes yeasts, molds, and mushrooms. Fungi have a cell wall that is composed of chitin, a polymer of beta glucose. Yeasts are very important in biotechnology. These fungal microorganisms were historically used to make bread and beer. In modern times, they are also used in protein production since they are inexpensive and easy to culture. They also have the unique ability to host extra artificially manufactured chromosomes, called yeast artificial chromosomes or YACs, to express recombinant eukaryotic genes. Many other fungi are used to produce enzymes commercially. *Aspergillus niger* is commonly used industrially to produce cellulases, pectinases, and amylases, which all have commercial applications. For example, amylases are used in detergents to remove stains from clothes.

Fungal enzymes are currently being investigated in biofuel research. As the decomposers in the food chain, fungi naturally have the ability to break down cellulose in plant cell walls. Much of the biofuel research is

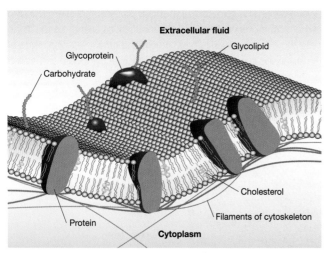

Figure 3.27. **Plasma membrane of a eukaryotic cell**. Proteins are embedded in a phospholipid bilayer. Glycoproteins and glycolipids reside only on the outside of the membrane and are used for cell recognition.

Organelles

Cell type	Organelle	Structure	Function
Animal	Lysosomes	Membrane-bound vesicles that contain hydrolytic enzymes (acid hydrolases)	Play a role in recycling worn-out organelles and digesting viruses or bacteria to recycle their nutrients for use in the cell. When cells die, lysosomes help break them down and recycle the nutrients back into the environment.
Both	Nucleus	A large double membrane compartment with numerous pores for the exchange of genetic material	Contains the genomic DNA and proteins that regulate gene expression.
	Nucleolus	Dense region of the nucleus	Where ribosomal RNA (rRNA) is transcribed.
	Mitochondria	Double membrane bean-shaped structure that contains 70S ribosomes and mitochondrial DNA (mtDNA)	Where cellular respiration produces ATP.
	Golgi apparatus	Stacks of flat membranous sacs	Contains enzymes that modify proteins by adding phosphate or carbohydrate groups; helps the cell distinguish between self and non-self (foreign) cells.
	Endoplasmic reticulum (ER)	Series of interconnected tubes and sacs made up of two types: smooth ER (SER) and rough ER (RER), which has 80S ribosomes attached to its surface	Internal transport system and site for protein translation, the production of new membrane, proteins, and lipids, and the detoxification of poisons. RER is where mRNA is translated into polypeptide chains during protein synthesis. Small vesicles often pinch off of the RER and merge with the Golgi apparatus for further processing. SER plays a role in lipid production and protein storage.
	Cytoskeleton	Microtubules, microfilaments, intermediate fibers	Provide support and aid in the movement of materials through the cytoplasm.
	Centrosome	Made up of cylindrical structures called centrioles near the nucleus in animal cells	Involved in the development of spindle fibers during cell division.
	Vesicles	Small membrane-bound compartments	Transportation, digestion, and containment of waste.
Plant	Chloroplast	Two external membranes with additional membranes inside that form stacks of thylakoid membranes called grana; area outside of thylakoids is called stroma	Light reactions of photosynthesis occur in thylakoid membranes; light-independent reactions (dark reaction or Calvin cycle) occur in the stroma, where sugar is produced.
	Large central vacuole (mature plant)	Large compartment that contains mostly water	Plays a role in storage of nutrients and poisons, and maintains the turgidity of the cell.

focused on finding ways to break down cellulose into sugars that are suitable for fermentation. Many species of fungi are being targeted and evaluated for their use in generating alternative energy sources.

Protists are a diverse group of eukaryotes. Many protists, including algae in the ocean, are photosynthetic primary producers. Protists are being investigated as potential sources of biofuels since they are so numerous and grow quickly. Microalgae secretes a compound that allows it to survive and reproduce in harsh environments, which is being used for anti-aging skincare products.

3.7 Uses of Eukaryotic Cells in Biotechnology

Fermentation

Yeast cells have been used for thousands of years in baking and brewing. When added to bread dough, yeast ferments the sugar in the dough, releasing CO_2 bubbles that get trapped in the dough. The build up of CO_2 causes the dough to rise during baking, producing a lighter, softer bread that is easier to chew. When used for alcohol production, yeast converts sugars from grains into alcohol because the yeast is forced to respire anaerobically. Historically, bakers and brewers generated different strains of yeast naturally; however more recently, specialized strains have been genetically engineered to create specific flavors or varieties of foods and wines.

Protein Production

Recombinant proteins are also made in eukaryotic cells. Some eukaryotic proteins cannot be made properly in bacteria. Eukaryotic proteins are more complicated than bacterial proteins and may not fold correctly or solubilize if they are made in bacteria. In addition, bacteria are unable to properly modify eukaryotic proteins after translation. **Posttranslational modification** includes glycosylation, a process in which carbohydrate molecules are attached to the newly synthesized protein.

Common eukaryotic cells used in cell culture for protein production include yeast and other fungi such as *Aspergillus*, insect cells such as the Sf9 **cell line** from the fall armyworm (*Spodoptera frugiperda*), and mammalian cell lines such as Chinese hamster ovary (CHO) cells.

The type of eukaryotic cell used for protein production depends on the complexity and specific requirements of the protein being produced. Yeast or other fungi are generally the system of choice since they are cheap and easy to culture. Many industrial enzymes, such as those in laundry detergents, are made using *Aspergillus*. However, depending on the recombinant protein being produced, these cells may not make the necessary posttranslational modifications, such as glycosylation. Insect cells are often used because they are less expensive to culture than mammalian cells and are much better than yeast at reproducing

the glycosylation of human proteins. However, insect cells also may not correctly perform all the required posttranslational modifications of proteins. Mammalian cells are the ideal system for protein production because they can perform most posttranslational modifications correctly. They can also make recombinant proteins that have multiple subunits, which cannot be achieved in other expression systems. However, mammalian cells are difficult and expensive to culture.

Antibodies are one class of recombinant proteins produced in eukaryotic cells for use as therapeutic drugs. As of January 2017, there were 68 therapeutic antibodies approved by the U.S. Food and Drug Administration (FDA) for a variety of disorders, including cancer and immune-related disorders. Antibodies are multisubunit proteins and therefore need to be made in mammalian cells. In 1998, Genentech developed an antibody drug called Herceptin that specifically binds and blocks signaling of a cell membrane protein called human epidermal growth factor receptor (HER2), which signals inappropriately in some cancers. If HER2 is blocked, it cannot send a signal, which slows the growth of the cancer. To produce Herceptin, the gene for the humanized antibody is cloned into CHO cells, which are cultured in a suspension culture in large bioreactors. The antibodies are then purified from the cells and used to treat cancer patients.

Stem Cells

Research into **stem cells** has dramatically increased in the last 20 years. Stem cells can differentiate into new cell types, such as brain, blood, or muscle cells, depending on the molecular signals they are given (see Figure 3.28). Because of this ability to become any tissue, stem cells hold the potential to cure diseases that are incurable today by replacing damaged or diseased tissue with brand new tissue.

Stem cell research is controversial because one source of stem cells is human embryos from in vitro fertilization clinics. **Embryonic stem cells** are derived from the inner cell mass (ICM) of a developing embryo and are pluripotent, meaning that they can differentiate into most types of tissue but cannot recreate an entire organism. The first human embryonic stem cells (hESC) were isolated in 1998. The first transplant using tissue derived from stem cells was performed in 2017. Currently there are multiple clinical trials in Phase 1, 2, or 3 to evaluate the safety and efficacy of stem-cell derived cells in patients, including stem-cell derived neural, adipose, and muscle cells.

Stem cells derived from adult tissues are multipotent, which means they can differentiate only into cells of their own lineage; for example, blood stem cells can differentiate only into specialized blood (hematopoietic) cells. In 2007, scientists induced human adult stem cells to become pluripotent or switch back to an embryonic ICM-type state. These **induced pluripotent stem cells (iPSC)** have huge potential since cells could be removed from a patient, induced into pluripotency, and then differentiated into the type of cells required to treat the patient's disease without any risk of immune rejection. Both hESC and iPSC lines are considered major research tools.

Bioethics
Should Human Embryos
Be Used for Research?

The use of human embryonic stem cells (hESCs) is highly controversial because the creation of hESCs destroys a human embryo. Many people feel that the destruction of human embryos for scientific research is wrong, while others feel the potential benefits to human health outweigh this ethical concern. Most of the candidate embryos used for research are left over from in vitro fertilization attempts and would be destroyed or stored until they were incapable of growth or development. Should these frozen embryos be used for potentially lifesaving research?

The alternative to hESCs, induced pluripotent stem cells (iPSCs), do not raise the same ethical considerations. However, iPSCs are not the same as hESCs, so hESCs are still in use as controls and as the standard against which new stem cell lines are evaluated. Some experts continue to study all stem cell types, since no one knows which will be the most useful. Research into how to best grow and stimulate hESCs to differentiate will also likely produce information that benefits the development of iPSCs. Finally, multiple hESC-based therapies are currently in clinical trials, whereas therapies based on iPSCs are years away. Should we halt research into hESC therapies to focus on advancing iPSC research?

The use of U.S. public funds for hESC research is hotly debated. Since 1995, U.S. law has prohibited the use of federal funds for the creation of human embryos for research purposes or for research in which human embryos are destroyed. In 2001, President George W. Bush restricted the use of federal funds for research into a specific set of existing hESC lines. In 2009, President Barack Obama lifted some of these restrictions, allowing public funds to be used for both existing and future hESC lines derived in the private sector. Should public funds be used for research that many people adamantly oppose?

Though iPSCs may be similar to hESCs, they cannot develop into a complete embryo. If iPSCs could be made to develop into embryos, they would be clones of their donor. Would they then lose their perceived ethical advantage over hESCs? What benefit would this technique provide? What ethical concerns would it raise? Should the laws affecting hESCs also cover other types of pluripotent stem cells? If so, what new legislation is needed?

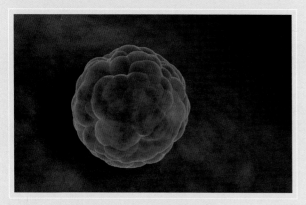

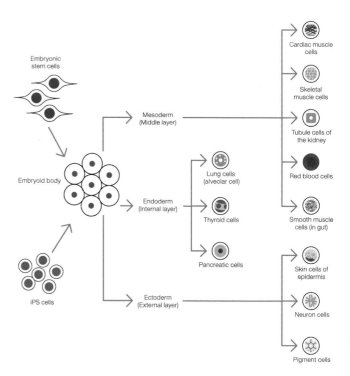

Figure 3.28. **Stem cells**. Embryonic stem cells or induced pluripotent stem cells are differentiated into different cell types by first forming an embryoid body. These cells then differentiate in response to various signals specific to the desired cell type. Stem cells can become mesodermal, endodermal, or ectodermal cells.

3.8 Eukaryotic Cell Culture

Many different methods are used for culturing eukaryotic cells. Yeast cells are cultured in a manner similar to bacteria: on agar plates or in liquid broth. The components of the media are specific to yeast. Yeast cells are very easy to culture compared to insect or mammalian cells.

Primary cells are used by researchers to investigate how normal cells behave. Primary cells are taken and cultured directly from an animal. However, primary cells can be cultured for only a short amount of time (a few days to a few weeks, depending on the cell type) before they stop growing. The short life span of primary cells makes them unsuitable for biotechnology companies that wish to manufacture therapeutic recombinant proteins. Biotechnology companies use established cell lines. Cell lines are cells that have changed in culture so that they grow indefinitely or are immortal. Most primary cells and cell lines need to adhere to tissue culture dishes or flasks while they are growing (**adherent cultures**) and stop growing when they come in contact with another cell (a phenomenon called contact inhibition). Some cell lines have lost the requirement for adhesion and can grow in suspension. **Suspension cultures** are much less expensive to culture than adherent cultures since more cells can be grown in less space and with much less media. Suspension cultures are grown in spinner flasks or large-scale bioreactors. Impellers or stirbars are used to keep the cells moving through the media.

Plant tissue culture is usually performed in biotechnology laboratories to introduce novel genes or DNA into plant cells with the purpose of regenerating an entire plant that expresses the novel DNA. Unlike mammalian cells, plant cells are totipotent, meaning a single cell can regenerate an entire plant. In plant tissue culture, almost any young piece of plant tissue can be cultured on a specialized medium and can revert back to an undifferentiated state. This forms a mass of undifferentiated cells called a **callus**. The callus can then be broken up into individual cells in a suspension culture or manipulated as a whole callus while the cells are transformed with novel DNA (see Chapter 5). Once plant cells are genetically engineered, they are given differentiation signals to produce roots and shoots, which grow into whole plants.

Growth Media

Each cell type has a specific type of medium in which it grows optimally. Eukaryotic cell culture media are a mixture of nutrients including amino acids, salts, glucose, vitamins, and iron. In addition to these nutrients, the media used for culturing mammalian cells typically contain a pH indicator called phenol red, which stains the media pink (see Figure 3.29). (Phenol red changes color if the pH of the media is not optimal.) Mammalian cells also usually require animal **serum** to grow. Serum is very expensive but provides the necessary hormones and growth factors. Serum is extracted from animal blood. Insect and plant cells do not require serum, which makes them much less expensive to culture.

Figure 3.29. **Mammalian cells in tissue culture**. Cells are grown in tissue culture dishes or flasks that contain a nutrient-rich medium. The cells are cultured in an incubator in the presence of a high level of CO_2. The medium typically contains phenol red, a pH indicator that stains the medium pink.

Environment

Similarly to bacteria, eukaryotic cells grow best at a specific temperature. Yeast cells grow best at 30°C, while mouse and human cells grow best at 37°C. Special incubators are needed for mammalian cells since they usually need to be grown in an environment with a high level of CO_2 to mimic the CO_2 levels in the body. Yeast, insect, and plant cells can be cultured in regular incubators. Plant cells are affected by light, so undifferentiated plant cells are often grown in the dark to discourage redifferentiation.

Biotech In The Real World

Artificial Life!

Synthetic biology combines biotechnology with other disciplines like chemistry, bioengineering, and molecular and evolutionary biology to build artificial biological systems for research, medical, and other applications. Synthetic life forms can theoretically be designed or engineered to exact specifications to perform specific functions.

In 2010, the first entirely synthetic bacterial cell was created at the J. Craig Venter Institute. An entire artificial genome composed of 1.08 million base pairs of DNA from *Mycoplasma mycoides* (*M. mycoides*) bacteria was produced synthetically in yeast, then isolated and inserted into a host *Mycoplasma capricolum* bacterial cell that had its own genome removed. The new cells produced only *M. mycoides* proteins, and not *M. capricolum* proteins. This first artificial cell was the first milestone in J. Craig Venter's mission to reduce life to its bare essentials in order to someday build life from scratch. (see Figure 3.30).

To reduce life to its bare essentials means understanding the minimum number and types of genes required for life. In 2016, Venter reported the creation of another synthetic cell, this one containing the smallest genome of any known independent organism. Using a 'design, build, and test' methodology, Venter's team dissected the *M. mycoides* genome into eight DNA segments and mixed and matched them to see which combinations produced living cells. Whatever they learned from doing this was applied to the next "design," and the cycle was repeated. Through this design process, the team identified DNA sequences that, among other things, do not encode genes but are still required to direct the expression of the critical genes. Eventually, they whittled the genome down to a 531,000-base pair, 473-gene design.

The new cell has all of its nutrients supplied through growth media, and its essential genes are those involved in making proteins, copying DNA, and building membranes. Interestingly, the functions of as many as 149 of its genes are still unknown, even though many are also found in other organisms, including humans.

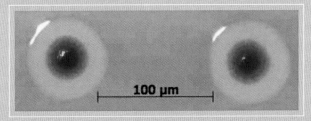

Figure 3.30. **Blue colonies of the first synthetic cell**. The J. Craig Venter Institute successfully manufactured a genome and placed it into a cell that had had its own genome removed. The synthetic cell successfully replicated itself millions of times. From Daniel G. Gibson et al. (2010). *Science* 329, 52. Reprinted with permission from AAAS.

Sterility

Eukaryotic cultures are more prone to contamination than bacterial cultures because they grow more slowly than bacteria. Class II biological safety cabinets (BSCs), commonly called tissue culture hoods, should be used to prevent contamination of eukaryotic cultures. These hoods have a front-access opening with carefully maintained inward airflow to protect the user from contamination. They also have HEPA-filtered, vertical, unidirectional airflow within the work area to protect the material from contamination and exhaust HEPA-filtered air to the room to protect the environment from the material. The level of protection required for the user, material, or environment determines the type of BSC that should be used. Class I BSCs do not protect the material in the cabinet from contamination in the room but protect the user and the environment. Traditional fume hoods protect just the user. BSCs should be sterilized after each use by wiping them down with 70% ethanol and then exposing them to shortwave ultraviolet germicidal light that kills bacteria. Aseptic technique must be used for all eukaryotic cell culture.

Visualization

Eukaryotic cells are much larger than bacteria, which means that they can be viewed using a microscope at relatively low power. Eukaryotic cells growing in tissue culture dishes can be viewed directly using an inverted microscope that has the objective lens under the stage (see Figure 3.31A). Cells can also be incubated with antibodies that bind to specific proteins on the cell surface, that helps identify the type of cell (see Figure 3.31B). The antibodies are linked to fluorescent dyes and imaged using a fluorescent microscope. Selective stains can be used to identify specific organelles using a compound microscope. For example, DNA-binding stains, such as Fast Blast™ DNA stain, UView™ 6x loading dye and stain, or methylene blue, are used to highlight the nucleus. Other stains react with specific enzymes in organelles. For example, Janus green B specifically stains the mitochondria.

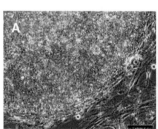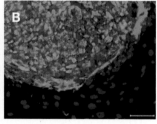

Figure 3.31. **Human-induced pluripotent stem cells growing in tissue culture**. iPSC are generated from an adult human fibroblast cell line. **A**, phase contrast microscopy image; **B**, immunofluorescence microscopy image. Cells were stained using a fluorescently labeled antibody specific for a human embryonic stem cell surface marker TRA1-60 (green). Nuclei are shown in blue. Image demonstrates that cells have been converted into an embryonic-like state. Images courtesy of Dr Miguel Esteban.

Chapter 3 Essay Questions

1. *E. coli* are an important model system used in the laboratory. Explain three reasons why *E. coli* are an ideal model organism.
2. How are antibodies used to treat cancer produced? Use a specific example.
3. Compare the process of making yogurt to that of making cheese and the role that microorganisms play in both processes.

Additional Resources

Further Reading:

Black J G and Black L J (2015). Microbiology: Principles and Explorations (Hoboken, NJ: Wiley Higher Ed).

Crawford D H (2009). Deadly Companions: How Microbes Shaped Our History (New York, NY: Oxford University Press).

Skloot R (2010). The Immortal Life of Henrietta Lacks (New York, NY: Crown Publishing).

Talaro K and Chess B (2015). Foundations in Microbiology, Basic Principles (Columbus, OH: McGraw-Hill Science/Engineering/Math).

Video lectures from Howard Hughes Medical Institute (HHMI):

HHMI (2006). Potent Biology: Stem Cells, Cloning, and Regeneration, hhmi.org/biointeractive/stemcells/lectures.html, Accessed October 17, 2017.

HHMI (1999). 2000 and Beyond: Confronting the Microbe Menace, hhmi.org/biointeractive/disease/lectures.html, Accessed October 17, 2017.

Interactive Tree of Life; an online phylogenetic tree viewer and Tree of Life resource:

Letunic and Bork (2016). Interactive Tree of Life (iTOLv3): an online tool for phylogenetic tree display and annotation. Nucleic Acids Research 44, W242-5. iTOL (2016). itol.embl.de/, Accessed October 17, 2017.

Online resources:

Centers for Disease Control and Prevention (2017). cdc.gov.

World Health Organization (2017). who.int/en/.

Activity 3.A Making Microbiology Media

Overview

Bacterial cell cultures require growth media. A common medium for culturing *E. coli* bacteria is lysogeny broth (LB), which is also known as Luria-Bertani broth and is commonly referred to as LB broth. This activity provides protocols for making LB media, including LB broth, LB agar plates, and LB agar deep tubes for stab cultures. In addition, instructions for modifying these basic media to add antibiotics and other additives are included.

LB broth is composed of water, tryptone (10 g/L), yeast extract (5 g/L), and sodium chloride. The concentration of NaCl depends on the formulation; LB, Lennox has 5 g/L, while LB, Miller has 10 g/L. There is little difference in the growth of *E. coli* with either formulation. Premixed, dry LB broth is purchased as premeasured capsules or tablets, each of which make 50 ml of medium, or as a ready-to-use powder, in which case, 20 g of LB, Lennox powder or 25 g of LB, Miller powder makes 1 L. LB agar has the same ingredients and composition as LB broth with the addition of agar, 15 g/L. It takes 35–40 g of premixed LB agar powder to make one liter, depending on whether the LB formulation is Lennox or Miller. Check the chemical container to find out the type and amount of medium that should be added to water.

Media frequently contain additional additives. In this course, sucrose, ampicillin, and arabinose are used to supplement the basic medium. LB/sucrose (LBS) agar is LB agar containing 1% sucrose. Sucrose is added prior to autoclaving the media. The sucrose in LBS agar promotes the growth of bacteria found in yogurt. Ampicillin selects for bacteria carrying a plasmid that codes for the β-lactamase enzyme, which breaks down ampicillin. Ampicillin is added to the LB broth and LB agar at a final concentration of 50–100 μg/ml. This medium is called LB/amp. Arabinose, which is used to induce the expression of a protein from a plasmid, is added to the LB broth and LB agar at a final concentration of 2 mg/ml. This media is called LB/ara. Arabinose will function between 1–3 mg/ml. Heat will degrade ampicillin and arabinose so they are added after autoclaving the media. LB medium containing both ampicillin and arabinose is called LB/amp/ara.

The protocols in this activity provide instructions for preparing the media required for the later activities. Prepare enough media for the activities to be performed within the next month. It is not advisable to make less than 50 ml of medium due to evaporation upon heating. Return to this section in the future to prepare media for other activities. The media required for one student workstation for each activity are listed next.

- **Activity 3.B** requires 3 ml of LB broth and one 100 mm LB agar plate
- **Activity 3.C** requires three 60 mm LBS agar plates
- **Activity 3.E** requires 10 ml of LB broth and seven 60 mm LB agar plates
- **Activity 5.A** requires 1 ml of LB broth, two 60 mm LB agar plates, two 60 mm LB/amp agar plates, and one LB agar deep tube (optional)
- **Activity 5.B** requires 1 ml of LB broth, two 60 mm LB agar plates, two 60 mm LB/amp agar plates, one 60 mm LB/amp/ara agar plate, and one LB agar deep tube (optional)
- **Activity 5.C** requires 10 ml of LB/amp broth
- **Activity 7.C** requires 4 ml of LB/amp/ara broth, one LB/amp agar plate, and one LB/amp/ara agar plate

Tips and Notes

- If an autoclave is not available for sterilizing media, use a microwave oven and ensure that all containers have been washed, bleached, and rinsed well

- Stock solutions of ampicillin and arabinose should be stored at –20°C and used within 1 year. If freeze-dried vials of ampicillin and arabinose are rehydrated with sterile water, they are ready to use without further sterilization. If ampicillin and arabinose stock solutions are made from powder, the stock solutions will need to be filter sterilized through a 0.22 μm filter

- Ampicillin is sensitive to heat and must not be added to hot agar. The agar must be allowed to cool to approximately 55°C before ampicillin is added. When media can be comfortably held in a bare hand for more than 10 sec, it is cool enough to add ampicillin

- Make every effort not to introduce bubbles into media in petri plates. If bubbles do occur they can be removed when the agar is still molten by flaming with a Bunsen burner, which pops the bubbles

- When pouring plates, ensure the agar has cooled to approximately 55°C so that the container is comfortable to hold. However, do not let the temperature of the molten agar drop below 50°C; otherwise, it will begin to solidify and will need to be remelted

- Use the lid of petri plates as a shield to prevent contaminants from falling into the agar while pouring plates

- While every effort may be made to keep growth media sterile, contamination can occur. Before using growth media, check that the broth is clear, not cloudy. Ensure that bacterial colonies or mold are not growing on or in the agar. Discard contaminated media as microbial waste

Activity 3.A Making Microbiology Media

Safety Reminder: Never use the autoclave unless under direct supervision of the instructor. Remember that once a medium is heated to its boiling point, it will pose a burn hazard. Molten medium containing agar can superheat and create flash boil conditions. Use heat-resistant gloves to minimize splash hazards when moving hot agar. Close-toed shoes should be worn when working in a laboratory to avoid burns due to splashing of molten agar. Wear appropriate PPE.

Objectives

For Chapter 3 activities:
- Make 50 ml of LB broth (see part 1)
- Make one 100 mm LB agar plate and seven 60 mm LB agar plates (see parts 4 and 5)
- Make three 60 mm LBS agar plates (see part 6)

For Chapter 5 activities:
- Make 50 ml of LB broth (see part 1)
- Make 50 ml of LB/amp broth (see part 2)
- Make four 60 mm LB agar plates, four 60 mm LB/amp agar plates, two 60 mm LB/amp/ara agar plates (see part 7)
- Make 2 LB agar deep tubes (optional, see part 8)

For Chapter 7 activities:
- Make 50 ml of LB/amp/ara broth (see part 3)
- Make one 60 mm LB/amp agar plate and one LB/amp/ara agar plate (see part 7)

Skills to Master

Refer to Laboratory Skills Assessment Rubric (Appendix E) and Laboratory Notebook Rubric (Appendix F) for more details.
- Record laboratory notebook entries
- Perform calculations necessary for growth media production
- Label agar plates and broth
- Make LB, LBS, LB/amp, and LB/amp/ara agar
- Use the autoclave to sterilize media
- Use aseptic technique
- Pour agar plates
- Make LB, LB/amp, and LB/amp/ara broth
- Make LB agar deep tubes

Student Workstation Materials

See the Materials list on the next page.

Prelab Focus Questions and Calculations

(Refer to the Preparing Solutions section in Chapter 2 for help with calculations)

1. If 35 g of LB agar powder is required to make 1 L of LB agar, how much LB agar powder is needed to make 100 ml of LB agar?
2. If 20 g of LB broth powder is required to make 1 L of LB broth, how much LB broth powder is needed to make 50 ml of LB broth?
3. If one capsule of LB broth powder makes 50 ml of LB broth, how many capsules are needed for 1 L of broth?
4. What information should be included when labeling a petri plate?
5. If the stock ampicillin solution is 10 mg/ml, how much of the stock solution should be added to 50 ml of LB broth for a final concentration of 100 µg/ml ampicillin?
6. If the stock arabinose solution is 200 mg/ml, how much of the stock solution should be added to 10 ml of LB/amp agar for a concentration of 2 mg/ml arabinose?
7. How much sucrose is needed to make 50 ml of 1% (w/v) sucrose?

Activity 3.A Making Microbiology Media

Activity Materials for Chapter 3, 5, and 7

Items	Quantity		
	Chapter 3	*Chapter 5*	*Chapter 7*
Water bath at 55°C (shared) (optional)	1	1	1
Autoclave and autoclave tape* (shared)	1	1	1
Microwave oven* (shared)	1	1	1
Balance (shared)	1	1	1
Weigh boats	3	3	2
100 ml graduated cylinder	1	1	1
Pipet pump or filler (optional)	0	1	0
5 ml sterile serological pipet (optional)	0	1	0
25 ml flask (optional)	1	1	1
100 ml reagent bottles or flasks	2	2	2
250 ml reagent bottle or flask	0	1	1
5–15 ml sterile screw cap culture tubes	0	1	0
60 mm petri dishes	10	10	2
100 mm petri dishes	1	0	0
Bunsen burner	0	1	1
Heat-resistant gloves (shared)	1 pair	1 pair	1 pair
Laboratory tape (optional) (shared)	1 roll	1 roll	1 roll
Foil (optional) (shared)	1 roll	1 roll	1 roll
Plastic wrap (optional) (shared)	1 roll	1 roll	1 roll
Laboratory marking pen	1	1	1
Distilled water (dH$_2$O)	250 ml	250 ml	250 ml
LB agar powder or capsules	6 g or 3 capsules	4.2 g or 2 capsules	2.1 g or 1 capsule
LB broth powder or capsules	1.25 g or 1 capsule	1.25 g or 1 capsule	1.25 g or 1 capsule
Sucrose	2 g	0	0
200x ampicillin (10 mg/ml)	0	0.5 ml	0.5 ml
100x arabinose (200 mg/ml)	0	0.25 ml	0.25 ml

* If an autoclave or a microwave oven is not available, a heated stir plate and stirbar can be used instead.

Activity 3.A Making Microbiology Media

Protocol

Part 1: Making LB Broth

1. Label a 100 ml bottle or flask with the name of the medium, date of preparation, and initials of the individual preparing the medium.

2. Measure 50 ml of dH$_2$O and add it to the labeled container.

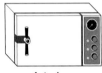

3. If using LB powder, measure the required mass of LB broth powder to make 50 ml of LB broth and add the LB powder to the container. Swirl to dissolve. If using LB capsules, add one capsule to the water, and microwave for 30 sec to 1 min to dissolve the capsule. Do not use foil in the microwave oven. Alternatively, use a stirbar in the container and heat the container on a stir plate until the capsule dissolves.

4. Place a cap **very** loosely on the container or put foil over the flask opening.

5. Place a piece of autoclave tape on the container and place the container in the autoclave.

6. Under direct supervision of the instructor, run the autoclave on the liquid cycle once all student groups have placed their media in the autoclave. Cap the bottles tightly once the media have cooled down.

   Note: If an autoclave is not available, microwave the container on medium power in 30 sec increments with swirling until all translucent particles have dissolved. Using an appropriate low power setting on the microwave oven, gently boil the LB broth for 5 min to sterilize.

7. Store LB broth at 4°C for up to 1 year.

Autoclave

Part 2: Making LB/Amp Broth

1. Calculate the volume of 10 mg/ml (200x) ampicillin stock solution that should be added to 50 ml of LB broth to make a final concentration of 50 µg/ml (1x) ampicillin.

2. Follow steps 1–7 of part 1 to make LB broth.

   Note: If LB broth is available, it does not need to be remade. The volume of ampicillin added to the LB depends on the volume of LB broth available.

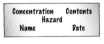

3. Relabel the bottle or flask containing LB broth **LB/amp broth** and add a new date on the container.

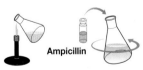

Ampicillin

4. Wait for the LB broth to cool to at least 55°C. Using aseptic technique, flame the neck of the container. Add the calculated amount of sterile ampicillin stock. Flame the container again, recap, and swirl to mix.

5. Store the LB/amp broth at 4°C and use within 2 months.

Activity 3.A Making Microbiology Media

Part 3: Making LB/Amp/Ara Broth

1. Calculate the volume of 10 mg/ml (200x) ampicillin and 200 mg/ml (100x) arabinose stock solutions that should be added to 50 ml of LB broth to make a final concentration of 50 µg/ml ampicillin and 2 mg/ml arabinose.

2. Follow steps 1–7 of part 1 to make LB broth.

   Note: If LB broth is available, it does not need to be remade. The volume of ampicillin and arabinose added to the LB depends on the volume of LB broth available.

3. Relabel the bottle or flask containing LB broth **LB/amp/ara broth** and add a new date on the container.

4. Wait for the LB broth to cool to at least 55°C. Using aseptic technique, flame the neck of the container. Add the calculated amount of sterile ampicillin and arabinose stock solutions. Flame the container again, recap, and swirl to mix.

5. Store the LB/amp/ara broth at 4°C and use within 2 months.

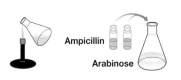

Ampicillin

Arabinose

Part 4: Making LB Agar

1. Calculate the amount of LB agar required based on the number and size of the petri plates to be poured. It is not advisable to make less than 50 ml of media.

   Note: One 100 mm plate requires around 20 ml of LB agar and one 60 mm plate requires around 8 ml of agar.

2. Label an appropriately sized bottle or flask with the name of the medium, date of preparation, and initials of the individual preparing the medium.

   Note: To make 50 ml of agar, use a 100 ml container. To make 100 ml of agar, use a 250 ml container.

3. Measure the required volume of dH$_2$O and pour it into the labeled container.

4. If using LB agar powder, add the required amount of powder. While swirling, slowly add the LB powder to the dH$_2$O, Swirl until no clumps exist. If using LB agar capsules, add the capsules directly to the water.

5. Place a cap **very** loosely on the bottle. (If using a flask, put an inverted 25 ml flask into the opening or loosely wrap plastic over the container opening.) Steam must be allowed to escape but too much evaporation will increase the concentration of the agar. Microwave the agar on a medium setting in 30 sec increments for 1–2 min until it is clear. Do not use foil in the microwave.

   Note: As an alternative, heat the container on a stir plate with a stirbar until the solution is clear, taking care that it does not boil over.

Microwave oven

Activity 3.A Making Microbiology Media

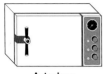
Autoclave

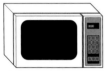
Microwave oven

6. When the medium is clear, loosely place a cap on the container or foil over the opening. Put a piece of autoclave tape on the lid of the container, ensuring that the lid is loose. Place the container into the autoclave.

7. Under direct supervision of the instructor, run the autoclave on the liquid cycle once all student groups have placed their media in the autoclave. Cap the bottles tightly once the media have cooled down.

    Note: If an autoclave is not available, microwave the agar at medium power in 30 sec increments with swirling until all translucent particles have dissolved. Using an appropriate low power setting on the microwave oven, gently boil the agar for 5 min to sterilize.

Part 5: Pouring Agar Plates

1. Label the bottom of petri plates with the name of the medium, date of preparation, and initials of the individual pouring the plates. Write the label around the outer edge of the plate.

    Note: To make agar plates for Activities 3.B and 3.E, label one 100 mm plate and seven 60 mm plates.

2. Retrieve the autoclaved agar.

3. If the agar solidifies before the plates are poured, loosen the cap on the container and microwave the agar for 1–2 min on medium power, checking every 30 sec until the agar is completely melted and clear with no lumps. Do not use foil in the microwave.

4. Before pouring the plates, cool the molten LB agar to around 55°C so that the container is comfortable to hold. The molten agar can be left to cool either on the laboratory bench or in a water bath set to 55°C. Do not let the agar cool so much that it starts to solidify; if this happens, remelt the agar.

    Note: Sterile antibiotics or other additives can be added at this stage; see part 7 for details.

5. Before pouring the plates, gently swirl the agar to mix. Swirl very gently and take care not to introduce bubbles into the medium.

6. Lift the lid of a petri plate and hold it over the plate as a shield while pouring the LB agar into the plate. Pour enough agar to cover about half of the bottom of the plate and then swirl the agar to cover the whole plate. Make sure the plate is about one-third full. Replace the lid.

    Note: A 100 mm plate requires approximately 20 ml of agar. A 60 mm plate requires approximately 8 ml of agar.

7. Pour the remaining plates and allow the plates to solidify undisturbed on the laboratory bench.

    Note: If agar deep tubes are going to be poured for stab cultures, place the molten agar in a 55°C water bath or proceed quickly to part 8 before the agar solidifies.

8. Dry the plates on the laboratory bench for 2 days. Then wrap them in plastic and store them upside down at 4°C. The plates should be used within 2 months.

Activity 3.A Making Microbiology Media

Part 6: Making LBS Agar

1. Label a 100 ml bottle or flask with the name of the medium, date of preparation, and initials of the individuals pouring the plates.

2. Pour 50 ml of dH₂O into the labeled container.

3. Measure the required mass of LB agar (if using LB agar powder) and sucrose required to make 50 ml of LBS agar.

4. While swirling, slowly add the LB agar and sucrose to the dH₂O. Swirl until no clumps of LB agar are present and the solution is cloudy. If using an LB agar capsule, add the capsule and sucrose directly to the water and proceed with the next step without swirling.

5. Follow steps 5–7 in part 4 and all of part 5 to make three 60 mm LBS agar plates.

Part 7: Making LB Agar, LB/Amp Agar, and LB/Amp/Ara Agar Plates Concurrently

1. Calculate the amount of LB agar required based on the number and size of the petri plates (and, if making, the number of LB agar deep tubes) to be poured. It is not advisable to make less than 50 ml of media.

   Note: One 60 mm plate requires approximately 8 ml of agar, while one LB agar deep tube requires 5 ml of agar.

2. Follow the instructions in part 4 to make molten LB agar.

3. Label the outside bottom of 60 mm petri plates with the date, your initials, the name of the medium, and the supplements added to it (for example, **LB**, **LB/amp**, or **LB/amp/ara**). Write the label around the outer edge of the plate.

   Note: If LB agar deep tubes are going to be made, see part 8.

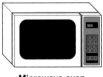

Microwave oven

4. If the agar solidifies before the plates are poured, loosen the cap and microwave the container for 1–2 min on medium power, checking every 30 sec until the agar is completely melted and clear with no lumps.

5. Before pouring the plates, cool the molten LB agar to around 55°C so that the container is comfortable to hold in your hand. Ampicillin is destroyed at temperatures above 60°C.

6. Follow part 5 to pour the LB plates.

   Note: If LB agar deep tubes are required, immediately proceed to part 8 to pour LB agar deep tubes.

Activity 3.A Making Microbiology Media

Ampicillin

Arabinose

7. Estimate the volume of the remaining LB agar using the volume measurements on the flask or bottle. Using sterile technique, add the appropriate amount of 10 mg/ml (200x) ampicillin for a final concentration of 50 µg/ml (1x) ampicillin. Swirl gently to mix the ampicillin with the LB agar and try not to introduce bubbles into the solution.

   For example, if there is approximately 50 ml of LB agar remaining, add 250 µl of 200x ampicillin. It is not necessary to be very accurate when measuring the volume of agar since ampicillin works across a wide range of concentrations.

8. Pour LB/amp plates.

9. Estimate the volume of the remaining LB/amp agar. Using sterile technique, add the appropriate amount of 200 mg/ml (100x) arabinose for a final concentration of 2 mg/ml (1x) arabinose. Swirl gently to mix. Pour LB/amp/ara plates.

   For example, if there is approximately 10 ml of LB/amp agar remaining, add 100 µl of 100x arabinose. It is not necessary to be very accurate when measuring the volume of agar.

10. Leave the plates undisturbed on the bench to allow the agar to solidify.

11. Dry the plates on the laboratory bench for 2 days. Then wrap them in plastic and store them upside down at 4°C. LB, LB/amp, and LB/amp/ara agar plates should be used within 2 months.

Part 8: Pouring LB Agar Deep Tubes for Stab Cultures

1. Label two sterile culture tubes **LB agar** and include the date and your initials.

2. If necessary, follow part 4 to make sterile molten LB agar.

3. Pour or pipet 5 ml of sterile molten LB agar into each sterile tube. Recap the tubes and let the agar solidify while the tubes are in an upright position. These tubes will be used for *E. coli* stab cultures.

4. LB agar deep tubes should be stored at 4°C and used within 1 year.

Postlab Focus Questions

1. Why is it important to store plates at 4°C and use them within one year?
2. Why is it important to cool agar prior to adding ampicillin?
3. Why is autoclave tape used?
4. How could you check that your answers to the calculations for ampicillin are accurate?
5. If a classmate were to make media for the first time, what are three key ideas or tips you would advise her of before she begins?

Self Assessment

Assess whether you have mastered the skills of this activity using the Laboratory Skills Assessment Rubric in Appendix E.

Assess your experimental write-up in your laboratory notebook using the Laboratory Notebook Rubric in Appendix F.

Activity 3.B Disk Diffusion Test (Modified Kirby-Bauer Test)

Overview

The Kirby-Bauer, or disk diffusion, test was developed as a standardized method to test the susceptibility of bacteria to antibiotics. This test is standardized by the Clinical and Laboratory Standards Institute. The size of the petri plate, depth of the agar, density of the inoculating culture, and quantity of antibiotic(s) are kept standard to yield reproducible results when testing the same bacteria. The zones of inhibition for known bacteria against different antibiotics are known, and unknown bacteria are measured against these data (see Figure 3.14).

Bacteria are becoming more and more resistant to antibiotics. MRSA is one example, but bacteria that cause multidrug-resistant tuberculosis (MDR TB) or drug-resistant pneumonia are also on the rise. Alternative antimicrobial substances are being investigated as viable treatments for these bacterial infections. Silver nanoparticles or colloids are at the intersection of biotechnology and nanotechnology. Silver nanoparticles (<100 nm in diameter) are thought to have antimicrobial properties and are currently added to antibacterial water filters and antimicrobial bandages for burns. Silver nanoparticles that are 10 nm in diameter have been shown to inhibit growth of MRSA bacteria.

Antimicrobial hand soaps are extremely prevalent in western society. A common antimicrobial ingredient is triclosan, a chlorinated aromatic compound that is present in many consumer goods, including soaps, deodorants, toothpastes, kitchen utensils, and toys.

In this activity, a modified Kirby-Bauer method will be used to test the sensitivity of E. coli HB101 to a variety of antimicrobial compounds. Ampicillin, a common antibiotic of the penicillin family, will be used as a positive control. The antimicrobial properties of silver colloids (sold as a natural remedy in health food stores) and an antibacterial household product (likely containing the antimicrobial triclosan) will be tested. A liquid culture of E. coli HB101 will be inoculated on an agar plate to create a bacterial lawn. ("Lawn" is a term used to describe bacteria completely covering the agar surface with no distinguishable colonies.) Paper disks impregnated with various antimicrobial agents will be placed in quadrants. A sterile disk will be placed on the plate as a negative control. The ability of the antimicrobial agent on the disk to inhibit the growth of E. coli will be determined by measuring the zone of inhibition surrounding each disk.

Tips and Notes

- When placing the disks onto the agar surface, remember to hold the lid directly above the plate as a shield. Always use sterile forceps to handle the disks. The forceps can be autoclaved or sterilized by flaming or soaking in 10% bleach and rinsing well in sterile water if an autoclave is not available. When placing the disks onto the surface of the plate, press them gently with sterile forceps to ensure that they stick when the plate is inverted

Safety Reminder: Inform your instructor if you are allergic to antibiotics and avoid contact with ampicillin. Use aseptic technique when handling E. coli bacteria and dispose of microbial waste properly. Wear appropriate PPE.

Research Question

- Which antimicrobial agent is most effective against E. coli HB101 bacteria?

Objectives

- If not already available, make one 100 mm LB agar plate
- Set up a disk diffusion assay using three chemicals and one negative control
- Evaluate the antimicrobial properties of three chemicals

Skills to Master

Refer to Laboratory Skills Assessment Rubric Appendix E) and Laboratory Notebook Rubric (Appendix F) for more details.
- Record laboratory notebook entries
- Perform a modified Kirby-Bauer test
- Use aseptic technique (Activity 3.A)
- Inoculate an agar plate for a bacterial lawn

Student Workstation Materials

Items	Quantity
Incubator at 37°C (shared)	1
Autoclave and autoclave tape (shared) (optional)	1
2–20 µl adjustable-volume micropipet and sterile tips	1
100–1,000 µl adjustable-volume micropipet and sterile tips, or 1 ml sterile graduated transfer pipet	1
Ruler	1
Sterile paper disks in a petri plate	4
Bunsen burner	1
Sterile inoculation loop or bacterial spreader	1
Sterile forceps	1
Laboratory marking pen	1
Pencil	1
Microbial waste container	1
Ampicillin (1 mg/ml)	20 µl
Silver colloid solution	20 µl
Antimicrobial product	20 µl
E. coli HB101 liquid culture	2 ml
100 mm LB agar plate (from Activity 3.A)	1

Prelab Focus Questions

1. What is a bacterial lawn?
2. Why would one disk have a greater zone of inhibition than another?
3. Why should you sterilize the forceps before starting this activity?
4. What would you expect to see if the chemical on the disk has no antimicrobial properties? What would you compare it to?

Activity 3.B Disk Diffusion Test (Modified Kirby-Bauer Test)

Activity

1. Place four sterile paper disks into an empty 60 mm petri plate. Using a pencil, label the disks **1**, **2**, **3**, and **4** for the following substances:

   Disk #1: negative control
   Disk #2: ampicillin
   Disk #3: silver colloid
   Disk #4: antimicrobial product

2. Using a fresh tip, pipet 10 µl of ampicillin solution onto the center of disk #**2**.

3. Using a fresh tip, pipet 10 µl of silver colloid solution onto the center of disk #**3**.

4. Using a fresh tip, pipet 10 µl of antimicrobial product onto the center of disk #**4**.

5. Close the lid of the petri plate containing the disks and incubate the plate in a 37°C incubator for 10–20 min to dry the disks.

6. Using a laboratory marking pen, divide the bottom of a 100 mm LB agar plate into four equal quadrants by drawing two perpendicular lines that form a large "X".

7. Label the quadrants **1**, **2**, **3**, and **4**. Label the plate with your initials and the date.

250 µl
E. coli

8. Using sterile technique, pipet 250 µl of *E. coli* HB101 culture to the middle of the LB agar plate.

9. Using a sterile inoculation loop, gently spread the bacterial culture over the entire surface in all directions, including the edge of the plate. If using a loop, skate it over the top of the plate and avoid digging into the agar. Spend approximately 1 min spreading the bacteria until most of the liquid has been absorbed by the plate. Remember to keep the lid over the plate as a shield. Close the lid and leave the bacterial culture to dry for 10–15 min.

10. Using sterile forceps, place disk #**1** (the negative control with no antimicrobial reagent) into the center of quadrant #**1** of the LB agar plate. Gently press the disk with the forceps to ensure that the disk makes contact with the agar. Take care not to contaminate the forceps. Do not lay them down on the benchtop; place them only on a sterile surface.

11. Place disk #**2** in quadrant #**2**.

12. Place disk #**3** in quadrant #**3**.

13. Place disk #**4** in quadrant #**4**.

14. Place the LB agar plate upside down in a 37°C incubator for 16–24 hr.

Activity 3.B Disk Diffusion Test (Modified Kirby-Bauer Test)

Results Analysis

Once the cultures have grown for 16–24 hr, turn the plates upside down and use a ruler to measure the diameter of the area around the disk where the bacteria did not grow. Do not open the plates. This area is called the zone of inhibition. The zone of inhibition for each antimicrobial compound on your agar plate should be compared to determine the relative effectiveness of each compound. Share the data with other teams and determine if your conclusions agree with those of the class.

Postlab Focus Questions

1. Which antimicrobial was the most effective against *E. coli* HB101 bacteria?
2. Which antimicrobial was the least effective against *E. coli* HB101 bacteria?
3. How could bacteria be tested for resistance to antibiotics?
4. What other compounds could you test using the disk diffusion method?
5. Consider the experimental design for this activity. Suggest three ways the test could be better standardized so that results would be reproducible from laboratory to laboratory.

Self Assessment

Assess whether you have mastered the skills of this activity using the Laboratory Skills Assessment Rubric in Appendix E.

Assess your experimental write-up in your laboratory notebook using the Laboratory Notebook Rubric in Appendix F.

Activity 3.C Microbes & Health: An Illustration of Koch's Postulates

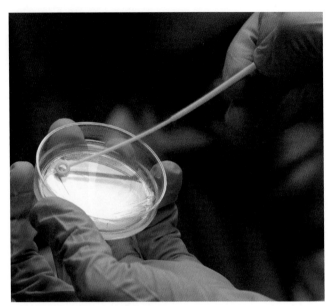

Figure 3.32. **Inoculation of LBS agar plate**.

Overview

To determine the microorganism that might be causing a disease, Robert Koch developed postulates to guide the process of assigning causation.

Koch's postulates are:

1. The microorganism must be found in abundance in all organisms suffering from the disease but should not be found in healthy animals.
2. The microorganism must be isolated from a diseased organism and grown in pure culture.
3. The cultured microorganism should cause disease when introduced into a healthy organism.
4. The microorganism must be re-isolated from the inoculated, diseased experimental host and identified as being identical to the original specific causative agent.

Scientists often use model systems to simulate diseases in humans. For example, a lot of medical research is performed on laboratory animals so that scientists can learn more about human diseases. In this activity, you will use a model to test Koch's postulates. Milk will represent a healthy individual and yogurt will represent a diseased individual. At times, milk will develop a condition that causes it to thicken and turn into yogurt; this condition will be referred to as the "yogurtness" disease. You will play the role of a medical investigator to determine the cause of yogurtness. You suspect that the yogurtness disease may be caused by something that is found in yogurt. You will use Koch's postulates to support or refute the hypothesis that microbes found in yogurt are the cause of the yogurtness disease. Of course, it is important to remember that this activity is a simulated investigation; real yogurt is a very healthy food. Microbes found in yogurt are harmless and do not cause disease in healthy humans.

Safety Reminder: Inform your instructor if you are allergic to antibiotics and avoid contact with ampicillin. Use aseptic technique when handling *E. coli* HB101 bacteria and dispose of microbial waste properly. Wear appropriate PPE. Do not eat or drink in the laboratory. Do not consume any of the yogurt produced in this activity or eat out of any container that is used in a laboratory.

Research Question
- What causes yogurtness?

Objectives
- If not already available, make three 60 mm LBS agar plates
- Describe the symptoms of yogurtness
- Make observations of milk and yogurt under the microscope
- Make streak plates and isolate bacteria from yogurt, milk, and *E. coli* HB101 cultures
- Describe bacterial colonies and compare the isolated bacteria microscopically with those in the original yogurt, milk, and *E. coli* HB101 samples
- Inoculate milk cultures using aseptic technique
- Describe symptoms of inoculated samples

Skills to Master
Refer to Laboratory Skills Assessment Rubric (Appendix E) and Laboratory Notebook (Appendix F) for more details.
- Record laboratory notebook entries
- Use aseptic technique (Activity 3.A)
- Streak bacteria to isolate single colonies
- Aseptically inoculate a plate with a liquid culture
- Aseptically inoculate a liquid culture from a plate
- Make a wet mount
- Observe bacteria using a microscope

Activity 3.C Microbes & Health: An Illustration of Koch's Postulates

Student Workstation Materials

Part 1: Identify Possible Pathogens (Postulate 1) and Isolate and Culture the Suspected Pathogen (Postulate 2)

Items	Quantity
Microscope and optional accessories, including immersion oil and lens cleaning tissue	1
Incubator at 37°C (shared)	1
Microscope slide	1
Coverslips	2
Bunsen burner*	1
Inoculation loop*	1–3
Sterile toothpicks or micropipet tips	3
Laboratory marking pen	1
Microbial waste container	1
Sterile water	1 ml
Fresh yogurt	5 ml
Fresh milk	5 ml
E. coli HB101 liquid culture (shared)	1 vial
pH paper	2 pieces
60 mm LBS agar plates (from Activity 3.A)	3

* If metal inoculation loops are available, one loop is sufficient and must be sterilized by flaming with a Bunsen burner. Alternatively, three disposable plastic loops can be used.

Part 2: Isolate and Culture the Suspected Pathogen (Postulate 2) and Infect a Healthy Individual with the Isolated Pathogen (Postulate 3)

Items	Quantity
Microscope and optional accessories, including immersion oil and lens cleaning tissue	1
Incubator at 37°C (shared)	1
20–200 µl adjustable-volume micropipet and sterile tips, or 1 ml sterile graduated transfer pipet	1
Microscope slides	2–4
Coverslips	4–8
Magnifying glass (optional)	1
Bunsen burner*	1
Inoculation loop*	1–3
Sterile toothpicks or micropipet tips	8
Parafilm (approx 10 cm strip) (shared)	1 roll
Laboratory marking pen	1
Microbial waste container	1
Sterile water	1 ml
Fresh yogurt	5 ml
Culture tubes with 5 ml of scalded milk	6
Ampicillin (10 mg/ml)	75 µl
LBS agar plates inoculated in part 1	3

* If metal inoculation loops are available, one loop is sufficient and must be sterilized by flaming with a Bunsen burner. Alternatively, three disposable plastic loops can be used.

Part 3: Isolate the Pathogen from the Newly Diseased Individual (Postulate 4)

Items	Quantity
Microscope and optional accessories, including immersion oil and lens cleaning tissue	1
Microscope slides	4
Coverslips	8
Sterile toothpicks or micropipet tips	8
Parafilm (approx 10 cm strip) (shared)	1 roll
Laboratory marking pen	1
Microbial waste container	1
Sterile water	1 ml
Milk cultures (from part 2)	6 tubes
pH paper	6 pieces
LBS agar plates with bacteria from yogurt (from part 1)	1

Prelab Focus Questions

1. Name three diseases that are caused by bacteria.
2. Name three diseases that are not caused by bacteria.
3. How are bacterial diseases treated?
4. Name two ways to prevent the spread of bacterial diseases.
5. Are all bacteria harmful? If not, describe the benefits of some bacteria.

Activity 3.C Microbes & Health: An Illustration of Koch's Postulates

Activity Protocol

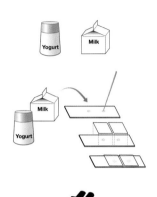

Part 1A: Identify Possible Pathogens (Postulate 1)

1. Compare yogurt and milk with respect to appearance, smell, and pH. Record your observations.

2. Obtain a microscope slide. Label the left edge of the slide **Yogurt** and right edge **Milk**.

3. Add a drop of sterile water to the left hand side of the slide. Dip a sterile toothpick in yogurt and transfer the toothpick into the drop of water on the slide. Swirl the toothpick to mix the yogurt sample with the water. Cover with a coverslip.

4. Add a drop of milk to the right hand side of the slide and cover with a coverslip.

5. Observe the yogurt and milk samples under the microscope. Describe and draw what you see.

Part 1B. Isolate and Culture the Suspected Pathogen (Postulate 2)

Milk   Yogurt   *E. coli*

6. Obtain three 60 mm LBS agar plates. If necessary, pour them by following part 6 of Activity 3.A. Label the bottom of each plate: one as **Milk**, one as **Yogurt**, and one as **_E. coli_**, along with your initials and the date.

7. Using the streak plate technique (described below), streak milk onto the plate labeled **Milk**, yogurt onto the plate labeled **Yogurt**, and *E. coli* HB101 bacteria on the plate labeled **_E. coli_**.

8. **Streak Plate Technique**

   Note: If a metal loop is used to spread the bacteria, sterilize it first by flaming. Alternatively, a sterile plastic inoculation loop can be used but should not be flamed.

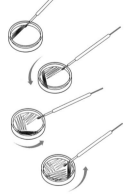

   a. Flame the metal loop by passing it back and forth in the Bunsen burner. Let the loop cool down. Dip the loop into the *E. coli* HB101 culture, milk, or yogurt, and obtain a film of sample across the loop.

   b. Gently rub the loop back and forth across the top left corner of the plate about 10 times. Stay in the top left quadrant of the plate and do not break the surface of the agar. Flame the loop again.

   c. Rotate the plate 45° and draw the loop through one end of the first streak and run back and forth in the second quadrant about 10 times. Flame the loop as before.

   d. Rotate the plate 45° and draw the loop through one end of the second streak and rub the loop back and forth in the third quadrant about 10 times. Flame the loop again.

   e. Repeat the process one final time in the forth quadrant. Flame and cool the loop after completing streak plating.

Activity 3.C Microbes & Health: An Illustration of Koch's Postulates

9. Invert the plates and place them in the incubator at 37°C for 24–48 hr.

10. Once the cultures have grown, wrap plates in Parafilm and store at 4°C until part 2.

Milk   Yogurt   *E. coli*

Part 2A: Isolate and Culture the Suspected Pathogen (Postulate 2)

1. Compare the number of colonies on the plates labeled **Milk**, **Yogurt**, and ***E. coli***. Record the results.

2. Observe colonies using a magnifying glass if one is available. Record the number of different types of colonies on each plate. Use a laboratory marking pen to circle one of each type of colony and label with a number on the bottom of the plate.

3. Describe the appearance of each numbered colony.

4. Label some slides according to your colony numbers. Use one slide for two samples.

5. Pick a circled, numbered colony from the yogurt plate. Prepare a wet mount by mixing the colony with a drop of water on the appropriately numbered slide. Cover with a coverslip.

6. Prepare a wet mount of the other numbered colonies from the plates labeled **Milk**, **Yogurt**, and ***E. coli***.

7. Observe the slides under the microscope. Draw and describe what you see.

8. Compare the bacteria in the above wet mounts with your descriptions of those observed in the yogurt and milk in postulate 1.

Part 2B: Infect a Healthy Individual with the Isolated Pathogen (Postulate 3)

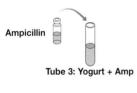

9. Label six tubes of scalded milk with your initials and as follows:

   Tube 1: Negative control
   Tube 2: Positive control (yogurt)
   Tube 3: Yogurt + amp
   Tube 4: Yogurt colony #1
   Tube 5: Yogurt colony #2
   Tube 6: *E. coli*

10. Add 50 µl or 1 drop of ampicillin to the tube labeled **Tube 3: Yogurt + amp**.

Ampicillin

Tube 3: Yogurt + Amp

11. Inoculate each tube with the appropriate sample as described below using aseptic technique. If a metal inoculation loop is used, sterilize the loop by flaming it between samples. Yogurt should be washed or wiped off the loop before flaming.

12. Dip the sterile loop in yogurt and inoculate the tube labeled **Tube 2: Positive control (yogurt)** with the yogurt sample.

Yogurt

Tube 2: Positive control

Yogurt

Tube 3: Yogurt + Amp

13. Dip the sterile loop in yogurt and inoculate the tube labeled **Tube 3: Yogurt + amp** with the yogurt sample.

Activity 3.C Microbes & Health: An Illustration of Koch's Postulates

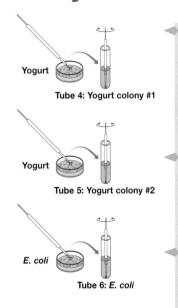

Yogurt

Tube 4: Yogurt colony #1

Yogurt

Tube 5: Yogurt colony #2

E. coli

Tube 6: E. coli

14. Identify two colonies on the plate labeled **Yogurt** (the colonies that were numbered in postulate 2). Using aseptic technique, gently touch one of the numbered colonies with a sterile loop and inoculate the tube labeled **Tube 4: Yogurt colony #1**.

    Note: If the colonies are too small to transfer with a loop, use a sterile toothpick or pipet tip.

15. Using aseptic technique, inoculate the tube labeled **Tube 5: Yogurt colony #2** with a second numbered colony from the plate labeled **Yogurt**.

16. Using aseptic technique, gently touch an *E. coli* HB101 colony from the plate labeled *E. coli* and inoculate the tube labeled **Tube 6: *E. coli***.

17. Place the inoculated milk cultures and the tube labeled negative control in the incubator at 37°C for 24–48 hr.

18. Wrap the plates in Parafilm and store them at 4°C until part 3.

Part 3: Isolate the Pathogen from the Newly Diseased Individual (Postulate 4)

1. Examine the milk cultures from the previous lesson. Describe each culture with respect to appearance, smell, and pH. Have any of the cultures contracted "yogurtness"?

2. Obtain four microscope slides. Label the first slide **Tube 1** on the right side and **Tube 2** on the left side.

3. Label the second slide **Tube 3** on the right side and **Tube 4** on the left side. Label the third slide **Tube 5** on the right side and **Tube 6** on the left side.

4. Prepare wet mounts of samples from all six tubes, two per slide.

5. Label the fourth slide **Yogurt colony #1** and **Yogurt colony #2**.

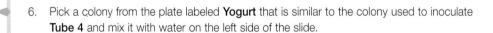

6. Pick a colony from the plate labeled **Yogurt** that is similar to the colony used to inoculate **Tube 4** and mix it with water on the left side of the slide.

7. Repeat to make a wet mount of the yogurt bacteria used to inoculate **Tube 5**.

8. Observe the four slides under the microscope and describe and draw what you see.

9. Wrap a strip of Parafilm around the edges of the LBS agar plates labeled *E. coli* and **Yogurt**, and store them at 4°C for use in Activity 3.D. Discard the plate labeled **Milk** as microbial waste.

Activity 3.C Microbes & Health: An Illustration of Koch's Postulates

Results Analysis

When observing the outcome of the experiment, it is important to reflect on Koch's postulates.

- Were microorganisms isolated from yogurt?
- Did they cause yogurtness when reinoculated into fresh milk?
- Did the organisms look the same when you looked at them again? Record detailed observations regarding appearance, smell, and pH of the cultures.
- Did the properties of the diseased culture differ from those of the original yogurt culture?

Postlab Focus Questions

1. What is the purpose of the tubes containing scalded milk alone, *E. coli* culture with ampicillin, and yogurt with ampicillin?
2. What conclusions can you make if you see more than one type of bacterium growing on an agar plate streaked with yogurt?
3. What can you conclude about the cause of yogurtness? Support your conclusion with data.
4. Which of Koch's postulates are tested by inoculating an agar plate with yogurt?
5. Do all bacteria cause milk to turn into yogurt? Which of the controls test for this?
6. Why did you add antibiotics to one of the tubes?

Self Assessment

Assess whether you have mastered the skills of this activity using the Laboratory Skills Assessment Rubric in Appendix E.

Assess your experimental write-up in your laboratory notebook using the Laboratory Notebook Rubric in Appendix F.

Activity 3.D Gram Staining

Overview

In 1882, a technique to discriminate between the two types of bacterial cell walls was invented by the Danish scientist Hans Christian Gram. This technique utilizes a four-step staining procedure with two different dyes and is still one of the first tests used when trying to identify unknown bacteria. Gram-positive bacteria have a very thick layer of peptidoglycan composed of layers of carbohydrates cross-linked with polypeptides. Crystal violet stain binds peptidoglycan very tightly and makes the bacteria a deep purple color. Gram-negative bacteria have a very thin layer of peptidoglycan in between two layers of phospholipid membrane. The crystal violet stain does not bind well and is washed out by decolorizer (alcohol). Safranin, which is used as a counterstain, makes gram-negative bacteria appear pink (see Figure 3.33).

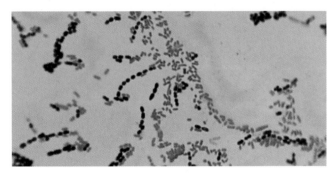

Figure 3.33. **A mixture of *E. coli* and bacteria from yogurt.** The mixture was Gram stained and viewed under an oil immersion lens at 1,000x magnification.

In this activity, you will perform Gram staining of *E. coli* HB101 and yogurt bacteria. You will observe the stained bacteria using a microscope and determine the shape of the bacteria and assess whether they are gram-positive or negative.

Tips and Notes

- Wear gloves to avoid staining your fingers. Have multiple beakers of water available for washing the stain from the slides during the Gram staining procedure. Wooden clothespins can be used to hold slides while staining and flaming; these clothespins can lower the chance of getting the stain on your fingers

Safety Reminder: Review the SDSs of all the stains used in this activity. Before placing any stains on the benchtop, ensure that flaming of slides is complete and that Bunsen burners are turned off. The decolorizer contains a high percentage of alcohol and is very flammable. Wear appropriate PPE.

Research Questions

- Are bacteria found in yogurt and *E. coli* HB101 bacteria gram-positive or gram-negative?
- What are the size and shape of *E. coli* HB101 and bacteria from yogurt?

Objectives

- Mount bacteria on a microscope slide using aseptic technique
- Perform Gram staining of bacteria
- Determine gram status of bacteria
- Determine cell shape of bacteria
- Determine size of bacteria

Skills to Master

Refer to Laboratory Skills Assessment Rubric (Appendix E) and Laboratory Notebook Rubric (Appendix F) for more details.

- Record laboratory notebook entries
- Perform Gram staining of bacteria
- Heat fix bacteria to a slide
- Observe bacteria using a microscope (Activity 3.C)
- Differentiate gram-positive bacteria from gram-negative bacteria
- Identify bacteria from cell shape
- Sketch microscopic details
- Estimate the size of cells using a microscope

Student Workstation Materials

Items	Quantity
Microscope and optional accessories, including immersion oil and lens cleaning tissue	1
Stage micrometer (optional)	1
Microscope slide	1
Bunsen burner	1
Inoculation loop*	1–4
Wooden clothespin (optional)	1
Tissue or paper towel	1
Wax pencil	1
Microbial waste container	1
Sterile water	1 ml
Beakers of tap water	3
Crystal violet stain	1
Gram's iodine	1
Decolorizer (alcohol)	1
Safranin stain	1
LBS agar plate with yogurt bacterial colonies (from Activity 3.C) or fresh yogurt	1
Agar plate with *E. coli* HB101 (from Activity 3.C)	1

* If metal inoculation loops are available, one loop is sufficient and must be sterilized by flaming with a Bunsen burner. Alternatively, four disposable plastic loops can be used.

Prelab Focus Questions

1. What part of the bacterium does the Gram staining procedure stain?
2. Describe the three basic shapes of bacteria and give the scientific terms used to describe these shapes.
3. How do you determine the magnification when using a microscope?

Activity 3.D Gram Staining

Protocol

Part 1: Heat Fix Bacteria to the Slide

1. Using a wax pencil, draw two circles about 1 cm in diameter at one end of a microscope slide. Label the left circle **Yogurt** and the right *E. coli*.

2. If using a metal inoculation loop, sterilize it by flaming. Sterile plastic loops should not be flamed.

3. Using aseptic technique, dip the loop in sterile water so that a film appears across the loop. Transfer the water into the circle drawn by the wax pencil. Flame the metal inoculation loop again or use a new sterile plastic loop, and transfer water into the second circle.

4. Flame the metal inoculation loop again or obtain a new sterile plastic loop. Use the loop to very lightly touch a yogurt colony from the **Yogurt** agar plate or touch the liquid on top of the fresh yogurt. Make sure to touch the bacterial colony or yogurt very lightly to avoid transferring too many bacteria.

5. Transfer the loop into the water in the left circle on the microscope slide and swirl the loop to mix the sample with the water.

6. Flame the metal inoculation loop again or obtain a new sterile plastic loop. Transfer a colony of *E. coli* HB101 bacteria growing on the *E. coli* agar plate. Make sure to lightly touch the bacterial colony to avoid transferring too many bacteria. Transfer the loop into the water in the right circle on the microscope slide and swirl the loop to mix the sample with the water.

7. Let the liquid completely evaporate, leaving behind dry spots on the slide. This will take at least 10 min.

8. Heat-fix the bacteria to the slide by quickly passing the slide through the flame. The flame should touch the bottom of the slide for about 1 sec during each of three passes.

Part 2: Gram Stain the Fixed Bacteria

1. Ensure all Bunsen burners are turned off before obtaining the set of Gram stains.

Crystal violet

2. Add a few drops of crystal violet stain to the bacteria until the wax circles are filled with stain. Let stand for 1 min.

3. Rinse the slide by dipping it in a beaker of water 2–3 times until most of the stain has been washed off.

Gram's iodine

4. Add Gram's iodine to cover the bacteria. Let stand for 1 min.

Activity 3.D Gram Staining

5. Rinse in a beaker of water.

6. Add drops of decolorizer (alcohol). Let the decolorizer flow over the bacteria at a 45° angle until it flows clear.

Safranin

7. Counterstain by adding safranin stain to cover the bacteria and let stand for 1 min.

8. Rinse the slide in a beaker of water.

9. Blot excess water from around the circles, making sure not to touch the inside of the circles. Let the slide air dry.

10. Once the slide is completely dry, observe the sample under the microscope, starting at low power and moving up to the higher-powered lenses. There is no need to add a coverslip. If oil immersion lenses are available, observe at 1,000x magnification.

Results Analysis

Once the slides are completely air dried, place them on the microscope and focus on the color visible at low power. Once focused on low power, move through the increasingly higher-power lenses, using only the fine focus to focus each time on the colored dots until you get to the highest power objective lens. If using an oil immersion lens, follow the directions of your instructor. Scan around the slide until you can see individual bacteria. This is often near the edge of the slide; the center of the slide often has too many bacteria, which appear as a dense mass of color. The determination of Gram status should be made while observing a single bacterium. Use a stage micrometer or information from your instructor on the size of the field of view to determine the size of the bacterium. Compare the bacteria found in yogurt to *E. coli* HB101 with respect to size, Gram stain result, and shape. Record all observations in your laboratory notebook and include sketches and the magnification used to view the bacteria.

Self Assessment

Assess whether you have mastered the skills of this activity using the Laboratory Skills Assessment Rubric in Appendix E.

Assess your experimental write-up in your laboratory notebook using the Laboratory Notebook Rubric in Appendix F.

Postlab Focus Questions

1. Are bacteria found in yogurt and *E. coli* HB101 bacteria gram-positive or gram-negative? What evidence do you have? What does the gram status tell you about the bacteria's cell wall structure?
2. What are the sizes of all the bacteria you observed under the microscope? What was the highest magnification you used to view the various bacteria?
3. What was the shape of each bacterium you observed?
4. Was there more than one type of bacterium present on any of the slides? If so, describe the bacteria and the evidence you have to prove that they are different from one another.

Activity 3.E Quantifying Bacterial Numbers

Overview

Bacteria are microscopic, and a bacterial culture contains billions of cells. There are several ways to determine the number of cells in a culture. The serial dilution and plate count method involves preparing serial dilutions of the bacterial culture to dilute the bacteria more than 1 million times. The individual bacteria are then grown into colonies on agar plates so that they can be seen and counted. The number of bacteria that formed colonies (colony forming units, or CFU) can then be used to calculate backward to the concentration of viable cells in the original culture.

In this activity, you will quantify the concentration of *E. coli* HB101 bacteria in a culture using the serial dilution and plate count method. The bacteria will first be diluted tenfold, and then that dilution will be diluted another tenfold to make a 100-fold dilution. The 100-fold dilution will be further diluted tenfold to make a 1,000-fold dilution, and so on, until the original culture has been diluted 10 million-fold. The amount that the culture is diluted by is called the dilution factor and can be stated in scientific notation as 10, 10^2, 10^3, etc. This dilution factor is used to calculate the number of bacteria in the original culture. Each dilution is plated and incubated overnight at 37°C. The CFU on the plates are then counted and used to calculate the number of bacteria per milliliter in the original culture.

Tips and Notes

- When doing serial dilutions, it is important to accurately measure the volume of cell cultures transferred into tubes and spread onto the plates. Solutions should be mixed well between each step to ensure that the cells are evenly distributed when diluting

Safety Reminder: Take care when opening microcentrifuge tubes that contain bacteria to avoid generating bacterial aerosols. Open containers very slowly. Use aseptic technique when handling *E. coli* HB101 bacteria and dispose of microbial waste properly. Wear appropriate PPE.

Research Question

- How many bacteria are in an *E. coli* HB101 culture?

Objectives

- If not already available, make 50 ml of LB broth and seven 60 mm LB agar plates
- Perform a tenfold serial dilution of bacteria seven times to dilute the original culture 10,000,000 times
- Spread each dilution on an LB agar plate and incubate overnight
- Count the colonies on each LB agar plate
- Calculate the concentration of bacteria in the original culture in CFU/ml

Skills to Master

Refer to Laboratory Skills Assessment Rubric (Appendix E) and Laboratory Notebook Rubric (Appendix F) for more details.
- Record laboratory notebook entries
- Perform a serial dilution of a bacterial culture
- Aseptically inoculate a plate with a liquid culture (Activity 3.C)
- Calculate concentration of bacteria in a culture

Student Workstation Materials

Items	Quantity
Incubator at 37°C (per class)	1
Vortexer (optional) (shared)	1
20–200 µl adjustable-volume micropipet and sterile tips	1
100–1,000 µl adjustable-volume micropipet and sterile tips	1
Sterile microcentrifuge tubes	7
Bunsen burner*	1
Inoculation loop or bacterial spreader*	1–8
Laboratory marking pen	1
Microbial waste container	1
LB broth (from Activity 3.A)	10 ml
E. coli HB101 culture	2 ml
60 mm LB agar plates (from Activity 3.A)	7

* If metal inoculation loops are available, one loop is sufficient and must be sterilized with a Bunsen burner. Alternatively, eight disposable plastic loops can be used.

Prelab Focus Questions

1. Why is it necessary to perform a serial dilution of the bacterial culture before plating each diluted sample on an LB agar plate?
2. Why is it important to thoroughly mix each dilution before pipetting it into the next dilution and before plating it?
3. If 60 colonies formed on an LB agar plate, how many bacteria were plated? How many CFU are there?
4. If you performed a tenfold serial dilution of a bacterial culture six times to achieve a millionfold dilution, and if 100 µl of that dilution yielded 60 CFU when plated, what is the concentration of bacteria in the original culture in CFU/ml?

Activity 3.E Quantifying Bacterial Numbers

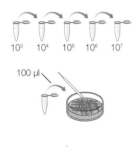

Protocol

1. If necessary, prepare seven 60 mm LB agar plates and 50 ml of LB broth using the instructions provided in Activity 3.A.

2. Label the LB agar plates on the bottom edge with the dilution factor (**10**, **10²**, **10³**, **10⁴**, **10⁵**, **10⁶**, and **10⁷**). Also add the name of the bacterium (*E. coli* HB101), your initials, and the date.

3. Label seven sterile microcentrifuge tubes with the dilution factor (**10**, **10²**, **10³**, **10⁴**, **10⁵**, **10⁶**, and **10⁷**).

4. Using aseptic technique, pipet 900 μl of LB broth into each microcentrifuge tube.

5. Using aseptic technique, pipet 100 μl of the *E. coli* HB101 culture into the microcentrifuge tube labeled **10**. Mix by vortexing or pipetting up and down at least five times.

6. Remove 100 μl of the diluted cell culture from the tube labeled **10** and add it to the tube labeled **10²**. Mix by vortexing or pipetting up and down at least five times.

7. Remove 100 μl of the diluted cell culture from the tube labeled **10²** and add it to the tube labeled **10³**. Mix by vortexing or pipetting up and down at least five times.

8. Continue to serially dilute the bacterial culture into each consecutive tube until the tube labeled **10⁷** is used.

9. Mix the diluted cell culture in the tube labeled **10** and, using aseptic technique, pipet 100 μl onto the LB agar plate labeled **10**. Sterilize an inoculation loop or bacterial spreader by flaming, and allow it to cool. Very gently spread the bacterial culture over the entire surface in all directions, including the edge of the plate. Do not use the streak plate technique — you are not isolating single colonies.

   Note: A sterile plastic inoculation loop or sterile spreader could also be used to spread the bacteria. Plastic tools are intended for single use and should not be flamed.

10. Plate 100 μl of each dilution on the matching plate using the instructions provided in step 9.

11. Stack the plates upside down and place in the incubator at 37°C for 16–24 hr.

Activity 3.E Quantifying Bacterial Numbers

Results Analysis

Retrieve the culture plates from the incubator and arrange them on the laboratory bench from least to most dilute. Record your observations, including the number of CFU across all plates.

- Plates that have bacterial growth so dense that no single colonies can be seen should be recorded as having a "lawn" of bacteria
- Plates where single colonies can be seen but are much too dense to actually count (more than 200 colonies per plate) should be recorded as "too many to count," or TMTC

To count colonies, use a laboratory marking pen and place a dot on the bottom of the plate in the middle of where each colony lies as it is counted. If you lose your place, you can wipe off the dots with alcohol and start again.

To accurately calculate the number of bacteria per milliliter in the original culture, count each plate that has between 30 and 200 CFU, and multiply the number of CFU by the dilution factor for that plate. You also need to adjust for the fraction of the dilution that was inoculated, since only 100 µl of the 1,000 µl dilution was spread on the plate. (The actual CFU in the dilution would have been 10 times higher if the entire 1,000 µl had been spread.)

$$\textit{Concentration of bacteria (CFU/ml)} = \textit{\# of CFU} \times \textit{dilution factor} \times \frac{\textit{1,000 µl}}{\textit{100 µl}}$$

Postlab Focus Questions

1. How many bacteria per milliliter of culture were in your original culture? How confident are you of this number?
2. Did each serially diluted sample have exactly 10 times the number of CFU as the previous dilution? If not, how different were the samples and why were they not exactly proportional?
3. How could you change the experiment to more accurately determine the number of CFU in the culture?

Self Assessment

Assess whether you have mastered the skills of this activity using the Laboratory Skills Assessment Rubric in Appendix E.

Assess your experimental write-up in your laboratory notebook using the Laboratory Notebook Rubric in Appendix F.

Activity 3.F Staining Eukaryotic Cells

Overview

Eukaryotic cells are larger than bacteria, which makes them much easier to observe using a light microscope. Contrast stains are used to see organelles and other cellular components in detail. Lugol's iodine is a general contrast stain that is used to observe all parts of the cell by increasing contrast and making shapes stand out. Nuclear stains such as Fast Blast™ DNA stain or methylene blue bind to DNA and turn it dark blue, which highlights cell nuclei; nuclear stains can also stain other DNA-containing organelles such as mitochondria. Janus green B stain is commonly used to observe the mitochondria in the cytoplasm. Janus green B stain turns from aqua blue to green then to yellow when exposed to the acidic environment in the mitochondria.

Safety Reminder: Review the SDSs of the stains. If any coverslips or slides are broken, place them into a sharps container, not in the trash can. Wear appropriate PPE and protect fingers and clothes from becoming stained. Properly discard the razor blade or cutting implement.

Research Questions

- What differences can be observed between plant and animal cells when they are stained with Lugol's iodine, Fast Blast DNA stain, or Janus green B stain?
- What is the size difference between plant and animal cells?

Objectives

- Stain onion and cheek cells using three different stains
- Estimate the size of cells using the field of view or stage micrometer
- Draw cells observed under the microscope

Skills to Master

Refer to Laboratory Skills Assessment Rubric (Appendix E) and Laboratory Notebook Rubric (Appendix F) for more details.
- Record laboratory notebook entries
- Make a wet mount (Activity 3.C)
- Observe eukaryotic cells using a microscope
- Sketch microscopic details (Activity 3.D)
- Estimate the size of cells using a microscope (Activity 3.D)

Student Workstation Materials

Items	Quantity
Microscope and optional accessories, including immersion oil and lens cleaning tissue	1
Stage micrometer (optional)	1
Microscope slides	6
Coverslips	6
Toothpicks	3
Razor blade, scalpel, or scissors	1
Laboratory marking pen or wax pencil	1
Waste container	1
Lugol's iodine stain	1–10 ml
100x Fast Blast DNA stain	1–10 ml
Janus green B stain	1–10 ml
Onion slice	~2 cm²

Prelab Focus Questions

1. What are two differences between eukaryotic cells and bacteria?
2. How do you estimate the size of an object using a microscope?
3. Name three major differences between animal and plant cells.

Activity 3.F Staining Eukaryotic Cells

Protocol

Part 1: Stain Cells with Iodine

1. Label one microscope slide **Cheek iodine** and another **Onion iodine**.

2. Place one drop of iodine stain in the middle of the slide labeled **Cheek iodine**, and use the blunt end of a toothpick to scrape the inside of your cheeks. Be gentle so you do not cause damage to your cheeks.

3. Place the toothpick into the drop of iodine stain on the slide and swirl it to release the cells into the stain.

4. Prepare a wet mount by lowering a coverslip over the stain, starting at a 45° angle. Avoid trapping bubbles under the coverslip.

5. Obtain a small piece of onion. Peel the inside epidermis off the onion leaf scale and cut it to roughly 0.5–1 cm^2.

6. Place a drop of iodine on the slide labeled **Onion iodine**, and lower the onion epidermis onto the stain. Keep the epidermis as flat as possible and avoid trapping bubbles under the sample.

7. Add another drop of iodine stain on top of the onion epidermis and then lower a coverslip into place.

8. Place a slide on the microscope stage, starting with the objective lens close to the slide and then moving away until the specimen is in the field of view. Use the coarse focus on low power until color is observed while looking through the eyepiece. Move to the fine focus once cells can be seen.

9. Move the slide around on the stage to look for an individual cell that can be sketched.

10. Switch to high power once the cell has been centered on the field of view.

11. Sketch and label a cheek cell and an onion cell, and note your observations. Use a stage micrometer or the diameter of your field of view to estimate the size of each cell.

Part 2: Staining Cells with Fast Blast DNA stain and Janus green B

1. Label four microscope slides:
    Cheek Fast Blast
    Cheek Janus green
    Onion Fast Blast
    Onion Janus green

2. Place one drop of Fast Blast DNA stain on the slide labeled **Cheek Fast Blast** and one drop of Janus green B stain on the slide labeled **Cheek Janus green**.

Activity 3.F Staining Eukaryotic Cells

3. Use the blunt end of a toothpick to gently scrape the inside of your cheeks. Place the toothpick into the stain on the slide labeled **Cheek Fast Blast** and swirl it to release the cells into the stain. Repeat, using a new toothpick to add cheek cells to the stain on the slide labeled **Cheek Janus green**. Prepare wet mounts using coverslips.

4. Obtain two new pieces of onion epidermis.

5. Place one drop of Fast Blast DNA stain on the slide labeled **Onion Fast Blast** and place one drop of Janus green B stain on the slide labeled **Onion Janus green**. Lower the onion epidermis onto the stain, trying not to trap bubbles. Keep the epidermis as flat as possible.

6. Add another drop of the appropriate stain on top of each sample of the onion epidermis and then lower a coverslip into place.

7. Observe each slide under the microscope as described earlier. Sketch your observations. Make a note of the differences between the onion epidermal cells and cheek cells and between the different stains.

Results Analysis

Sketch your observations and make comparisons between cheek cells and onion epidermal cells, keeping in mind the different features observed with different stains. Estimate the cell sizes using either a stage micrometer or the diameter of the field of view of the microscope provided by your instructor. It is more difficult for stains to enter the onion epidermis; therefore, scan the edges of the tissue to find cells that have been stained. Janus green B should stain mitochondria, which should show up as greenish or yellowish dots in the cell cytoplasm.

Postlab Focus Questions
1. What cell structure takes up most of the space in an onion epidermal cell?
2. Which cells are larger, cheek cells or onion epidermal cells?
3. Why do onion cells not have chloroplasts?

Self Assessment

Assess whether you have mastered the skills of this activity using the Laboratory Skills Assessment Rubric in Appendix E.

Assess your experimental write-up in your laboratory notebook using the Laboratory Notebook Rubric in Appendix F.

Chapter 3 Extension **Activities**

1. Perform Gram staining of the different types of colonies isolated from yogurt. Alternatively, stain bacteria from different brands of yogurt.
2. Use a spectrophotometer to make a standard curve of the optical density (OD) at 600 nm of bacterial cultures against the cell number as determined by a serial dilution and plate count. Then use the standard curve to estimate the number of cells in a bacterial culture of unknown density. Check the estimation by performing a serial dilution and plate count.
3. Estimate the number of bacteria in a container of yogurt, or quantify bacteria in a specific volume of potting soil or meat from the grocery store.
4. Use the disk diffusion method to assay the antimicrobial properties of spices or medicinal plants.
5. Perform plant or animal tissue culture using a kit or online protocol.

DNA Structure and Analysis

Summary

Molecular biology is the study of biology at the molecular level. The term also refers to a set of tools and techniques that have revolutionized biology. The central dogma of molecular biology states that DNA is transcribed into RNA, which is then translated into proteins that produce traits, or phenotypes. Restriction enzymes, which cut DNA, and ligases, which join DNA fragments back together, are the foundational tools in molecular biology for manipulating genetic information. CRISPR technology is a powerful new technology in molecular biology that allows precise changes in DNA and has many potential applications. These tools have been used to develop genetically modified plants with resistance to agricultural pests and genetically engineered tissue culture cells that produce therapeutic drugs for breast cancer just to name two. Once DNA has been cut with enzymes, the fragments need to be examined and isolated. This is done using horizontal agarose gel electrophoresis, in which DNA is separated according to size in a gel matrix using an electric current. This chapter provides background into fundamental molecular biology techniques and practice with restriction enzyme digestion and agarose gel electrophoresis.

4.1 Molecular Biology

Molecular biology bridges the disciplines of **genetics** and **biochemistry**, which include the study of gene and protein function, respectively (see Figure 4.1).

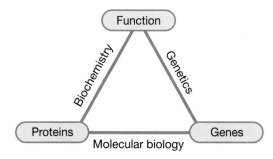

Figure 4.1. **The relationship among biochemistry, genetics, and molecular biology**.

Molecular biology also refers to an array of techniques that relate specifically to the manipulation of DNA and is used in that context in Chapters 4–6. Molecular biology includes techniques such as restriction enzyme digestion, ligation, bacterial transformation, the polymerase chain reaction (PCR), and recently CRISPR-Cas9 technology.

Using molecular biology tools, scientists have been able to free wrongfully accused individuals who have spent more than half of their lives in prison. Scientists have also used molecular biology tools to trace strains of cultured microorganisms back to their source, solving mysteries such as the anthrax poisonings of 2001. Billions of dollars are generated each year by the pharmaceutical industry through the production of recombinant proteins for medicine using molecular biology tools. For example, companies such as Genentech and Amgen have used molecular biology to develop drugs like Humulin to treat diabetes, Herceptin to treat cancer, and Protropin to alleviate deficiencies in human growth hormone. The tools, techniques, and applications of molecular biology continue to evolve.

The Central Dogma of Molecular Biology

Molecular biology research tackles questions of DNA and ribonucleic acid (RNA) structure and function, including **replication**, **transcription**, and **translation**. Francis Crick coined the phrase the "central dogma of molecular biology" in 1956, referring to the normal flow of genetic information (see Figure 4.2). The core of the central dogma is that genetic information contained in DNA is transmitted to RNA and then to **protein** (see orange arrows in Figure 4.2). The information in DNA is transferred to new DNA through DNA replication. In DNA replication, an **enzyme** called **DNA polymerase** copies both DNA strands and makes a brand new copy of the DNA (see Chapter 6). The information in DNA is transferred to RNA through transcription. In transcription, an enzyme called RNA polymerase copies one DNA strand into **messenger RNA (mRNA)**. The information in RNA is transferred to

proteins through translation. In translation, the sequence of bases on the mRNA is read and the information is converted into a **polypeptide** chain composed of **amino acids**. The polypeptide chain folds into a protein (see Chapter 7). Proteins then determine the **phenotype**, or traits, of the organism. The central dogma is summarized by the phrase "DNA > RNA > protein > trait."

There are special circumstances in which genetic information flows differently (see grey arrows in Figure 4.2). For example, RNA viruses, such as the influenza virus, can replicate their own RNA using a virus-encoded RNA-dependent RNA polymerase. Other viruses, such as the human immunodeficiency virus (HIV), can copy RNA into DNA. Retroviruses like HIV use reverse transcriptase, an enzyme that has become an important tool in cloning.

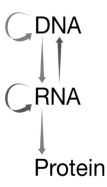

Figure 4.2. **The central dogma of molecular biology**. DNA is copied through replication. The information in DNA is transferred into RNA through transcription, and the information in RNA is transferred into protein through translation. In special cases, RNA is replicated into RNA and reverse transcribed into DNA.

4.2 DNA Structure

DNA is a molecule present in all living things, including bacteria, plants, and animals. DNA carries genetic information that is inherited, which is passed down from parents to offspring. Since 1953, when James Watson and Francis Crick deciphered the structure of DNA, scientists have been studying how its structure leads to its function. This research has been the foundation of the recombinant DNA revolution, leading to the development of genetic engineering and the polymerase chain reaction (PCR).

DNA strands are polymers of **nucleotides**, consisting of a sugar, a phosphate group, and one of four nitrogenous bases: adenine (A), thymine (T), guanine (G), and cytosine (C) (see Figure 4.3). Guanine and adenine have two nitrogenous rings and are purines, while thymine and cytosine have a single nitrogen ring and are pyrimidines.

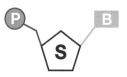

Figure 4.3. **Nucleotide structure**. P represents the phosphate group, S the sugar, and B one of the four nitrogenous bases.

Bioethics

Personal Genetic Information

As the cost of DNA sequencing continues to drop, the genetic testing industry is expanding rapidly. Nearly 100 companies, like 23andMe, Ancestry, and MyHeritage, now provide some form of direct-to-consumer genetic testing, and personal genetic tests are available online.

In most cases, these companies sell kits with materials and instructions for customers to send saliva samples to a commercial laboratory where the DNA is extracted and sequenced. Those sequence data are analyzed to determine the customer's risk and carrier status for conditions like Alzheimer's disease, breast or lung cancer, and diabetes as well as many other traits.

These genetic tests pose no physical risk. But as private companies store and share personal genetic information, ethical questions arise. How might people interpret disease risk results and react to news that they are at higher risk for disease? Would they understand the correlation between a positive test result and the probability of developing the disease? Some companies require doctor approval prior to testing, but others, like 23andMe, do not, and offer contacts to genetic counselors instead. Could identifying risk inspire customers to change their lifestyle? If a woman learns she is at high risk for breast cancer, for example, how would she know whether to take radical action, such as a double mastectomy to eliminate the risk altogether, or to get more frequent mammograms instead?

Genetic information applies not only to the customer but also to their relatives. If someone shares information about their personal cancer risk with others, they may also be sharing information about the cancer risk of their siblings, children, or even grandchildren. In 2008, President George W. Bush signed into law the Genetic Information Nondiscrimination Act (GINA), which prohibits discrimination on the basis of information derived from genetics tests. Nevertheless, what types of consent should someone be required to get from family members before sharing that type of information online?

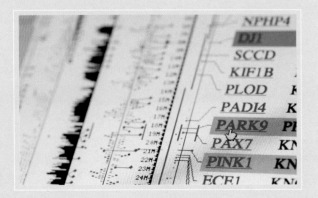

DNA molecules are formed by single strands of DNA twisting together to form a double helix (see Figure 4.4). The two strands are held together by hydrogen bonds between complementary pairs of nucleotides: A with T and G with C (see Figure 4.5). These pairings of A–T and G–C are called **base pairs**, and it is the **complementarity** between bases that is the basis of PCR and DNA sequencing.

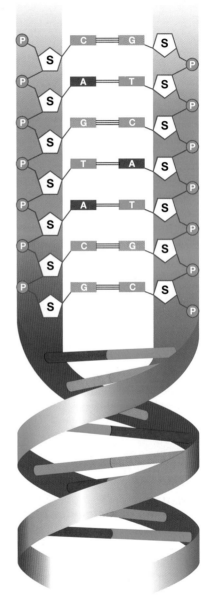

Figure 4.4. **Double helix structure of DNA**. The two strands of DNA are connected by hydrogen bonds between the bases. Cytosine and guanine are connected with three hydrogen bonds, and thymine and adenine with two hydrogen bonds.

Each sugar residue in the backbone of a DNA strand has five carbons, making it a pentose. The carbons are numbered based on their location in the pentose (see Figure 4.6). The phosphate group of the nucleotide is attached to carbon 5 and the base is attached to carbon 1 (see Figure 4.6). The sugar in DNA is a deoxyribose, as there is a hydrogen atom on carbon 2.

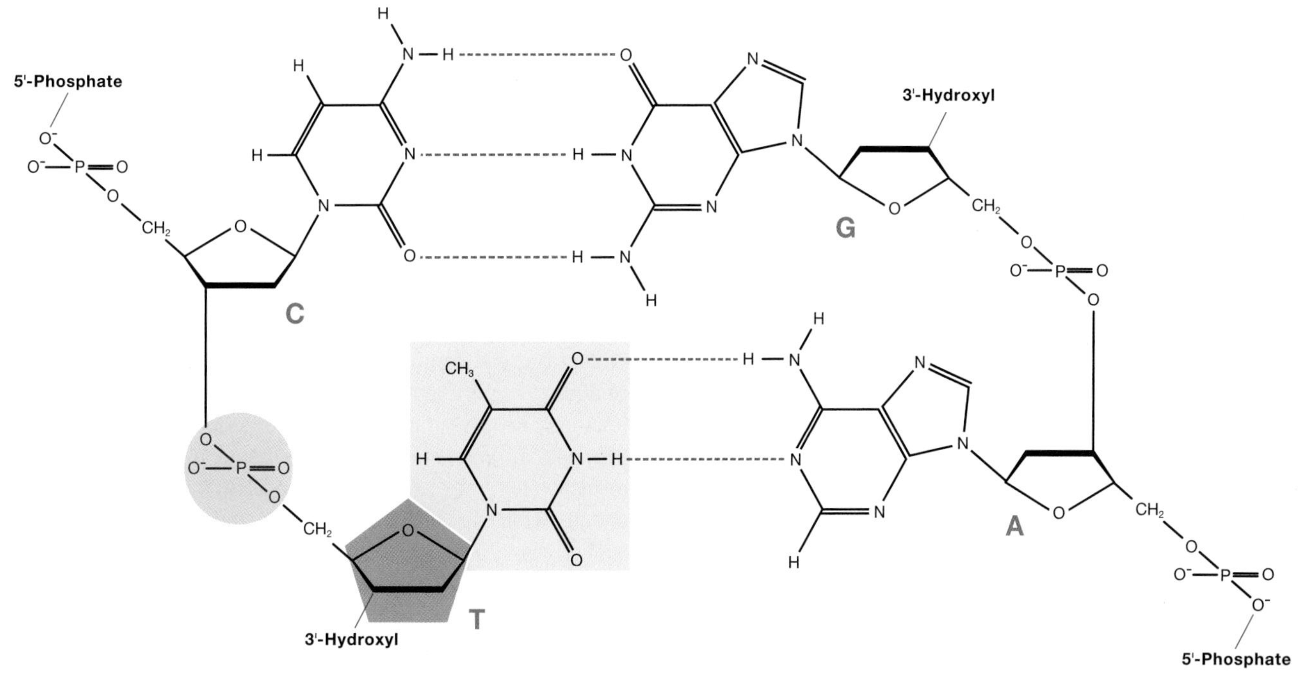

Figure 4.5. **DNA base pair interactions**. Double stranded DNA has two sugar-phosphate backbones (purple and blue) in opposite orientations that are joined by hydrogen bonds between the base pairs. Guanine and adenine are purines made up of two nitrogenous rings, and thymine (yellow) and cytosine are pyrimidines made up of a single nitrogen ring.

The two strands of a DNA molecule are oriented in opposite directions. By looking at the structure of each strand, it can be determined to have either 5' to 3' (five prime to three prime) orientation or 3' to 5' (three prime to five prime) orientation based on the attachment points to the sugar. For example, in Figure 4.4, the directionality of the strand on the left is 5' to 3', as the first attachment is that of a phosphate group to carbon 5 of the sugar.

Nucleic acids are synthesized **in vivo** only in a 5' to 3' direction. As the strand grows, each new nucleotide is added to the 3'-hydroxyl group (attached to carbon 3) of the sugar. As a standard practice, the sequences of single strands of DNA and RNA are written in the 5' to 3' direction. When discussing specific regions on the strand, such as genes or regulatory elements, the relative positions are usually noted as being either upstream (toward the 5' end) or downstream (toward the 3' end) from one another.

Two other terms used to describe the strands of DNA are sense and antisense. The sense strand has nucleotides running 5' to 3', while the antisense strand has the complementary nucleotides

running 3' to 5' (see Figure 4.5). When DNA is transcribed, the antisense strand serves as the template for the synthesis of mRNA. mRNA has the same sequence as the sense strand of DNA but with the base uracil instead of thymine.

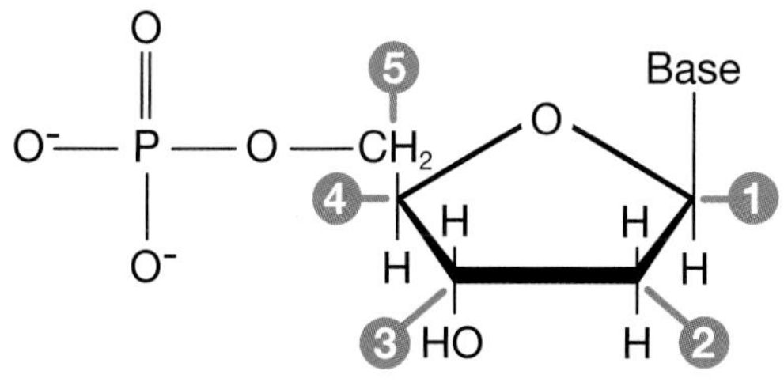

4.3 Recombinant DNA Technology

Modern biotechnology is based on the premise that DNA is virtually the same in all cells and can be read and acted on (or expressed) in all cells. In other words, DNA from one organism can still be read and expressed when transferred into another organism. This means that scientists can cut out a DNA fragment from one organism and insert it into a different organism, and the recipient organism carries out the genetic instructions provided by the newly acquired DNA. This process is called genetic engineering. DNA that has been assembled by linking fragments of DNA from different organisms is called **recombinant DNA**.

Any organism can be genetically engineered. For example, cotton plants have been genetically engineered to express a bacterial gene that makes them resistant to agricultural pests. In the pharmaceutical industry, recombinant proteins are produced in bacteria and eukaryotic cells and sold as therapeutic drugs. In medicine, people with some genetic diseases can be helped by gene therapy where a recombinant gene is expressed in specific tissues to counter a genetic defect. The first gene therapy to treat a form of leukemia was approved in the United States in 2017 and others are in the clinical trial stage.

Figure 4.6. **Chemical structure of a deoxyribonucleotide**. The numbering and location of the five carbons on the deoxyribose sugar are indicated in orange.

Careers In Biotech

Elisa Ciullo

Sr. Supervisor, Clinical Filing Operations
Genentech, South San Francisco, CA

Photo courtesy of Elisa Ciullo

Elisa Ciullo has always been fascinated with all the things that science can accomplish. "In a high school classroom, you can dice and splice a DNA sequence. On a larger scale, you can genetically engineer an antibody to harness your natural immune system to resolve a health problem. How could you not find that interesting?" she asks.

Elisa discovered her aptitude for math and science in high school. Once she heard that engineers earned good salaries, she chose to combine her interests and study bioengineering at the University of California, San Diego (UCSD). She earned a B.S. in 2010 and then her M.S. in 2011, both in bioengineering. Throughout her years at UCSD she worked in a variety of research environments, but by the end of her M.S. work, she was sure that laboratory research was not for her. Instead, the biotechnology industry seemed a better fit.

Genentech was the first place to offer her a position, as an Associate Manufacturing Technical Specialist. She now works at Genentech as a Clinical Filing Supervisor.

Elisa believes one of the most promising areas of biotechnology is personalized medicine. This field is developing at a rapid pace, and so it constantly presents new challenges for the biotechnology industry. New globalized manufacturing models must be developed, and new standards for licensing strategies have to be created to accommodate all the novel technologies, tests, and drugs hitting the market. Her advice to someone starting out in biotechnology is to never limit yourself by assuming a career in biotechnology means you will be tied to research and benchtop experiments. Biotechnology is a massive industry with so much to offer. The sooner you can learn about all the possibilities it holds, the quicker you, as an individual with unique skills and aptitudes, can make your mark on the industry as it evolves.

Restriction Enzymes

Genetic engineering relies on the ability to move a gene from one location to another. In the late 1960s, bacterial enzymes that cut DNA in very predictable locations were discovered. These enzymes, called **restriction enzymes**, have enabled scientists to move genes from the genome of one organism into the genome of another. The discovery of restriction enzymes by Werner Arber at the University of Basel, Switzerland, and Hamilton Smith and Dan Nathans at Johns Hopkins University, Baltimore, MD, started the DNA revolution.

In nature, restriction enzymes evolved as a defense mechanism to protect bacteria from invading viruses called **bacteriophages**; or **phages**. Phages inject their DNA into bacteria and then use the bacterial cellular machinery to reproduce more copies of themselves, frequently killing the bacteria. Bacteria produce restriction enzymes to cut the phage DNA, thereby destroying the phage. (Restriction enzymes were so named because they limit or restrict the growth of phages; at the time, scientists did not understand how these enzymes functioned.) Since the bacteria's own DNA may contain the same restriction sites, bacteria protect their DNA from being cut by adding methyl groups to the DNA bases. The presence of the methyl groups changes the bacterial DNA conformation and prevents restriction enzymes from recognizing and cutting the restriction sites on the DNA.

Specific restriction enzymes are named for the bacteria from which they were originally isolated. For example, EcoRI was isolated from *Escherichia coli*. The enzyme name is derived from the bacterial genus, species, and strain. The name uses the first initial of the genus and the first two letters of the species name of the bacterium, followed by the order in which the enzymes were discovered in that particular strain.

EcoRI: The first restriction enzyme isolated from *Escherichia coli* RY13 strain.

HindIII: The third restriction enzyme isolated from *Haemophilus influenzae* Rd strain.

PstI: The first restriction enzyme isolated from *Providencia stuartii* bacteria.

Restriction enzymes, more accurately called restriction endonucleases ("*endo*" means within and "*nuclease*" means an enzyme that cuts nucleic acids), cut DNA from any source. They are thought to associate with the DNA molecule and slide along the helix until they recognize a specific sequence of base pairs that signal the enzymes to stop sliding. Restriction enzymes then cut, or chemically separate, the DNA molecule at that site, called a restriction or recognition site. Enzymes hydrolyze the sugar-phosphate bond between two specific nucleotides in the restriction site on each strand of DNA, thus breaking of the DNA molecule in a predictable location.

Some enzymes cut across both DNA strands in the middle of the restriction site, leaving "blunt" ends with no unpaired bases (see Figure 4.7). Other enzymes make staggered cuts in the DNA, leaving a short length of unpaired bases, called a "sticky" end. Also, depending on where the cut occurs, the sticky end can have a 5' or a 3' overhang (see Figure 4.8). In general, restriction sites are palindromic, meaning the sequence of bases reads the same forward as it does backward on the opposite DNA strand (see Figures 4.7 and 4.8).

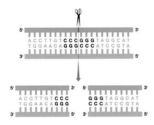

Figure 4.7. **Blunt ends generated by SmaI enzyme**.

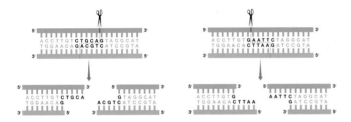

Figure 4.8. **Sticky ends generated by PstI and EcoRI enzymes**. PstI generates a 3' overhang while EcoRI generates a 5' overhang.

If a specific restriction site occurs in more than one location on a DNA molecule, a restriction enzyme makes a cut at each of those sites, resulting in multiple fragments. Therefore, if a linear piece of DNA is cut at two restriction sites, three fragments of different lengths are generated. If a circular piece of DNA is cut at two restriction sites, only two fragments of different lengths are generated. (Imagine a piece of string versus a rubber band being cut in two places.) The length of each fragment generated depends upon the location of the restriction sites on the DNA molecule. DNA that has been cut with restriction enzymes is separated using a process known as agarose gel **electrophoresis**, which will be described later. Restriction sites are usually 4 to 6 bases long but they can also be longer than 12 bases. By chance, a 4-base sequence occurs every few hundred base pairs in a DNA sequence, while a 6-base sequence occurs every several thousand base pairs. Therefore, a human genome of 3×10^9 base pairs (bp) contains approximately 1 million restriction sites of any 6-base restriction enzyme, while a **plasmid** (a circular piece of DNA that is commonly used to clone genes in the laboratory) of 6,000 bp (6 kilobases or kb) should have 1–3 restriction sites.

Ligases

Once restriction enzymes cut the DNA of interest from one location, the fragment can be moved to a new location and joined to the new genome by an enzyme called **ligase**. Ligases join (ligate) two pieces of DNA together by reforming the phosphate bonds that were broken by the restriction enzymes. In nature, ligase is used in DNA replication to ligate the Okazaki fragments made on the lagging strand of DNA (see Chapter 6). This enzyme catalyzes the formation of a phosphodiester bond between the free 5'-phosphate end on the backbone of a DNA fragment and the 3'-hydroxyl group of another DNA fragment. If a restriction enzyme has created DNA fragments with complementary sticky ends, the ligation reaction is helped by hydrogen bonds forming between the complementary base pairs, which hold the DNA fragments in position while the ligase works to fuse the backbone (see Figure 4.9). If there are no complementary sticky ends and the DNA fragments are blunt-ended, ligation will still occur but is less efficient.

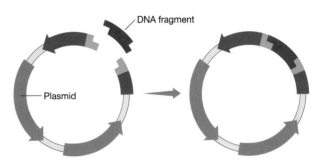

Figure 4.9. **Ligation of a DNA fragment into a plasmid**. In this instance, the DNA fragment (shown in red) has sticky ends with a 5' overhang; the plasmid also has a 5' sticky end.

Plasmids are circular pieces of DNA that scientists use to carry DNA fragments, including genes, between organisms and are discussed further in Chapter 5. When ligating DNA fragments into plasmids, there is a higher probability of the plasmid ligating to itself and recircularizing than there is of it incorporating the new fragment. This is because the two ends of the opened plasmid are in close proximity to each other, while the DNA fragments are floating around the solution. To prevent plasmids from self-ligating, the phosphate on the 5' end of the plasmid DNA is removed using an enzyme called phosphatase in a process called dephosphorylation. Dephosphorylating the plasmid prevents ligase from forming a phosphodiester bond, thereby preventing self-ligation of the plasmid. The DNA fragments still have a 5' phosphate and so the plasmid can re-circularize only if the DNA fragment is inserted. The DNA fragments can also ligate to one another. To reduce this occurrence, the molar ratio of plasmid to DNA fragment is controlled during a ligation reaction and is usually, but not always, 1:1. After the ligation is complete, the ligated DNA is transformed into bacteria (see Chapter 5).

Nontraditional Cloning Techniques

The use of restriction enzymes and ligases to cut and paste pieces of DNA into plasmids has been the standard technique for cloning. The technique, however, relies on the presence of restriction sites in the DNA sequences to be cloned. It also requires several separate reactions, so it can be time-consuming, error-prone, and inefficient. Other cloning techniques have been developed, which overcome some of these challenges. Three of these approaches are described below.

How To...

Set Up a Restriction Digest

1. To set up a restriction digest, the components and quantities need to be determined. Calculate the necessary volume of DNA. This depends on the concentration of the DNA sample. For example, if 1 µg of DNA is to be digested in a 20 µl reaction volume and the DNA stock is 0.1 µg/µl, then 10 µl of DNA should be used.

2. Calculate the amount of restriction enzyme required.
This depends on the concentration of the enzyme, which typically is printed on the enzyme tube label. In a digest, 10 U of enzyme are commonly used per µg of DNA. For example, if the enzyme concentration is 10,000 U/ml, then 1 µl of enzyme contains 10 U, and 1 µl of enzyme should be added to the reaction. **Note:** When working with DNA and restriction enzymes, use a fresh, clean pipet tip for each addition to avoid contaminating the DNA sample with enzymes or the enzymes with DNA.

3. Determine the type and volume of restriction digestion buffer. Most restriction buffers are provided at a 10x concentration and are generally packaged with the enzyme. If a digestion using two enzymes is to be performed, look up the manufacturer's recommendation to determine the appropriate buffer for your application.

4. Calculate the amount of water required to bring the reaction up to the final volume.

To set up the reaction, add the calculated volume of the components to a microcentrifuge tube in the following order: water, buffer, DNA, and, last, enzyme. When possible, always keep enzymes on ice. Mix the components by pipetting up and down or gently flicking the tube, and pulse-spin the tube in a microcentrifuge to collect the contents at the bottom. Incubate the reactions at 37°C in a water bath or dry bath for 30–60 minutes. Place the tube at 4°C until analysis on an agarose gel.

Table 4.1. **Components required for a typical 20 µl restriction digestion reaction.**

Concentration of Stock	Quantity or Concentration	Volume in Digestion Reaction
DNA stock (0.1 µg/µl)	1 µg	10 µl
10x reaction buffer	1x	2.0 µl
Enzyme (10 U/µl)	10 U	1 µl
Molecular biology grade water	x µl (to bring up to final volume)	7 µl
	Total	**20 µl**

Golden Gate Assembly

Golden Gate assembly is similar to traditional cloning in that it uses restriction enzymes and DNA ligases. Instead of standard restriction enzymes like EcoRI, however, this method uses Type IIs restriction enzymes. Type IIs restriction enzymes (for example, BsaI) cut DNA outside of their recognition sites. This special feature of Type IIs enzymes means that the sticky ends generated are unique in sequence because they are not part of a recognition site.

With Golden Gate assembly, the fragments and plasmid assemble in a directional manner as a result of the unique sequences in the sticky ends. The reaction can occur within a single tube with only one restriction enzyme and one ligase.

Gibson Assembly

Gibson assembly does not involve restriction enzymes at all. Instead, long stretches of DNA are copied, so that when they are assembled, adjacent regions have 20–40 base pairs of overlapping sequence. This overlap is necessary for the DNA fragments to assemble in a designated order. The DNA fragments are mixed with several enzymes that cleave, repair, and join the pieces of DNA in a single reaction. This allows the fragments to be joined properly and yield a single piece of DNA with the desired final sequence.

With Gibson assembly, no restriction sites are required within a sequence, and since the process involves fewer steps, fewer reagents, and less time than Golden Gate assembly, it is very efficient. Multiple DNA fragments can be combined simultaneously in a single reaction. Using this technique, scientists can overlap more than ten DNA fragments, each hundreds of kilobases long.

TOPO (TA) Cloning

TOPO (TA) cloning, unlike all the other cloning techniques described, does not involve a restriction enzyme or a DNA ligase. Instead, the technique exploits the biological activity of DNA topoisomerase I. The topoisomerase I cuts DNA at a specific sequence much like a restriction enzyme. It then stays with the DNA (for example, the plasmid) and ligates any DNA fragment with compatible overhangs to create a recombinant molecule.

While traditional restriction enzyme/ligation methods can take from four hours to overnight, TOPO (TA) cloning can be completed in under an hour. In addition, plasmids are available that allow TOPO methods to be used on DNA sequences with blunt ends.

Using Enzymes in the Laboratory

Enzymes used for genetic engineering are produced and sold by many companies. Enzyme concentration is often given in units of activity (U) rather than the concentration of protein. The definition of the unit is usually provided by the supplier. Most enzymes are also supplied with reaction buffers that provide the optimal reaction conditions.

Restriction enzymes are used to specifically cut, or digest, DNA. Setting up a restriction digestion reaction requires that the concentrations of buffer, DNA, and enzymes be adjusted to the optimal levels. A typical restriction digest is conducted in 10–20 μl final volume and uses approximately 1 μg of DNA, 5–10 U of enzyme, and 2 μl of a 10x reaction buffer. Any remaining volume is made up with purified water. A unit of restriction enzyme is usually defined as the amount of enzyme needed to digest 1 μg of lambda phage DNA in 1 hour at 37°C. Common restriction enzymes have a concentration of 10,000 or 20,000 U/ml, which translates to 10 or 20 U/μl, respectively. Most restriction enzymes are supplied with a concentrated reaction buffer that provides the optimal salts and pH for that specific enzyme. Some restriction enzymes also require additives such as bovine serum albumin (BSA). When two restriction enzymes are used to digest DNA in the same tube, it is called a double digest. In this case, consult a chart from the enzyme supplier(s) to determine the reaction buffer that provides the optimal conditions for both enzymes.

Many enzymes are temperature sensitive and lose activity when heated, so they should be stored at the recommended temperature, usually –20°C. Enzymes are usually supplied in glycerol so that they do not freeze at –20°C. Repeated freeze-thawing of enzymes should be avoided since it reduces enzyme activity. Some enzymes are supplied freeze-dried and must be hydrated before use. Freeze-dried enzymes are designed for a single use; however, if necessary, they can be rehydrated in a small volume of 50% glycerol and stored at –20°C for future use. These enzymes require further dilution before use to reduce the glycerol concentration to less than 5% in the final reaction. When working with enzymes in the laboratory, they should be kept on ice or at 4°C and returned to the freezer as soon as possible. The freezer should be open only long enough to take out the required enzyme; the entire container of enzymes should not be removed when retrieving the required enzyme from the freezer.

CRISPR-Cas9 Technology

In the late 1980s, genome sequencing was coming into the spotlight and many bacterial and viral genomes were being sequenced. Scientists noticed a DNA sequence in bacteria that was repeated many times, with unique sequences between the repeats. They named these repeated sequences "**clustered regularly interspaced short palindromic repeats**," or **CRISPR**. In a palindromic repeat, the sequence of nucleotides is the same in both directions, and in CRISPR, each palindromic repeat is followed by short segments of spacer DNA. Surprisingly, these spacer sequences matched DNA sequences found in the genomes of viruses known to infect bacteria. Small clusters of CRISPR-associated (Cas) genes were also found next to CRISPR sequences. The **Cas genes** encode enzymes that cut DNA in specific places.

Eventually, the role of CRISPR sequences in an adaptive "immune" response was realized. When a bacterium or archaeon is infected by a virus, the microbe captures some of the viral DNA and inserts it into a CRISPR sequence as a spacer (see Figure 4.10). The cut and capture is carried out by two microbial Cas enzymes, Cas1 and Cas2. In the second phase, the spacers are transcribed and each binds to a Cas9 protein to form a "search and destroy complex" in the cell. Should a matching virus infect daughter cells descended from this bacterium or archaeon, the spacer RNA transcript and Cas9 complex would bind and cleave the viral DNA to prevent it from replicating. This would halt the viral infection. A bacterium can collect sequences from many different infecting viruses using this system to create a sort of "library." Since the CRISPR sequence is contained in genomic DNA, it is passed on to each generation, and the library continues to grow over time.

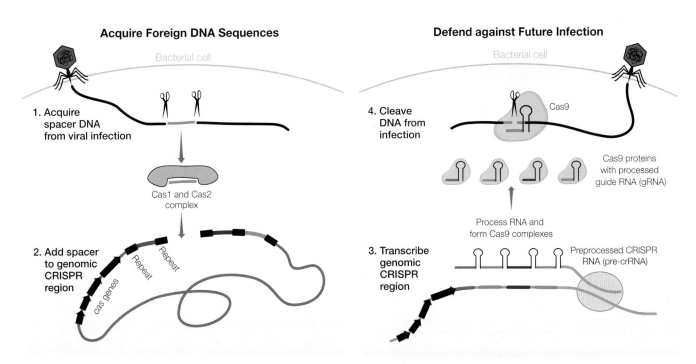

Figure 4.10. **The CRISPR-Cas9 system as a microbial defense system.** 1, the Cas1-Cas2 enzymes of the microbe recognize foreign DNA and cut out a segment; 2, the Cas1-Cas2 enzymes insert the DNA segment into the CRISPR region of its own genome as a spacer; 3, a spacer sequence is transcribed and then linked to a Cas9 protein; 4, upon reinfection by the same invader, the CRISPR-Cas9 complex can recognize the foreign DNA sequence and cut it to prevent complete infection.

How CRISPR technology works

Genetic engineering is based on the ability to cut and paste together (ligate) DNA. Traditional genetic engineering methods, though powerful, are often imprecise. For example, generating mutations through radiation or chemical exposure is a random process; it is impossible to control the number or location of the mutations created. Similarly, when introducing new genes by transformation, the new genes may insert themselves in a way that disrupts other genes, affects their gene expression, or causes other unintended effects. **CRISPR-Cas9 technology** allows scientists to manipulate genes and gene expression precisely and in the least intrusive manner possible.

To introduce precise changes in DNA, the CRISPR system consists of two key molecules:

- Cas9 enzyme (**Cas9**) — endonuclease that cuts double-stranded DNA at a site dictated by the particular guide RNA that is bound to the Cas9

- Guide RNA (**gRNA**) — RNA approximately 100 nucleotides long that can form a complex with Cas9. A 20 nuleotide region at the end of the gRNA contains a spacer sequence complementary to the target DNA sequence

1. Cas9 with guide RNA binds to the target DNA sequence

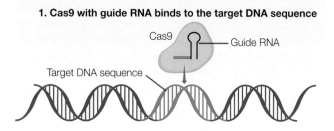

2. Cas9 cleaves the target DNA sequence

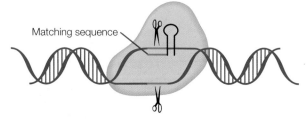

3. DNA repair mechanisms insert donor DNA to form a new DNA sequence

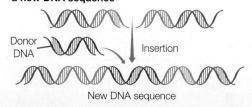

Figure 4.11. **The CRISPR-Cas9 system for targeted gene editing.** The CRISPR system has two components: a Cas9 enzyme and guide RNA. 1, the guide RNA includes a spacer sequence of about 20 nucleotides that recognizes and binds to a target DNA sequence; 2, the Cas9 enzyme cuts the double-stranded DNA (dsDNA); 3, the cell's own natural repair enzymes then go to work repairing the region that was cut. This repair is error prone and provides the opportunity to insert donor DNA to create new DNA sequences.

When Cas9 cuts the target DNA, the cell recognizes the DNA is damaged and starts to repair it. Scientists take advantage of the ways that cellular DNA repair machinery works to introduce changes to one or more genes (Figure 4.11). Gene sequences can be mutated by the deletion, insertion, or substitution of DNA sequences that result in destruction, addition, or change in gene function.

Applications of CRISPR technology

CRISPR is a uniquely powerful gene editing technique particularly for eukaryotic organisms. With CRISPR, targeted disruptions of any gene — in most organisms — are now possible. It allows scientists to modify genomic DNA with precision that is not possible using other gene editing techniques. That precision ensures that no other genes or sequences are unintentionally disrupted. CRISPR technology is easier, faster, and less expensive than other gene editing techniques. In addition, multiple genes can be edited at the same time in a single cell using CRISPR. Finally, CRISPR requires the introduction of only one protein (Cas9) and a gRNA into a cell. Such a powerful technology can be expected to have a vast range of applications.

Agriculture

CRISPR technology is expected to accelerate the development of new, improved crops. The technology has produced crops and livestock with desirable traits such as faster growth, higher nutrient content, and disease resistance. And, since CRISPR technology can modify genes without introducing new genes, CRISPR-modified plants may be subjected to lighter regulations than other genetically modified crops.

Industry

Scientists have used a modified CRISPR-Cas9 system to create a yeast strain to produce lipids and polymers. These molecules could be useful in the development of biofuels, adhesives, and fragrances. Currently these lipids and polymers are made synthetically from non-renewable petroleum-based materials that are more expensive and could present safety risks.

Public Health

Scientists are experimenting with using CRISPR to engineer "gene drives" to spread specific genes through a population of insect pests that cause them to die or become infertile. This technique is being considered to eradicate the mosquitoes carrying human pathogens like malaria parasites or Zika virus.

Medicine

Researchers are looking to CRISPR as a technique for editing out genetic defects that result in sickle cell disease, cystic fibrosis, hemophilia, and muscular dystrophy, and for developing more targeted and effective cancer treatments. One study showed that adult rats engineered to have a genetic form of blindness could be treated using CRISPR gene therapy. The goal is to someday have patients' diseased cells removed, "fixed" with CRISPR, and then returned to their bodies to treat various conditions, or have diseased organs be treated directly with CRISPR.

The potential for using CRISPR to change genetic traits in humans has raised serious concerns, however, about possible unintended effects, as well as ethical questions. The ease of applying CRISPR has caused worry about potential misuse of the technology. Despite these concerns, CRISPR is revolutionizing many aspects of biotechnology and scientific research.

4.4 DNA Analysis Techniques

Horizontal Agarose Gel Electrophoresis

DNA fragments need to be analyzed before they are used in genetic engineering. Agarose gel electrophoresis separates DNA fragments by size using an electric field. The term "*electrophoresis*" means to carry with electricity. DNA fragments are loaded into an agarose gel, which is placed into a chamber filled with a conductive buffer solution. A direct current is passed between wire electrodes at each end of the chamber. Since DNA fragments are negatively charged, they are repelled from the negative electrode (**cathode**) and toward the positive electrode (**anode**) when placed in an electric field. The agarose gel matrix acts as a molecular sieve through which smaller DNA fragments can move more easily than larger ones. Therefore, the rate at which a DNA fragment migrates through the gel is inversely proportional to its size in base pairs. Over a period of time, smaller DNA fragments travel farther than larger ones. DNA fragments of the same size stay together and migrate as a single band. The agarose gel is stained with an appropriate dye to make the DNA bands visible.

Agarose Gels and Running Buffer

DNA electrophoresis is generally performed on an **agarose** gel. The agarose gel is formed by pouring molten agarose into a gel tray whose ends have been blocked to contain the agarose. A comb is inserted into the gel tray before the molten agarose is poured (see Figure 4.12), and the wells formed by the comb are later loaded with DNA.

Figure 4.12. **Agarose gels**. From left to right, an agarose gel containing a comb and an agarose gel with the comb removed.

Electrophoresis buffer is used to make the gel, and the same buffer is used in the electrophoresis chamber to conduct the electric current. The most common electrophoresis buffers for DNA electrophoresis are Tris/acetic acid/EDTA (TAE) and Tris/boric acid/EDTA (TBE); the same buffer should always be used to make the gels and to fill the electrophoresis chamber. These buffers are supplied as concentrated stock solutions and diluted to 1x working solutions before use. Agarose gels are made by mixing agarose powder with 1x electrophoresis buffer, then heating the mixture to melt the agarose, similar to making Jell-O. The molten agarose is cooled to around 55°C before it is poured into the gel tray.

The sizes of the DNA fragments being separated determine the concentration of agarose needed to separate the DNA. If relatively large fragments (>1 kb) are being separated, the percentage of agarose should be 0.7–1% (m/v). Fragments <1 kb are usually separated with a higher percentage of agarose. There are also different types of agarose. Good quality molecular biology agarose is suitable for most purposes. Specialty agaroses are available for separating very small fragments. Low melting–point agarose is recommended for DNA fragment recovery. High-strength agarose is available for separating very large DNA fragments.

Standards for DNA Electrophoresis

To determine the size of unknown DNA fragments on a gel, scientists use a standard, or control, sample that contains DNA fragments of known sizes. **DNA size standards** are commonly referred to as standards, ladders, rulers, or markers. The original DNA standards were made from bacteriophage or plasmid DNA cut with specific restriction enzymes that yielded DNA fragments of known number and size (for example, bacteriophage lambda DNA cut with HindIII restriction enzyme). More recently, DNA size standards have been made from engineered DNA that forms a regular "ladder" of DNA with precisely defined size intervals, such as a 1 kb ladder or ruler that contains fragments of 1 kb, 2 kb, 3 kb, and so on (See Figure 4.13). Lambda phage DNA is useful as a standard because it covers a broad range of DNA sizes when cut with HindIII enzyme. Lambda phage DNA is 48,502 bp long and the fragments generated by the HindIII digestion range from 125–23,130 bp (see Figures 4.14 and 4.15). DNA standards with more focused ranges are useful when the sizes of the expected fragments are known and lie within the range of the ladder

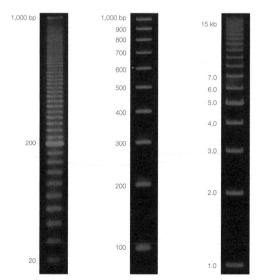

Figure 4.13. **Example DNA size standards**. From left to right, 20 bp molecular ruler, 100 bp molecular ruler, and 1 kb molecular ruler.

Biotech In The Real World

Fighting Crime with DNA

DNA fingerprinting was first used in a criminal case in 1986 to exonerate a suspect. The case involved the rape and murder of two schoolgirls near Leicester in the United Kingdom. Interestingly, DNA fingerprinting was used first to exonerate the man who had confessed to the crime and later to screen the village and ultimately identify the culprit. Since 1986, DNA evidence has evolved to enable more powerful crime fighting, even allowing use of microscopic samples to resolve cases that occurred decades ago.

Back in the 1980s and 1990s, samples collected from crime scenes had to be at least the size of a nickel to be of use in DNA analysis, and even as late as 2005, they had to be visible. Now forensic specialists can work from samples no larger than the head of a pin. The polymerase chain reaction (PCR, see Chapter 6) is at the heart of these improvements. Using heat and enzymes to copy DNA, PCR doubles the DNA in a sample many times to generate enough DNA for the creation of a profile.

It is also now possible to use chemical techniques to separate human DNA from microbial DNA. This means forensic scientists can learn about the habits of suspects from the microbes they leave behind. Bacteria are often specific to certain environments, and we all leave bacteria behind when we touch door knobs, cell phones, keyboards, or even knives or bullets. By identifying the bacteria you leave behind, scientists can know where you have been or what you ate recently. Such clues can help narrow down to a single suspect.

Another tool with a promising future is DNA phenotyping. DNA phenotyping leverages knowledge about how DNA determines hair and eye color, face shape, skin tone, height, and even freckles with DNA from crime scenes to create a theoretical image of the suspect. The technique led Ryan Derek Riggs to confess to a 2016 murder in Texas after a DNA-generated image of him was circulated.

Advances in microfluidics and automation are greatly increasing the speed of DNA analysis. While the typical process for analyzing suspect DNA can take up to 8 hours, the RapidHIT ID box from IntegenX (Pleasanton, CA) uses tiny robots to move the DNA from extraction to PCR amplification to analysis within 90 minutes. This process allowed police to apprehend accused murderer Christopher Jacquell Williams just hours after he committed a crime in Pennsylvania. Use of this technology will create a constantly growing database of DNA profiles. Instead of taking fingerprints, police in the future may take cheek swabs and analyze them using the RapidHIT ID box or similar technology.

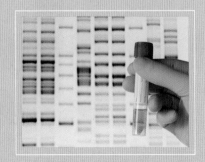

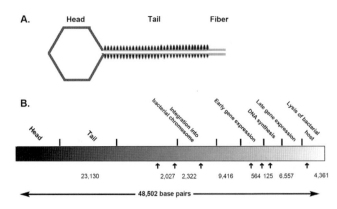

Figure 4.14. **Bacteriophage lambda**. A, bacteriophage lambda capsid consists primarily of a head, which contains the genomic DNA, and a tail that is involved in phage attachment to bacterial cells. B, lambda genomic DNA with functions of important gene clusters indicated. Arrows mark the sites where HindIII enzyme cuts the DNA, and the numbers indicate the number of bases in each fragment.

(for example, PCR gels with expected fragments of just a few hundred base pairs). To accurately predict DNA sizes, a standard curve based on the control is generated and used to estimate the size of the unknown bands.

DNA standards can also be used to estimate the quantity of DNA in each band. The intensity or thickness of a DNA band is proportional to the total amount of DNA in the band. When the quantity of the DNA in a control band is known, the intensity or thickness of the unknown band can be compared to the intensity of bands in the DNA control to provide an estimate of DNA quantity. Controls for DNA mass can be phage or plasmids digested with enzymes to produce fragments whose mass can be calculated. Alternatively, commercial **molecular mass rulers** that are engineered to provide bands of evenly distributed molecular mass can be used (see Figure 4.13).

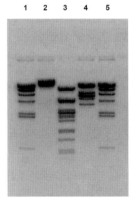

Figure 4.15. **Agarose gel with digested lambda DNA samples**. Lane 1, lambda HindIII DNA standard; lane 2, uncut lambda DNA; lane 3, lambda DNA cut with PstI; lane 4, lambda DNA cut with EcoRI; lane 5, lambda DNA cut with HindIII.

Equipment

Horizontal gel electrophoresis requires specialized equipment to hold the agarose gels and electrophoresis buffer and to provide the electric current that drives the DNA through the gel matrix.

Horizontal Gel Electrophoresis Chambers

Agarose gels are run in **horizontal gel electrophoresis chambers**, which are also called gel tanks or gel boxes (see Figure 4.16). Horizontal gel electrophoresis chambers are usually made from acrylic plastic and have a positive electrode (anode) at one end of the chamber and a negative electrode (cathode) at the other end. The electrodes are attached to red (positive) and black (negative) cables that will plug into a power supply. To prevent electrical shocks, the electrophoresis chamber should have a lid that cannot be removed without breaking the current. Different sizes of electrophoresis chambers are available. Some chambers, like those used in this course, hold a 7 x 7 cm gel for running 8 or 12 samples. Electrophoresis chambers that run very large gels (for example, a 25 x 25 cm gel for running 192 samples) are also available.

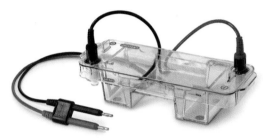

Figure 4.16. **Mini-Sub® cell horizontal agarose gel electrophoresis chamber**.

Power Supplies

Electrophoresis needs electricity, which is regulated by a **power supply**. There are many different types of power supplies, most of which run two to four electrophoresis chambers at a time (see Figure 4.17). Simple power supplies have fixed voltages. Most power supplies allow either the voltage or the current to be set by the user, and many have a timer that automatically turns the power off when the run is complete. The choice of power supply depends on the application. Agarose gel electrophoresis requires a standard power supply that can run voltage and current in the mid range. Western blotting (see Chapter 8) requires a power supply that can run at high current, while sequencing gels require a power supply with very high voltage. Western blotting and sequencing applications use specialized high current or high voltage power supplies that produce 3,000 mA or 5,000 V, respectively.

Figure 4.17. **Power supply**. Bio-Rad's PowerPac Basic power supply is commonly used for gel electrophoresis.

Running an Agarose Gel

DNA samples need to be mixed with **sample loading buffer** prior to being loaded into an agarose gel. Sample loading buffer contains an agent such as glycerol or Ficoll to increase the density of the sample so that it sinks into the well, and a dye to monitor migration of the samples. Different dyes co-migrate with different lengths of DNA and the migration of dyes is highly dependent on the percentage of the agarose used (see Table 4.2). Sample loading buffer is usually supplied as a 5x concentrate; therefore, 1 μl of sample loading buffer should be added to every 4 μl of sample, so that the final loading buffer is at 1x concentration.

Table 4.2. **Dyes and their relative migration in different percentages of agarose**.

Dye	Size of DNA fragments that comigrate with dye	
	1% agarose gel	**3% agarose gel**
Xylene cyanol	~4,000 bp	~800 bp
Bromophenol blue	~400 bp	~100 bp
Orange G	~50 bp	<50 bp

To run an agarose gel (see Figure 4.18), place the agarose gel (in its gel tray) in the electrophoresis chamber, making sure that the sample wells are near the cathode, which is usually color-coded black. Pour 1x TAE or 1x TBE electrophoresis running buffer (the same buffer as the one used to make the gel) into the chamber until it covers the gel by approximately 2 mm. Load the DNA samples into the wells of the agarose gel using a micropipet. The volume loaded depends on the size of the comb used and the depth of the gel poured. A typical load volume is 20 μl.

To load the sample in a well of an agarose gel, hold the tip of the micropipet just under the surface of the electrophoresis buffer and directly over the well (see Figure 4.19). Slowly press down on the pipet plunger until it stops at the first stop to release the sample from the pipet tip; the sample will slowly flow into the well due to the increased density provided by the sample loading buffer. Remove the pipet tip from the buffer before releasing the plunger to avoid sucking the liquid back into the pipet.

Once all wells are loaded, place the lid on the electrophoresis chamber, matching the red and black jacks on the lid with the red and black plugs on the chamber. Plug the electrical cables leading from the lid into a power supply by attaching the red cable to the red socket and the black cable to the black socket.

Agarose gels are typically run at a constant voltage of 100 V for 30 minutes to 1 hour depending on the desired separation of the DNA fragments. Bio-Rad has devised a method to run gels faster using a modified buffer (see Appendix A for details). Since DNA is colorless and therefore invisible in the gel, the migration of DNA in the gel is monitored through the movement of the dye in the sample loading buffer (see Table 4.2). Once electrophoresis is complete, turn off the power supply and remove the lid of the electrophoresis chamber.

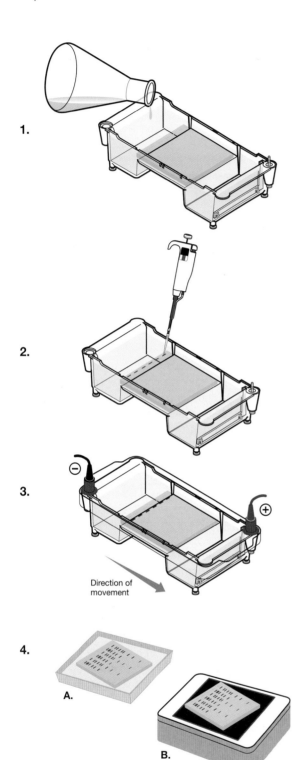

1.

2.

3.

Direction of
movement

4.

A.

B.

Figure 4.18. **Running an agarose gel** 1. the prepared agarose gel is placed into the electrophoresis chamber and running buffer is added; 2. DNA samples are loaded into the sample wells; 3. the gel is run at a constant voltage until the electrophoresis is complete; 4. the gels are visualized using a) a positively-charged stain or b) a fluorescent dye with a UV light source.

Agarose Gel Staining and Imaging

DNA is colorless and cannot be seen in an agarose gel without staining. DNA can be visualized in an agarose gel using a positively charged **DNA stain** that is colored and binds to the outside of the DNA molecule. Examples of positively charged stains are Bio-Rad's Fast Blast™ DNA stain and methylene blue. Positively charged stains must be used after the gel has been run because the dyes would disrupt the migration of the DNA in the gel. Alternatively, SYBR® Safe DNA stain, UView™ 6x loading dye and stain, and ethidium bromide are fluorescent stains that fit, or intercalate, between the base pairs of DNA. Fluorescently stained DNA is visualized by exposing the gel to ultraviolet (UV) or blue light, depending on the stain. When ethidum bromide or SYBR safe DNA stain is added to the molten agarose gel matrix, it will stain the DNA during electrophoresis and allow the DNA to be visualized immediately following electrophoresis. No additional staining is required when using UView 6x loading dye and stain as it is added to the DNA before loading on the gel. Post-electrophoresis staining of gels with these stains can also be performed. Different protocols for staining agarose gels are available in Appendix B.

Ethidium bromide is still the most common DNA stain used in research and industry. It is a known mutagen and possible carcinogen. For safety reasons, this stain, and the gels and buffer treated with this stain must be used and disposed of properly. In addition, visualization of DNA fragments on gels stained with ethidium bromide requires the use of a UV light, which is also hazardous and requires UV protective shielding. UV light can also damage the DNA that is being viewed. SYBR® Safe DNA stain and UView 6x loading dye and stain are newer stains that are much less harmful than ethidium bromide. DNA fragments stained with these stains can also be visualized with UV light. DNA fragments stained with SYBR® Safe DNA stain can also be seen with a special, non-hazardous blue light source and orange filter. If fluorescent DNA stains are used, a **gel documentation system** consisting of a transilluminator, blue light (see Figure 4.20), and an

Figure 4.19. **Students loading DNA samples into an agarose gel for electrophoresis**.

image capturing system such as a camera is required to view and record the gel results. Images of gels stained with positively-charged stains can also be digitally captured using a white-light transilluminator.

Most research or industrial laboratories use more complex imaging systems, such as Bio-Rad's ChemiDoc™ Imaging systems, which are capable of imaging many different types of gels and stains at high resolution. Such imaging systems have specialized software that allows quantitation of band intensity in all gel types.

Figure 4.20. **UView Mini Transilluminator** can be used as a UV source to visualize fluorescent DNA stains.

4.5 Restriction Enzymes as Forensic Tools

Restriction Fragment Length Polymorphisms
Restriction enzymes are the workhorses of molecular biology. These enzymes are used extensively in recombinant DNA technology, which involves moving DNA fragments from one location to another. DNA profiling is another use for restriction enzymes. Invented in 1985 by the English geneticist Alec Jeffreys, DNA profiling is the foundation of modern forensics (see Biotech In the Real World) and is based on variances in the DNA sequences among individuals. Genetic variations among individuals sometimes include mutations in restriction sites such that the sites are no longer recognized by the restriction enzyme. These genetic differences, discovered in 1978, are called **restriction fragment length polymorphisms** (**RFLP** or "rif-lips"). RFLP analysis generates a banding pattern unique for each individual. This DNA fingerprint can be used to include or exclude a suspect in a criminal investigation.

Southern Blotting
When performing RFLP analysis on genomic DNA, the DNA is digested with a restriction enzyme and the fragments generated are analyzed by agarose gel electrophoresis. Because there are thousands of restriction enzyme sites within a genome (resulting in thousands of DNA fragments being generated in the digestion), the fragments appear as a smear rather than as distinct bands on the agarose gel. To create a useful banding pattern and detect RFLPs, only a select number of DNA fragments are visualized using a technique called **Southern blotting**, named after Edwin Southern who invented it in 1975.

RFLP analysis and Southern blotting were widely used through the 1980s and 1990s to exclude suspects in criminal investigations; however, PCR testing is now more commonly used for forensic investigations. (PCR will be covered in more detail in Chapter 6.) Southern blotting requires a large amount of DNA sample and is time-consuming; however, there is no risk of contamination with this technique and a positive result is virtually indisputable. The evidence needed for DNA fingerprinting can be obtained from any biological material that contains DNA, including body tissues, body fluids (blood and semen), and hair follicles. DNA analysis can even be performed on dried material, such as blood stains or mummified tissue.

Chapter 4 Essay Questions

1. Discuss the current status and implications of direct-to-consumer (home) genetic tests.

2. Investigate the potential benefits of CRISPR to humans.

3. Explain whether forensic evidence is sufficient for a criminal investigation.

Additional Resources

Additional information on restriction enzymes:

Roberts R J (2005). How restriction enzymes became the workhorses of molecular biology. Proc Natl Acad Sci USA 102, 5905–5908.

New England BioLabs' current product catalog and website (neb.com), Accessed June 18, 2018.

Collins M et al. (2009). DNA staining using Fast Blast™ DNA stain or SYBR® Safe stain and digital analysis of gel images using Logger Pro software. Bio-Rad Bulletin 5914.

Vincze T et al. (2003). NEBcutter: a program to cleave DNA with restriction enzymes. Nucleic Acids Res 31, 3688–3691.

Profile on Alec Jeffreys:

Zagorski N (2006). Profile of Alec J. Jeffreys. Proc Natl Acad Sci USA 103, 8918–8920.

Video of a lecture by Professor Alec Jeffreys at the Royal Society. The first-half of the video discusses the history of DNA fingerprinting and the second half discusses Professor Jeffreys' latest genetic research (67 min):

Jeffreys A (2010). Prize Lecture: Croonian Lecture: Genetic Fingerprinting and Beyond. youtube.com/watch?v=O28XJQCrlJg, Accessed November 14, 2017.

Activity 4.A Restriction Site Prediction Using NEBcutter®

Overview

This activity will use an online tool, NEBcutter, to predict the DNA fragments that will be generated from a specific restriction digestion of DNA from a bacteriophage called lambda. The virtual DNA fragments will then be separated on a virtual gel. Tools such as NEBcutter are useful for planning DNA cloning experiments.

New England BioLabs (NEB), which was established in the mid-1970s, is one of the first companies to commercially sell restriction enzymes. NEB has been responsible for the discovery of hundreds of restriction enzymes, greatly assisting the molecular biology community. Its catalog and website are great resources for information on all aspects of restriction enzyme use.

NEBcutter is a free online tool developed by NEB to assist experimental design. This tool contains a vast database of DNA sequences, including bacteriophage and plasmid reference DNA sequences, which can be selected and virtually cut with any restriction enzyme. NEBcutter can also be used to virtually cut a DNA fragment with multiple enzymes at the same time, and the results of a compound digestion will be shown on the virtual gel.

Research Questions

- How many times does a given restriction enzyme cut lambda DNA?
- When cutting lambda DNA with a given restriction enzyme, what pattern will the resulting DNA fragments make on a gel?

Objectives

- Use NEBcutter to predict where restriction enzymes will cut the bacteriophage lambda DNA sequence
- Produce a virtual gel image of lambda DNA cut with specific restriction enzymes

Skills to Master

Refer to Laboratory Skills Assessment Rubric (Appendix E) and Laboratory Notebook Rubric (Appendix F) for more details.
- Use NEBcutter software to predict restriction sites
- Record detailed software instructions into a laboratory notebook

Student Workstation Materials

Items	Quantity
Computers with Internet connection	1–4
Printer (shared)	1

Prelab Focus Questions

1. If a linear piece of DNA has three recognition sites, how many bands do you expect to see on the gel after digestion with a restriction enzyme that recognizes those sites?
2. If a circular piece of DNA has three recognition sites, how many bands do you expect to see on the gel after digestion with a restriction enzyme that recognizes those sites?
3. When electrophoresed, will the largest DNA fragment be nearest to or farthest from the wells of the gel?
4. If a linear piece of DNA is cut exactly in the middle, how many fragments do you expect to see on the gel?

Activity 4.A Restriction Site Prediction Using NEBcutter

Protocol

1. Go to the New England BioLabs website (neb.com) and locate NEBcutter under the NEB tools banner.

   Note: Be sure to write detailed notes on how to use this online tool in your laboratory notebook.

2. Click the **# Viral + phage** box and select **Lambda** on the dropdown menu for **Standard sequences** (in the upper right area of the gray box).

3. Ensure that the radio button at **The sequence is** section is selected for **Linear** and click **Submit**. Once submitted, the window will refresh and show all NEB enzymes that cut the sequence.

4. Select **Custom digest** under **Main options** in the lower left corner.

5. Select one restriction enzyme from the list by clicking the checkbox to the left of the enzyme's name. Your instructor may tell you which enzymes to use.

   Note: The table shows the number of sites, recognition sequence, and percent activity of the restriction enzyme in different NEB reaction buffers.

6. Click **Digest** at the bottom of the window. The results will show the restriction enzyme name below the linear representation of the lambda DNA.

7. Click **Fragments** under the heading **List**. This will open a new window showing the number of fragments and the length of each fragment. Copy this table into your laboratory notebook.

8. Click **View gel** on the main results page under **Main options**. This will open a new window with the fragments on a virtual agarose gel.

9. Adjust **Gel type** to **1% agarose**, **Marker** to **Lambda - HindIII Digest**, and **DNA type** to **Unmethylated**. Print the virtual gel and paste it into your laboratory notebook.

10. Repeat steps 2–10 using a second enzyme. Repeat again using both enzymes for a double digestion.

Results Analysis

Count the number of restriction sites for the selected enzymes on the lambda DNA map.

Count the number of fragments in the virtual gel and confirm that the number matches the information in the data table.

Add up the lengths of the fragments in the virtual gel and verify that the total length matches that of the uncut lambda DNA.

Postlab Focus Questions

1. State one experiment for which a DNA researcher would use NEBcutter or a similar online tool.
2. What can you conclude if the prediction from NEBcutter does not match the results of the actual digestion?

Self Assessment

Assess whether you have mastered the skills of this activity using the Laboratory Skills Assessment Rubric in Appendix E.

Assess your experimental write-up in your laboratory notebook using the Laboratory Notebook Rubric in Appendix F.

Activity 4.B Casting Agarose Gels

Overview

This activity will provide a procedure for casting an agarose gel (of any percentage/concentration and volume) for use in horizontal gel electrophoresis. This procedure will use a microwave oven to melt the agarose. (Agarose can also be melted using a hot plate.) The protocol describes how to tape the ends of the gel casting tray with laboratory or painter's tape. If casting gates or dams are available, they can be used instead to block ends of the gel tray.

The percentage of agarose in the gels depends on the task: high-percentage agarose gels are used to separate small molecules, while low-percentage gels are used to separate large molecules. If small PCR fragments are going to be separated, 2–3% agarose gels are ideal. If digested lambda DNA is to be electrophoresed, 0.7–1% agarose gels work best.

The volume of molten agarose required depends on the size of the gel tray. A 7 x 10 cm gel tray holds approximately 50 ml of agarose, while a 7 x 7 cm tray holds approximately 40 ml. To determine the volume of a gel tray, tape up the ends of the tray, add water until it is approximately 0.75 cm high, and then measure the water in a graduated cylinder.

Once the volume of agarose solution has been determined, calculate the required amount of agarose powder by using the formulae in the Preparing Solutions section of Chapter 2, page 42.

Tips and Notes

- Two combs can be added to a 7 x 10 cm gel tray to create two rows of wells to allow more samples to be run simultaneously. When using two combs, place the first comb at one end of the gel and the other comb in the middle of the gel. During electrophoresis, care should be taken not to let the upper set of samples run into the lower set of wells. If only one-half of the gel is to be used, use the lower half and leave the upper half free to use later

- It is important to cool the molten agarose to approximately 55°C in a water bath or on a heat-resistant pad. Excessive heat from the molten agarose will damage the acrylic gel trays and may melt the tape adhesive. Cooler agarose is also less likely to leak because the solution is more viscous and does not flow through tiny spaces. Do not shake or vigorously swirl the molten agarose to avoid introducing bubbles, which could affect the migration of DNA fragments. Any bubbles should be removed before the gel solidifies

- Solidified agarose can be remelted in the microwave oven, but agarose mixed with fluorescent DNA stains should never be reheated

Safety Reminder: Molten agarose is very hot and should be handled carefully with heat-resistant gloves. Swirl the molten agarose before it is removed from the microwave oven to ensure that it will not flash boil. All appropriate PPE should be worn.

Objectives

- Calculate the amount of agarose powder needed to make 50 ml of 1% agarose
- Prepare gel trays for pouring the molten agarose
- Prepare and cast a 1% agarose TAE gel
- Remove the gel comb and store the agarose gel

Skills to Master

- Prepare agarose gels

Student Workstation Materials

Items	Quantity
Water bath at 55°C (optional)	1
Microwave oven (shared)	1
Balance (shared)	1
Weigh boat	1
100 ml graduated cylinder	1
25 ml flask or bottle cap (optional)	1
150–250 ml flask or reagent bottle	1
Gel tray	1
Gel comb	1
Leveling bubble (optional)	1
Heat-resistant gloves (shared)	1 pair
Heat-resistant pad (optional)	1
Plastic wrap or small plastic bag (optional)	1
Laboratory tape	20 cm
Laboratory marking pen	1
1x TAE electrophoresis buffer (from Activity 2.D)	50 ml
Agarose	0.5 g

Prelab Focus Questions

1. Calculate the amount of agarose needed to make 50 ml of 1% TAE agarose solution.
2. Describe how to make 50 ml of molten 1% TAE agarose that is ready to pour into a gel casting tray.
3. Once poured, why should the agarose gel be opaque before removing the comb from the gel casting tray?
4. Why do scientists use gels containing different percentages of agarose?

Activity 4.B Casting Agarose Gels

Protocol

1. Determine the volume of molten agarose required for the gel(s) to be poured. Calculate the mass of agarose required for the volume and agarose percentage to be made.

2. Measure the required quantity of agarose powder using a weigh boat and balance.

3. Measure the required volume of 1x TAE buffer and pour it into an appropriate heat-resistant bottle or flask. Do not fill more than one-third of the bottle or flask. Label the container appropriately. Include the percentage of the agarose, the buffer used, the date, and your initials.

4. Add the agarose powder to the container of buffer while swirling. Place a cap **very** loosely on the bottle or put an inverted 25 ml flask into the flask opening. Alternatively, loosely wrap plastic over the container opening. Steam must be allowed to escape but too much evaporation will increase the concentration of agarose.

   Safety note: Agarose can superheat, resulting in flash boiling. The overflowing agarose may cause burns. Ensure proper PPE, heat-resistant gloves, and safety goggles are worn when boiling agarose. Wait until the agarose stops bubbling and gently swirl to check for superheating before removing it from microwave oven.

5. Microwave the agarose solution on a medium setting in 30 sec increments. Allow the solution to stop bubbling and gently swirl to check whether agarose particles have dissolved. Continue heating and swirling every 30 sec until no translucent particles remain.

6. Cool the molten agarose to approximately 55°C in a water bath or on a heat-resistant pad.

7. Tape both ends of a gel tray with laboratory or painter's tape, making sure to securely seal the edges of the tray. Label the tape with your initials. Place the comb into the comb guides on the gel tray. For DNA electrophoresis, place the comb at one end of the gel tray. For dye electrophoresis, place the comb in the middle of the tray. Ensure the gel tray is horizontal using a leveling bubble.

8. (Optional) If SYBR® Safe DNA stain or ethidium bromide is to be added to the agarose, add the concentrated stain to the molten agarose (once the agarose is cooled to 55°C), and gently swirl to mix.

9. Once the molten agarose is cooled to 55°C, pour it into the gel tray until the agarose comes to within 2–4 mm of the top of the teeth of the comb.

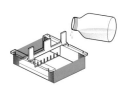

10. Let the gel cool and solidify, it will become opaque.

11. When the gel is cold and opaque, remove the comb by pulling it gently straight up and out of the wells. If the comb is removed before the gel is set, the wells will collapse.

Activity 4.B Casting Agarose Gels

12. If the gel is to be used immediately, remove the tape from both ends of the tray and continue with the activity. If the gel is to be stored for later use, keep the gel in the gel tray (do not remove the tape from the tray) and store it at 4°C in a small plastic bag or in plastic wrap. Label the wrapped gel with your initials, the date, and type of gel.

Postlab Focus Questions

1. Explain three key things that must be done in order for a gel to be correctly cast and run using electrophoresis.
2. Why is it necessary to store the gel in a plastic bag or in plastic wrap?

Self Assessment

Assess whether you have mastered the skills of this activity using the Laboratory Skills Assessment Rubric in Appendix E.

Assess your experimental write-up in your laboratory notebook using the Laboratory Notebook Rubric in Appendix F.

Activity 4.C Dye Electrophoresis

Overview

This activity is a good introduction to electrophoresis and provides a simple and fun way to learn how to load and run agarose gels. Different dyes will be loaded into agarose gels and electrophoresed. The dyes should move toward either the positive electrode (anode) or the negative electrode (cathode) based on their chemistries. Dyes are available not only in molecular biology laboratories but also in materials we encounter every day, including many foods and candies we consume. Food dyes can be obtained from food coloring or from candies dissolved in liquid. Unlike DNA, the distance dyes migrate is mainly based on their charge rather than size because they are much smaller than DNA molecules. In this activity, extracts from colored food samples will be compared after electrophoresis to reference dyes commonly used during food production. The net charge as well as the identities of the food dyes present in the food samples can be determined from these results.

Safety Reminder: Appropriate PPE should be worn at all times, and attention should be given when using electricity in the presence of liquids. Safety measures built into the equipment should prevent accidental exposure; however, should any leakage occur, notify your instructor before proceeding.

Research Questions

- What charge do dyes have?
- Are samples composed of multiple dyes?

Objectives

- Load and run dyes on an agarose gel
- Determine the charge of each electrophoresed dye
- Determine the composition of each colored sample

Skills to Master

Refer to Laboratory Skills Assessment Rubric (Appendix E) and Laboratory Notebook Rubric (Appendix F) for more details.

- Record laboratory notebook entries
- Load an agarose gel
- Perform agarose gel electrophoresis
- Use a power supply

Student Workstation Materials

Items	Quantity
Horizontal gel electrophoresis chamber	1
Power supply (shared)	1
2–20 µl adjustable-volume micropipet and tips	1
Microcentrifuge tube rack	1
Waste container	1
1x TAE electrophoresis buffer (from Activity 2.D)	300 ml
Food coloring in dye extraction solution (optional)	1–4 tubes
Candy dyes in dye extraction solution (optional)	1–5 tubes
Fast Blast stain in dye extraction solution (optional)	1 tube
1x sample loading buffers (optional)	1–2 tubes
Set of four reference dyes (Blue 1, Red 40, Yellow 5, Yellow 6)	4 tubes
1% agarose TAE gel with wells formed in the middle of the gel (from Activity 4.B)	1

Prelab Focus Questions

1. If a dye is negatively charged and is loaded in lanes 1–3, in which direction will it migrate when electrophoresed? Copy the gel below and place an arrow to indicate the direction of the expected movement.

2. If a dye is positively charged and loaded in lanes 4–6, in which direction will it migrate when electrophoresed? Place an arrow to indicate the direction of the expected movement.

3. Why is dye extraction solution added to dyes before they are loaded into the gel?

Activity 4.C Dye Electrophoresis

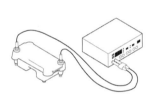

Protocol

1. Place the agarose gel (on the gel tray) into the electrophoresis chamber. Add sufficient 1x TAE buffer to the chamber to cover the gel by approximately 2 mm.

2. Record the sample names in your laboratory notebook. Make a table listing the order in which the samples will be loaded into the gel and write down the well (lane) number on the gel and the volume to be loaded (20 µl).

3. Set a 2–20 µl adjustable-volume micropipet to deliver 20 µl. Place a tip on the micropipet, and draw 20 µl of the first dye into the pipet tip.

4. Hold the micropipet tip directly over the well and just break the surface tension of the buffer over the well with the tip (**Note:** Do not put the tip into the well to avoid breaking the gel.) Slowly depress the plunger to deliver the dye into lane 1. Stop depressing at the first (soft) stop of the micropipet. Keeping the plunger depressed, lift the pipet tip out of the buffer.

5. Remove the tip, and follow steps 3 and 4 to load the next dye into lane 2 according to your table. Continue until all the dyes have been loaded on the gel according to your table.

6. Place the lid on the electrophoresis chamber. Connect the electrical leads into the power supply, red to red and black to black. Record the orientation of the gel with respect to the positive and negative electrodes in your laboratory notebook.

7. Set the power supply to run the gel at 100 V for 20 min, and start the electrophoresis. Watch for bubbles at both electrodes to confirm that the electrophoresis has started. It is important to monitor the dye migration on the gel, although condensation on the lid may sometimes prevent viewing.

8. After 20 min, turn off the power supply and remove the lid of the chamber. Quickly record the results in your notebook with a summary and a sketch, as the dyes will begin to diffuse quite rapidly.

Postlab Focus Questions

1. Which dyes were negatively charged?
2. Which dyes were positively charged?
3. Which samples were composed of multiple dyes? Record the colors of the dyes.
4. State one way this experiment might be improved.

Self Assessment

Assess whether you have mastered the skills of this activity using the Laboratory Skills Assessment Rubric in Appendix E.

Assess your experimental write-up in your laboratory notebook using the Laboratory Notebook Rubric in Appendix F.

Activity 4.D Restriction Digestion and Analysis of Lambda DNA

Overview

In this activity, DNA from bacteriophage lambda will be digested with PstI, EcoRI, and HindIII restriction enzymes. The digested DNA will be analyzed by horizontal agarose gel electrophoresis on a 1% agarose gel. The gel will be stained and the banding pattern will be recorded. A standard curve of the molecular size standard will be plotted and used to estimate the lengths of the various lambda DNA fragments generated by the digests.

Tips and Notes

- When working with DNA and restriction enzymes, use a fresh, clean pipet tip for each addition to avoid contaminating the DNA sample with the incorrect enzymes or the enzymes with DNA

- Thaw the tubes of DNA and restriction buffer on ice

- Keep restriction enzymes on ice at all times to maintain their activity

Safety Reminder: Appropriate PPE should be worn at all times and attention should be given when using electricity in the presence of liquids. Safety measures built into the equipment should prevent accidental exposure; however, should any leakage occur, notify your instructor before proceeding. If ethidium bromide, fluorescent stains, or UV light are used, special precautions are required.

Research Questions

- How many HindIII, EcoRI, and PstI recognition sites are present on lambda DNA?
- How long are DNA fragments generated by digestion of lambda DNA with these enzymes?

Objectives

- Digest lambda DNA with HindIII, EcoRI, and PstI restriction enzymes
- Load and run digested DNA on a 1% agarose TAE gel
- Determine the number of recognition sites for various enzymes on lambda DNA
- Generate a standard curve based on lambda HindIII DNA standard
- Determine the length of all DNA fragments by comparing with the standard curve

Skills to Master

Refer to Laboratory Skills Assessment Rubric (Appendix E) and Laboratory Notebook Rubric (Appendix F) for more details.
- Record laboratory notebook entries
- Perform a restriction digest
- Perform agarose gel electrophoresis (Activity 4.C)
- Analyze an agarose gel
- Generate a standard curve using a DNA size standard
- Determine DNA fragment sizes using a standard curve

Student Workstation Materials

Part 1: Setting Up the Digestion Reactions

Items	Quantity
Water bath, dry bath, or incubator at 37°C (shared)	1
Microcentrifuge or mini centrifuge (optional) (shared)	1
2–20 µl adjustable-volume micropipet and tips	1
Colored microcentrifuge tubes (yellow, violet, green, and orange)	4
Microcentrifuge tube rack	1
Floating tube rack (optional)	1
Container of ice	1
Laboratory marking pen	1
Waste container	1
Lambda DNA on ice	25 µl
2x restriction buffer on ice	60 µl
EcoRI enzyme on ice	5 µl
HindIII enzyme on ice	5 µl
PstI enzyme on ice	5 µl

Part 2: Running the Gel

Items	Quantity
Horizontal gel electrophoresis chamber	1
Power supply (shared)	1
Digital imaging system (optional) (shared)	1
Computers with graphing software (optional)	1–4
Water bath or dry bath at 65°C (optional) (shared)	1
Microcentrifuge or mini centrifuge (optional) (shared)	1
UV light source (optional) (shared)	1
Rocking platform (optional) (shared)	1
0.5–10 µl adjustable-volume micropipets or 2–20 µl adjustable-volume micropipets and tips*	1
Rulers	4
Microcentrifuge tube rack	1
Gel staining tray	1
Agarose gel support film (optional)	1
Semilog graph paper	4
Laboratory marking pen	1
Waste container	1
DNA digest samples from part 1	4
5x sample loading buffer (SLB)	30 µl
or UView 6x loading dye and stain (UView)	10 µl
Lambda HindIII DNA standard (λ Std)	15 µl
1x TAE electrophoresis buffer (from Activity 2.D)**	300 ml
1x Fast Blast DNA stain***	100 ml
1% agarose TAE gel†	1

* 0.5–10 µl micropipets are ideal for measuring 1 µl volumes of enzyme; however, 2–20 µl micropipets can be used to measure 1 µl with sufficient accuracy for this experiment.

** If necessary, make 1x TAE electrophoresis buffer (see part 1 of Activity 2.D).

*** Alternative staining methods may be used. See Appendix B for details.

† If necessary make a 1% agarose TAE gel according to the protocol in Activity 4.B. This may require extra 1x TAE electrophoresis buffer.

Prelab Focus Questions

1. Why do restriction enzymes need to be kept on ice?
2. What order should the DNA, enzyme, water, and buffer be added to the microcentrifuge tube for a restriction digest?
3. If lambda DNA is linear, how many times would the enzyme have to cut the DNA to generate five DNA fragments?
4. Would a shorter DNA fragment move faster or slower through the agarose gel than a longer fragment? Why?

Activity 4.D Restriction Digestion and Analysis of Lambda DNA

Protocol

Part 1: Setting Up the Digestion Reaction

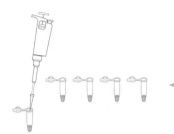

1. Label colored microcentrifuge tubes as below:

yellow tube	**L** (lambda DNA)
violet tube	**P** (PstI lambda digest)
green tube	**E** (EcoRI lambda digest)
orange tube	**H** (HindIII lambda digest)

    Label all the tubes with your initials and date, and place them in the microcentrifuge tube rack.

2. Using a fresh tip for each reagent and sample, pipet the lambda DNA, 2x restriction buffer and enzymes into each tube according to the table below. Make sure the reagents are added in the order indicated: DNA, followed by buffer, and last, enzyme.

Tube	DNA	Buffer	PstI	EcoRI	HindIII
L	4 µl	6 µl	—	—	—
P	4 µl	5 µl	1 µl	—	—
E	4 µl	5 µl	—	1 µl	—
H	4 µl	5 µl	—	—	1 µl

3. Tightly cap the tubes and mix the components by gently flicking the tubes with your finger. Collect the sample at the bottom of each tube by tapping the tube gently on the bench or by pulse-spinning it in a microcentrifuge.

4. Incubate digestion reactions for 30 min at 37°C or overnight at room temperature.

5. After the incubation, store the samples at 4°C until the next laboratory period. Samples can be stored for 1 month at 4°C. If there is sufficient time, proceed to running the gel.

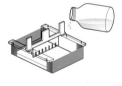

6. 1x TAE electrophoresis buffer and a 1% agarose TAE gel are required for the next part of the activity. If necessary, prepare 1x TAE (refer to part 1 of Activity 2.D) and a 1% agarose TAE gel (refer to Activity 4.B).

Part 2: Running the Gel

1. If condensation has collected on the lids of the tubes, collect the samples at the bottom of the tubes by tapping them gently on the bench or by pulse-spinning them in a microcentrifuge.

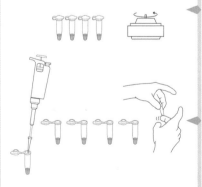

2. Using a fresh tip each time, pipet 2.5 µl of 5x sample loading buffer (**SLB**) or 2 µl of UView™ 6x loading dye and stain (**UView**) into each tube. Cap the tubes and mix the contents by gently flicking the tubes with your finger. Collect the sample at the bottom of each tube by tapping the tube gently on the bench or by pulse-spinning it in a microcentrifuge.

Activity 4.D Restriction Digestion and Analysis of Lambda DNA

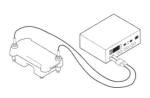

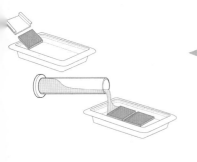

3. (Optional) Heat the DNA samples at 65°C for 5 min and then put them on ice to improve band separation.

4. Place a 1% agarose TAE gel into the electrophoresis chamber. Fill the electrophoresis chamber with sufficient 1x TAE buffer to cover the gel by approximately 2 mm.

5. Check that the wells of the agarose gels are near the black (–) electrode or cathode and the bottom edge of the gel is near the red (+) electrode or anode.

6. Using a fresh tip for each sample, load 10 µl of each sample into 5 wells of the gel in the following order:

    | Lane 1 | **λ Std** (lambda HindIII DNA standard) |
    | Lane 2 | **L** (uncut lambda) |
    | Lane 3 | **P** (PstI lambda digest) |
    | Lane 4 | **E** (EcoRI lambda digest) |
    | Lane 5 | **H** (HindIII lambda digest) |

7. Place the lid on the electrophoresis chamber. Connect the electrical leads to the power supply in the following orientation: red to red and black to black.

8. Turn on the power and run the gel at 100 V for 30 min.

    Note: If using the fast gel running buffer (0.25x TAE), run the gel at 200 V for 20 min (see Appendix A).

9. When the electrophoresis run is complete, turn off the power and remove the lid from the chamber. Carefully remove the gel and tray from the electrophoresis chamber. Be careful — the gel is very slippery. If using UView 6x loading dye and stain or other fluorescent stains, use a UV light source to visualize DNA bands. Otherwise, slide the gel into the staining tray and carefully pour a sufficient volume of 1x Fast Blast DNA stain to completely submerge the gel.

    Note: For alternative DNA staining options, see Appendix B.

10. Stain the gel overnight to visualize the DNA bands and then record the results. (Optional) A rocking platform can be used to enhance results.

11. (Optional) Trim away any unloaded lanes and air-dry on a piece of agarose gel support film. Tape the dried gel into your laboratory notebook.

Activity 4.D Restriction Digestion and Analysis of Lambda DNA

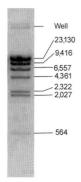

Figure 4.21. Lambda HindIII DNA standard. Labels indicate the size of the DNA bands in base pairs.

Results Analysis

Record the banding pattern on the gel either by tracing the wells and bands onto acetate sheets, photocopying or photographing the gel, or using a digital gel imaging system. Once the gel image is obtained, label the sizes of the bands in the lambda HindIII DNA standard (see Figure 4.21). Then roughly estimate the sizes of the bands from the EcoRI and PstI digests by visually comparing them with the bands from the lambda HindIII DNA standard. (If a band whose size is not known migrates between two bands of the DNA standard, the size of the unknown band must be between the sizes of the two known bands.) Record your estimates in a data table.

To determine the sizes of the unknown DNA fragments, a standard curve will be generated by plotting the distance each fragment of the lambda HindIII DNA standard traveled on the agarose gel against the known size of the fragment. The standard curve will then be used to estimate the sizes of the DNA fragments generated by the restriction digest.

First, determine the distance traveled by each fragment of the lambda HindIII DNA standard by measuring the distance from the bottom of the well to the bottom of each DNA band. Record your measurements in your laboratory notebook in a table similar to Table 4.3. Use either semilog graph paper provided by your instructor or graphing software such as Microsoft Excel to generate a standard curve. Each of these methods is described below. Second, measure the distance traveled by each unknown band in each restriction digestion and record these measurements in your table. Third, use your standard curve to determine the sizes of the unknown fragments. Digital gel imaging software that analyzes the gel automatically and provides band sizes without the need to measure and graph can also be used. However, it is important to acquire the skills needed to determine the sizes of unknown DNA fragments using a standard curve and to understand what the software is doing behind the scenes. Information on using digital gel imaging systems is described below.

Once the analysis is complete, compare the data obtained using the standard curve with the rough estimates you made at the beginning of the gel analysis.

Table 4.3. **Results table for lambda DNA digestions.**

Largest fragment first ↓	M = DNA Marker		L = Uncut Lambda No Enzyme		P = PstI Restriction Digest of Lambda DNA		E = EcoRI Restriction Digest of Lambda DNA		H = HindIII Restriction Digest of Lambda DNA	
	Distance (mm)	Actual base pairs	Distance (mm)	Estimated base pairs	Distance (mm)	Estimated base pairs	Distance (mm)	Estimated base pairs	Distance (mm)	Estimated base pairs
Band 1		23,130								
Band 2		9,416								
Band 3		6,557								
Band 4		4,361								
Band 5		2,322								
Band 6		2,027								

Estimating DNA Sizes Using Semilog Graph Paper

Using the semilog graph paper provided by your instructor, plot the bands from the lambda HindIII DNA standard. Plot the distance (in mm) that the fragments moved on the x-axis and the number of base pairs of the DNA fragments on the y-axis. Add a line of best fit through the data points. Do not include the 23,130 bp data point in the line of best fit since it is out of the linear range (see Figure 4.22). Once the measurements of the lambda HindIII DNA standard have been plotted on the semilog graph paper, the length of the unknown bands on the gel can be estimated. Using the graph, find the distance traveled for each band on the x-axis and read up until the line of best fit is reached on the y-axis. Then read across to find the estimated number of base pairs. Repeat until the approximate length of all unknown bands has been estimated.

Activity 4.D Restriction Digestion and Analysis of Lambda DNA

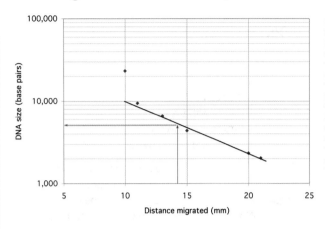

Figure 4.22. **Standard curve of lambda HindIII DNA fragments.** The size of the DNA fragments in base pairs was plotted on a logarithmic scale against the distance the fragments migrated in the agarose gel (in mm). A line of best fit was drawn through the data points. (**Note:** The data point for the largest DNA lambda HindIII fragment, 23,130 bp, should not be included in the line of best fit since it is outside the linear range.) To determine the sizes of unknown DNA fragments, read up from the distance the fragment migrated and across to the DNA size, as shown by the arrows on the graph.

Estimating DNA Sizes by Graphing

The sizes of the unknown bands in the various digested samples can be estimated by plotting the results using a graphing software tool such as Microsoft Excel or Google Sheets. Create a new spreadsheet and enter the distance traveled (in mm) by the various bands of the lambda HindIII DNA standard in column A and the number of base pairs of the bands in column B. Highlight both columns and create an XY scatter plot. Once the graph has been produced, click the data points and insert a line of best fit. There will be an option to display the equation of a line. To estimate the fragment lengths using the line of best fit, add the values for "x" (which represents the distance measurements) into the equation and solve for "y" (which represents the estimated number of base pairs).

Estimating DNA Sizes Using a Digital Gel Imaging System

A digital gel imaging system can also be used to estimate the sizes of the unknown bands in the DNA digestions. Place the gel onto the transilluminator and launch the imaging software on the computer attached to the transilluminator. Acquire the gel image using the system camera. You will need to identify the lane of the gel containing the DNA size standard. You may also need to enter the base pair sizes for each band of the lambda HindIII DNA standard. Depending on the imaging system, the software will either automatically identify the DNA bands or you will need to click each band to identify it for the software. The software will generate a standard curve and estimate the size of each band (in bp) in a data table. Detailed procedures will be provided by your instructor if a digital imaging system is going to be used.

Postlab Focus Questions

1. If an enzyme cuts lambda DNA seven times, how many fragments would you see on the gel?
2. How many HindIII, EcoRI, and PstI sites are present in the lambda DNA sequence?
3. Why does DNA move from the negative electrode (cathode) to the positive electrode (anode)?
4. If the gel is electrophoresed for too long, what will happen to the DNA fragments?
5. Why do you add sample loading buffer to the DNA samples?
6. Why do you need to stain the DNA in the gel?

Self Assessment

Assess whether you have mastered the skills of this activity using the Laboratory Skills Assessment Rubric in Appendix E.

Assess your experimental write-up in your laboratory notebook using the Laboratory Notebook Rubric in Appendix F.

Activity 4.E Forensic DNA Fingerprinting

Overview

In this activity, restriction fragment length polymorphism (RFLP) analysis will be performed on samples that represent DNA taken from a crime scene and DNA from five suspects in a criminal investigation.

Technicians working in forensic labs are often asked to perform DNA profiling, or fingerprinting, to analyze evidence in law enforcement cases. DNA evidence is used to exclude suspects from investigations. DNA fingerprinting is also used for many non-criminal applications, such as paternity testing, environmental sampling, and ancestry analysis. Most DNA profiling nowadays is performed using polymerase chain reaction (PCR), which can analyze minute samples of DNA; however, RFLP analysis, the original method for DNA fingerprinting, still has its place in scientific research. RFLP analysis requires much more DNA than PCR analysis and uses restriction enzymes to cut the DNA, followed by analysis of the generated fragments. If two DNA samples are from the same individual, they will have the same DNA sequence and therefore the same restriction enzyme recognition sites. Thus, when cut with the same enzymes, these two DNA samples will generate the same banding pattern on a gel. Samples from different individuals will produce different patterns on a gel.

Tips and Notes

- When working with DNA and restriction enzymes, use a fresh, clean pipet tip for each addition to avoid contaminating the DNA sample with the incorrect enzymes or the stock enzymes with DNA

- Keep restriction enzymes on ice at all times to maintain their activity

Safety Reminder: Appropriate PPE should be worn at all times and attention should be given when using electricity in the presence of liquids. Safety measures built into the equipment should prevent accidental exposure; however, should any leakage occur, notify your instructor before proceeding. If ethidium bromide , fluorescent stains, or UV light are used, special precautions are required.

Research Question

- Which suspects can be eliminated from the investigation?

Objectives

- Digest DNA samples from the crime scene and the five suspects with EcoRI and PstI restriction enzymes
- Load and run samples on a 1% agarose TAE gel
- Generate a standard curve based on lambda HindIII DNA standard
- Determine the length of all DNA fragments by comparing to the standard curve
- Determine if the DNA from the crime scene matches the DNA from any of the suspects

Skills to Master

Refer to Laboratory Skills Assessment Rubric (Appendix E) and Laboratory Notebook Rubric (Appendix F) for more details.
- Record laboratory notebook entries
- Perform a restriction digest (Activity 4.D)
- Perform agarose gel electrophoresis (Activity 4.C)
- Analyze an agarose gel (Activity 4.D)
- Generate a standard curve using a DNA size standard (Activity 4.D)
- Determine DNA fragment sizes using a standard curve (Activity 4.D)

Activity 4.E Forensic DNA Fingerprinting

Student Workstation Materials

Part 1: Setting Up the Digestion Reaction

Items	Quantity
Water bath, dry bath, or incubator at 37°C (shared)	1
Microcentrifuge or mini centrifuge (optional) (shared)	1
2–20 µl adjustable-volume micropipet and tips	1
Colored microcentrifuge tubes (green, blue, orange, violet, pink, and yellow)	6
Microcentrifuge tube rack	1
Floating tube rack (optional)	1
Container of ice	1
Laboratory marking pen	1
Waste container	1
DNA samples, 10 µl of each (CS, S1, S2, S3, S4, and S5)	6
Restriction enzymes (**ENZ**) (mixture of PstI and EcoRI) on ice	80 µl

Part 2: Running the Gel

Items	Quantity
Horizontal gel electrophoresis chamber	1
Power supply (shared)	1
Digital imaging system (optional) (shared)	1
Computers with graphing software (optional)	1–4
Microcentrifuge or mini centrifuge (optional) (shared)	1
UV light source (optional) (shared)	1
Rocking platform (optional) (shared)	1
2–20 µl adjustable-volume micropipet and tips	1
Rulers	4
Microcentrifuge tube rack	1
Gel staining tray	1
Agarose gel support film (optional)	1
Semilog graph paper	4
Laboratory marking pen	1
Waste container	1
DNA digest samples from part 1	6
5x sample loading buffer (**SLB**)	30 µl
or UView 6x loading dye and stain (**UView**)	24 µl
Lambda HindIII DNA standard (**λ Std**)	15 µl
1x TAE electrophoresis buffer (from Activity 2.D)*	300 ml
1x Fast Blast DNA stain**	100 ml
1% agarose TAE gel***	1

*   If necessary, make 1x TAE electrophoresis buffer (see part 1 of Activity 2.D).
** Alternative staining methods may be used. See Appendix B for details.
*** If necessary make a 1% agarose TAE gel according to the protocol in Activity 4.B. This may require extra 1x TAE electrophoresis buffer.

Prelab Focus Questions

1. What does RFLP analysis test for?
2. If an RFLP result from a suspect does not match that of the DNA found at the crime scene, what does that mean for the suspect? What does it mean if the suspect's DNA matches the DNA found at the crime scene?
3. If two samples of DNA with the same sequence are cut with the same restriction enzyme, will the resulting banding patterns match?
4. If two samples of DNA with different sequences are cut with the same enzyme, will the resulting banding patterns match?
5. If two DNA samples with the same sequence are each cut with a different enzyme, will the resulting banding patterns match?

Activity 4.E Forensic DNA Fingerprinting

Protocol

Part 1: Setting Up the Digestion Reactions

1. Label six colored microcentrifuge tubes:

green tube	**CS** (crime scene)
blue tube	**S1** (suspect 1)
orange tube	**S2** (suspect 2)
violet tube	**S3** (suspect 3)
pink tube	**S4** (suspect 4)
yellow tube	**S5** (suspect 5)

   Label all the tubes with your initials and date, and place them in the microcentrifuge tube rack.

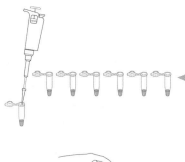

2. Using a fresh tip for each sample, pipet 10 µl of each DNA sample into the matching colored microcentrifuge tube. Make sure each sample is transferred to the bottom of the tube.

3. Using a fresh tip each time, pipet 10 µl of enzyme mix (**ENZ**) into the bottom of each tube. Pipet up and down carefully to mix.

4. Tightly cap the tubes and mix the components by gently flicking the tubes with your finger. Collect the samples at the bottom of the tubes by tapping the tubes gently on the table or by pulse-spinning them in a microcentrifuge.

5. Incubate the digestion reactions for 45 min at 37°C or overnight at room temperature.

6. After the incubation, store the samples at 4°C until the next laboratory period. Samples can be stored for 1 month at 4°C. If there is sufficient time, proceed to running the gel.

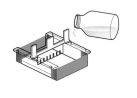

7. 1x TAE electrophoresis buffer and a 1% agarose TAE gel are required for the next part of the activity. If necessary, prepare 1x TAE (refer to part 1 of Activity 2.D) and a 1% agarose TAE gel (refer to Activity 4.B).

Part 2: Running the Gel

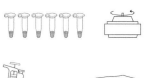

1. If condensation has collected on the lids of the tubes, collect the samples at the bottom of the tubes by tapping them gently on the bench or by pulse-spinning them in a microcentrifuge.

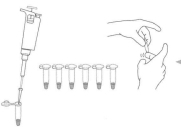

2. Using a fresh tip for each sample, pipet 5 µl of the 5x sample loading buffer (**SLB**) or 4 µl of UView™ 6x loading dye and stain (**UView**) into each tube. Cap the tubes and mix by gently flicking the tubes with your finger. Collect the samples at the bottom of the tubes by tapping them gently on the bench or by pulse-spinning them in a centrifuge.

Activity 4.E Forensic DNA Fingerprinting

3. Place a 1% agarose TAE gel into the electrophoresis chamber. Fill the electrophoresis chamber with sufficient 1x TAE buffer to cover the gel by approximately 2 mm.

4. Check that the wells of the agarose gels are near the black (–) electrode or cathode and the bottom edge of the gel is near the red (+) electrode or anode.

5. Using a fresh tip for each sample, load the indicated volume of each sample into seven wells of the gel in the following order:

Lane 1	**λ Std** (lambda HindIII DNA standard)	10 µl
Lane 2	**CS** (crime scene digest)	20 µl*
Lane 3	**S1** (suspect 1 digest)	20 µl*
Lane 4	**S2** (suspect 2 digest)	20 µl*
Lane 5	**S3** (suspect 3 digest)	20 µl*
Lane 6	**S4** (suspect 4 digest)	20 µl*
Lane 7	**S5** (suspect 5 digest)	20 µl*

* If using UView 6x loading dye and stain, load 10 µl of each sample.

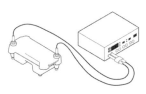

6. Place the lid on the electrophoresis chamber. Connect the electrical leads to the power supply, red to red and black to black.

7. Turn on the power and run the gel at 100 V for 30 min.

   Note: If using the fast gel running buffer (0.25x TAE), run the gel at 200 V for 20 min (see Appendix A).

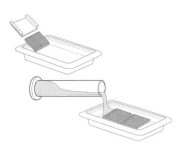

8. When the electrophoresis run is complete, turn off the power and remove the lid of the chamber. Carefully remove the gel and tray from the electrophoresis chamber. Be careful — the gel is very slippery. If using UView 6x loading dye and stain or other fluorescent stains, use a UV light source to visualize DNA bands. Otherwise, slide the gel into the staining tray and carefully pour a sufficient volume of 1x Fast Blast DNA stain to submerge the gel.

   Note: For alternative DNA staining options, see Appendix B.

9. Stain the gel overnight to visualize the DNA bands and then record the results. (Optional) A rocking platform can be used to enhance results.

10. (Optional) Trim away any unloaded lanes and air-dry on gel support film. Tape the dried gel into your laboratory notebook.

Activity 4.E Forensic DNA Fingerprinting

Results Analysis

Record the banding pattern of DNA fragments on the gel either by tracing the wells and bands onto acetate sheets, photocopying or photographing the gel, or using a digital imaging system.

Label the sizes of the bands of the lambda HindIII DNA standard on your gel image as shown in Figure 4.21 in Activity 4.D. Then estimate the sizes of the bands in the digested DNA samples by comparing the bands in these samples with the bands from the DNA standard. Record your estimations in a data table.

To accurately determine the sizes of the unknown DNA fragments, generate a standard curve using the known sizes of the lambda HindIII DNA standard and the distance they traveled on the agarose gel. The standard curve can be plotted using semilog graph paper, a graphing software tool, or a gel documentation system and software. The distance traveled for each band is measured and recorded in a table similar to Table 4.3 in Activity 4.D. The standard curve is then used to determine the size of each DNA fragment in the digested samples. See the Results Analysis section of Activity 4.D if more assistance is needed with this task.

Postlab Focus Questions

1. Which suspect(s) can be excluded from the investigation? Why?
2. Which suspect(s) cannot be excluded? Why?
3. How many DNA fragments do the crime scene and suspect DNA sample(s) share? How many base pairs are estimated for each of these DNA fragments?

Self Assessment

Assess whether you have mastered the skills of this activity using the Laboratory Skills Assessment Rubric in Appendix E.

Assess your experimental write-up in your laboratory notebook using the Laboratory Notebook Rubric in Appendix F.

Activity 4.F Plasmid Mapping

Overview

In this activity, you will use logic to determine the number and relative positions of the recognition sites of two restriction enzymes on the suspect 3 (S3) plasmid from Activity 4.E. You will be provided with a gel image, from which you will draw a map of the S3 plasmid showing the relative positions of EcoRI and PstI restriction sites.

Plasmids are circular pieces of DNA that molecular biologists use to clone and express genes. Plasmids are described in detail in Chapter 5. If the sequence of a plasmid is not known, it is important that molecular biologists find the location of the restriction sites on the plasmid. This allows scientists to determine which restriction enzyme(s) they should use to cut out the piece of DNA of interest from the plasmid. The process of finding out where the restriction sites are in a plasmid is called plasmid mapping since it involves generating a map of the restriction sites on a plasmid.

Plasmid mapping can be performed using a combination of simple experiments and logic. The general procedure is to cut (digest) a plasmid with two restriction enzymes separately (two single digests), followed by a digestion of the plasmid with both enzymes together (a double digest). Sizes of the resulting DNA fragments are determined, and then logic is used to determine the relative location of the restriction sites.

The most challenging part of plasmid mapping is using logic to overlap the results of two single digests with the results from a double digest. To overlay the cuts from one restriction enzyme with cuts from a second restriction enzyme, check if any of the fragments digested with the first enzyme remain uncut with the second enzyme. In addition, determine if the sizes of any of fragments from the double digest add up the size of a fragment from a single digest. Do any of the fragments seem to remain the same after being cut by the second enzyme? A simple example of plasmid mapping is shown in Table 4.4.

Table 4.4. **DNA fragments from a 1,000 bp plasmid cut with different restriction enzymes**.

DNA Size Standard	Undigested	Digested with Enzyme 1	Digested with Enzyme 2	Digested with Enzyme 1 & 2
1,000 bp 700 bp 500 bp 300 bp 200 bp	1,000 bp	700 bp 300 bp	1,000 bp	500 bp 300 bp 200 bp

Note that the 1,000 bp fragments from the undigested plasmid and the plasmid digested with Enzyme 2 (see Table 4.4) might not migrate exactly the same distance on an agarose gel. There is a difference in migration among fragments of the same size, depending on whether the fragments are uncut (circular), cut (linear), or uncut and twisted (supercoiled). In addition, fragments that are very similar in size may migrate together as a single band in an agarose gel, making it impossible to distinguish among them. Furthermore, fragments that are very small may not be detected by the DNA stain or may run off the end of the gel.

Tips and Notes

- Remember that a shorter DNA fragment migrates ahead of a longer one and that two DNA fragments of exactly the same size migrate together. Fragments with significant differences in size may migrate less than 1 mm from one another; therefore, it is important to study the actual gel image as well as the migration measurements to determine the size of DNA fragments

Research Question

- Where are the PstI and EcoRI restriction sites on the S3 plasmid?

Objectives

- Generate a standard curve based on the lambda HindIII DNA standard
- Determine the length of all DNA fragments on the provided gel image by comparing them with the standard curve
- Determine the number and relative location of EcoRI and PstI restriction sites on the S3 plasmid

Skills to Master

Refer to Laboratory Skills Assessment Rubric (Appendix E) and Laboratory Notebook Rubric (Appendix F) for more details.

- Record laboratory notebook entries
- Analyze an agarose gel (Activity 4.D)
- Generate a standard curve using a DNA size standard (Activity 4.D)
- Determine DNA fragment sizes using a standard curve (Activity 4.D)
- Construct a plasmid map from restriction digestion data

Student Workstation Materials

Items	*Quantity*
Rulers	4
Semilog graph paper or computer with graphing software	1–4

Activity 4.F Plasmid Mapping

Prelab Focus Questions

1. If a restriction enzyme cuts a circular plasmid twice, how many fragments would you see on a gel?
2. How would you estimate the total number of base pairs in a plasmid by looking at the DNA fragments of the digested plasmid on a gel?
3. If a linear 1 kb DNA fragment has a restriction site that is located 50 bp from one end of the plasmid, what would you expect to see if the digested and undigested DNA samples were run on a 1% agarose TAE gel?

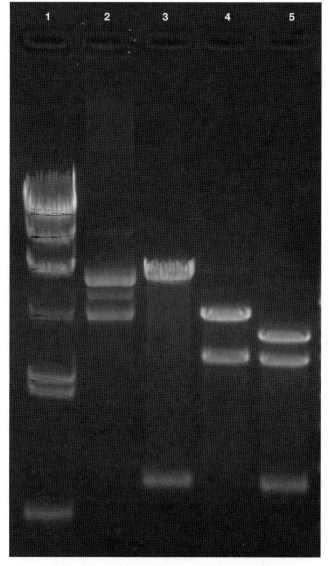

Figure 4.23. **DNA fragments resulting from digestion of suspect 3 (S3) plasmid with EcoRI, PstI, or both restriction enzymes**.

Lane 1, lambda HindIII DNA standard;
Lane 2, undigested S3 plasmid;
Lane 3, S3 plasmid digested with EcoRI;
Lane 4, S3 plasmid digested with PstI;
Lane 5, S3 plasmid double digested with EcoRI and PstI.

Note that the uncut S3 plasmid in lane 2 appears smaller than the EcoRI large band in lane 3 due to supercoiling of the uncut plasmid. In addition, the multiple bands in lane 2 indicate that the uncut plasmid has been nicked, resulting in some unwinding of the DNA fragment.

Activity 4.F Plasmid Mapping

Protocol

1. Generate a standard curve using the lambda HindIII DNA standard in lane 1 of the gel in Figure 4.23 (refer to Activity 4.D if you need help with this). Measure the distance from the bottom of the wells of the gel to the bottom of each of the DNA bands to an accuracy of 0.5 mm. See Figure 4.21 for the sizes of the bands in the lambda HindIII standard.

2. Using the standard curve, determine the length in base pairs of each fragment in lanes 3–5 of the gel.

3. Using the base pair estimates for the plasmid digested with EcoRI, draw a plasmid map in your laboratory notebook with the estimates for the location of the EcoRI restriction sites.

4. Using the base pair estimates for the plasmid digested with PstI, draw a plasmid map in your laboratory notebook with the estimates for the location of the PstI restriction sites.

5. Estimate the total number of base pairs in the S3 plasmid.

6. Using both maps, determine where the two restriction enzymes digest the S3 plasmid at the same time. Draw an overall plasmid map showing all the EcoRI and PstI restriction sites.

Self Assessment

Assess whether you have mastered the skills of this activity using the Laboratory Skills Assessment Rubric in Appendix E.

Assess your experimental write-up in your laboratory notebook using the Laboratory Notebook Rubric in Appendix F.

Postlab Focus Questions

1. How many times does the EcoRI enzyme digest the plasmid? What evidence do you have to support your findings?
2. How many times does the PstI enzyme digest the plasmid? What evidence do you have to support your findings?
3. Estimate the number of base pairs in this plasmid and explain your reasoning. How confident are you of your answer?
4. When the S3 plasmid is digested with both EcoRI and PstI enzymes, how many bands are visible on the gel? How many times is the S3 plasmid cut by the EcoRI and PstI enzymes? What data can you use to explain your conclusions?

Chapter 4 Extension **Activities**

1. Compare various food dyes from similar colored candies to determine whether the same dyes are used.
2. Use NEBcutter to predict the restriction fragments that are generated from the digestion of lambda DNA with various enzymes. Then perform a restriction digestion reaction with lambda DNA using the same enzymes to verify your predictions.
3. Explore the minimum concentration and the minimum or maximum temperature at which a restriction enzyme can still function. (Note: Enzymes will exhibit differences in temperature sensitivity.)
4. Digest the suspect 3 (S3) plasmid with EcoRI and PstI enzymes, and compare the results with the plasmid map from Activity 4.F.
5. Identify another restriction enzyme site on the S3 plasmid map by performing the appropriate digests. (Note: HindIII, ScaI, and PvuII all cut the S3 plasmid at least once.)
6. Ligate DNA fragments into a plasmid and transform the plasmid into bacteria.

Bacterial Transformation and Plasmid Purification

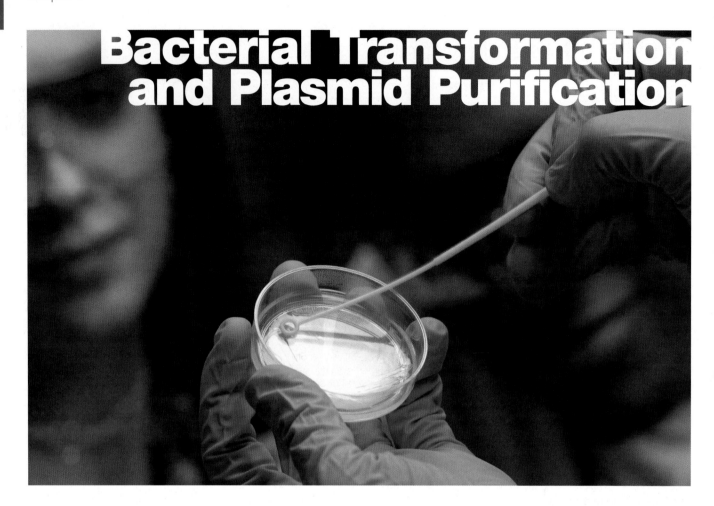

Summary

Discovered by Joshua Lederberg in the 1950s, the small extragenomic circular loops of DNA called plasmids have revolutionized the world of biotechnology and become a fundamental tool in genetic engineering. Plasmids allow a gene to be moved from one organism and expressed in another. Plasmids occur naturally, but are engineered by scientists to serve many research and medical goals. They can be used with bacterial, plant, and even mammalian cells. For example, plasmids are the starting point for recombinant protein production in drug development. In fact, the gene that codes for human insulin was initially genetically engineered into a plasmid and expressed in bacteria in 1978 by scientists at Genentech. This was the first-ever recombinant protein developed for a pharmaceutical application.

Genetic engineering requires that scientists introduce new DNA into cells. Plasmids and viruses are used to introduce DNA using various techniques, such as transformation, transfection, and infection. Scientists amplify (propagate) plasmids by growing them in *E. coli*,

purify them using various purification techniques, and then assess the quantity and quality of the plasmids produced. Production of plasmids on a small scale is commonly referred to as a miniprep. Activities in this chapter include genetically engineering bacteria to express a jellyfish gene, performing minipreps to purify plasmid DNA, and quantitating plasmid DNA.

Chapter 5: Overview

Chapter 5: Laboratory Activities

5.1 History of Bacterial Transformation and Plasmids

Bacteria naturally transfer DNA using a variety of mechanisms, including **transformation, conjugation, and transduction**. These mechanisms were discovered in the last century, starting with transformation. In the 1920s, pneumonia was frequently fatal and many scientists were studying *Streptococcus pneumoniae* (pneumococci) to understand more about these deadly bacteria. A British medical officer named Frederick Griffith found that there were two forms of pneumococci bacteria. One form of the bacteria was encapsulated in a polysaccharide coat and was referred to as smooth, or S, strain. The second form of the bacteria was not encapsulated with a coat and was referred to as rough, or R, strain. The R strain was not lethal and did not cause the disease when injected into mice. In contrast, when the S strain was injected into mice, the mice died. When the S strain was killed by boiling, it no longer killed the mice. However, when the dead S strain bacteria were mixed with live R strain bacteria and then injected into mice, they died and their blood contained live encapsulated S strains. Griffith referred to the new encapsulated bacteria as being "transformed." However, the mechanism by which the bacteria were transformed was not determined until the 1940s when Oswald Avery, Colin MacLeod, and Maclyn McCarty discovered that the "transforming principle" responsible for the phenomenon observed by Griffin was DNA from the dead S strain bacteria being transferred into the live R strain and expressed by the R strain to form a capsule.

Bacterial conjugation (see Figure 5.1) was discovered in 1946 by Joshua Lederberg and Edward Tatum. In bacterial conjugation, *E. coli* shuttle DNA across a bridge that forms between the cells. This is dependent on a fertility, or F, factor, and only F+ bacteria can transfer DNA in this manner.

Transduction was discovered in 1951 by Lederberg and Norton Zinder. Lederberg and Zinder were testing whether salmonella bacteria could conjugate like *E. coli*. They discovered that the bacteria did not need to physically contact each other to transfer genetic information, indicating that bacteria do not build a bridge as they do as in conjugation. After further investigation, these researchers discovered that DNA was transferred by viruses called bacteriophages that infect bacteria.

Plasmids were discovered by scientists investigating the process of bacterial conjugation. In conjugation, it was discovered that it is not chromosomal DNA that is transferred from one cell to another, but a type of extrachromosomal DNA. This discovery was made independently by William Hayes and Lederberg, and, in 1952, Lederberg proposed the name "plasmids" for these extrachromosomal pieces of DNA. In 1961, Tsutomu Watanabe and Toshio Fukasawa were investigating Shigella, a bacterium that was causing dysentery in Japanese hospitals. They found that

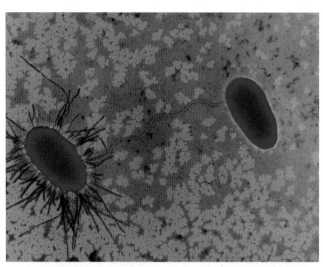

Figure 5.1. **Bacterial conjugation.** An F+ bacteria conjugating with an F– (hairless) bacteria. The genes for the hairs (pili) are coded on the F factor being transferred. The scanning electron microscope image is at 3,645x magnification. Science Photo Library/Dennis Kunkel Microscopy.

some Shigella strains carried plasmids that conferred resistance to antibiotics and that these plasmids could be transferred to non-drug-resistant strains. During this early research, plasmid DNA was assumed to be linear; however, in 1962, Allan Campbell correctly proposed that plasmid DNA was circular.

After the discovery of restriction enzymes in the late 1960s, the concept of using restriction enzymes and plasmids to recombine DNA was proposed by Peter Lobban, a graduate student at Stanford University, Palo Alto, CA. Prior to this proposal, research into plasmids had been curiosity-driven and focused on the interesting natural phenomena. In 1973, Stanley Cohen, Annie Chang, Herbert Boyer, and Robert Helling published a paper describing how to artificially construct a biologically functional plasmid and began the age of recombinant DNA technology. Herb Boyer and venture capitalist Robert Swanson founded Genentech in 1976 and used recombinant plasmids to produce insulin. Roche purchased Genentech in 2009 for $46 billion.

Bacterial transformation and plasmid technology have been used to **clone** countless genes for the production of thousands of proteins, which have affected all aspects of life, from laundry detergents to dairy farms to breast cancer treatments. These technologies have also been used to sequence thousands of whole genomes, from archaea in the deepest parts of the ocean to pufferfish and people.

Plasmid Structure

Most plasmids are extrachromosomal circles of DNA that can be replicated in the cytosol of bacteria. Plasmids are represented using plasmid maps, and the name of each plasmid usually starts with a lower case "p," which stands for plasmid. Plasmid maps can be simple or complex and show different levels of detail (see Figure 5.2 for a simple plasmid map).

Bioethics

A World without Antibiotics?

Antibiotics have revolutionized medical practice. The first major use of commercially manufactured antibiotics was to treat injured soldiers in World War II, a little more than a decade after Alexander Fleming noticed that *Penicillium* mold inhibited bacterial growth and named the first antibiotic penicillin. Prior to the discovery of antibiotics, infections were frequently fatal, and surgeries carried huge risks of infection and subsequent complications. Postoperative infections became rare with the use of antibiotics.

Antibiotic-resistance genes, or R-factors, are often carried on plasmids and evolve in bacteria under selective pressure when antibiotics are used. These R-factors are naturally passed between bacteria, generating bacterial strains that are resistant to antibiotics. The increase in antibiotic-resistant pathogenic bacteria, such as methicillin-resistant *Staphylococcus aureus* (MRSA), is raising the question of how long we have until pathogens evolve to become resistant to all known antibiotics. We have come one step closer to this frightening future with the emergence of gram-negative enterobacteriaceae (gut bacteria), such as pathogenic forms of *E. coli* that are resistant to almost all of the last line of powerful antibiotics. Most worrying is that the resistance gene is carried on a natural plasmid that can be easily transferred between enterobacteriaceae. This gene is called *New Delhi metallo-β-lactamase (NDM-1)* and is named after the Indian city where it was first isolated.

A recently identified contributor to antibiotic resistance is the use of antibiotics intended for humans to treat and prevent diseases among chickens, cattle, pigs, and other livestock to treat and prevent diseases in the animals, and to help them gain weight. The FDA has issued guidelines in recent years to limit the use of antibiotics in farm animals.

What should the medical community do with the small number of antibiotics that remain effective against enterobacteriaceae that carry *NDM-1*? Should the use of these antibiotics be restricted in an attempt to slow down the evolution of pathogenic bacteria that will eventually become resistant to even the most effective antibiotic treatments? Who would be eligible for treatment in such a case and how would such eligibility be regulated in the global community?

Pharmaceutical companies will likely continue the race to produce new antibiotics as bacteria evolve resistance, but there is not much profit to be made investing billions of dollars to develop new antibiotics. People use them only for short-term treatments, and once bacteria evolve to be drug resistant, there is no longer a market for these drugs. What should the government's role? Should we invest in research into nonantibiotic-dependent antimicrobial therapies such as phage therapy? Consider this the next time you have a minor infection — in a world without antibiotics, it could be fatal!

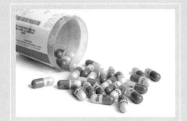

Features of Plasmids

To copy themselves in the cytosol, plasmids must have a starting point for replication, which is called the origin of replication, or "ori" (see Figure 5.2). The ori has recognition sites for DNA polymerases that replicate DNA and enable the plasmid to be cloned (copied) as the bacterium divides. When bacteria divide during fission, each new bacterium gets a share of the plasmids. The number of plasmids per bacterial cell is referred to as the copy number and can vary from 5 to more than 1,000. The ori determines the copy number of plasmids. Plasmids that are present in relatively low numbers in the cells are referred to as low copy number plasmids, while plasmids that are present in much higher numbers are referred to as high copy number plasmids.

Plasmids also contain genes. Many natural plasmids have genes that code for enzymes that confer resistance to antibiotics. For example, β-lactamases are a group of enzymes that have been found to break down antibiotics that have a β-lactam ring as part of their molecular structure, including penicillin and ampicillin. The *β-lactamase* gene is abbreviated as *bla* or *amp^r* (for ampicillin resistant) (see Figure 5.2). If bacteria contain a plasmid with an antibiotic-resistance gene, they will grow in the presence of that antibiotic, while bacteria without the plasmid will not grow. Using an antibiotic to allow only bacteria that contain a plasmid to grow is called **selection**. Early genetic engineers used the genes and regulatory sequences of natural plasmids as the basis for recombinant plasmids to produce recombinant proteins, including insulin. Antibiotic- resistance genes were vital to these recombinant plasmids because they provide a mechanism for scientists to separate bacteria containing recombinant plasmids from those that do not.

Figure 5.2. **pGLO plasmid map**. The plasmid map shows the location of the origin of replication (ori) and the *green fluorescent protein (GFP), β-lactamase (bla),* and *araC* genes.

Genes are found on either strand of the plasmid DNA double helix. RNA polymerases read in opposite directions for genes on opposite strands. To indicate the strand the genes are on, a plasmid map often has arrows. If the arrows point in opposite directions, the genes are on opposite strands. If the arrows point in the same direction, the genes are on the same strand of the double helix.

To express a gene, a plasmid must have a **promoter** DNA sequence before the coding region of the gene. A promoter is a sequence of DNA that is located before the protein-coding sequence, and provides a landing site for RNA polymerase so that the gene can be transcribed and the protein can be made. RNA polymerase also needs a signal to stop transcribing the gene so that the messenger RNA (mRNA) is the correct length. This signal is called a **terminator** and is a DNA sequence that is present at the end of a gene in a plasmid.

Using Plasmids in Biotechnology

Scientists have used the natural abilities of plasmids and engineered them to create useful products. Plasmids are often manipulated to perform two basic functions: to express recombinant proteins and to carry or house genes that have been cloned (see Figure 5.3). A plasmid used to express recombinant proteins is referred to as an expression plasmid, while a plasmid used to house genes is referred to as a cloning plasmid. A cloning plasmid is required because a stretch of linear DNA containing a gene is not very stable. Plasmids are very stable, and millions of copies can be made by growing them in bacteria. Also, DNA can be manipulated much more easily in a plasmid.

To insert a specific gene into a plasmid, the gene of interest first needs to be obtained using a process called **cloning**. Cloning involves making an exact copy of the DNA sequence of the gene. (The term cloning is also used in cell biology when an exact copy of a cell is made.) Historically, genes were cloned using restriction enzymes that break DNA at specific locations to cut out a target gene from the original organism. Now, the **polymerase chain reaction (PCR)** is used to make copies of the gene directly from the genome of the original organism. PCR is discussed further in Chapter 6. The plasmid is then cut with restriction enzymes to open the circle of DNA. The gene of interest may be blunt-ended or have sticky ends from the restriction digest (see Figures 4.7 and 4.8). The open ends of the plasmid must also be blunt or have complementary sticky ends; therefore, they should be cut with the same restriction enzyme(s). The gene of interest is then linked, or ligated, into the plasmid using ligase. During ligation, the sugar phosphate bonds are reformed between the base pairs and the plasmid again forms a circle of DNA.

An entire gene is not always cloned from the original organism into the final plasmid all at once. The gene can be cloned in pieces using multiple steps and multiple plasmid intermediates.

Modern lab-generated plasmids are designed to make cloning as easy as possible. These plasmids have an area called a **multiple cloning site (MCS)**, which is a string of unique restriction enzyme recognition sites. The MCS is used to open up the plasmid so that it is ready to receive the gene of interest. Because there are many restriction sites in the MCS, it is likely that there will be a site that matches the sticky ends of the DNA fragment or the gene of interest. If an expression plasmid is used, the MCS will be downstream of the promoter and will provide options for getting the gene of interest into the plasmid in the correct direction and in frame so that it is transcribed and translated correctly.

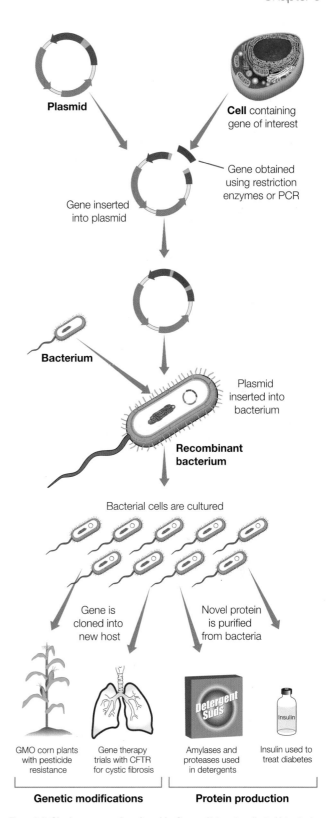

Figure 5.3. **Cloning genes using plasmids**. Genes of interest are ligated into cloning or expression plasmids to make more copies of the gene or to express proteins, respectively. Recombinant proteins can be used in detergents and as pharmaceutical drugs. Cloned genes can be used to make genetically modified crops and for gene therapy trials.

If a gene is to be expressed from a plasmid (that is, a protein is to be made from the gene), the DNA sequence that encodes the protein needs to be cloned downstream of a promoter and upstream of a terminator. The protein-coding sequence also needs to be cloned in frame, meaning that the first **codon** of the mRNA is positioned to be read correctly by the **ribosome** and is not shifted by one or two bases, which would result in a different amino acid sequence. Transcription and translation are discussed further in Chapter 7. These details are not important if the gene is only being housed in a cloning plasmid and is not to be expressed.

Transcriptional Regulation of Plasmids

Transcribing and translating genes takes up energy; therefore, only necessary proteins are expressed by cells at any one time. This means that cells must regulate when and to what level they transcribe their genes. Some genes need to be expressed all the time, while others need to be expressed only at certain times or in certain environments. Genes that are always expressed are referred

to as **constitutive** genes, while genes that are transcribed only when needed are referred to as **facultative** genes. Bacteria regulate expression of some of their facultative genes using operons. **Operons** are naturally occurring control units in bacterial chromosomal DNA that consist of one promoter, multiple genes, and a single terminator. One mRNA molecule that encodes multiple proteins is produced from an operon. This means that all the proteins encoded by the operon are made in exactly the same proportions and are all made at the same time. This is beneficial for bacteria that can instantly produce multiple proteins with a single switch for fast use of a new food source. Only prokaryotes have operons. Eukaryotes have more complex transcriptional regulation in which each gene is thought to be regulated individually.

The mode of action of the *lac* operon was discovered by François Jacob and Jacques Monod in 1961. This discovery was important because it was the first time that scientists found out how genes are regulated. The *lac* operon controls the production of three enzymes

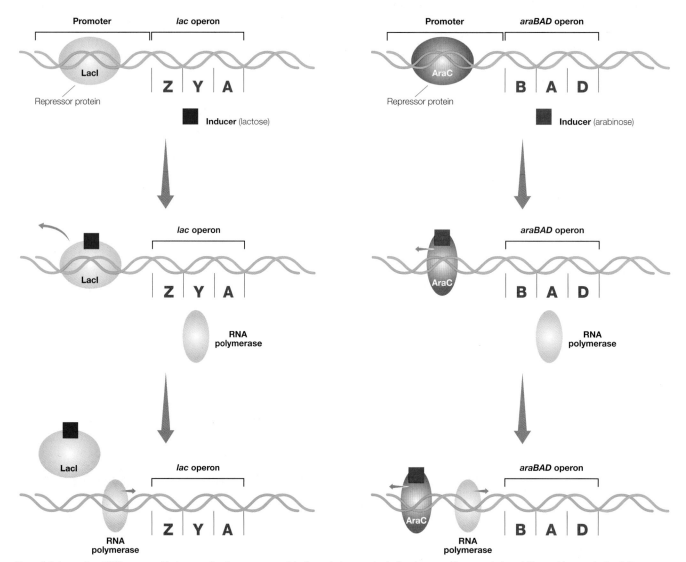

Figure 5.4. *lac* and *araBAD* operons. Each operon has three genes regulated by a single promoter. In the absence of the sugar inducer (either arabinose or lactose), the repressor protein binds near the promoter and blocks the RNA polymerase from binding to the promoter. In the presence of the sugar, the repressor protein either moves or changes shape and allows the RNA polymerase to transcribe the genes.

that are involved in the metabolism of lactose in *E. coli*, while the *araBAD* operon controls the production of enzymes involved in the metabolism of another sugar, arabinose. Operons are controlled by **repressor** proteins that are encoded by an additional gene. Repressor proteins bind to the operator region near the operon and block RNA polymerase from transcribing. To turn the operon on, the sugar (lactose or arabinose, depending on the type of operon) binds to the repressor and relieves the block, thereby allowing transcription (see Figure 5.4). The sugar is called an **inducer**. Inducible operons can turn on genes that are normally off, while repressible operons can turn off genes that are normally on.

Genetically Engineering the pGLO Plasmid

Scientists took inducible operons from bacterial chromosomal DNA and genetically engineered them into plasmids. In the case of the pGLO plasmid, a team of scientists made a plasmid called pBAD18 by cloning the *araBAD* promoter and the gene for the AraC repressor protein into a parental plasmid containing an ori

and the antibiotic-resistant *bla* gene. Another team cloned the *green fluorescent protein* (*GFP*) gene from the jellyfish *Aequorea victoria* into a different plasmid. A third group of scientists then used restriction enzymes to clone the *GFP* gene into the pBAD18 plasmid downstream of the *araBAD* promoter. As a result, the *araBAD* operon and the AraC protein regulate the expression of GFP. Using this system GFP is expressed only in the presence of arabinose (see Figure 5.5).

Plasmids for Eukaryotic Expression

Bacteria and eukaryotic cells use different types of promoters. Therefore, there are different expression plasmids depending on the plasmid's uses. Plasmids are generally made and amplified (propagated) in bacteria. If the plasmids are destined to be used in eukaryotic cells, they must carry features that allow them to be used in both bacteria and eukaryotes. Plasmids that have both properties are called **shuttle plasmids**. For example, a yeast shuttle plasmid would have features that allow for replication and selection in both

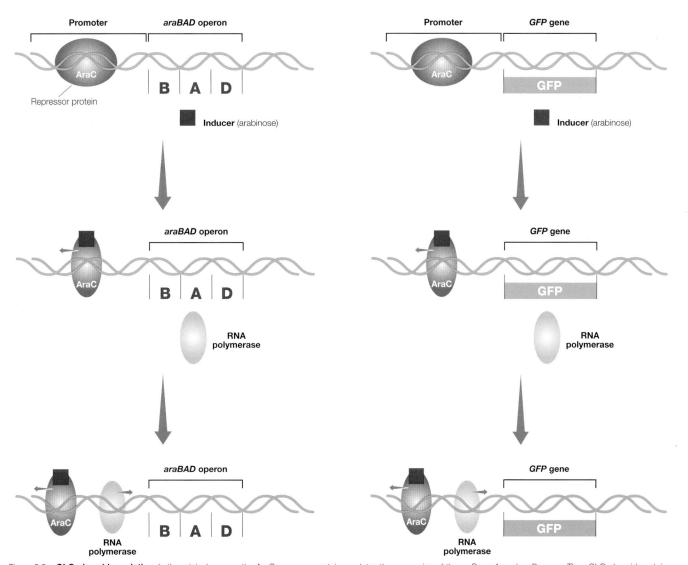

Figure 5.5. **pGLO plasmid regulation**. In the original operon, the AraC repressor protein regulates the expression of the *araB*, *araA*, and *araD* genes. The pGLO plasmid contains a modified *araBAD* operon that expresses GFP instead.

E. coli and yeast cells. For propagation in *E. coli*, the yeast shuttle plasmid would have an origin of replication and a selectable marker (for example, the *bla* gene). For expression in yeast, the yeast shuttle plasmid would have the sequences necessary for replication in yeast and a different selectable marker. Since the gene of interest contained in the plasmid is destined to be expressed in yeast, a yeast promoter would be used to drive the gene.

Shuttle plasmids also exist for expression in mammalian cells. For propagation in *E. coli*, the mammalian shuttle plasmid would contain an ori and a *bla* gene, similar to the yeast shuttle plasmid. For expression in mammalian cells, the main shuttle plasmid would have a selectable marker, such as a neomycin-resistance gene, and a promoter from a mammalian virus to drive gene expression, such as the CMV promoter from the human cytomegalovirus. Viral promoters are ideal for driving expression in many cell types because viruses evolved to use the resources in the cells they infect. Mammalian shuttle plasmids also have a polyadenylation sequence at the end of the gene of interest, so a poly A tail is added to the mRNA when it is transcribed. Polyadenylation is discussed in Chapter 7. Interestingly, mammalian shuttle plasmids do not replicate independently in mammalian cells; they are either expressed transiently or stably integrated into the mammalian genome.

A naturally occurring plasmid called Ti (for tumor-inducing) occurs in plants. This plasmid is derived from a pathogenic plant bacterium called *Agrobacterium tumefaciens* that causes crown gall disease (see Figure 5.6). When a plant is infected with *A. tumefaciens*, a specific fragment of the plasmid DNA called T-DNA gets integrated into the plant's genomic DNA (gDNA). In the laboratory the Ti plasmid is engineered to remove the tumor-inducing genes and foreign genes are inserted into the T-DNA. The modified plasmid is then propagated in *A. tumefaciens* using the natural regulatory sequences of the bacterium, and the genes of interest are engineered with plant viral promoters driving their expression. Terminator sequences are also required and are frequently cloned from genes that are found in the natural Ti plasmid.

Figure 5.6. **Crown gall disease**.

5.2 Transforming Cells

To genetically engineer a cell, foreign DNA must be placed into the cell. The cell receiving the foreign DNA is said to be transformed (see Figure 5.7). There are many ways to introduce foreign DNA into a cell, depending on the cell type and the purpose of the experiment. The most common methods used to transform bacteria are calcium chloride transformation and **electroporation**. The success of a transformation procedure is measured by determining the **transformation efficiency**, which is calculated by quantitating the number of bacteria that were successfully transformed per microgram (μg) of DNA. The number of transformed bacteria is determined by counting the number of colonies (CFU) the bacteria form when spread on an agar plate.

In calcium chloride transformation, actively dividing bacteria are repeatedly washed in an ice-cold calcium chloride solution. The calcium chloride makes the cells "chemically competent" and more

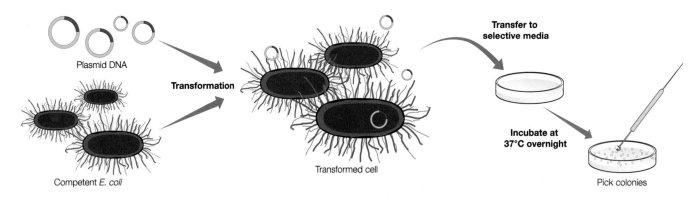

Figure 5.7. **Bacterial transformation**. Plasmids are mixed with the bacteria and enter cells through a variety of techniques. The cells are then plated on a selective medium so that only the bacteria that received plasmids will grow.

permeable to DNA. Chemically **competent cells** can be flash frozen and stored at –80°C for several months. Chemically competent bacteria produced in the laboratory have transformation efficiencies ranging from 10^4–10^6 CFU/µg of DNA. Alternatively, highly competent cells with efficiencies of 10^7–10^9 CFU/µg of DNA are commercially available.

To transform chemically competent bacteria, plasmid DNA is added to the cells and the mixture is incubated on ice. The cells are heat shocked in a 42°C water bath for 50 seconds and then immediately placed back on ice. Heat shock is thought to temporarily open up, or fracture, the cell membranes, allowing the plasmid DNA to enter the bacterial cells. Non-selective growth medium is then added to the cells and they are incubated and allowed to recover at 37°C with shaking for 30 minutes to 1 hour. The cells are then plated and grown overnight on a selective medium. A modified version of the calcium chloride transformation method will be used in Activities 5.A and 5.B.

Electroporation, another common transformation method, uses electricity to disrupt cell membranes and temporarily increase the permeability of the cells, which allows plasmid DNA to enter the cells. Electroporation of bacteria is described in How To... Transform *E. coli* Using Electroporation. Electroporation can be used to transform both bacteria and eukaryotic cells.

Eukaryotic cells can also be transformed using a method called **transfection**. In chemical-based transfection, plasmid DNA is first sealed in tiny oil bubbles called lipid vesicles and incubated in the culture medium. The vesicles then fuse with the cell membranes and deliver the DNA into the cells.

Plants have cell walls that create a barrier to most delivery methods. To penetrate plant cell walls, biolistics is the method of choice. Biolistics uses small gold or tungsten particles coated with DNA that are shot into cells under helium pressure. A handheld device called a Helios® gene gun was developed for delivery of DNA into plant cells. A larger system, the PDS-1000/He™ was also developed for transfecting cells, tissues, or organelles (see Figure 5.8). Biolistics is also used to deliver DNA into other eukaryotic cell types, including animal cells and fungi.

Figure 5.8. **Biolistics**. Left, the Helios gene gun. Right, the PDS-1000/He system. Both systems are used to shoot microparticles coated with DNA into various cell types.

Transform *E. coli*
Using Electroporation

Electroporation uses an electric pulse between two metal plates to disrupt cell membranes, increasing their permeability. Bacteria, yeast, and mammalian cells are commonly transformed by this method. Electroporation of bacteria is much faster and often more efficient than calcium chloride transformation and must be done in an electroporator, such as Bio-Rad's MicroPulser™ electroporator (see Figure 5.9).

Electroporation requires electrocompetent bacteria. Electrocompetent cells are commercially available from various suppliers. Alternatively, electrocompetent cells can be prepared by washing actively growing *E. coli* cells multiple times with ice-cold nonionic solutions such as 10% glycerol. (Electroporation requires that all reagents be ice-cold.) Electrocompetent cells can be flash frozen and stored at –80°C.

To electroporate *E. coli*, thaw the electrocompetent *E. coli* cells on ice and pipet them into chilled microcentrifuge tubes on ice. Add salt-free plasmid (1–50 ng/µl) and incubate for 1 minute. It is important to add the smallest volume possible of DNA since buffer from the DNA can affect the resistance of the electroporation solution and reduce transformation efficiency. Transfer the mixture to a chilled electroporation cuvette and keep the cuvette on ice until ready to electroporate.

Place the cuvette into the electroporator and pulse once. Immediately remove the cuvette, and add 1 ml of SOC medium. (SOC medium is a rich culture broth similar to LB nutrient broth that nourishes the cells.) Transfer the culture to a culture tube and incubate at 37°C with shaking for 1 hour. Finally, spread the culture on LB agar with the appropriate antibiotics and incubate overnight at 37°C.

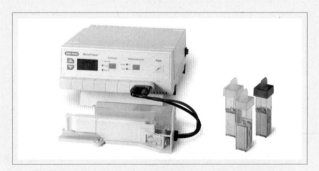

Figure 5.9. **Bio-Rad's MicroPulser electroporator and electroporation cuvettes**. Electroporation cuvettes have two metal plates on either side of a narrow sample space. During electroporation, an electrical pulse is passed between the plates through the sample of cells.

Genes can also be delivered into plants using a genetically modified Ti plasmid from *A. tumefaciens* that integrates foreign genes into the plant genome. Plant cells are transformed by being infected with cultures of Agrobacterium containing genetically engineered Ti plasmids. For example, *Arabidopsis thaliana*, a plant frequently used in the laboratory, can be genetically engineered by dipping the flowers in a solution of Agrobacteria. This is sufficient to transform the Arabidopsis seeds, which can then be propagated in a selective medium. Alternatively, Ti plasmids can be introduced into plant cells using a biolistic system.

Viruses can also be used to deliver genes into cells. A process called **transduction** is used by bacteriophages to deliver genes into bacterial cells. (Bacteriophages, or phages, are viruses that attack bacteria.) Scientists genetically engineer bacteriophages, such as lambda phage, to encode the DNA of interest and then infect bacteria with the phage. The infected bacteria then express the recombinant DNA. Baculoviruses, which naturally infect insect larval cells, are often engineered to express recombinant genes. Baculoviruses are widely used to infect insect cells for recombinant protein production. Retroviruses are similarly used in mammalian cells. These RNA viruses convert their RNA genome into DNA and then insert this DNA into the host cell's genome.

Selection of Transformed Cells

Not every cell in a transformation reaction will be transformed and contain the plasmid. Depending on the efficiency of the transformation, 1 in 100 cells or 1 in 1 million cells may take up the plasmid. How do you find that one cell in the mass of untransformed cells? The process of identifying the transformed cells is called selection. Since most plasmids contain an antibiotic resistance gene, selection uses an antibiotic to identify bacteria that contain the plasmid. Cells that are transformed contain the plasmid; therefore, they can express the gene that confers resistance to the antibiotic. When the cells are grown in the presence of the antibiotic, only those that contain the plasmid will grow.

Transformation is used not only to propagate whole plasmids but also to propagate newly formed plasmids. New plasmids are usually generated by ligating a new DNA fragment into the MCS of an existing plasmid backbone. Scientists have devised systems to select for plasmids that contain the new fragment of DNA over plasmid backbones that recircularized without the new DNA fragment being inserted (see Figure 5.10). For example, the pUC18 plasmid contains the *lacZ* gene, which encodes the enzyme β-galactosidase. β-galactosidase cleaves X-gal, an artificial substrate and lactose analog. In the presence of β-galactosidase, X-gal turns blue and the bacterial colonies containing the plasmid also turn blue. Scientists have carefully engineered a MCS into the *lacZ* gene without disrupting its function. However, once a DNA fragment is ligated into the MCS, it disrupts the *lacZ* gene. This means that β-galactosidase cannot be expressed and the cells do not turn blue in the presence of X-gal. So bacteria containing the recircularized plasmid will be

blue, while bacteria containing plasmids that have ligated a DNA fragment will be white (see Figure 5.10). This selection method is referred to as blue-white screening. Developed in the early 1980s, this method is still commonly used today. However, this method is not perfect and often produces white colonies that are actually negative; such colonies are referred to as false positives. Blue-white screening also requires that the bacterial host not have a functioning chromosomal *lacZ* gene. For example, *E. coli* strain HB101 has a functioning *lacZ* gene and cannot be used for blue-white screening.

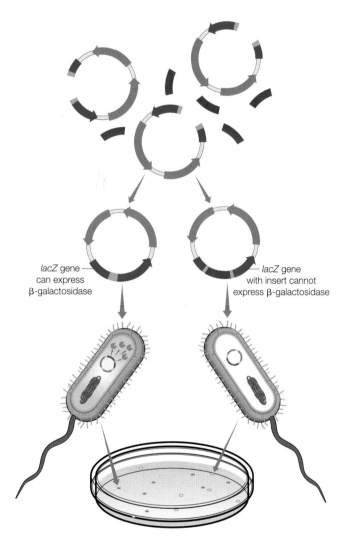

lacZ gene can express β-galactosidase

lacZ gene with insert cannot express β-galactosidase

Figure 5.10. **Blue-white screening**. The pUC18 plasmid is used to confirm that genes of interest have been cloned successfully. The resulting bacterial colonies are white if they have been successfully transformed with the gene of interest. Colonies that do not contain the gene of interest turn blue.

Positive selection is another method used to identify bacterial colonies that have been transformed with the gene of interest. This method prevents unsuccessful ligations from growing altogether. For example, the pJET1.2 plasmid has been engineered to insert a MCS into the *Eco47IR* gene without disrupting the function of the gene. The *Eco47IR* gene codes for the Eco47IR restriction enzyme, which is toxic to *E. coli* when expressed. Ligation of DNA fragments into the plasmid inactivates the gene and allows bacteria to grow, while plasmids that recircularize express the toxic gene and kill their bacterial hosts (see Figure 5.11). Since pJET1.2 cannot be propagated in *E. coli* due to the toxicity of the *Eco47IR* gene, the plasmid is available commercially only as a pre-opened or linearized plasmid.

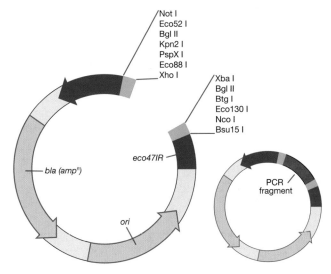

Figure 5.11. **pJET1.2 plasmid**. In the presence of an insert, the gene for the Eco47IR restriction enzyme is disrupted. If no insert is present, the plasmid recircularizes and the restriction enzyme is expressed, thereby digesting the bacterium's own DNA and killing the bacterium.

Transformation Efficiency

Transformation efficiency is a measurement of the number of transformed bacteria that can grow into colonies (CFU) per µg of plasmid DNA. Transformation efficiency depends on the bacterium's growth phase, the amount of plasmid used and whether it is supercoiled, the type of bacterial cell, the temperatures used to heat shock, and the method used to transform the cells. Scientists often purchase commercially available bacterial cells that have been made either chemically competent or electrocompetent. These cells need to be stored at −80°C. Not only are these competent cells conveniently packaged but they also have transformation efficiencies that are much greater than can be obtained using chemically competent cells prepared in the laboratory (10^8 vs. 10^5 CFU/µg of DNA).

Electroporation is usually more efficient than calcium chloride transformation and transformation with plasmid DNA is much more efficient than transformation with a ligation reaction. Successful transformation with a ligation reaction requires competent cells that have high transformation efficiencies (>10^6 CFU/µg of DNA), while intact plasmid DNA can be effectively transformed using bacteria of low competency. In a ligation reaction, the majority of the opened plasmids will not ligate fragments, resulting in a very small number of correctly reformed plasmids available to transform bacteria.

To calculate transformation efficiency, the following information is required: the number of CFU, the volume of bacterial culture plated

Biotech on the Pharm

Using microbes to manufacture drugs like insulin can be expensive. The technique requires large cell culture facilities as well as systems for extracting and purifying the drugs, and it often results in relatively low yields. In addition, some protein drugs cannot be made using microbes because they require modifications that occur after translation, using processes that only animal cells possess. For these reasons, scientists have looked to other methods and systems for producing therapeutic drugs.

One system that has already given great results is the mammary gland — a gland whose purpose is to produce a solution (milk) filled with proteins. Some of the proteins that can be created in milk are difficult or impossible to create in tissue culture–based systems. The Massachusetts-based GTC Biotherapeutics Company (now rEVO Biologics) genetically engineered goats to produce foreign proteins (such as therapeutic drugs) in their milk. In 2009, the U.S. Food and Drug administration (FDA) approved the sale of the first such therapeutic drug, ATryn, a recombinant human antithrombin protein produced in goat milk and used to break up blood clots.

GTC Biotherapeutics used PCR to clone the antithrombin gene from human DNA and regulated its expression by coupling it to a goat milk protein promoter. This ensures the drug is made only in the goat mammary glands and is secreted into the milk. GTC uses goats because goats reproduce more rapidly than cattle and produce more milk than rabbits or mice. To create the transgenic goats, GTC transfected the expression vectors into single-celled goat embryos, which were then implanted into female goats that carried the embryos to term. The first transgenic goats that resulted are called founder animals, and they are bred to produce more transgenic goats. All animals produced in this manner are tested by PCR to confirm the presence of the foreign gene. A single mature female goat produces 3 liters of milk per day and about 3 kg of recombinant protein per year, about ten times the amount that could be produced in cell culture.

Though it is also expensive to make transgenic goats, the cost associated with making recombinant antithrombin from them is significantly less than with traditional methods. The drug is produced only in mammary glands and does not cause any known health problems for the goats. But how might making transgenic animals and using them as drug factories be controversial?

Photo courtesy of the Agricultural Research Service, USDA

as a fraction of the overall final transformation volume, and the mass of DNA used in the transformation. The transformation efficiency can then be calculated as the number of CFU per µg of DNA.

For example, 50 ng of plasmid DNA is transformed into a final transformation volume of 500 µl, and 10 µl of this volume is spread on an agar plate. Let's assume that 60 CFU are observed on the agar plate. **Note:** 1 µg is 1,000 ng, so 50 ng = 0.05 µg of DNA.

1. Count the number of colonies growing on the LB/ampicillin (LB/amp) agar plate. In this case, the CFU is 60.

2. Determine the amount of plasmid DNA (in µg) spread on the LB/amp agar plate. In this example, only 10 µl of a 500 µl transformation was spread on the plate.

$$\text{DNA spread on the plate (µg)} = \frac{\text{Volume spread (µl) x DNA in transformation (µg)}}{\text{Total volume of transformation (µl)}}$$

$$= \frac{10 \text{ µl x 0.05 µg}}{500 \text{ µl}}$$

$$= 0.001 \text{ µg}$$

3. Calculate the transformation efficiency by dividing the number of colonies on the plate (in CFU) by the amount of DNA (in µg) spread on the plate.

$$\text{Transformation efficiency} = \frac{60 \text{ CFU}}{0.001 \text{ µg}}$$

$$= 60,000 \text{ CFU/µg}$$

$$= 6.0 \times 10^4 \text{ CFU/µg}$$

5.3 Plasmid Purification and Quantitation

Once a plasmid has been transformed into competent bacteria and the cells have been selected on an agar plate, the next step is to grow the transformed bacteria in a liquid culture and purify the plasmid out of the bacteria. The bacteria most commonly used to produce plasmids are *E. coli*. The process of growing a small culture of bacteria and purifying the plasmid from the bacteria is commonly referred to as a **miniprep**. A miniprep serves two purposes: 1) it purifies the plasmid out of the bacteria and away from the bacterial **genomic DNA** (**gDNA**), the full complement of DNA within a cell that makes up its genome, enabling analysis of the plasmid to check if a ligation was successful, and 2) it amplifies the plasmid and provides more DNA for subsequent experiments.

Growing Bacteria in a Liquid Culture

To start the liquid culture, a single colony is inoculated into a culture medium containing the appropriate antibiotic. For optimal growth, *E. coli* need nutrients, oxygen, and a warm temperature. For example, LB broth provides the necessary nutrients, while a shaking incubator provides warmth and oxygenation (see Figure 5.12). It is important to oxygenate the bacteria by shaking them to help the aerobic *E. coli* grow. If the culture is not shaken, a poor plasmid yield may result.

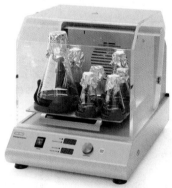

Figure 5.12. **A shaking incubator** has an orbiting platform with clamps to hold various sizes of liquid culture flasks and tubes. The chamber is kept at a constant temperature and the platform orbits to swirl the liquid to increase oxygenation.

Bacterial growth follows a regular growth cycle with four phases (see Figure 5.13). After the culture medium is inoculated with the bacteria, there is a lag phase before the cell numbers begin to increase. After the lag phase, the cells begin an exponential growth phase. Rates of division depend on the growth conditions and the bacterial species. Growth rates of species vary greatly. For example, under optimal conditions, *E. coli* divide every 17 minutes. In contrast, *Mycobacterium tuberculosis*, the bacteria that cause tuberculosis, divide approximately every 15 hours. When the nutrients in the growth medium are consumed, the culture enters a stationary phase, during which the number of bacteria remains the same. Finally, the culture enters the death phase, in which the number of cells decreases. Cells are typically harvested late in the growth phase, when the bacterial number is high and the cells are still healthy and dividing. For minipreps, *E. coli* are optimally grown for 16–24 hours at 37°C with shaking at 200–250 rpm.

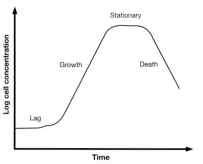

Figure 5.13. **Bacterial growth curve**. Bacterial growth follows a regular cycle with four phases.

Purifying Plasmid DNA from a Culture

Transformed bacteria that have grown overnight are harvested by centrifugation, and the plasmid DNA is purified. Purification is a multistep process designed to separate plasmid DNA from gDNA and other cellular components (see Figure 5.14).

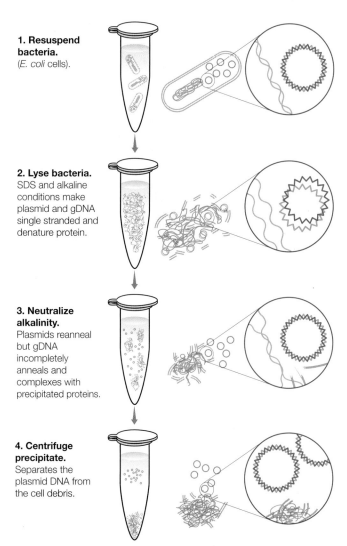

1. Resuspend bacteria.
(*E. coli* cells).

2. Lyse bacteria.
SDS and alkaline conditions make plasmid and gDNA single stranded and denature protein.

3. Neutralize alkalinity.
Plasmids reanneal but gDNA incompletely anneals and complexes with precipitated proteins.

4. Centrifuge precipitate.
Separates the plasmid DNA from the cell debris.

Figure 5.14. **Plasmid purification**. The bacterial pellet is first resuspended. When the bacteria are lysed, gDNA and plasmid DNA become single-stranded, but the two single strands of the plasmid DNA remain linked. Once the lysate is neutralized the plasmids reanneal but the gDNA aggregates with precipitated proteins. When the lysate is centrifuged, the gDNA and precipitated proteins are pelleted, leaving the plasmid DNA in solution in the supernatant.

1. The bacteria must be removed from the LB broth. After the culture is grown, the culture is cloudy due to the billions of suspended bacteria. To separate the bacteria, the culture is spun in a **microcentrifuge** (see How To... Use a Microcentrifuge on the next page). The centrifugal force pulls the bacteria to the bottom of the tube to form a dense mass referred to as a **pellet**. Following centrifugation, the culture liquid becomes clear because it no longer contains floating bacteria. The liquid is removed by pipetting or using a vacuum source. (The liquid fraction resulting from centrifugation is referred to as the **supernatant**.) The pellet is resuspended in a buffered solution.

2. The bacteria must be broken open, or lysed, to release the cell contents (including the plasmid DNA), but the treatment cannot be so harsh that all the cell components are broken down. Lysing bacteria requires disrupting or weakening both the cell wall and the cell membrane. Bacteria are lysed with an alkaline solution (pH 12–12.5) containing sodium hydroxide, sodium dodecyl sulfate (SDS), and ethylenediaminetetraacetate (EDTA). SDS, an ionic detergent, disrupts the cell membranes and denatures proteins to release DNA from the cell. EDTA, a chelating agent, removes magnesium ions, destabilizes the bacterial cell wall, and inactivates enzymes such as DNases that might destroy DNA. When the cell wall is weakened, the cell bursts and the alkaline conditions break the hydrogen bonds that link the strands of double-stranded DNA. This causes the double helix to unwind and the two strands to separate. Hence, the gDNA is denatured but the molecules remain in high molecular–weight form since the strands themselves are intact. The hydrogen bonds connecting the strands of the smaller circular plasmid DNA molecules are also broken under alkaline conditions; however, the single strands stay linked together because plasmid DNA molecules are circular and supercoiled.

3. A low pH, high salt solution, such as 3–5 M potassium or sodium acetate, pH 5.5, neutralizes the alkaline pH of the lysate. The high salt concentration causes proteins to precipitate, and the gDNA partially renatures and aggregates with the precipitated proteins due to its length. Because plasmids are small and their single strands remained linked in the lysis step, they reanneal, or reconnect, into soluble, double-stranded molecules.

4. The cellular debris, including gDNA and proteins, is removed by centrifugation and the supernatant contains the soluble plasmid DNA.

The cleared lysate has a high salt concentration that would interfere with subsequent experiments, so the final step usually involves purification to remove salts and other minor contaminants. This step has the added benefit of concentrating the DNA into a much smaller volume. Historically, alcohol precipitation was used to desalt and concentrate the DNA but many researchers now use chromatography columns from commercial kits to purify the plasmid DNA.

How To...

Use a Microcentrifuge

A variable-speed microcentrifuge is designed to hold microcentrifuge tubes. The number of holes, or wells, in the rotor (the part that rotates) determines the number of tubes a microcentrifuge can accommodate, typically 18–24 (see Figure 5.15). The speed of a microcentrifuge usually ranges from 1,000–14,000 revolutions per minute (rpm) and can be digitally or manually adjusted. The duration of centrifugation is set by a dial or digital display. There is also a pulse or quick spin button. The pulse button is used if a brief spin of 5 or 10 seconds is required to pull reagents to the bottom of a tube.

When using a microcentrifuge, it is important to balance the microcentrifuge tubes so that each tube has another tube of the same mass on the exact opposite side of the rotor. The balance tube can have the same components as the first tube or can contain water to make it the same mass as the experimental tube. If necessary, the matching tubes should be weighed to ensure that they are actually the same mass.

The speed settings for most microcentrifuges are in rpm, yet many protocols require speeds in relative centrifugal force (RCF), which is the number of times the force of gravity (g force) the centrifuge generates as it spins. RCF is measured in units of g and is constant among centrifuges. RCF is dependent on the radius of the rotor; therefore, the rpm to produce a specific RCF varies among centrifuges since the radii of rotors vary. The rpm values can be converted to RCF or vice versa using either the formula below or an Internet converter. Alternatively, a nomograph, a form of graphical calculator, can be used (see Figure 5.16).

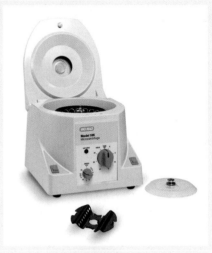

Figure 5.15. **Bio-Rad's Model 16K microcentrifuge**.

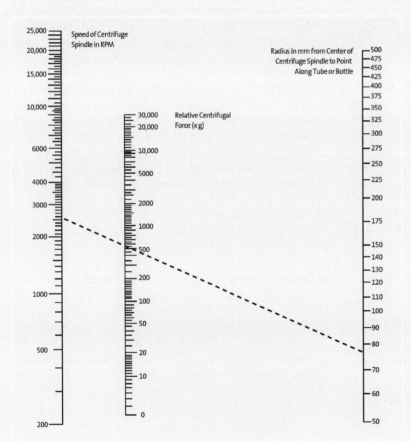

$$RCF\ (g) = (1.12 \times 10^{-6})(s)^2 r$$

or

$$S\ (rpm) = \sqrt{\frac{RCF\ (g)}{1.12 \times 10^{-6} \times r}}$$

Where r is the radius of the rotor, measured in millimeters from the center of the rotor to the bottom of the microcentrifuge tube, and s is the speed of the rotor in rpm.

Figure 5.16. **Nomograph.** A nomograph is a common chart used to convert from RCF to rpm and vice versa. A straight line is drawn through two points on the nomograph: the radius of the centrifuge rotor and the desired RCF. The line will cross the rpm scale at the rotor speed necessary to generate the desired RCF. Chart provided by Corning Incorporated, Life Sciences, Lowell, MA USA. (corning.com/lifesciences)

Alcohol Precipitation: The cleared lysate is mixed with absolute ethanol, incubated at –20°C to precipitate the DNA, and then centrifuged. The supernatant is removed, and the pellet of DNA is washed in 70% alcohol to remove salts, centrifuged again, and then air dried. Finally, the pellet is hydrated in water or in Tris-EDTA (TE) buffer.

Commercial Column Purification: Following lysis, the supernatant containing the plasmid DNA is added to a chromatography column. DNA binds strongly to silica in high salt conditions. Contaminants, such as salts, do not bind to silica and are washed from the column when a column wash solution that contains ethanol is added and the column is spun in a centrifuge. The DNA remains bound to the column as the contaminants flow through. The DNA is released (eluted) from the column in a low salt elution solution.

DNA Quantitation

Once plasmid DNA has been purified, the concentration of the DNA needs to be determined. Plasmid purification can yield very different quantities of DNA depending on whether a high or low copy number plasmid is used.

DNA can be quantitated in different ways depending on the sample and the technology available. Three common ways of DNA quantitation are gel quantitation, spectrophotometric quantitation, and fluorometry. Gel quantitation provides an estimate of DNA concentration, while spectrophotometric and fluorometry DNA quantitation provide more accurate measurements. DNA quantitation using real-time PCR will be discussed in Chapter 6.

Gel Quantitation

The concept behind gel quantitation is that the intensity (darkness) of a band of DNA on a gel is proportional to the amount of DNA in the band. DNA stains attach to DNA proportionally so DNA bands of similar intensity should contain similar quantities of DNA. An estimate of DNA quantity can be obtained by running DNA of unknown concentration alongside a DNA standard with known quantities of DNA in the bands. The estimate can then be used to calculate the DNA concentration of the sample. Mass rulers with known concentrations of DNA in each band are commercially available, such as Bio-Rad's precision molecular mass ruler that has DNA fragments of 1,000, 700, 500, 200, and 100 bp. Each band has a corresponding mass dependent on the volume loaded into the gel. For example, when 5 µl of the mass ruler is loaded, the 1,000 bp band contains 100 ng of DNA, the 700 bp band contains 70 ng of DNA, and so on (see lane 2 of the gel in Figure 5.17).

Linear DNA runs proportionally to its size on an agarose gel but circular DNA, like plasmids, does not. Plasmids coil up (supercoil) due to the tension from the DNA double helix, which makes them run faster than linear DNA through a gel. If one strand of the double helix has been cut (nicked) or if the plasmid is uncoiled for other reasons, these plasmids will also run at different rates. Therefore, DNA size standards cannot be used with accuracy to gauge plasmid size, and a pure plasmid preparation may form multiple bands on a gel because the plasmid DNA has several different levels of coiling.

When quantitating a circular plasmid using gel quantitation, use the most intense plasmid band for the comparison.

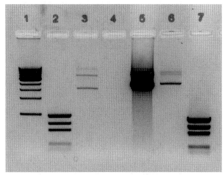

Figure 5.17. **Example of gel quantitation results.** Lane 1, 1 kb molecular ruler; lane 2, 5 µl of precision molecular mass ruler (DNA quantity from top to bottom: 100, 70, 50, 20, and 10 ng); lane 3, plasmid 1; lane 4, plasmid 2; lane 5, plasmid 3; lane 6, plasmid 4; lane 7, 20 µl of precision molecular mass ruler (DNA quantity from top to bottom: 400, 280, 200, 80, and 40 ng). Only plasmids 1 and 4 can be quantified using this gel. Plasmid 2 in lane 4 is not visible and plasmid 3 in lane 5 is more intense than any of the mass ruler bands.

Spectrophotometric Quantitation

To quantitate DNA using a spectrophotometer, the instrument must have an ultraviolet (UV) light source (see chapter 7). DNA absorbs light in the UV range at 260 nanometers (nm). A 50 µg/ml solution of double-stranded DNA has an absorbance of 1 at 260 nm. The absorbance at 260 nm is abbreviated as A_{260}. If the A_{260} is 0.5, the DNA concentration would be 25 µg/ml.

Standard plastic or glass cuvettes cannot be used to measure DNA concentration because they do not transmit UV light. Instead, quartz cuvettes that allow UV light transmission are best. However quartz cuvettes are expensive and fragile. Cuvettes made of plastic resin that are specially designed for DNA quantitation allow transmission of UV light and are cheaper and less fragile.

Since DNA samples are usually precious, cuvettes that use small volumes are often used to conserve the samples. These are called microcuvettes. Microcuvettes are available in different sizes. To conserve DNA samples, the DNA is often first diluted 10 to 100 times. Therefore, this dilution factor must be factored into the calculation when the DNA concentration is determined.

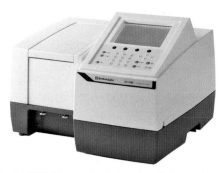

Figure 5.18. **A UV/Visible spectrophotometer.** Photo courtesy of Shimadzu Corporation, Kyoto, Japan.

Careers In Biotech

Denise Gangadharan, PhD
Associate Director for Science
Centers for Disease Control and Prevention
Atlanta, GA

Denise is the Associate Director for Science at the Centers for Disease Control and Prevention (CDC) in Atlanta, Georgia. She works in the Division of Select Agents and Toxins, which regulates the possession, use, and transfer of select agents and toxins, as well as the importation of etiological agents, hosts, and vectors of human disease to protect public health in the United States. Specifically, Denise works in the Science Office, where she provides technical help to the CDC inspectors as well as the regulated community. She is responsible for training and outreach, and for providing technical advice to the Director to help guide the creation of regulations, policy, and guidance documents. She enjoys applying the knowledge she gained in college and graduate school to her rapidly changing field.

Photo courtesy of Denise Gangadharan

Denise became interested in biology during high school. She recalls the excitement of pouring and loading her first gel as a pivotal moment, and after high school, she chose to study microbiology and molecular genetics at the University of California, Los Angeles (UCLA). While at UCLA, Denise developed a particular interest in immunology as she worked as a research technician investigating B cells. After taking some upper division immunology courses, she decided to pursue graduate studies in that subject. So, after graduating from UCLA in 1999, she enrolled as a graduate student at the University of California, San Diego School of Medicine, where she earned her PhD. Her thesis research investigated the development of CD8 intraepithelial T cells. Denise then went to Emory University for a post-doctoral fellowship and studied CD8 T cell memory. After her postdoc, she realized she wanted to shift her career into public health, and so she accepted a position at the CDC. Her scientific interests now are mainly in the study of disease and the interplay between host and microbe.

Over the next ten years, Denise sees her field of science regulation and policy growing to include the regulation of synthetic biology and of synthetic organisms, both of which are rapidly growing fields. Her advice to anyone interested in a career in biotechnology is to pursue an education in the field that interests them most, and include graduate study in the education plan. There are many options besides academia for those with advanced degrees in biology, and it's worth the effort to look into all the options while working toward a degree.

For example, 1 µl of DNA is added to 99 µl of distilled water (1:100 dilution) and A_{260} of the sample is 0.45. **Note:** a 50 µg/ml solution of DNA has an A_{260} of 1.

$$\text{Concentration of DNA} = A_{260} \times \frac{50 \text{ µg}}{1 \text{ ml}} \times \text{dilution factor}$$

$$= 0.45 \times \frac{50 \text{ µg}}{1 \text{ ml}} \times 100$$

$$= 2{,}250 \text{ µg/ml}$$

Single-stranded DNA and RNA absorb UV light differently. A 33 µg/ml solution of single-stranded DNA (for example, a PCR primer) and a 40 µg/ml solution of RNA both have an A_{260} of 1.

A spectrophotometer can also be used to assess the purity of the DNA samples using the ratio of absorbances at 260 and 280 nm (expressed as $A_{260}{:}A_{280}$) as an indicator of protein contamination. (The aromatic rings of proteins absorb light at 280 nm, while DNA absorbs light at 260 nm.) Most spectrophotometers have a setting to automatically measure the $A_{260}{:}A_{280}$ ratio, or this ratio can be calculated from the raw absorbance data. A pure DNA solution has an $A_{260}{:}A_{280}$ ratio >1.8, while a pure RNA sample has a ratio of >2.0. An $A_{260}{:}A_{280}$ ratio <1.7 implies that there is protein contamination. If the ratio is well below 1.7, the DNA must be purified before continuing to the next activity.

The NanoDrop spectrophotometer is a specialized, small-volume spectrophotometer that measures a much smaller sample size (typically 1–2 µl) without using cuvettes. This instrument can read A_{260} and also provide the $A_{260}{:}A_{280}$ ratio.

Quantitation of DNA with a Fluorometer
A fluorometer can measure DNA at a very low concentration and can specifically measure the concentration of double-stranded DNA. This is useful if the DNA sample is contaminated with RNA. A spectrophotometer requires approximately 1 µg of DNA for accurate quantitation, while a fluorometer requires only nanograms of DNA. A fluorometer requires the DNA be stained with a dye such as Hoescht 33258. This dye fluoresces only when it is bound to double-stranded DNA; therefore, the amount of fluorescence is proportional to the amount of DNA. A standard curve is drawn using known concentrations of DNA, and then the fluorescence reading of the unknown concentration is determined from the standard curve. Fluorescent molecules are excited by light at one wavelength and then produce (emit) light at a different wavelength. Hoechst 33258 is excited at 350 nm and emits at 450 nm. If the concentration of RNA or single-stranded DNA is required, dyes that excite and emit at different wavelengths are used.

Chapter 5 Essay Questions

1. Compare calcium chloride transformation, electroporation, and transfection. State the benefits and drawbacks of each method and describe the situations in which each method may be used.

2. What are the challenges of using biolistics for genetically modifying organisms? Give one case study on the use of biolistics.

3. Explain the use of *Agrobacterium tumefaciens* in genetically modifying crops.

Additional Resources

Papers on derivation of pGLO plasmid (also called pBADGFPuv):

Crameri A et al. (1996). Improved green fluorescent protein by molecular evolution using molecular DNA shuffling. Nat Biotech 14, 315–319.

Guzman L et al. (1995). Tight regulation, modulation, and high-level expression by vectors containing the arabinose PBAD promoter. J Bacteriol 14, 121–130.

Chalfie M et al. (1994). Green fluorescent protein as a marker for gene expression. Science 263, 802–805.

Papers and commentary describing worldwide incidence of antibiotic resistance and NDM-1:

Ventola, C L (2015). The Antibiotic Resistance Crisis: Part 1: Causes and Threats. Pharmacy and Therapeutics, 40(4), 277-283.

Kumarasamy K et al. (2010). Emergence of a new antibiotic resistance mechanism in India, Pakistan, and the UK: a molecular, biological, and epidemiological study. Lancet Infect Dis 10, 597–602.

Pitout J (2010). The latest threat in the war on antimicrobial resistance. Lancet Infect Dis 10, 578–579.

Activity 5.A Bacterial Transformation with S3 Plasmid

Overview

Bacteria can naturally take up DNA from their environment by transformation. Scientists have developed various protocols that greatly increase the rate and probability of transformation. One of these protocols is calcium chloride transformation. A modified version of this protocol will be used in this activity.

In this activity, *E. coli* HB101 bacteria will be transformed with the S3 plasmid (suspect 3 from Activity 4.E). (see Figure 5.19). The transformation efficiency will then be calculated. The transformation protocol will involve placing bacteria in an ice-cold environment surrounded by calcium ions and then adding the plasmid DNA. The mixture will be incubated on ice for 10 min. The bacteria will be heat shocked in a 42°C water bath for 50 sec, then immediately be placed back on ice. (Heat shock is thought to create holes or fractures in the bacterial membrane that allows passage of DNA inside. The DNA has been partially neutralized by the calcium ions, allowing easier passage through the hydrophobic lipid bilayer.) The bacteria will be incubated in LB broth for a recovery period to allow expression of the antibiotic resistance gene (*β-lactamase*), which is needed for the bacteria to grow on the selectable media. The bacteria will then be plated onto a selective medium containing ampicillin. In the presence of ampicillin, bacteria that contain the plasmids grow, while those that do not have the plasmid are inhibited from growing.

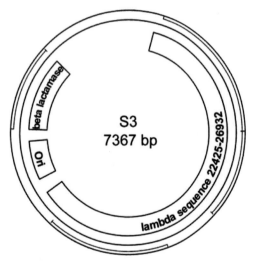

Figure 5.19. **Suspect 3 (S3) plasmid map.** S3 plasmid contains a 4.5 kb fragment of lambda phage DNA and the *β-lactamase* (*bla*) gene that confers resistance to ampicillin. The map also shows positions of various restriction sites on the plasmid.

In Activity 5.C, bacteria transformed with the S3 plasmid will be used to grow cultures. The S3 plasmid will then be purified from the bacterial cultures.

In Activity 5.D, the purified S3 plasmid will be quantitated by gel and spectrophotometric analyses.

Tips and Notes

- It is important to let the bacteria cool down on ice when performing the transformation protocol. It is also vital that the heat shock step be performed immediately (with no delay) after the incubation on ice. If necessary, take the ice bucket to the water bath

- Giving the cells a recovery period following the heat shock increases the success rate. If calculating the transformation efficiency, the cells should be plated after the 10 min recovery to accurately reflect the number of transformants per μg of DNA used. The bacteria have a chance to divide if they are left for more than 20 min

- Sterile 0.5 M calcium chloride solution made in Activity 2.D can be diluted to 0.05 M with sterile water and used for this activity. A total of 500 μl of 0.05 M CaCl₂ is needed for this activity. Alternatively, transformation solution can be used

- Two LB agar plates, two LB/amp agar plates, 1 ml of LB broth, and 1 LB agar deep tube (optional) are needed for this activity. These media should be made according to protocols in Activity 3.A. If three activities (Activities 5.A, 5.B, and 5.C) are to be performed, four LB agar plates, four LB/amp agar plates, one LB/amp/ara agar plate, two LB agar deep tubes, 2 ml of LB broth, and 10 ml of LB/amp broth in total are required

Safety Reminder: Follow all laboratory safety procedures. Inform your instructor if you are allergic to antibiotics and do not handle ampicillin. Use aseptic technique when handling bacteria and dispose of microbial waste properly. Wear appropriate personal protective equipment (PPE).

Research Questions

- Can *E. coli* HB101 be genetically engineered to be resistant to an antibiotic?
- What is the transformation efficiency of *E. coli* HB101 when using a calcium chloride transformation protocol?

Objectives

- Make growth media
- Streak an LB agar starter plate for single colonies
- Transform *E. coli* HB101 bacteria with S3 plasmid
- Inoculate LB and LB/amp plates with transformed bacteria
- Calculate the transformation efficiency

Activity 5.A Bacterial Transformation with S3 Plasmid

Skills to Master

Refer to Laboratory Skills Assessment Rubric (Appendix E) and Laboratory Notebook Rubric (Appendix F) for more details.

- Record laboratory notebook entries
- Streak bacteria to isolate single colonies (Activity 3.C)
- Transform bacteria by calcium chloride transformation
- Aseptically inoculate a plate with a liquid culture (Activity 3.C)
- Calculate transformation efficiency
- Use aseptic technique (Activity 3.A)

Student Workstation Materials

Two LB agar plates, two LB/amp agar plates, 1 ml of LB broth, and one LB agar deep tube (optional) are needed for this activity. If these media are not available, they should be made according to the protocols in Activity 3.A.

Part 1: Make Streak Plates

Items	*Quantity*
Incubator at 37°C (shared)	1
Bunsen burner*	1
Inoculation loop*	1
Laboratory tape (optional) (shared)	1
Laboratory marking pen	1
Microbial waste container	1
E. coli HB101 (shared)	1 vial
LB agar plate (for starter colonies)	1

* If sterile disposable inoculation loops are used, a Bunsen burner is not required.

Part 2: Transform Bacteria

Items	*Quantity*
Incubator at 37°C (shared)	1
Water bath at 42°C (shared)	1
2–20 µl adjustable-volume micropipet and sterile tips*	1
20–200 µl adjustable-volume micropipet and sterile tips**	1
100–1,000 µl adjustable-volume micropipet and sterile tips**	1
Sterile microcentrifuge tubes	2
Microcentrifuge tube rack	1
Floating microcentrifuge tube rack	1
Bunsen burner***	1
Inoculation loop***	1
Parafilm (shared)	1
Laboratory tape (optional) (shared)	1 roll
Container of ice	1
Laboratory marking pen	1
Microbial waste container	1
Transformation solution (0.05 M CaCl$_2$)	600 µl
S3 plasmid (0.3 µg/µl)	10 µl
LB broth	1 ml
LB agar plate	1
LB/amp agar plates	2
LB agar deep tube (optional)	1

* Disposable 10 µl inoculation loops can be used to measure 10 µl of plasmid. A liquid film over the loop is 10 µl.

** Sterile 1 ml graduated transfer pipets can be used instead of 20–200 µl and 100–1,000 µl micropipets.

*** If sterile disposable inoculation loops are used, a Bunsen burner is not required.

Activity 5.A Bacterial Transformation with S3 Plasmid

Protocol

1. If necessary, 3 days prior to the activity make two LB agar plates, two LB/amp agar plates, one LB agar deep tube (optional), and 50 ml of LB broth using the instructions provided in Activity 3.A.

2. One day prior to the activity, streak one LB agar plate for single colonies with *E. coli* HB101 bacteria (see step 8 of Activity 3.C). Incubate the plate overnight at 37°C. This is the starter plate for step 5.

3. On the day of the activity, label one microcentrifuge tube **+S3** and another **–S3**. Label both tubes with your initials.

4. Using aseptic technique, pipet 250 μl of transformation solution (0.05 M CaCl$_2$) into each tube. Place the tubes on ice.

5. Flame the inoculation loop and let it cool down. Scrape 2–4 single colonies from the surface of the starter plate so that they collect on the loop. Transfer the loop into the **+S3** tube and swirl it in the transformation solution to disperse the bacteria. Close the tube and place it back on ice.

   Note: A sterile plastic inoculation loop can also be used to collect bacterial colonies, but should not be flamed.

6. Flame the loop and repeat step 5 to transfer bacteria to the **–S3** tube.

7. Pipet 10 μl of S3 plasmid into the **+S3** tube. Do not add any plasmid into the **–S3** tube. Mix by pipetting gently up and down.

8. Incubate both tubes on ice for 10 min, making sure that the tubes are in full contact with the ice.

9. Label one LB agar plate and two LB/amp agar plates with your initials and the date. Label the plates as follows:

   LB/amp +S3
   LB/amp –S3
   LB –S3

Activity 5.A Bacterial Transformation with S3 Plasmid

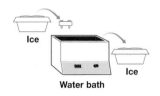

Ice

Ice

Water bath

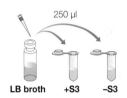

250 µl

LB broth   **+S3**   **−S3**

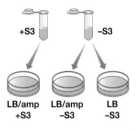

+S3   **−S3**

LB/amp   **LB/amp**   **LB**
+S3   **−S3**   **−S3**

10. After 10 min, transfer the **+S3** and **−S3** tubes directly from the ice into a 42°C water bath for exactly 50 sec, then immediately place them back on ice. Incubation at 42°C is used to heat shock the bacteria.

    Note: Make sure the tubes are in full contact with the 42°C water. For best results, do not exceed the recommended incubation time on ice and at 42°C.

11. Incubate tubes on ice for 2 min.

12. Remove the tubes from the ice and pipet 250 µl of LB broth into each tube using aseptic technique. Incubate the tubes for 10 min at room temperature.

13. Mix the tubes by inverting. Pipet 100 µl of each transformation mixture onto the appropriately labeled LB agar plate or LB/amp agar plate using a sterile pipet tip each time.

14. Flame the inoculation loop and let it cool down. Spread the bacteria over the entire surface of the plate in all directions. Flame the loop.

    Note: A sterile plastic inoculation loop can also be used to spread the bacteria but should not be flamed.

15. Repeat for each plate.

16. Stack the plates and tape them together with the agar side up. Incubate the plates in a 37°C incubator for 16–24 hr.

17. Once the cultures have grown for 16–24 hr, remove the plates from the incubator. Wrap the plates in Parafilm and store them at 4°C for up to 2 weeks.

18. (Optional) If your instructor advises you to do so, make a stab culture of the S3 transformed bacteria by picking one colony with a sterile inoculation loop or needle and stabbing it down into a labeled LB agar deep tube (see part 8 of Activity 3.A). Seal the tube and incubate it at 37°C for 24–48 hr. The tube can be stored at room temperature for up to 1 year.

Results Analysis

Record your observations and include sketches of the agar plates and the number of colonies on each plate. Plates that have too many single colonies to count should be recorded as too many to count (TMTC). Plates that have bacterial growth so dense that single colonies cannot be seen should be recorded as having a lawn. Calculate the transformation efficiency in CFU per µg of DNA. The concentration of the S3 plasmid is 0.3 µg/µl unless otherwise specified by your instructor. Compare the transformation efficiencies obtained by each group in the class.

Activity 5.A Bacterial Transformation with S3 Plasmid

Determining transformation efficiency (see page 156 for an example):

Step 1: Count the number of colonies on the LB/amp agar plate.

Step 2: Determine the amount of plasmid DNA spread on the LB/amp agar plate.

$$\text{DNA spread on plate (µg)} = \frac{\text{Volume spread (µl) x DNA in transformation (µg)}}{\text{Total volume of transformation (µl)}}$$

Step 3: Calculate the transformation efficiency by dividing the CFU by the amount of DNA spread on the plate.

$$\text{Transformation efficiency (CFU/µg of DNA)} = \frac{\text{CFU}}{\text{DNA spread on plate (µg)}}$$

Postlab Focus Questions

1. Have the E. coli been genetically engineered to be resistant to an antibiotic? What is your evidence?
2. What does the LB agar –S3 plate control for?
3. What does the LB/amp agar –S3 plate control for?

Self Assessment

Assess whether you have mastered the skills of this activity using the Laboratory Skills Assessment Rubric in Appendix E.

Assess your experimental write-up in your laboratory notebook using the Laboratory Notebook Rubric in Appendix F.

Activity 5.B Bacterial Transformation with pGLO Plasmid

Overview

Expressing proteins takes effort; therefore, protein expression is tightly regulated so that only needed proteins are expressed. Bacteria use operons to regulate proteins used in the metabolism of sugars like lactose (milk sugar) and arabinose (plant sugar). Scientists have designed operons to control the bacterial expression of proteins that have been genetically engineered.

The pGLO plasmid has been designed to express green fluorescent protein (GFP), under the control of the arabinose operon (araBAD) (see Figure 5.20). The *GFP* gene was isolated from the *Aequorea victoria* jellyfish, and the three scientists who discovered and worked on GFP were awarded the Nobel Prize in Chemistry in 2008. The discovery of GFP has revolutionized cell biology. For example, GFP can be fused to other proteins to enable scientists to localize proteins in live cells and organisms, something that had been impossible before the discovery of GFP.

GFP expression from the pGLO plasmid is prevented (repressed) by AraC, a protein encoded by the pGLO plasmid. The expression of GFP is activated (induced) by arabinose, a sugar that causes AraC to stop blocking transcription. The pGLO plasmid also encodes the *β-lactamase (bla)* gene that confers resistance to ampicillin.

Figure 5.20. **pGLO plasmid**. The plasmid map shows the location of the origin of replication (ori) and the *GFP*, *β-lactamase (bla)*, and *araC* genes.

In this activity, *E. coli* HB101 will be transformed with the pGLO plasmid. The bacteria will then be grown on a non-selective medium, as well as on a medium containing ampicillin, arabinose, or both.

In Activity 5.C, bacteria transformed with pGLO plasmid will be used to grow cultures. The pGLO plasmid will then be purified from the bacteria.

In Activity 5.D, the purified pGLO plasmid will be quantitated by gel and spectrophotometric analyses.

In Activity 7.C, GFP protein will be purified from bacteria transformed with pGLO plasmid. Stab cultures will be prepared to keep the transformed bacteria viable for a few months.

Tips and Notes

- It is important to let the bacteria cool down on ice when performing the transformation protocol. For the heat shock step, bacteria must be transferred directly from ice to the 42°C water bath for 50 sec and immediately back to ice again with no delay. If necessary, take the ice container to the water bath

- Giving the cells a recovery period following the heat shock increases the success rate. If calculating the transformation efficiency, the cells should be plated after the 10 min recovery to accurately reflect the number of transformants per µg of DNA used. The bacteria have a chance to divide if left for more than 20 min

- Expression of GFP from the pGLO plasmid can be controlled by the addition of arabinose. Colonies on the LB/amp agar plate can be induced to express GFP by adding 5–10 µl of arabinose (200 mg/ml) directly to one or more colonies on the plate. The colonies should glow green the next day

- Sterile 0.5 M calcium chloride ($CaCl_2$) solution made in Activity 2.D can be diluted to 0.05 M with sterile water as an alternative to using transformation solution for this activity. A total of 500 µl of 0.05 M $CaCl_2$ solution is needed for this activity

- Two LB agar plates, two LB/amp agar plates, and one LB/amp/ara agar plate are needed for this activity and should be made according to protocols in Activity 3.A. If three activities (Activities 5.A, 5.B, and 5.C) are to be performed, four LB agar plates, four LB/amp agar plates, one LB/amp/ara agar plate, two LB agar deep tubes, 2 ml of LB broth, and 10 ml of LB/amp broth in total are required

- UV lamps should be long-wave

Safety Reminder: Follow all laboratory safety procedures. Inform your instructor if you are allergic to antibiotics and do not handle ampicillin. Use aseptic technique when handling bacteria, and dispose of microbial waste properly. Wear appropriate PPE. Special precautions are required when using UV lights.

Activity 5.B Bacterial Transformation with pGLO Plasmid

Research Questions
- Can bacteria be genetically engineered to produce a jellyfish protein?
- Can the expression of GFP be regulated?
- What is the transformation efficiency of *E. coli* HB101 when using a calcium chloride transformation protocol?

Objectives
- Make growth media
- Streak an LB agar plate for single colonies
- Transform *E. coli* HB101 bacteria with the pGLO plasmid
- Grow transformed bacteria on a selective and inducible medium
- Calculate the transformation efficiency
- Control the expression of GFP with arabinose

Skills to Master
Refer to Laboratory Skills Assessment Rubric for more details.
- Streak bacteria to isolate single colonies (Activity 3.C)
- Transform bacteria by calcium chloride transformation (Activity 5.A)
- Aseptically inoculate a plate with a liquid culture (Activity 3.C)
- Calculate transformation efficiency (Activity 5.A)
- Use aseptic technique (Activity 3.A)

Student Workstation Materials
Two LB agar plates, two LB/amp agar plates, one LB/amp/ara agar plate, one deep agar tube (optional), and 1 ml of LB broth are needed for this activity. If these media are not available, they should be made according to the protocols in Activity 3.A.

Part 1: Make Streak Plates

Items	Quantity
Incubator at 37°C (shared)	1
Bunsen burner*	1
Inoculation loop*	1
Laboratory tape (optional) (shared)	1
Laboratory marking pen	1
Microbial waste container	1
E. coli HB101, rehydrated (shared)	1 vial
LB agar plate (for starter colonies)	1

* If sterile disposable inoculation loops are used, in a Bunsen burner is not required.

Part 2: Transform Bacteria

Items	Quantity
Incubator at 37°C (shared)	1
Water bath at 42°C (shared)	1
2–20 µl adjustable-volume micropipet and sterile tips*	1
20–200 µl adjustable-volume micropipet and sterile tips**	1
100–1,000 µl adjustable-volume micropipet and sterile tips**	1
Sterile microcentrifuge tubes	2
Microcentrifuge tube rack	1
Floating microcentrifuge tube rack	1
Bunsen burner***	1
Inoculation loop***	1
UV light	1
Parafilm (shared)	1
Laboratory tape (optional) (shared)	1 roll
Container of ice	1
Laboratory marking pen	1
Microbial waste container	1
Transformation solution (0.05 M CaCl$_2$)	600 µl
pGLO plasmid (0.08 µg/µl)	10 µl
LB broth	1 ml
LB agar plate	1
LB/amp agar plates	2
LB/amp/ara agar plate	1
LB agar deep tube (optional)	1
LB agar starter plate with *E. coli* (HB101) colonies from part 1	1

* Disposable 10 µl inoculation loops can be used to measure 10 µl of plasmid. A liquid film over the loop is 10 µl.
** Sterile 1 ml graduated transfer pipets can be used instead of 20–200 µl and 100–1,000 µl micropipets.
*** If sterile disposable inoculation loops are used, a Bunsen burner is not required.

Prelab Focus Questions
1. Why is arabinose present in the LB/amp/ara agar plates?
2. Would the pGLO plasmid glow green when exposed to UV light? Explain why or why not.
3. Predict the phenotype of pGLO-transformed bacteria under UV light when growing on LB/amp agar plates and LB/amp/ara agar plates. Explain your prediction.

Activity 5.B Bacterial Transformation with pGLO Plasmid

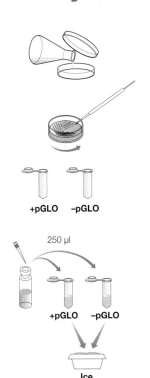

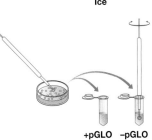

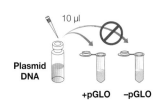

Protocol

1. If necessary, 3 days prior to the activity make two LB agar plates, two LB/amp agar plates, one LB/amp/ara plate, and one LB agar deep tube (optional) by following the instructions in Activity 3.A.

2. One day prior to the activity, streak one LB agar plate for single colonies with *E. coli* HB101 bacteria (see step 8 of Activity 3.C). Incubate the plate overnight at 37°C. This is the starter plate for step 5.

3. On the day of the activity, label one microcentrifuge tube **+pGLO** and another **–pGLO**. Label both tubes with your initials.

4. Using aseptic technique, pipet 250 µl of transformation solution (0.05 M CaCl₂) into each tube. Place the tubes on ice.

5. Flame the inoculation loop and let it cool down. Scrape 2–4 single *E. coli* colonies from the surface of the starter plate so they collect on the loop. Transfer the loop into the **+pGLO** tube and swirl it in the transformation solution to disperse the bacteria. Close the tube and place it back on ice.

   Note: A sterile plastic inoculation loop can also be used to pick the colonies but should not be flamed.

6. Flame the loop and repeat step 5 to transfer bacteria to the **–pGLO** tube.

7. Pipet 10 µl of pGLO plasmid into the **+pGLO** tube. Do not add any plasmid into the **–pGLO** tube. Mix by pipetting gently up and down.

8. Incubate the tubes on ice for 10 min, making sure the tubes are in full contact with the ice.

9. Label one LB agar plate, two LB/amp agar plates and one LB/amp/ara plate with your initials and the date. Label the plates as follows:

   LB/amp +pGLO
   LB/amp/ara +pGLO
   LB/amp –pGLO
   LB –pGLO

.B/amp +pGLO   LB/amp /ara +pGLO   LB/amp –pGLO   LB –pGLO

Activity 5.B Bacterial Transformation with pGLO Plasmid

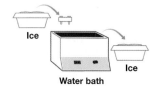

Ice

Ice

Water bath

250 µl

LB broth   +pGLO   –pGLO

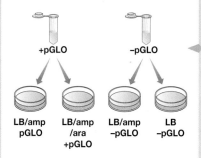

+pGLO   –pGLO

LB/amp   LB/amp   LB/amp   LB
pGLO   /ara   –pGLO   –pGLO
          +pGLO

10. After 10 min, transfer the **+pGLO** and **–pGLO** tubes directly from the ice into a 42°C water bath for exactly 50 sec, then immediately place them back on ice. Incubation at 42°C is used to heat shock the bacteria.

    Note: Make sure the tubes are in full contact with the 42°C water. For best results, do not exceed the incubation time on ice and at 42°C.

11. Incubate the tubes on ice for 2 min.

12. Remove the tubes from the ice and pipet 250 µl of LB broth into each tube using sterile technique. Incubate the tubes at room temperature for 10 min.

13. Mix the tubes by inverting. Pipet 100 µl of each transformation mixture onto the appropriately labeled LB agar plate, LB/amp agar plate, or LB/amp/ara agar plate using a sterile pipet tip each time.

14. Flame the inoculation loop and let it cool down. Spread the bacteria over the entire surface of the plate in all directions. Flame the loop.

    Note: A sterile plastic inoculation loop can also be used to spread the bacteria but should not be flamed.

15. Repeat for each plate.

16. Stack the plates together with the agar side up. Incubate the plates in a 37°C incubator for 16–24 hr.

17. Once the cultures have grown for 16–24 hr, remove them from the incubator. Wrap the plates in Parafilm, and store them at 4°C for up to 2 weeks.

18. (Optional) If your instructor advises you to do so, make a stab culture of the pGLO transformed bacteria by picking one colony with a sterile inoculation loop or needle and stabbing down into a labeled LB agar deep tube (see part 8 of Activity 3.A). Seal the tube and incubate it at 37°C for 24–48 hr. The tube can be stored at room temperature for up to 1 year.

Activity 5.B Bacterial Transformation with pGLO Plasmid

Results Analysis

View the plates under a long-wavelength UV light. Record your observations and include sketches of the agar plates. Record the number and color of the colonies on each plate; record if the colonies are too many to count (TMTC) or if a lawn of bacteria is present on the plate. Calculate the transformation efficiency in CFU/µg of DNA. (If necessary, see the Results Analysis section in Activity 5.A for more detail.) The concentration of the pGLO plasmid is 0.08 µg/µl unless otherwise specified your instructor. Compare the transformation efficiencies obtained by each group in the class.

Postlab Focus Questions

1. What is the transformation efficiency of *E. coli* HB101 when using the calcium chloride transformation protocol?
2. Did any of the colonies glow green under UV light? If so, why?
3. How could colonies on the LB/amp +pGLO plate be made to glow green?

Self Assessment

Assess whether you have mastered the skills of this activity using the Laboratory Skills Assessment Rubric in Appendix E.

Assess your experimental write-up in your laboratory notebook using the Laboratory Notebook Rubric in Appendix F.

Activity 5.C Purification of S3 and pGLO Plasmids

Overview

Bacteria that have been transformed with a plasmid can be used as biological machines to produce more plasmids. Depending on the copy number, each bacterium can contain 5–1,000 identical copies of the plasmid. To analyze or manipulate the plasmid DNA, many billions of copies are required. Since growing bacteria in liquid culture is the most efficient way to obtain large numbers of bacteria, a colony of bacteria that contain the plasmid is first grown in liquid culture. The plasmid DNA is then purified from the bacterial proteins and genomic DNA (gDNA). When plasmid DNA is purified on a small scale (as performed in this activity), it is called a miniprep. The culture volume used for a miniprep ranges from 2–5 ml. Larger scale plasmid purifications can be performed in the laboratory; these purifications are called midipreps and maxipreps and use cultures of approximately 25 ml and 100 ml, respectively.

In this activity, *E. coli* HB101 containing S3 and pGLO plasmids (from Activities 5.A and 5.B) will be grown in liquid culture. The plasmids will then be extracted using the alkaline lysis method. Finally, column chromatography will be performed to purify the plasmid DNA.

The amount of DNA obtained from transformed bacteria depends on the copy number of the plasmid, which in turn depends on the sequence of the origin of replication. A low copy number plasmid yields 10–100 times less DNA in a miniprep than a high copy number plasmid. Plasmid yield will be investigated in Activity 5.D.

A total of 10 ml of LB/amp broth is needed for this activity. LB/amp broth should be made according to protocols in Activity 3.A.

Tips and Notes

- When adding the lysis solution, gently invert or turn the tubes upside down. Do not shake or vortex. Shaking or vortexing will shear the gDNA into small pieces and make it hard to separate from the plasmid DNA

- Make sure that the centrifuge is balanced at all times. Once tubes have been centrifuged, take care not to accidentally bump or shake the tubes to avoid resuspending the pellets. Centrifuge the tubes again if the pellets become dislodged

Safety Reminder: Follow all laboratory safety procedures. Inform your instructor if you are allergic to antibiotics and do not handle ampicillin. Use aseptic technique when handling bacteria and dispose of microbial waste properly. The lysis solution contains sodium hydroxide, which can cause burns. Handle with care and read the SDS for complete safety information. Wear appropriate PPE.

Objectives

- Make 10 ml of LB/amp broth
- Grow a pure culture of *E. coli* HB101 transformed with S3 plasmid
- Grow a pure culture of *E. coli* HB101 transformed with pGLO plasmid
- Purify pGLO and S3 plasmids from *E. coli* HB101

Skills to Master

- Use aseptic technique (Activity 3.A)
- Aseptically inoculate a liquid culture from a plate (Activity 3.C)
- Purify plasmid DNA (perform a miniprep)
- Use a microcentrifuge

Activity 5.C Purification of S3 and pGLO Plasmids

Student Workstation Materials

A total of 10 ml of LB/amp broth is needed for this activity. If LB/amp broth is not available, it should be made according to protocols in Activity 3.A.

Part 1: Inoculating Cultures

Items	Quantity
Shaking incubator, shaking water bath, or incubator with mini-roller (shared)	1
Pipet pump or filler	1
5 ml sterile serological pipet	1
14 ml culture tubes	2
Bunsen burner*	1
Inoculation loop*	1
Laboratory marking pen	1
Microbial waste container	1
LB/amp broth	10 ml
LB/amp agar plate with colonies transformed with S3 plasmid (Activity 5.A)	1
LB/amp agar plate or LB amp/ara agar plate with colonies transformed with pGLO plasmid (Activity 5.B)	1

* If sterile disposable inoculation loops are used, a Bunsen burner is not required.

Part 2: Purifying Plasmid DNA

Items	Quantity
Vortexer (optional) (shared)	1
Microcentrifuge (shared)	1
20–200 µl adjustable-volume micropipet and sterile tips	1
100–1,000 µl adjustable-volume micropipet and sterile tips	1
1.5 ml microcentrifuge tubes	4
2 ml capless collection tubes	2
Microcentrifuge tube rack	1
Laboratory marking pen	1
Microbial waste container	1
Overnight cultures from part 1	2
Resupension solution	600 µl
Lysis solution	600 µl
Neutralization solution	1 ml
5x wash solution	2 ml
Elution solution	250 µl
Plasmid purification mini columns	2

Prelab Focus Questions

1. Why should bacterial cultures be grown in LB broth containing ampicillin?
2. Why do bacterial cultures need to be shaken to grow most efficiently?
3. It is important to keep track of the location of the plasmid at each stage of plasmid purification. For the following steps of the protocol, state whether the plasmid is in the pellet, in the supernatant, bound to the column, or in the flowthrough:
   a. After centrifuging the bacterial culture (pellet or supernatant).
   b. After centrifuging the neutralized cell lysate (pellet or supernatant).
   c. After centrifuging the supernatant through the column (column or flowthrough).
   d. After centrifuging the wash solution through the column (column or flowthrough).
   e. After centrifuging the elution solution through the column (column or flowthrough).

Activity 5.C Purification of S3 and pGLO Plasmids

Protocol

Part 1: Inoculating Cultures

1. Label two culture tubes with your initials and the date. Label the tubes as follows:

   S3 mini
   pGLO mini

2. Use aseptic technique to transfer 4 ml of LB/amp broth to each tube.

3. Flame the loop and let it cool down. Using the sterile loop, scrape one colony transformed with the S3 plasmid from the surface of the **LB/amp +S3** plate. Transfer the loop into the culture tube labeled **S3 mini**. Swirl the loop to release the bacteria. Recap the tube.

   Note: A sterile plastic inoculation loop can also be used to spread the bacteria but should not be flamed.

4. Repeat step 3 to transfer one colony transformed with the pGLO plasmid from the **LB/amp +pGLO** plate into the culture tube labeled **pGLO mini**. Recap the tube.

5. Place the culture tubes into a shaking incubator, shaking water bath, or tube roller in an incubator. Grow the culture overnight at 37°C.

   Note: If a 37°C incubator is not available, incubate the samples at room temperature for 48 hr with shaking or rolling; however, expect a lower plasmid yield. This protocol is not recommended if the samples cannot be shaken or rolled while incubating at room temperature.

6. After an overnight incubation, the cultures should appear cloudy.

   (Optional) Measure the optical density (OD) of the cultures at 600 nm using a spectrophotometer.

Part 2: Purifying Plasmid DNA

1. Label two 2 ml capless collection tubes with your initials. Label the tubes as follows:

   S3 mini
   pGLO mini

2. Label two plasmid mini columns on the top ridge with the same information as above. Place each column in its corresponding capless tube.

3. Label two 1.5 ml capped microcentrifuge tubes with your initials and the same information as above.

+S3   +S3

+pGLO   +pGLO

Activity 5.C Purification of S3 and pGLO Plasmids

1.5 ml

4. Transfer 1.5 ml of each culture into the corresponding 1.5 ml capped microcentrifuge tube by pipetting 750 µl twice. Close the tubes.

5. Place the tubes in the microcentrifuge in a balanced pattern. Put the hinges of the tubes facing outward so that the pellet will be easy to locate after centrifugation. Centrifuge the tubes containing the cultures for 1 min at top speed (≥12,000 rpm) to pellet the bacteria.

   Note: Make sure the microcentrifuge is balanced and accommodate tubes of classmates to ensure economic use of the microcentrifuge.

6. Locate the bacterial pellet (it should be below the hinge of the tube). Remove the liquid (supernatant) from each tube using a 100–1,000 µl micropipet or a vacuum source. Avoid touching the pellet. Discard the supernatant.

1.5 ml

7. Add another 1.5 ml of each bacterial culture to the corresponding 1.5 ml capped microcentrifuge tube that already contains the bacterial pellet. Spin again for 1 min. Locate the pellet and remove the supernatant without disturbing the pellet. Discard the supernatant.

250 µl

Resuspension solution

8. Using a fresh pipet tip each time, pipet 250 µl of resuspension solution into each tube. Resuspend the bacterial pellet by pipetting up and down or by vortexing. Ensure no bacterial clumps remain.

250 µl

Lysis solution

9. Pipet 250 µl of lysis solution into each tube and mix by gently inverting 6–8 times. Do not pipet or vortex.

350 µl

Neutralization solution

10. Pipet 350 µl of neutralization solution into each tube and mix by gently inverting 6–8 times. A precipitate should form.

    Note: Neutralization solution should be added within 5 min of adding the lysis solution.

11. Spin the tubes in the microcentrifuge for 5 min at top speed.

    Note: Remember to orient the tube hinges outwards. The precipitate should form a large, white pellet.

12. Pour or pipet the supernatant from the centrifuged tubes into the corresponding plasmid mini column.

    Note: Avoid transferring any precipitate or cellular debris. If necessary, centrifuge the tubes again to pellet cellular debris.

Activity 5.C Purification of S3 and pGLO Plasmids

13. Spin the columns in the capless collection tubes in the microcentrifuge for 1 min at top speed.

    Note: The inner lid of the microcentrifuge may not fit with the columns in place, in which case centrifuge the tubes without the inner lid.

14. Take the column out of the capless tube and discard the liquid (flowthrough) from the collection tube. Replace the column in the collection tube.

Wash solution

750 µl

15. Pipet 750 µl of wash solution into each column.

    Note: Ensure that ethanol is already added to the wash solution.

16. Spin the columns in the capless collection tubes in the microcentrifuge for 1 min at top speed.

17. Discard the flowthrough from the collection tube.

18. Place the columns back into the collection tubes and spin for 1 min to dry out the columns.

19. Label two clean 1.5 ml capped microcentrifuge tubes with your initials and the date. Label the tubes as follows:

    S3 plasmid
    pGLO plasmid

Elution solution

100 µl

20. Transfer each column to the corresponding clean 1.5 ml capped microcentrifuge tube. Discard the capless collection tubes. Pipet 100 µl of elution solution directly onto the column bed of each column.

21. Allow the elution solution to absorb into the column for 1–2 min.

22. Spin the columns in the clean 1.5 ml capped microcentrifuge tubes for 2 min at top speed. Since the tube caps are not closed, the microcentrifuge may need to be run without closing the inner lid.

23. Discard the columns and cap the tubes.

24. Measure the volume of solution eluted from the column using a 20–200 µl adjustable-volume micropipet. This volume will be used to calculate the plasmid yield in Activity 5.D.

25. Store miniprep plasmid DNA at 4°C for up to 1 month or at –20°C for 1 year.

Activity 5.C Purification of S3 and pGLO Plasmids

Results Analysis
The concentration of the purified plasmids will be determined in Activity 5.D using gel or spectrophotometric quantitation.

Postlab Focus Questions
1. Describe the appearance of the bacterial culture after the addition of lysis solution and gentle inversion of the tube.
2. What did you observe once neutralization solution was added?
3. What did you observe after the neutralized lysate was centrifuged?
4. Where was the plasmid located after the neutralized lysate was centrifuged?

Self Assessment
Assess whether you have mastered the skills of this activity using the Laboratory Skills Assessment Rubric in Appendix E.

Assess your experimental write-up in your laboratory notebook using the Laboratory Notebook Rubric in Appendix F.

Activity 5.D DNA Quantitation

Overview

The amount of DNA obtained from transformed bacteria depends on the copy number of the plasmid, which in turn depends on the sequence of the origin of replication of the plasmid. Low copy number plasmids yield 10–100 times less DNA in a miniprep than a high copy number plasmid. This activity will measure the amount of plasmid generated from the minipreps in Activity 5.C.

There are many ways to quantify DNA; the two methods that will be used in this activity are gel and spectrophotometric quantitation. In part 1 of this activity, the amount of DNA in the purified plasmid preparations will be estimated using gel quantitation. A molecular mass ruler that has known quantities of DNA in each band will be used as the control. Purified plasmid preparations vary greatly in DNA concentration, and the amount of plasmid in the band on the gel should lie in the range of the mass ruler (see Figure 5.17). Two different amounts of mass ruler will be loaded to provide a range of concentrations from 10 to 400 ng. In addition, different amounts of the two plasmids will be loaded on the gel.

Bio-Rad's 1 kb molecular ruler will also be used to estimate the sizes of the plasmids. (This standard will not be used to estimate DNA quantity.) Remember that plasmid DNA runs faster than linear DNA because of supercoiling; therefore, the measurement is just a rough estimate.

In part 2, the amount of DNA in the purified plasmid preparation will be estimated using a spectrophotometer. The plasmid preparations will first be diluted 10 times. DNA absorbs UV light at 260 nm, and 50 µg/ml of DNA has an absorbance of 1.0. This information can be used to calculate the concentration of the plasmid DNA from its absorbance. The amount the sample is diluted will need to be factored into the calculation or input into the spectrophotometer prior to measuring. Spectrophotometers also allow the purity of the DNA to be measured by looking at the ratio of the absorbance at 260 nm against the absorbance at 280 nm. Pure DNA has an $A_{260}:A_{280}$ ratio of ≥1.8. A ratio <1.7 indicates impure DNA.

Tips and Notes

- When quantitating DNA it is important to mix the solutions well to ensure the DNA is evenly distributed throughout the solution. A vortexer is useful for mixing solutions

- Ensure that all cuvettes or containers are very clean. Cuvettes are specialized for quantitating DNA and are made of quartz or plastics that do not absorb UV. Quartz cuvettes are very expensive and breakable so take extra care when using them

- Be sure to orient the cuvette correctly in the spectrophotometer so that light passes through the sample in a 1 cm path length. Do not handle the cuvette where the light passes through, fingerprints can interfere with the reading. Wipe off fingerprints with a laboratory tissue if necessary

Safety Reminder: Appropriate PPE should be worn at all times and attention should be given when using electricity in the presence of liquids. Safety measures built into the equipment should prevent accidental exposure; however, should any leakage occur, notify your instructor before proceeding. If ethidium bromide, fluorescent stains, or UV light is used, special precautions are required.

Research Questions

- How much plasmid DNA was produced in the miniprep?
- Which plasmid purification yielded the most DNA?

Objectives

- Make a 1% agarose TAE gel and 300 ml of 1x TAE buffer (if necessary)
- Electrophorese S3 and pGLO plasmids on a 1% agarose gel alongside a DNA mass standard
- Analyze the gel and estimate the quantity of S3 and pGLO plasmids in the gel
- Measure the absorbance of S3 and pGLO plasmid preparations at 260 nm (A_{260}) and 280 nm (A_{280})
- Calculate the concentration of DNA in S3 and pGLO plasmid preparations
- Calculate the $A_{260}:A_{280}$ ratio to determine the purity of DNA

Skills to Master

Refer to Laboratory Skills Assessment Rubric (Appendix E) and Laboratory Notebook Rubric (Appendix F) for more details.

- Record laboratory notebook entries
- Perform agarose gel electrophoresis (Activity 4.C)
- Quantitate DNA using gel analysis
- Analyze an agarose gel (Activity 4.D)
- Use a spectrophotometer
- Quantitate DNA using a spectrophotometer
- Determine DNA purity by the $A_{260}:A_{280}$ ratio

Activity 5.D DNA Quantitation

Student Workstation Materials

One 1% agarose TAE gel and 300 ml of 1x TAE electrophoresis buffer are required for this activity. If not available, refer to Activities 4.B and 2.D, respectively, to prepare these reagents.

Part 1: Gel Quantitation

Items	*Quantity*
Horizontal gel electrophoresis chamber	1
Power supply (shared)	1
Vortexer (optional)	1
Digital imaging system (optional) (shared)	1
Microcentrifuge or mini centrifuge (optional) (shared)	1
Rocking platform (optional)	1
UV light source (optional) (shared)	1
2–20 µl adjustable-volume micropipet and tips	1
20–200 µl adjustable-volume micropipet and tips	1
Microcentrifuge tubes	8
Microcentrifuge tube rack	1
Gel staining tray	1
Agarose gel support film (optional)	1
Laboratory marking pen	1
Waste container	1
Sterile water	100 µl
S3 plasmid (Activity 5.C)	12 µl
pGLO plasmid (Activity 5.C)	12 µl
1 kb molecular ruler (1 KB)	8 µl
Precision molecular mass ruler (MMR)	30 µl
5x sample loading buffer (SLB)	15 µl
or UView 6x loading dye and stain (UView)	12 µl
1x TAE electrophoresis buffer*	300 ml
1x Fast Blast DNA stain**	100 ml
1% agarose TAE gel***	1

* If necessary, make 1x TAE electrophoresis buffer (see part 1 of Activity 2.D).
** Alternative staining methods may be used. See Appendix B for details.
*** If necessary, make a 1% agarose TAE gel according to the protocol in Activity 4.B. This may require extra 1x TAE electrophoresis buffer.

Part 2: Spectrophotometric Quantitation

Items	*Quantity*
Spectrophotometer (shared)	1
Vortexer (optional)	1
2–20 µl adjustable-volume micropipet and sterile tips	1
20–200 µl adjustable-volume micropipet and sterile tips	1
Microcentrifuge tubes	2
Quartz or other UV-transparent cuvettes	1–3
Laboratory marking pen	1
Waste container	1
Sterile water	250 µl
S3 plasmid (Activity 5.C)	10 µl
pGLO plasmid (Activity 5.C)	10 µl

* If using quartz cuvettes, it is likely you will have just one or two cuvettes, which are shared with others. You will need to wash out each cuvette with sterile water between samples. If the quartz cuvettes require volumes > 50 µl you may also need to prepare a larger volume of diluted plasmid for measurement.

Activity 5.D DNA Quantitation

Protocol

Part 1: Gel Quantitation

1. If a 1% agarose TAE gel is not available, follow the procedures in Activity 4.B to prepare the gel.

2. If 1x TAE running buffer is not available, make 300 ml of buffer using the procedures in Activity 2.D.

3. Label two microcentrifuge tubes as follows: **1/25 pGLO**, and **1/25 S3**.

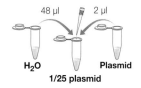

4. Dilute each plasmid 1/25 by pipetting 48 µl of sterile water and 2 µl of the corresponding plasmid DNA into each tube. Vortex or pipet up and down to thoroughly mix. These are the diluted plasmids.

5. Label four microcentrifuge tubes as follows: **pGLO gel**, **1/25 pGLO gel**, **S3 gel**, and **1/25 S3 gel**.

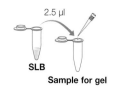

6. Pipet 2.5 µl of 5x sample loading buffer (**SLB**) or 2 µl UView 6x loading dye and stain (**UView**) into each tube.

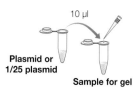

7. Pipet 10 µl of each undiluted or diluted plasmid into the corresponding microcentrifuge tube containing sample loading buffer. Collect the sample at the bottom of the tube by tapping the tube gently on the table or pulse spinning it in a microcentrifuge to collect the sample at the bottom of the tube.

8. Place a 1% agarose TAE gel into the electrophoresis chamber. Fill the chamber with sufficient 1x TAE running buffer to cover the gel by approximately 2 mm.

9. Check that the wells of the agarose gel are near the black negative electrode, or cathode, and the bottom edge of the gel is near the red positive electrode, or anode.

10. Using a fresh tip for each sample, load the samples in the following order with the indicated volumes:

Lane	Sample	Volume
1	**1 KB**	5 µl
2	**MMR**	5 µl
3	**pGLO gel**	12.5 µl*
4	**1/25 pGLO gel**	12.5 µl*
5	**S3 gel**	12.5 µl*
6	**1/25 S3 gel**	12.5 µl*
7	**MMR**	20 µl*

    * If using UView 6x loading dye and stain, load 12 µl of each sample.

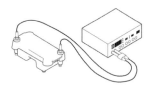

11. Place the lid on the electrophoresis chamber. Connect the electrical leads to the power supply in the following orientation: red to red and black to black.

Activity 5.D DNA Quantitation

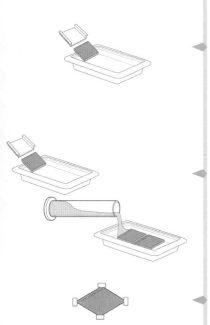

12. Turn on the power and run the gel at 100 V for 30 min.

    Note: If using the fast gel running buffer (0.25x TAE), run the gel at 200 V for 20 min (see Appendix A).

13. When the electrophoresis run is complete, turn off the power and remove the lid from the chamber. Carefully remove the gel and tray from the electrophoresis chamber. Be careful—the gel is very slippery. If using UView 6x loading dye and stain or other fluorescent stains, use a UV light source to visualize DNA bands. Otherwise slide the gel into the staining tray and carefully pour a sufficient volume of 1x Fast Blast DNA stain to completely submerge the gel.

    Note: For alternative DNA staining options, see Appendix B.

14. Stain the gel overnight to visualize the DNA bands, then record the results. (Optional) A rocking platform can be used to enhance results.

15. (Optional) Trim away any unloaded lanes and air-dry on a piece of agarose gel support film. Tape the dried gel into your laboratory notebook.

Part 2: Spectrophotometric Quantitation

1. Approximately 10 min before starting the spectrophotometric quantitation, turn on the spectrophotometer so that the UV lamp has time to warm up.

2. Label two microcentrifuge tubes as follows: **1/10 pGLO**, and **1/10 S3**.

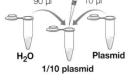

3. Dilute each plasmid 1/10 by adding 90 µl of sterile water to each tube and then pipetting 10 µl of each plasmid to the corresponding tube. Mix the solution by vortexing, thoroughly pipetting up and down, or flicking the tubes.

4. When using quartz cuvettes, follow your instructor's guidelines on labeling and proper use.

5. Add 50 µl of sterile water to the cuvette labeled **Blank**.

6. Quartz cuvettes should be rinsed with water after each sample is read in the spectrophotometer. Transfer the sample from the cuvette back to its original tube before rinsing the cuvette. Remove as much water as possible from the cuvette and then add the second sample and repeat the reading.

Activity 5.D DNA Quantitation

Blank

S3

pGLO

7. Set the spectrophotometer to its DNA quantitation setting or set it to read two wavelengths: 260 nm and 280 nm. Follow the directions on the spectrophotometer to set up the instrument.

    Note: If possible and desired, a dilution factor may also be entered into the spectrophotometer at this stage. Remember, the samples were diluted tenfold.

8. Insert the cuvette labeled **Blank** (containing the water) into the spectrophotometer. Take the reading of the blank cuvette or set the instrument to 100% transmittance.

9. Remove the cuvette labeled **Blank** and insert the cuvette labeled **S3** into the spectrophotometer.

10. Read the absorbance of the sample. The instrument should automatically read the absorbance at 260 nm and 280 nm.

11. If the spectrophotometer has the capability of providing the $A_{260}{:}A_{280}$ ratio, record this reading. If not, calculate this ratio from the raw absorbance data when analyzing results.

12. Repeat for the cuvette labeled **pGLO** and record the absorbance values.

13. Once the DNA concentration has been calculated (see Results Analysis), label each tube containing the purified plasmid with the DNA concentration. Store the purified plasmid at −20°C for up to 1 year.

Activity 5.D DNA Quantitation

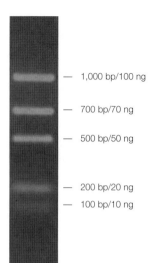

— 1,000 bp/100 ng

— 700 bp/70 ng

— 500 bp/50 ng

— 200 bp/20 ng
— 100 bp/10 ng

Figure 5.21. **Precision molecular mass ruler**. The mass standard has bands of specific length and quantity that can be compared to unknown bands. The quantities shown correspond to a 5 µl load.

Results Analysis

Part 1: Gel Quantitation

Step 1: Calculate the volume of the plasmid solution that was loaded into the gel.
The volume will be 10 µl if the purified plasmid was loaded. If the 1/25 dilution was loaded, the volume of plasmid will be 10 µl x 1/25 or 0.4 µl.

Step 2: Compare sample band intensities with standards.
Compare the intensity of the plasmid band with the mass standard and estimate the band of the standard that is closest in intensity to the plasmid DNA. The intensity (brightness) of the band corresponds to the quantity of DNA. Each band of the precision molecular mass ruler contains a specific quantity of DNA (see Figure 5.21). The DNA quantities when 5 µl and 20 µl volumes of the molecular mass ruler are loaded are shown in Table 5.1.

DNA Length	Mass of DNA in Various Bands	
	5 µl	**20 µl**
1,000 bp	100 ng	400 ng
700 bp	70 ng	280 ng
500 bp	50 ng	200 ng
200 bp	20 ng	80 ng
100 bp	10 ng	40 ng

Table 5.1. **Mass of DNA in the precision molecular mass ruler**. Each time a different volume is loaded using the mass standard, the bands contain a different quantity of DNA. The table shows the DNA quantities for the two load volumes used in this experiment.

Step 3: Calculate the DNA concentration.
To convert the estimated quantity into a concentration, divide the quantity of DNA in the band by the volume of plasmid DNA (from the plasmid stock) loaded on the gel (10 µl or 0.4 µl). Present the concentration in units of ng/µl. For example, if 10 µl of DNA was loaded into the gel and the quantity of DNA in the band was estimated to be 70 ng, the DNA concentration of the sample would be 70 ng/10 µl or 7 ng/µl.

Step 4: Calculate the total yield of plasmid DNA from each purification.
The yield is the total quantity of DNA (in ng or µg) that was purified from 3 ml of culture. Calculate yield by multiplying the concentration of DNA by the volume of plasmid stock in the microcentrifuge tube, which was measured at the end of Activity 5.C.

Part 2: Spectrophotometric Quantitation

Results analysis will depend on the format of the data obtained from the spectrophotometer. It is important to know how to calculate DNA concentration from raw absorbance data. You can verify your manual calculation by comparing it to automatic data from the spectrophotometer.

Step 1: Calculate the DNA concentration from the A_{260} reading.
To calculate the DNA concentration from the A_{260} reading, you need the following information: the absorbance value at 260 nm, the dilution factor, and the absorbance of double-stranded DNA at 50 µg/ml (which is 1).

$$\text{Dilution factor} = \frac{\text{Vol of DNA (µl)} + \text{Vol of water (µl)}}{\text{Vol of DNA (µl)}}$$

Activity 5.D DNA Quantitation

Concentration of DNA sample = A_{260} x 50 µg/ml x dilution factor

Step 2: Calculate the total yield from the plasmid purification.
The yield is the total quantity of DNA (in ng or µg) that was purified from 3 ml of culture. Calculate yield by multiplying the concentration of DNA by the volume of plasmid stock in the microcentrifuge tube, which was measured at the end of Activity 5.C.

Step 3: Calculate the A_{260}:A_{260} ratio to assess DNA purity.
To calculate the A_{260}:A_{280} ratio, you need the absorbance values at 260 nm and 280 nm.

Ratio of absorptions $= \dfrac{A_{260}}{A_{280}}$

A pure DNA sample has a ratio of >1.8, while a sample is considered contaminated with protein if the ratio is <1.7. How does your sample compare?

Comparison of Results
Compare the results of the two methods used to quantitate DNA. The results should be converted so that they are given in the same units of measure (µg/µl or ng/µl).

Self Assessment
Assess whether you have mastered the skills of this activity using the Laboratory Skills Assessment Rubric in Appendix E.

Assess your experimental write-up in your laboratory notebook using the Laboratory Notebook Rubric in Appendix F.

Postlab Focus Questions
1. Using the experimental data from the gel quantitation and spectrophotometric analysis, state which of these methods provides the most accurate determination of DNA concentration. Explain your reasoning.
2. Which plasmid yield was higher, S3 or pGLO plasmid?
3. Speculate on whether S3 and pGLO plasmids are high or low copy number plasmids and give your reasoning.
4. What was the A_{260}:A_{280} ratio for each plasmid preparation? Were the ratios ≥1.8 (pure DNA), <1.7 (impure), or 1.7–1.8 (borderline)?

Chapter 5 Extension **Activities**

1. During the plasmid purifications, the cultures can be measured by either optical density (OD) readings or by serial dilution to quantitate the number of bacteria used in the plasmid purification. The quantity of DNA per bacterium can then be calculated.
2. Optimize transformations by varying plasmid concentration and bacterial concentration. Use Bio-Rad's Ligation and Transformation module to make competent cells and compare highly efficient competent bacteria with the modified method used in Activities 5.A and 5.B.
3. Map the S3 and or pGLO plasmids with restriction enzymes using the skills you learned in chapters 4 and 5.
4. Purify plasmids, such as pBR322, that contain dual resistance against ampicillin and tetracycline.
5. Make plasmids for restriction digestion experiments with younger students by performing plasmid purifications and determining the plasmid concentrations.
6. Make stab cultures and glycerol stocks of *E. coli* HB101 that contain plasmids.
7. Perform ligation of PCR products or restriction digest fragments.

The Polymerase Chain Reaction

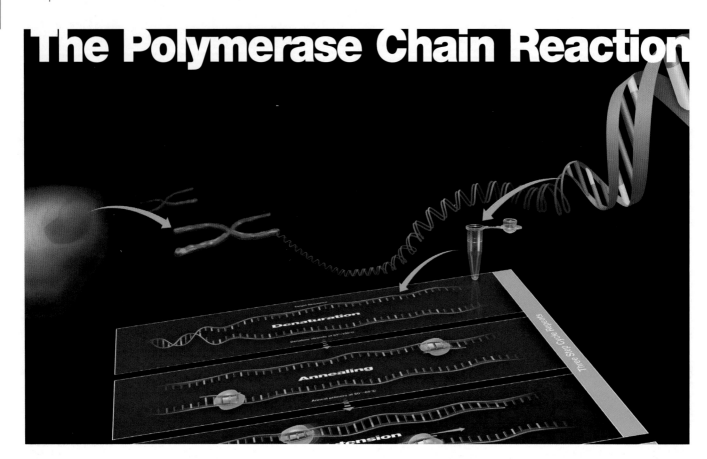

Summary

The polymerase chain reaction (PCR) has revolutionized the study of living things. Invented by Kary Mullis in 1983, PCR has been a springboard for molecular biology research. It has been the basis of the Human Genome Project, modern forensic analysis, and genetic engineering. Using PCR, a small DNA sequence consisting of just a few hundred base pairs can be found within a genome of billions of base pairs. Billions of copies of the sequence can be generated, making the DNA sequences available for study and manipulation. Agriculture has been transformed by PCR with the advent of genetically modified crops. Cows and goats have been genetically engineered to produce pharmaceutical drugs in their milk, creating an industry called biopharming. PCR has made forensic analysis cheap, fast, and extremely accurate. Today's DNA profiles have less than a one in a trillion chance of matching another random individual, providing law enforcement with a powerful tool to fight crime. PCR has also been used to compare Neanderthal and human DNA to provide insights into how these populations interacted tens of thousands of years ago. The activities in this chapter use PCR to investigate DNA profiling, to detect genetic modifications in food, and to study human ancestry.

6.1 Invention of PCR

It was on Highway 128 in California at mile marker 46.58 in April 1983 that Kary Mullis (see Figure 6.1) had an epiphany. He pulled off the road and sketched out the process that would later be known as the polymerase chain reaction (PCR). He envisioned the use of small pieces of DNA to bracket and replicate a section of DNA. Mullis was a chemist working at Cetus, one of the first biotech companies in the U.S. (Cetus was acquired by Chiron Corporation in 1991, and Chiron was acquired by Novartis International AG in 2006.) Mullis ran a laboratory that made oligonucleotides (short, single strands of DNA) and was interested in methods for sequencing DNA. After reporting his theory to the company, he was placed on the project full time. In December 1983, Mullis got the process to work and generated millions of copies of the target DNA sequence. Mullis was given a $10,000 bonus at the time of his discovery. He left Cetus in 1986 and won the Nobel Prize in Chemistry in 1993 for his invention. After much controversy regarding the patents for PCR, they were sold to Hoffman-LaRoche for $300 million in 1992.

Figure 6.1. **Kary Mullis.** Mullis won the Nobel Prize in 1993 for the development of PCR.

The Nobel Prize was given to Mullis because of the impact PCR has had on the world. PCR revolutionized molecular biology and affected research in almost all fields of biology and beyond. PCR made gene cloning and DNA fingerprinting accessible and affordable to most research laboratories, whereas these technologies previously could be performed only by specialists at great expense and effort. Even more important, PCR paved the way for brand new technologies such as automated sequencing, which allowed the Human Genome Project and enabled whole new research areas in genomics.

6.2 What Is PCR?

PCR is a simplified version of bacterial DNA replication that copies a specific sequence of DNA (the target sequence) so that it is amplified. The target sequence is replicated again and again to make millions or billions of copies. Copies produced by PCR are called **PCR products** or **amplicons**.

The strength of PCR lies in its ability to specifically target a section of DNA within a much larger quantity of DNA, such as a whole genome. The sequence is targeted with short, single strands of DNA, called **primers**, which are designed to match and bind to each end of the target sequence. The first primer, called the forward primer, **anneals** at the beginning of the targeted region of DNA, and the second primer, called the reverse primer, is designed to bind at the end of the targeted region (see Figure 6.2). Primers provide the specificity of PCR, selecting the region to be amplified.

Special DNA polymerases are used in PCR. They are stable at high temperatures (thermophilic) so they are not denatured by the 94°C heat that is necessary to separate the DNA strands. DNA polymerases are derived from thermophilic bacteria that exist in hot environments (see Figure 6.3).

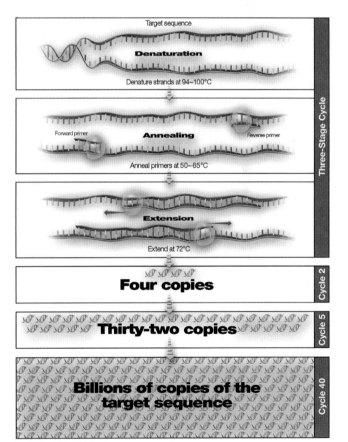

Figure 6.2. **The polymerase chain reaction.** One cycle of PCR consists of separation of strands (denaturation), binding of primers to the single-stranded DNA in a specific location to bracket the target sequence (annealing), and extension of the primers by DNA polymerase, which reads the sequence from the template DNA strand and adds complementary nucleotides to the 3' end of the primers (extension). These three steps are repeated and the number of copies of the target sequence doubles each cycle. After cycle 2, there are 4 copies, by cycle 5 there are 32 copies, and by cycle 40 there are billions of copies.

PCR has revolutionized DNA analysis because it provides a way to make a lot of DNA with specificity and speed. PCR has made many techniques possible that were previously unimaginable.

Figure 6.3. **Geyser at Yellowstone National Park.** Taq DNA polymerase was the first thermophilic DNA polymerase used in PCR. This enzyme was isolated from *Thermus aquaticus* bacteria found in hot springs in Yellowstone National Park. Taq is still the most common DNA polymerase used in PCR.

Three Stages of PCR

A PCR reaction is comprised of three stages: **denaturation**, **annealing**, and **extension**, which are repeated again and again (see Figure 6.4). Each stage occurs at a different temperature because DNA changes its state, from double-stranded to single-stranded, depending on the temperature.

Denaturation

DNA must be single-stranded before it can be replicated, so the first step of PCR is to denature the **template DNA** that will be copied in the PCR reaction by heating the reaction to 94°C. The high temperature causes the DNA double helix to separate by breaking hydrogen bonds between base pairs, resulting in single-stranded DNA.

Annealing

Once the DNA is single-stranded, the next step of the PCR cycle is for the primers to anneal, or bind to the target sequence. The PCR reaction is cooled to a temperature that allows hydrogen bonds to form between the primers and the single-stranded template DNA. The temperature at which primers bind to the template is called the annealing temperature. The optimal annealing temperature is specific for each pair of primers (usually 50–60°C) and depends on the length and sequence of the primers. This temperature is low enough for the primers to bind, but high enough to discourage reforming of the double-stranded template DNA.

Extension

The next step of the PCR cycle is for a DNA polymerase to extend the primers by reading the complementary strands and by adding matching **nucleotides** to the 3' end of each primer. The DNA polymerase binds to the region of double-stranded DNA created by the binding of the primer to the template DNA. The DNA polymerase

reads the template strand in the 3' to 5' direction and adds nucleotides or writes the new strand extending in the 3' direction. (The mantra "read 3' 5' and write 5' 3'" helps with remembering the direction of DNA replication.) The most common DNA polymerase used in PCR, Taq DNA polymerase, functions best at 72°C, so the extension step of PCR is usually conducted at 72°C.

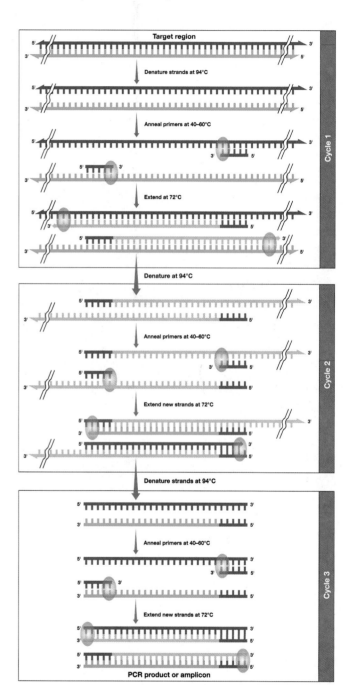

Figure 6.4. **PCR requires three cycles to form the PCR product.** In the first cycle (labeled Cycle 1), the original template DNA is read and the newly replicated DNA strand starts at the primer sequence and extends beyond the boundary of the other primer. In the second cycle (labeled Cycle 2), DNA made from the template strand during the first cycle is the length of the target DNA but is only single-stranded. In the third cycle (labeled Cycle 3), DNA made from the template strand during the third cycle is the exact complement of the Cycle 2 strand and these two strands form a DNA double helix that is the length of the target sequence; the PCR product.

The three stages of PCR—denaturation, annealing, and extension—comprise one cycle of PCR, which is then repeated 25–40 times to amplify the target sequence. A double-stranded PCR product is not formed until the third cycle of PCR (see Figure 6.4). The PCR product, or amplicon, is the DNA sequence from the 5' end of one primer to the 5' end of the other. Theoretically after 25 rounds of PCR, there are 2^{22}, or 4 million, times more copies, and after 35 rounds of PCR, there are 2^{32}, or 4 billion, times more copies of the target sequence than were present at the beginning of the reaction. It is this power of amplification that enables forensic scientists to amplify the DNA remaining in a bloody fingerprint at a crime scene so that a suspect can be identified, and makes it possible for fertility clinic technicians to screen the DNA of a single cell of a preimplantation embryo.

PCR and DNA Replication

When the DNA within a bacterial cell is replicated, the entire DNA genome (4.6 million base pairs (bp) in *E. coli*) is copied to make two genomes of DNA in a directed process involving many different enzymes. One copy of the genome is packaged into the original cell and the other is packaged into the daughter cell. When DNA is replicated in a tube using PCR, a short segment of DNA composed of a few hundred to a few thousand bp is selected from

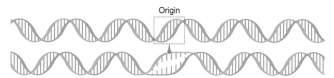

Figure 6.5. **Origin of replication.** DNA replication starts at the origin of replication, which forms a "bubble" of separated DNA. Each side of the bubble is a replication fork, where DNA replication will start and progress in both directions.

a much longer piece of DNA (often an entire genome) and replicated again and again to make millions or billions of copies of that segment of DNA using only one type of enzyme.

DNA replication in a cell or in a tube uses the same fundamental process but with some significant differences (see Table 6.1). In both scenarios, double-stranded DNA is unwound and then separated into single template strands. Then short complementary strings of nucleotides (primers) are bound to the template DNA strands so that a DNA polymerase can bind to the double strands and start to copy the template DNA. A DNA polymerase adds nucleotides to the 3' end of the primer to extend the new, complementary DNA strand (see Figure 6.6).

Table 6.1. **Comparison of PCR replication and in vivo replication.**

Common Processes	PCR	Replication *In Vivo*
Separation of double-stranded DNA	The DNA template is heated at temperatures >94°C to break the hydrogen bonds holding the double helix together, which makes all the DNA single stranded (see Figure 6.2).	Replication initiator proteins separate the DNA at origins of replication (ori) (see Figure 6.5) and recruit the other proteins involved in replication. The two strands of DNA remain joined together ahead of the replication fork. Topoisomerase gradually unwinds the DNA helix ahead of the replication fork (see Figure 6.6). Helicase unzips the double-stranded DNA by enzymatically breaking the hydrogen bonds between the DNA strands. Single-stranded binding proteins prevent the DNA from reannealing.
Annealing of primers to the single-stranded DNA	DNA primers are designed based on the sequence of the target DNA. Primers are chemically synthesized.	RNA primers are made by primase, which reads the template DNA.
Extension of DNA by DNA polymerase that reads the template strand in the 3' to 5' direction and adds nucleotides to the 3' end of the primer in the 5' to 3' direction	Heat-stable (thermophilic) DNA polymerase extends the DNA primers. The nucleotides used to make the new DNA are added to the reaction by the researcher.	DNA polymerase III extends the RNA primer. Because DNA is made at a replication fork, the leading strand of DNA is extended as a continuous piece, while the lagging strand is replicated in discontinuous fragments (Okazaki fragments), using multiple RNA primers. Once the new DNA strand has been replicated, RNA nucleotides from the RNA primer are exchanged for DNA nucleotides by DNA polymerase I. Ligase joins the discontinuous Okazaki fragments. The nucleotides used to make the new DNA are made by metabolic reactions in the cell.
DNA replication complete	A specific section of the template DNA (the target sequence) has been made. The process is started all over again.	Two complete copies of the template DNA have been made and one copy is packaged into a daughter cell.

DNA polymerases extend DNA by adding a nucleotide to an existing DNA molecule. DNA polymerases read the complementary DNA strand and select a nucleotide that is complementary. They then catalyze a reaction that links the phosphate group attached to the 5' carbon of one nucleotide to the hydroxyl group on the 3' carbon of the next nucleotide. This reaction releases pyrophosphate (see Figure 6.7).

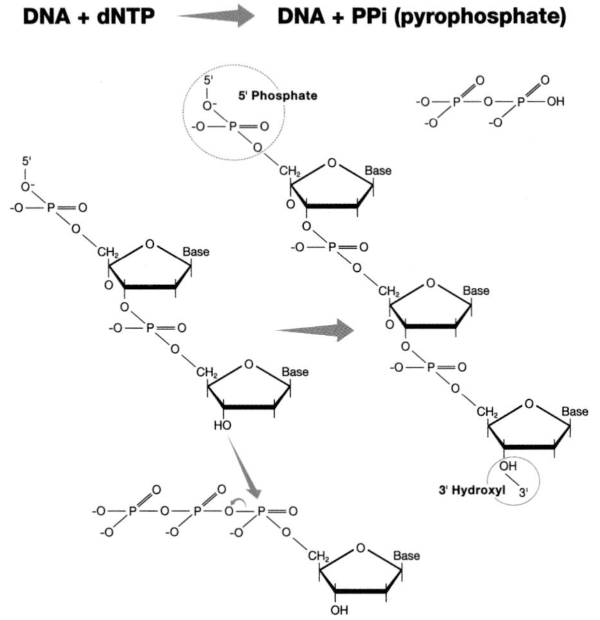

Components of a PCR Reaction

PCR reactions require specific components:

- Template DNA
- Nucleotides (dNTPs)
- PCR buffer
- Magnesium chloride (MgCl₂)
- Water
- Forward and reverse primers
- DNA polymerase

The template DNA is often genomic DNA (gDNA), but it may also be plasmid DNA or **complementary DNA** (**cDNA**). cDNA is a complementary copy of mRNA. Primers are designed based on the sequences at each end of the target DNA. Primer sequences must be designed specifically for the PCR reaction and then optimized. Primers are chemically synthesized by specialized laboratories. When primers are ordered, they arrive freeze-dried and must be hydrated to produce concentrated stock solutions that are stored at –20°C. **Deoxyribonucleoside triphosphates** (**dNTPs**, or nucleotides) are the building blocks of the new strands of DNA (see Figure 6.8) and provide the power source for the reaction. dNTPs are added as a mixture of deoxyadenosine triphosphate (dATP), deoxythymidine triphosphate (dTTP), deoxyguanosine triphosphate (dGTP), and deoxycytidine triphosphate (dCTP); all of the dNTPs are added at the same concentration. The PCR buffer provides the optimal ionic concentration of monovalent and divalent cations and buffers to maintain pH for optimal enzyme activity. MgCl₂ is an essential cofactor for DNA polymerases. The first and most common DNA

DNA + dNTP ⟶ **DNA + PPi (pyrophosphate)**

Figure 6.7. **DNA extension reaction.** A nucleotide is added to a DNA molecule to form DNA with one extra nucleotide and a pyrophosphate (PPi) molecule. Hydrolysis of the pyrophosphate releases energy to drive DNA synthesis.

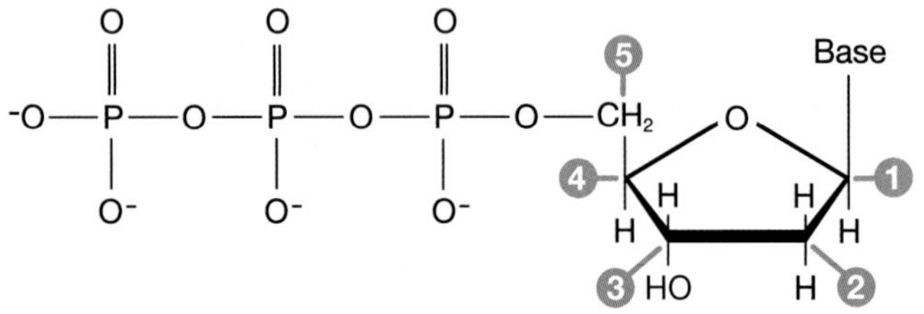

Figure 6.8. **Deoxyribonucleoside triphosphate.** The nucleoside portion is composed of the sugar (deoxyribose) and the nitrogenous base (adenosine, thymidine, guanosine, or cytidine). The nucleoside is bonded to three phosphate groups at the 5' position.

polymerase used in PCR is Taq DNA polymerase, derived from *Thermus aquaticus*, thermophilic bacteria found in geysers in Yellowstone National Park (see Figure 6.3).

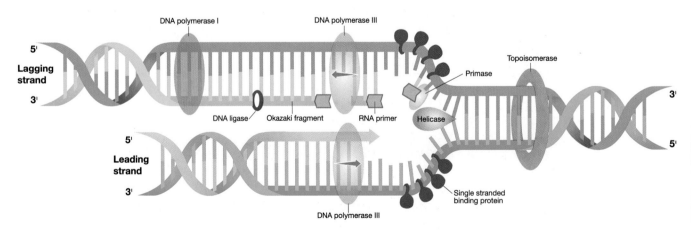

Figure 6.6. **In vivo replication of DNA.** Topoisomerase unwinds DNA and helicase separates the DNA strands that are then kept separate by single-stranded binding proteins. Primase adds RNA primers that are extended by DNA polymerase III on the leading and lagging strands. The lagging strand is made discontinuously and fragments are joined by ligase. DNA polymerase I replaces the RNA primers with DNA.

Setting Up a PCR Reaction

Specialized tubes, plates, and tips are needed to set up reactions, and special precautions are necessary to prevent contamination or suboptimal results. PCR reactions can be set up in individual PCR tubes, in PCR tube strips of 8 or 12 tubes, or in plates that have 48, 96, or 384 wells to hold reactions (see Figure 6.9). These tubes, strips, and plates have specially designed thin walls to efficiently transfer heat from the thermal cycler (an instrument that automates PCR cycles) to the reactions.

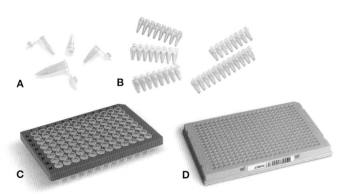

Figure 6.9. **PCR tubes, strips, and plates.** A, 0.2 ml and 0.5 ml thin-walled PCR tubes; B, 8-well and 12-well PCR tube strips; C, 96-well PCR plate; D, 384-well PCR plate.

Due to its sensitivity, PCR is very prone to contamination. For example, reagents that have aspirated into the pipet barrel can be transferred into the reaction, which can lead to amplification of the wrong DNA and false results. To prevent contamination, aerosol-barrier pipet tips containing filters, screw-cap tubes, and gloves are used, and PCR reactions are often set up in designated workstation cabinets that have UV lights installed to destroy DNA and decontaminate the work area between amplifications. In addition, all experiments have negative controls to ensure cross-contamination has not occurred.

When setting up PCR reactions, it is important to keep the reagents on ice until the reaction is ready to be placed into the thermal cycler. This prevents non-specific amplification of the template DNA caused by primers binding inappropriately to the template or to themselves.

Master Mixes

DNA polymerase, dNTPs, primers, and reaction buffer are often combined into a **master mix**. A master mix is a bulk preparation of enzymes and buffers required in each of a series of reactions. Use of a master mix is highly recommended because it eliminates the need for transferring small volumes (some <1 μl) of each component to each reaction tube, which can introduce errors. A master mix also ensures that the components in the reaction are present in exactly the same concentrations in each reaction. Master mixes can be purchased (without primers) or produced in the laboratory.

Preparing a master mix requires that the amount of each component be calculated for the required number of PCR reactions. Each component is then added to the master mix tube so that it is at the appropriate concentration in the final reaction (see Table 6.2).

Table 6.2. **Components of a PCR reaction.**

Component	Common Stock Concentration	Final Concentration or Amount per Reaction
DNA template	N/A	Genomic DNA: 50–500 ng, plasmid DNA: 50 pg–50 ng
dNTPs	10 mM	0.2 mM
$MgCl_2$	50 mM	1–6 mM
PCR buffer	10x	1x
Forward primer	100 μM	0.1–1 μM
Reverse primer	100 μM	0.1–1 μM
DNA polymerase	Varies depending on the polymerase and the manufacturer	Usually <1 μl

To set up a PCR reaction, a master mix containing the DNA polymerase, dNTPs, primers, and reaction buffer is added directly to template DNA in the PCR tube. Once the tube is capped, the samples are ready to be placed in the thermal cycler.

The amount of template DNA varies depending on the PCR reaction. For example 50–500 ng of gDNA is usually used for PCR, while 50 pg to 50 ng of plasmid DNA is sufficient. A higher amount of gDNA is required compared to plasmid DNA because there is only one target sequence in almost 3 billion bp of gDNA. In contrast, plasmid DNA has one target in a few thousand bp. PCR can also be optimized to amplify much lower amounts of DNA (see PCR Optimization).

Analysis of PCR Products

After PCR, products need to be assessed. PCR products can be analyzed using horizontal agarose gel electrophoresis as described in Chapter 4. PCR products are usually smaller than plasmid DNA (often between 100 and 3,000 bp), so they must be analyzed on gels made with a higher percentage of agarose than gels used to analyze most plasmid DNA. PCR products less than 1,000 bp are analyzed on 2–4% agarose gels—the smaller the expected PCR product, the higher the agarose percentage needed.

Careers In Biotech

Dora Barbosa

Research Associate
University of California
San Francisco, CA

Dora's interest in biotechnology began when she was 10 years old, when her mother was diagnosed with advanced lymphoma. Constantly overhearing the doctors speak about things neither she nor her parents could understand made her want to learn more about the disease, how it had impacted her mother, and what could be done

Photo courtesy of Dora Barbosa

about it. During her high school biology class in Tracy, CA, her innate curiosity was reignited when she first held a pipet and performed an experiment. Curiosity is still at the heart of all she does. "There is something spectacular about knowing that there is an infinite amount of things you can not only learn from others, but that you can also discover yourself," says Dora.

After graduating from Tracy High School, Dora attended the University of California, Berkeley, where she studied within the Biology Scholars Program.

For those interested in a career in biotechnology, Dora's advice is to obtain hands-on experience early on, as this is the best way to figure out whether you are choosing the right field. "There are so many opportunities out there, all you have to do is look for them. It took me six years and countless jobs to get the position I have always wanted," she says. "Remember that not everyone takes the same path, but when opportunities come be ready to take them on."

Dora believes that the biotechnology industry is going to be in for some unbelievable discoveries and boundary breaking in the next ten years. Many opportunities are opening up to individuals of various backgrounds, and she believes we will continue to see the value of diversity and collaboration.

6.3 Thermal Cyclers

For a PCR reaction to work, the reaction must be transferred between three different temperatures quite rapidly. When PCR was developed, samples were manually transferred between three separate water baths, each at the temperature required for a PCR step. This laborious process was soon superseded by the invention of thermal cyclers. A thermal cycler is an electronically controlled heat block that can quickly change from one temperature to another and can be programmed to hold and change temperature according to the required cycling parameters (see Figure 6.10). Thermal cyclers are also referred to as PCR machines.

Figure 6.10. **Thermal cycler.** Bio-Rad's T100™ thermal cycler amplifies 96 samples.

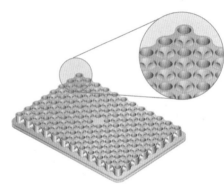

Figure 6.11. **An example 96-well heating block.**

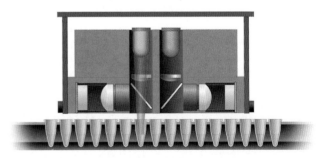

Figure 6.12. **Optics shuttle schematic of a real-time thermal cycler.** The optics shuttle travels across the plate of samples and measures the concentration of amplified DNA. A source light is used to excite the fluorescent dye and the fluorescence is detected. Different wavelength source lights can be used to detect different dyes.

Most thermal cyclers have the same basic features:

- Heat block with holes, or wells, hold 0.2 or 0.5 ml PCR tubes (see Figure 6.11). The number of wells ranges from 8 to 384 per block, with 96 wells being the most common. Some thermal cyclers have multiple heat blocks that can be programmed differently, allowing different PCR reactions to be run concurrently. The speed at which the block can change from one temperature to another is called the ramping speed. The temperature is changed quickly by using electricity (the Peltier effect) or by blowing air over the heating element. The faster the ramping time, the quicker the PCR can be completed. Some heat blocks can generate a thermal gradient, which allows different annealing temperatures to be tested

- Heated lids prevent the reactions from evaporating at 94°C and then condensing on the colder tube cap. When using a cycler that does not have a heated lid, a layer of mineral oil is applied to the top of the samples in the PCR tubes prior to the reaction

- Graphical interfaces allow for intuitive programming. Some advanced thermal cyclers are controlled by specialized software on an accompanying computer. The software regulates multiple parameters, provides detailed information on the run for quality control purposes, and controls multiple thermal cyclers or heat blocks on one thermal cycler

There are two main categories of thermal cyclers: conventional PCR instruments and **real-time PCR** instruments. Real-time PCR thermal cyclers have the same features as conventional thermal cyclers, but they also have an optical module that detects increases in fluorescence as PCR products are being made in real time, allowing the original template to be quantified (see Figure 6.12).

6.4 Types of PCR

Since the invention of PCR, researchers have developed many variations of the method. One of the most recent advancements in PCR has been digital PCR (dPCR). Other types of PCR include real time or quantitative PCR (qPCR), reverse transcription PCR (RT-PCR), multiplex PCR, and nested PCR. Different types of PCR are frequently combined to achieve the goals of the test being performed.

Real-Time, or Quantitative, PCR

In real-time, or **quantitative**, **PCR** (**qPCR**), the amount of PCR product is measured as each cycle is completed in real time and used to deduce the amount of input DNA. In conventional PCR, PCR products can be analyzed only at the end of the reaction, and inferences on amount of starting material cannot be made. Real-time PCR enables researchers to quickly measure DNA and gene expression levels, which previously required the use of complicated protocols that involved radioactivity.

Real-time PCR detection systems detect increases in fluorescence as new PCR products are made (see Figure 6.12). DNA is detected by a few different methods. One common method uses fluorescent DNA stains such as EvaGreen® Dye or SYBR® Green supermix that

specifically bind double-stranded DNA. During the extension step, the fluorescent stain binds to the double-stranded DNA. Following each extension step, the PCR reaction is illuminated at a specific wavelength and the fluorescence that is emitted is recorded. To quantify the amount of template, the level of fluorescence of real-time PCR reactions is compared to controls. Using the concentration of control samples, the concentration of the unknown sample can be determined from a standard curve (see Figure 6.13). Real-time PCR is used for detection of some pathogens in human serum or urine including the Zika virus.

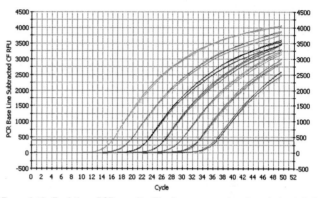

Figure 6.13. **Real-time PCR results.** The fluorescence at each cycle is plotted against the cycle number for each PCR reaction. The cycle at which the fluorescence passes a threshold (called C_q) is determined and used to compare different reactions. In the example, the differently colored lines from left to right represent PCR reactions in triplicate with tenfold reductions in the quantity of starting DNA. The horizontal red line represents the threshold, and the C_q value for each reaction is the cycle where each curve intersects the threshold.

Digital PCR

Digital PCR (**dPCR**) builds on traditional PCR and fluorescent probe–based detection but allows more sensitive detection and direct measurement of nucleic acid concentration without the use of standard curves. Because of its sensitivity, dPCR is used for detecting rare or low-abundance gene targets and small changes in gene expression.

The key to digital PCR is the division of a sample into many separate samples that are small enough to contain either zero or one (or a few) template molecules. Each sample is then amplified separately by PCR. Samples that contain amplified product are considered positive (1, fluorescent), and those without product are negative (0, little or no fluorescence), hence the name "digital" PCR. The ratio of positives to negatives in each sample is the basis for quantitation (see Figure 6.14).

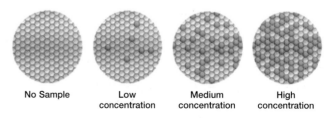

| No Sample | Low concentration | Medium concentration | High concentration |

Figure 6.14. **Digital PCR.** Each droplet in each sample undergoes PCR amplification and analysis separately. The droplets are then individually counted and scored as positive (green) or negative (white, or blank) for fluorescence. The ratio of positive to negative droplets is used to determine the concentration of template in each sample.

How To...

Program a Thermal Cycler

Thermal cyclers can save a number of programs for various uses (see Figure 6.15). Each thermal cycler is different, so refer to the instruction manual provided by the manufacturer for details on programming the cycler.

1. Write up the cycling parameters in a table similar to Table 6.3. The duration of each step in your program will depend on several factors including your template length, primer annealing temperatures, and even the thermal cycler.

2. Create a new program and name it. To set up a new program, select the option to create a new program. The thermal cycler may have saved templates or model programs that can be modified as needed. The program may also be developed from scratch.

3. If necessary, add a new cycle. Enter the initial denaturation cycle by entering the temperature in degrees Celsius (°C) and duration in minutes. The initial denaturation is usually performed at 94–96°C for 2–5 minutes. This cycle will be run once.

4. If necessary, add a new cycle and three steps. Enter the temperature and duration of each of the three steps of the main thermal cycle. Enter the number of times to repeat this cycle (this is usually 25–40).

5. If necessary, add a new cycle. Enter the final extension cycle by entering the temperature and duration. For Taq DNA polymerase, the final extension cycle is usually performed at 72°C for 5–10 minutes. This cycle will be run once.

6. If necessary, add a new cycle. Enter the final hold by entering the temperature and choosing the infinity symbol or the "forever" option on the thermal cycler. The thermal cycler will hold that temperature until the user stops the program. Incubation at 15°C is adequate for the final hold step.

7. Name the program (if not already done in step 2), and save it.

Table 6.3. **Example of the cycling parameters for a three-step PCR program.**

Cycle	Step	Temperature	Duration	No. of Repeats of Each Cycle
Initial denaturation	Denature DNA	94°C	5 min	1x
Thermal cycling	Denature	94°C	1 min	40x
	Anneal primers	52°C	1 min	
	Extend	72°C	2 min	
Final extension	Extend	72°C	6 min	1x
Hold	Hold	15°C	∞	1x

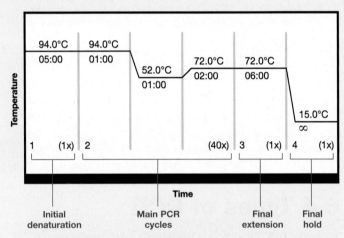

Figure 6.15. **A PCR program on a thermal cycler display**. In the PCR program shown here, the cycles and steps are separated by vertical lines. The horizontal line provides a visual representation of changes in temperature. Cycle 1, the initial denaturation, has a single hold step at 94°C for 5 minutes. Cycle 2 has three steps: 94°C for 1 minute, 52°C for 1 minute, and 72°C for 2 minutes. The steps in cycle 2 are repeated 40 times. Cycle 3, the final extension step, is a single step of 72°C for 6 minutes. Finally, the reaction is held indefinitely (∞) at 15°C to maintain the PCR products.

Reverse Transcription PCR

Reverse transcription PCR (RT-PCR) uses the retroviral enzyme reverse transcriptase to reverse transcribe mRNA into DNA before the actual PCR reaction begins. The resulting DNA is cDNA. The cDNA is used as a template for PCR. The mRNA is reverse transcribed during a first cycle, when the reaction is held at approximately 42°C for 30–60 minutes, after which the reaction is conducted as a normal PCR reaction. RT-PCR is most frequently used in combination with real-time PCR to measure gene expression levels. RT-PCR has been used to detect RNA viruses in blood and was instrumental in diagnosing cases of influenza A H1N1 virus that caused a pandemic in 2009. (**Note**: RT-PCR is often confused with real-time PCR due to its abbreviation, real-time PCR is abbreviated to qPCR.)

Multiplex PCR

Multiplex PCR reactions simultaneously detect multiple target sequences on the same starting material in the same reaction tube. In multiplex PCR, multiple sets of primers each target and amplify a different sequence. Uses of multiplex PCR include forensic DNA profiling, where multiple sequences are identified in a single reaction, and detection of genetically modified organisms (GMOs) in food samples, in which the specific genetic modification of the originating plant may be unknown. Activity 6.B uses multiplex PCR to test for GMOs in food.

Degenerate PCR

Degenerate PCR can amplify DNA when the information about the target sequence is limited or when the same PCR needs to work with DNA templates from different species. Degenerate primers are sets of primers that have almost the same sequence as each other, differing in 1–3 bases, and made to compensate for the differences in sequence of the target DNA. Use of degenerate primers improves the likelihood of successful PCR and removes the need to design individual primers and to test each one separately. A set of degenerate primers usually has 1–3 different base pair positions, which can mean 24 different primers in the reaction—12 forward and 12 reverse. Activity 6.D uses degenerate PCR to identify a variety of possible fish species with DNA barcoding.

Nested PCR

When PCR conditions are suboptimal, primers may anneal inappropriately or nonspecifically and therefore amplify incorrect targets as well as the desired target. This will produce unwanted PCR products as well as the desired amplicon. A nested PCR reaction uses a second round of PCR and a second set of primers that bind within the PCR product produced during the initial round of PCR. A nested set of primers ensures that the product generated is specific, since the chance of nested primers binding nonspecifically to a PCR product from the initial round (rather than to the original full-length template) is minimal.

Fast PCR

The total run time for traditional three-cycle PCR is relatively long, 2.5–4 hours. In fast PCR, the total run time is 0.5–1.5 hours. This is achieved by using specially designed primers and a two-step PCR program. The primers anneal at a similar temperature to the extension temperature (68–72°C), so that the annealing and extension steps are combined into a single step. The reaction times at each step are also shortened. The denaturation step is reduced to 1–5 seconds and is run at a slightly lower temperature than in standard PCR (92°C rather than 95°C), so that less ramping time is required. An example of a fast PCR reaction is an initial denaturation step at 98°C for 30 seconds, 35 cycles of 92°C for 1 second followed by 70°C for 15 seconds, and then a final extension for 1 minute at 72°C; the entire protocol takes less than 30 minutes to complete. Fast PCR must be optimized and does not work for all reactions.

Isothermal PCR

Isothermal DNA amplification techniques are carried out at a single temperature instead of in a cycle of three temperatures, as in traditional PCR. These techniques are typically faster than traditional PCR and are great for mobile labs. Loop mediated isothermal amplification (LAMP), an example of isothermal amplification, requires multiple sets of carefully designed primers and a polymerase that displaces a strand of the DNA as it polymerizes. Other types of isothermal amplification use different strategies for separating the strands of dsDNA to complete polymerization. These techniques are primarily used for detection of particular sequences because the products cannot be used in cloning.

Random Amplification of Polymorphic DNA

Random amplification of polymorphic DNA (RAPD, pronounced "rapid") is a method of DNA fingerprinting. It is used when the genomic sequence of a particular organism is not known. RAPD identifies differences among similar genomes and is often used to differentiate closely related plant species or bacterial strains. In RAPD PCR, a mix of random primers is used to amplify the template DNA. The primers are decamers (ten nucleotides long), and four to eight primers are used in a single PCR reaction (multiplex PCR). The sequences of the primers are random, meaning that they are not based on any specific sequence; however, the same primers are used to compare samples. The primers are short enough to recognize multiple sites on the template DNA. If two primers anneal closely enough to one another in the correct orientation, they will generate a PCR product. The number and size of the PCR products generated can be used to compare species. Species with the exact same genome will generate the exact same number of PCR products, whereas different species will produce different PCR products.

6.5 PCR Optimization

Each PCR reaction is different. The components and conditions of each reaction are specific to the target region, the primers used, the source of DNA template, and the goal of the PCR. The sequences at the ends of the target region determine the choice of primers. The optimal types of reaction components need to be selected based on the purpose of the PCR assay. For example, if the goal of the PCR is to clone a gene, it is important that the PCR products are free of errors. In this case, a thermally stable DNA polymerase that can proofread the DNA, such as Pfu isolated from the bacterium *Pyrococcus furiosus,* may be chosen. In cases where the goal of PCR is only to determine the presence of a DNA sequence, a basic Taq DNA polymerase is appropriate.

The quality of the template DNA influences the success of the PCR. For example, naturally occurring compounds from the original tissue sample and impurities in the template DNA preparation can inhibit PCR. Additionally, PCR is very sensitive so steps should be taken to reduce the risk of sample contamination. In industry, thermal cyclers are sometimes given their own hood or room to reduce the chances of cross contamination.

Each PCR reaction must be optimized; in other words, the best combination of components and conditions needs to be determined. Each factor that can influence the reaction when changed is called a **variable**. The process of optimization tests the variables in a PCR reaction to determine the optimal reaction conditions. When optimizing, only one variable at a time should be changed so that the effect of the change can be determined.

Quality of Template DNA

PCR reactions ideally have template DNA comprised of pure, unbroken DNA strands at an optimal concentration. Working with DNA templates of low quantity, quality, or purity is very difficult but sometimes it is unavoidable. For example, ancient DNA from archeological finds or DNA from processed foods can be very degraded, meaning the DNA is broken up into small fragments. PCR can be optimized to work on low-quality DNA but the target sequences must be short (about 100 bp) to improve the chances of the primers finding a fragment containing the complete target. Some PCR reactions can be optimized to work on minute quantities of DNA, although the techniques for optimizing such reactions are challenging. Too much template DNA can inhibit a PCR reaction, while too little can generate too little product.

A relatively crude genomic **DNA extraction** can be performed using a **chelating agent** such as InstaGene™ matrix. Samples are boiled in InstaGene matrix to break open cells and release the DNA. The Chelex® resin in InstaGene matrix binds divalent cations such as Ca^{2+} and Mg^{2+} to prevent cellular DNases (enzymes that degrade DNA) from destroying the DNA. (Cellular DNases need divalent ions as cofactors.) Although the DNA extract still contains the cellular proteins and other molecules, it is sufficient for many PCR reactions. DNA purified with this method will be used in Activities 6.B and 6.C.

If a purer sample is needed, cells are lysed with special detergents and the DNA is purified by passing the cell lysate over a silica spin column. DNA binds to the silica in the column in the presence of alcohol, while the other cellular material passes through the column. Pure DNA is then collected from the column in an aqueous solution. The best method for DNA extraction for each sample should be determined experimentally.

Primer Design

Two primers are specifically designed for each PCR reaction to amplify the target sequence. Primers are short, single-stranded pieces of DNA that are usually designed to bind to one sequence in the entire template. Primers can vary in sequence and length. Many factors influence primer design:

- Primer sequence must match sequences at each end of the target region

- The **melting temperature (T_m)** of each primer in the pair must be within a few degrees Celsius of each other

- Primers should be designed without intra- or intercomplementary sequences to minimize the formation of primer secondary structures called primer-dimers

The options for primer sequences are dependent on the target sequence. If the sequences at the edges of the region of interest do not provide good options, the target sequence may sometimes need to be expanded by approximately one hundred base pairs to provide good candidate primer sequences. However, depending on the application, primer choice may be limited to suboptimal primers and the PCR reaction must be optimized around the suboptimal primers.

The T_m is the temperature at which half of the primers disassociate from the template DNA. The T_m provides a guide for determining the annealing temperature, which is then programmed into the thermal cycler for the annealing step of PCR. The T_m of the two primers needs to be similar, since only one annealing temperature can be used. The T_m should ideally be 50–65°C, while the annealing temperature is usually 3–5°C lower than the T_m.

The T_m is dependent on the length and base pair composition of the primers. The longer the primer sequence, the less likely that the sequence will occur anywhere else in the template and the more hydrogen bonds there are to hold the primer in position. (Longer primers have higher T_m values.) Similarly, Gs and Cs have three hydrogen bonds compared to two hydrogen bonds with As and Ts; therefore, primers with a higher proportion of Gs and Cs have higher T_m values. The length of a primer should be 17–28 bp and the GC content should be 40–60%. In addition, primers are often designed with GC clamp, a term that refers to the presence of 1 to 3 G or C bases at the 3' end, which increases specific binding of the primer.

Different formulas can be used to calculate the T_m. These formulas take into account the length of the primer and the number of As, Ts, Cs, and Gs in the primer. (Many online primer calculators determine the T_m using complex formulas). The most basic formula is:

$$T_m = [4°C \times (G+C)] + [2°C \times (A+T)]$$

where G, C, A, and T represent the number of each base in the primer. The optimal annealing temperature must be tested experimentally. If the annealing temperature is too low, primers may bind non-specifically, and if it is too high, primers may not bind well enough to be properly extended into PCR products. Many thermal cyclers have a gradient function that generates a thermal gradient, allowing multiple annealing temperatures to be tested at one time (see Figure 6.16).

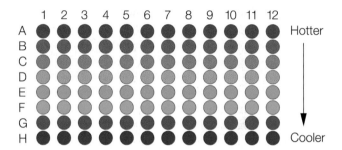

Figure 6.16. **Thermal gradient on a 96-well heat block.** Row A can be set at a higher temperature than row H and the rows in between will form a gradient between the temperatures of A and H, allowing multiple annealing temperatures to be tested with a single PCR run.

Primer design must also take into account the possible formation of primer secondary structure. Intercomplementarity occurs when a section of the primer sequence is complementary to its paired primer, resulting in the two primers binding to each other and forming primer-dimers, which are short PCR products based on the primer sequence. Intracomplementarity occurs when a primer binds to a section of its own sequence, forming structures such as hairpin loops (see Figure 6.17). Intracomplementarity changes the sequence of the primer available to bind to the template, leading either to the amplification of a non-specific product or to no PCR product being produced. (Primer design software available on the Internet helps identify primer secondary structure(s).) The formation of primer-dimers or hairpin loops also reduces the quantity of primers available for amplification.

Primer   T A C A C G A A G T G C T G C T T C G G A T C

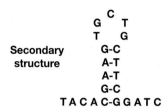

Figure 6.17. **Primer secondary structure.** The primer shown here has intracomplementarity, which results in two complementary sections of the sequence annealing to each other to form a hairpin loop. Primers should be designed without intracomplementary sequences so hairpin loops do not form.

Cycling Parameters

The cycling parameters are temperature, length of time for which the temperature is held for at each step of the PCR program, and the number of cycles. All of these parameters are variables that need to be optimized:

- The annealing temperature is affected by the T_m of the primers and must be determined experimentally

- The temperature of the extension step is determined by the type of DNA polymerase used for the reaction

- The length of time for which the temperature is held at each step is dependent on the length of the target sequence and the processing speed of the DNA polymerase. For example, Taq processes 1 kb of target sequence in 1 minute. Short target sequences usually have shorter incubation times than long target sequences

- The number of cycles is dependent on the quality and amount of starting template. Low-quality or scarce template requires more cycles to generate a visible PCR product than a high-quality template that is present in abundance. The number of cycles is usually between 25 and 40. The polymerase degrades during the reaction so later cycles are less efficient. Therefore, performing more than 40 cycles usually does not generate more product. The total cycling time for PCR can range from 30 minutes to 4 hours depending on the type and purpose of the PCR and the cycling parameters used

Magnesium Concentration

The cofactor magnesium is necessary for DNA polymerases to function and is included at a minimal level in most PCR reaction buffers. Varying the concentration of Mg^{2+} can improve PCR results. To optimize the Mg^{2+} concentration, a series of PCR reactions is set up with all components and conditions held constant except the $MgCl_2$ concentration, which ranges from 1–6 mM. The $MgCl_2$ concentration that results in the best amplification is chosen for the reaction.

6.6 Techniques Based on PCR

PCR has become an integral part of many modern biotechnological techniques such as DNA microarrays and **DNA sequencing**.

DNA Microarrays

DNA microarrays can be used to determine which genes in a cell have been turned on or turned off. Information like this can help cancer researchers understand why cancerous cells behave abnormally, or provide clues as to how different cells respond to different drugs. DNA microarray assays allow scientists to compare the levels of DNA or RNA among samples. For example, researchers using microarrays can compare the mRNA levels of specific genes between cancerous cells and healthy cells.

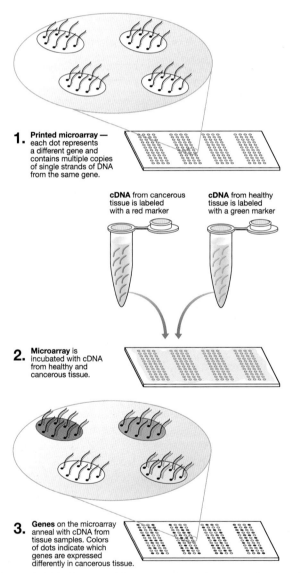

1. **Printed microarray —** each dot represents a different gene and contains multiple copies of single strands of DNA from the same gene.

cDNA from cancerous tissue is labeled with a red marker

cDNA from healthy tissue is labeled with a green marker

2. **Microarray is** incubated with cDNA from healthy and cancerous tissue.

3. **Genes** on the microarray anneal with cDNA from tissue samples. Colors of dots indicate which genes are expressed differently in cancerous tissue.

Figure 6.18. **DNA microarrays.** Microarrays can be used to compare the expression levels of genes between two tissue samples (for example, healthy and cancerous tissues).

A microarray is a microscope slide or other solid support (for example, a chip) that has an array of spots composed of hundreds or thousands of different single-stranded pieces of DNA—each representing a different gene sequence—printed in a regular matrix (see Figure 6.18). The process of making a microarray slide is called printing because when the process was invented scientists modified commercial ink-on-paper printers for use in spotting microarrays. Today, specialized microarray printers are used. The single-stranded DNA used for each spot of the array is amplified using PCR from a plasmid template, and the PCR products are then purified. Hundreds or thousands of different plasmids are used to generate sufficient PCR products to print an array. Microarrays can be made by a researcher or purchased from a supplier.

To compare cancerous cells and healthy cells using a microarray, mRNA from each sample is first reverse transcribed into cDNA. The cDNAs are tagged with specific colors. Typically red and green are used, however other colors are possible. The tagged cDNAs are then incubated on a microarray. cDNAs that are complementary to the PCR products printed on the array will bind (hybridize), and the bound spots on the array will fluoresce. If the cDNA from cancerous cells is tagged red and the cDNA from healthy cells is tagged green, a red spot represents a gene sequence that is expressed more highly in cancerous cells. If a spot is green, the expression of that particular gene is higher in normal cells than in cancerous cells. If the spot is yellow, the expression level is the same in both cancerous and healthy cells (red and green fluorescence combine to appear yellow). If there is no color, the gene is not expressed in either cancerous or healthy cells (see Figure 6.18).

DNA Sequencing

Advances in PCR have made it possible to sequence entire genomes of multiple organisms including humans. The elucidation of the human genome is revolutionizing health care by enabling therapies to be designed for an individual's genetic makeup. Before PCR, DNA sequencing required the use of radioactively labeled nucleotides and a large amount of template DNA, and the data took a long time to obtain and interpret. PCR has made sequencing faster and more sensitive, and eliminated the need for radioactivity. A cycle sequencing reaction is similar to a PCR reaction but uses a single primer and a small number of specialized dNTPs called **dideoxynucleotide triphosphates** (**ddNTPs**) in addition to dNTPs. ddNTPs cannot be extended by a DNA polymerase because they do not have a 3'-hydroxyl (3'-OH) group; therefore, incorporation of a ddNTP prematurely terminates the new DNA strand (see Figure 6.19).

Primer Sequences

In cycle sequencing, each ddNTP (ddATP, ddTTP, ddCTP and ddGTP) is linked to a differently colored fluorescent molecule. The sequencing reaction is performed in a thermal cycler for at least 50 cycles. During each round of cycling, the primer is extended, copying the template DNA. Most of the time, dNTPs are incorporated as the strand is extended; however, there is a chance that a terminating ddNTP will be incorporated into the DNA strand, terminating it prematurely and tagging it with a fluorescent dye that

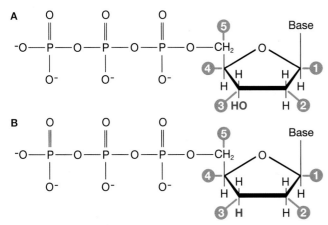

Figure 6.19. **Structure of dNTPS and ddNTPs**. A, dNTPs have a 3'-hydroxyl (3'-OH) group, which is necessary for elongation of DNA. B, ddNTPs do not have a 3'-OH; instead, the 3' position has been modified to have a hydrogen (-H) at that position. When a ddNTP is incorporated into a DNA molecule, the synthesis ends at that nucleotide and the DNA chain is terminated.

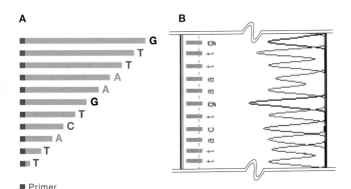

■ Primer

Figure 6.20. **Cycle sequencing**. A, a single primer is extended by copying the template DNA in each round of cycling and may randomly incorporate a fluorescently tagged ddNTP, terminating the strand extension and generating a mixture of fragment sizes; B, when the fragments are electrophoresed, they pass a laser that records the color of the tag on each fragment and presents the results in a chromatogram (also known as an electropherogram). The sequence is determined by reading the terminating nucleotides in order.

indicates the last nucleotide incorporated. With more than 50 cycles, every possible fragment length (up to approximately 800 bp) is generated, and each fragment ends with a fluorescent dye that indicates the last nucleotide incorporated (see Figure 6.20).

After cycling, the tagged DNA fragments are separated by size through capillary electrophoresis (gel electrophoresis in a tiny tube) on a sequencing instrument. As the DNA fragments separate, each fragment passes by a laser, which records the fragment as a colored peak on a chart called a chromatogram. Each peak represents a single fragment and the color of its fluorescent tag (see Figure 6.20). The order of the peaks represents the order in which the fragments passed by the laser, which is the DNA sequence from smallest to largest fragment.

Next-Generation Sequencing

New ways to sequence DNA (called next-generation sequencing, or NGS) are being developed to sequence entire genomes more quickly, more accurately, and at lower cost. Based on PCR and cycle sequencing methods, NGS technologies can analyze millions, if not billions, of sequences at the same time.

Several different NGS methods are available (see Table 6.4), but all comprise the same three general steps (see Figure 6.21):

- **Library preparation**: breaking genomic DNA into smaller pieces (100s–1,000s of base pairs long) and attaching them to a solid surface, like a microchip, slide, or bead. This library of DNA fragments makes it possible to analyze long stretches of genomic DNA as millions of smaller stretches, all at the same time. This is the key to quicker and more efficient coverage

- **Library amplification**: performing PCR amplification of the bound genomic DNA fragments to make clusters of identical sequences. This ensures the signal generated during sequencing will be large enough to be detected accurately

- **Sequencing**: amplified DNA fragments are sequenced by one of several different NGS technologies

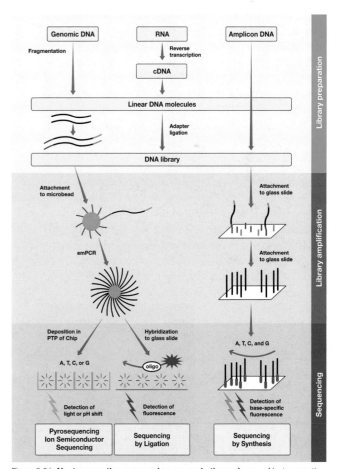

Figure 6.21. **Next-generation sequencing occurs in three phases.** Next generation sequencing occurs in three phases: library preparation, library amplification, and sequencing. During library preparation, different starting materials can be used including genomic DNA, RNA, or amplicon DNA from a previous PCR reaction. Image credit: Knief, C. (2014) Front. Plant Sci. 5:219 licensed under CC BY 3.0, recreated and modified.

Biotech In The Real World

A Universe of Viruses

On an average day, it's estimated that 800 million virus particles fall onto every square meter of our planet, according to a 2018 study by Curtis Suttle and team. Some virus particles are swept up by dust storms, but most enter the atmosphere by sea spray and travel long distances before settling. Viruses prosper in just about every environment on Earth, yet we know surprisingly little about them and the "virosphere" — fewer than 2,200 virus genomes have been sequenced compared to more than 50,000 bacterial genomes.

Viruses have generally been difficult to study because of the challenges of isolating and culturing them in labs. However, newly developed sequencing methods have allowed viruses to be identified in their natural environments. Metagenomics is the study of genetic material recovered directly from environmental samples, which allows thousands of sequences to be obtained from a single sample. Frederik Schulz, a postdoctoral researcher at the Department of Energy Joint Genome Institute in Walnut Creek, California, and colleagues used a metagenomics approach to study the microbes living in sludge from a water treatment plant in Klosterneuburg, Austria. The team discovered a new type of giant virus they named Klosneuvirus. It is larger even than many microbes, and its genome is much more cell-like than those of any other viruses. Genome sequencing revealed that it has the genes for the enzymes needed to synthesize all 20 standard amino acids, the building blocks of life. The genetic complexity found within these newly discovered giant viruses has led some scientists to question the definition of viruses as small and without genes involved in replication and metabolism.

Students are also helping to discover and characterize previously unknown viruses through the Science Education Alliance Phage Hunters Advancing Genomic and Evolutionary Science (SEA-PHAGES) program. Students use metagenomics and other methods to sequence bacteriophages (viruses that infect bacteria) from local environments and submit the annotated sequences to the National Center for Biotechnology Information GenBank database. Students then attend the SEA Symposium to share and discuss their discoveries. In 2016–17, more than 4,100 students from 100 different colleges and universities participated in the program and generated more than 20 peer-reviewed publications related to the discovery of new viruses.

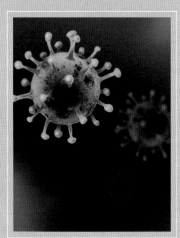

Each sequencing method has its benefits and drawbacks. Whereas cycle sequencing can give much longer sequence reads, the parallel nature of NGS methods allows longer reads to be constructed from many adjoining short reads. This parallel analysis has been key to the improved accuracy and speed of NGS, as well as its reduced manpower requirements and cost. Sequencing a genome has never been so simple or inexpensive (see Table 6.4).

Sequencing by Synthesis

This method is perhaps the most similar to cycle sequencing, in that it uses DNA polymerase to incorporate fluorescent dNTPs one at a time. Each dNTP is labeled with a different color, so the sequence can be determined as the strand is assembled.

Pyrosequencing

Pyrosequencing is based on the release of pyrophosphate (PPi) by DNA polymerase during DNA extension (Figure 6.7). Each dNTP is added one at a time, and when one is added to the growing DNA strand, the PPi produced is detected through chemical reactions that generate light.

Ion Semiconductor Sequencing

This approach is based on the release of protons (H+) during the DNA polymerase reaction. The sequencing microchips position the amplified DNA fragments directly above a semiconductor transistor that detects changes in the pH of the solution whenever a dNTP is incorporated.

Sequencing by Ligation

This method uses DNA ligase instead of DNA polymerase during sequencing. Various fluorescently labeled sequences are added with the ligase, and because ligases are highly specific, the DNA sequence can be determined from whether those sequences are ligated with each sequencing cycle.

Table 6.4. **Next-generation sequencing methods.**

Method	Read Length	Benefits	Disadvantages
Sequencing by synthesis	100–150 bp	High throughput	Expensive, requires high concentrations of DNA
Pyrosequencing	1,000 bp	Fast, long reads	Expensive; errors with runs of the same nucleotide
Ion semiconductor sequencing	2,000 bp	Less expensive, fast	Errors with runs of the same nucleotide
Sequencing by ligation	<100 bp	Less expensive	Slower; may have issues with palindromic sequences

Sequence Data Analysis Using Bioinformatics

Bioinformatics is the use of information technology and computer science to perform biological research. Bioinformaticians apply the power of computer processing to data, such as DNA sequences, protein sequences, and protein structures, to generate new information. Bioinformatics is also essential to help sort and evaluate data, such as the vast amounts of information generated by next-generation DNA sequencing and microarray technologies, data sets far too large to be processed manually. Bioinformatics is used in many different areas of research, including protein modeling, evolutionary biology, and genome mapping. Molecular biologists also need to know some bioinformatics to perform their research.

Bioinformaticians use databases that contain data derived by scientists over many years. One of the largest databases is GenBank, a searchable database that hosts billions of DNA and protein sequences. GenBank is operated by the National Center for Biotechnology Information (NCBI) and is funded by the U.S. National Institutes of Health (NIH). Each sequence record is associated with a unique identifier composed of letters and numbers called an accession number. Scientists use accession numbers to find DNA or protein sequences in GenBank.

Bioinformatic methods are also used to map genomes, that is, to identify important elements such as genes within DNA sequences. DNA sequences are scanned using computer algorithms that identify potential regulatory elements such as promoters, terminators, and stop/start codons that may indicate the presence of a gene.

6.7 Real-World Applications of PCR

PCR in Medicine

PCR is an everyday tool in biomedical research. Most biomedical scientists use PCR to detect, identify, sequence, or clone genes under investigation. Production of biological drugs such as therapeutic antibodies rely on PCR to clone the genes and construct recombinant expression vectors.

Reducing the time it takes to diagnose a patient improves the outcome for the patient and real-time PCR (qPCR) has greatly improved the accuracy and speed of disease diagnosis. If a specific disease is suspected, a real-time PCR reaction that uses primers specific to the suspected disease microorganism can be performed. The reaction may or may not amplify a product, producing a positive or negative diagnosis, respectively. For example, the influenza A H1N1 outbreak in 2009 and the Zika virus outbreak in 2015 were confirmed by real-time RT-PCR reactions.

PCR in Agriculture

Genetically modified (GM) plants have revolutionized modern agriculture and PCR is one of the techniques used to create GM organisms (GMOs). Genetic modification of plants uses components derived from the common soil bacterium *Agrobacterium tumefaciens*.

Forensic DNA Databases—
Is Your Privacy Protected?

Many countries have forensic DNA databases that contain the DNA profiles of people who have been convicted of crimes. The rights of individuals whose DNA is in the databases and the ways in which these databases are used raise concerns about the infringement of civil rights.

The U.K. has one of the world's largest DNA databases per capita, with over 5 million individuals (~8% of the country's population) on file as of 2018. Established in 1995, the database used to include the profiles of all individuals who were arrested, and of individuals, including minors, who voluntarily provided a DNA sample in order to be excluded as a suspect in an investigation. As a result of the Protection of Freedoms Act of 2012, however, the DNA data of individuals not charged or not found guilty cannot be kept beyond a specified period of time. If adults are arrested for a serious offense but not convicted, their profiles are kept for up to three years with a possible two-year extension. In addition, as a result of the law, volunteer profiles are only checked against the relevant crime scene profiles. Should DNA be retained in the database at all when a person is found not guilty?

In the U.S., The National DNA Index System, which is part of the Combined DNA Index System (CODIS), contains more than 13 million profiles, most of which are profiles of individuals convicted of state or federal crimes. The profiles contain data not only about the criminal but also the criminal's family, since families share many of the same short tandem repeat (STR) markers. When police cannot find a perfect match for crime scene DNA in the database, they can search for partial matches, actively looking for relatives of the perpetrator who are in the database. This is called familial searching. As of 2010, the FBI does not use familial searching. Some states forbid the use of familial searches while others allow it in certain circumstances. Is it fair that people who have not committed a crime or given permission for their profiles to be in the database are effectively searchable through a criminal database?

A. tumefaciens naturally contains Ti (tumor inducing) plasmids, which can insert their DNA into the chromosomal DNA of a plant to cause tumors. For example, *Bacillus thuringiensis* (Bt) bacteria produce a delta endotoxin (Cry1) that is lethal to specific species of moth larvae. Scientists used PCR to genetically engineer the gene for Cry1 into a Ti plasmid, then used the engineered plasmid to genetically modify crop plants such as corn and soy. The GM plants produce Cry1, which makes them resistant to moth larvae such as corn borers. There are dozens of genetically modified crops that contain different kinds of genetic modifications (for example, herbicide resistance, pest resistance, and frost resistance).

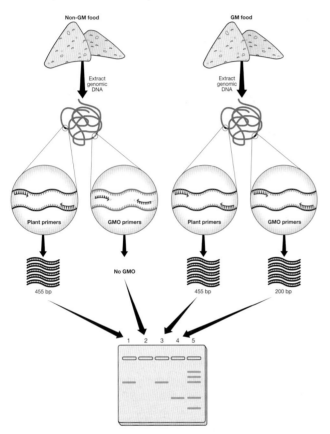

Figure 6.22. **Detection of GMO foods.** When assessing genetic modification in food, PCR controls are important. Primers specific to the genetic modifications (GMO primers) determine whether the food is genetically modified. Primers that amplify any plant DNA (plant primers) act as a positive control to ensure that intact plant DNA has been extracted from the food.

PCR can also identify GM plants or foods. This is necessary on farms to determine if GM crops are growing outside of their boundaries. PCR is also used in food testing to ensure that food meets governmental regulations on GM food. The DNA sequences used to genetically modify plants are derived from other organisms; these sequences serve as templates for the PCR assays designed to detect genetic modifications. While the actual genes inserted into crops are often quite different, the promoters and terminators used to drive the expression of the foreign genes are similar. These regulatory sequences are usually used to identify the presence of genetically modified material. Proper controls must be performed to verify that the PCR assay is working. For example,

DNA sequences that are present in any food (non-modified or genetically modified) can be amplified and used as a control for the actual PCR reaction (see Figure 6.22). While conventional PCR can be used to identify GM foods, real-time PCR is typically used to quantitate the percentage of GM material in foods. Since different countries have different regulations regarding the permissible level of GM material in food, real-time PCR plays an important role in food exports.

Just as PCR has improved the diagnosis of human diseases, it is used similarly in veterinary medicine. PCR is also used in a new type of farming—biopharming—in which transgenic animals are created to produce pharmaceutical products such as therapeutic drugs expressed in milk (see Biotech in the Real World: Biotech on the Pharm in Chapter 5).

PCR in Forensics

Television crime shows have increased public awareness of new forensic techniques. Forensic scientists use biological samples (containing DNA) found at crime scenes to identify suspects who could have been present at the scene. The source of DNA can be large (for example, pools of blood and tissues) or small (for example, DNA in a blood spatter).

Many different techniques exist to fingerprint biological samples (see Figure 6.23). These methods have different abilities to differentiate among individuals or to correctly identify one person from the population. ABO blood typing is the oldest technique. Although this technique is very fast, it has a very low **power of discrimination**; it can differentiate only between large groups, not individuals. Conversely, restriction fragment length polymorphism (RFLP) analysis has a very high power of discrimination but is time-consuming. The current method of choice for forensics investigators is **short tandem repeat (STR)** analysis. STR analysis is based on PCR, so it is fast and highly discriminatory.

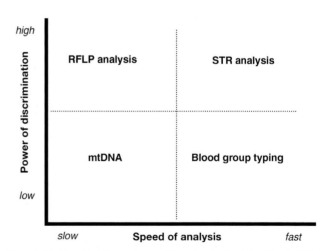

Figure 6.23. **Biological fingerprinting.** Different methods of identifying individuals that rely on biological factors are positioned based on the power of discrimination and the speed of analysis.

STR and variable number tandem repeats (VNTR) are specific regions (loci) of chromosomal DNA that have repeating sequences. STRs are repeating units of 2–6 bp, while VNTRs can be up to 80 bp. The number of repeating units varies among individuals. Each chromosome of an individual usually has a different number of repeats, since each member of a chromosome pair is inherited from a different parent. The number of repeats on each chromosome is called an **allele**; for some loci there can be as many as 30 repeats. An allele refers to a specific **genotype** at a **locus** of DNA, and each diploid individual has two alleles for each locus of DNA.

The ability to distinguish between any two individual DNA profiles increases with the number of loci tested. If only one locus is examined, many people would have the same genotype, leaving many potential suspects in a criminal case. As more loci are added to the test, fewer people would have the same profile. The more loci analyzed, the higher the power of discrimination. Given that each person has two alleles, if one STR has eight different possible alleles that each have equal frequency, the probability of a person's allele combination randomly matching another person's is 1 in 8 multiplied by 1 in 8 for a probability of 1 in 64. If two more different loci are tested and if each locus has the same random match probability of 1 in 8, the probability of a random match (after testing all three loci) will be reduced to 1 in 64 multiplied by itself three times, which is 1 in 262,144.

Allele frequencies are not equal; they vary among different ethnic groups, with some alleles more common in certain groups than in others. The frequencies of occurrence of various alleles have been published for different ethnic groups. For example, in Caucasians, the frequency of eight repeats at the TPOX locus is 0.535. This frequency can be referred to as p, so p = 0.535. This means that the chance for any Caucasian having the TPOX 8 allele is 53.5%. Similarly, there is a 4.1% chance that a Caucasian has 12 repeats at the TPOX locus. This frequency can be referred to as q, so q = 0.041. The probability that a person received the 8 allele from

his/her mother and the 12 allele from his/her father is represented as pq, and the opposite—that he/she received the 8 allele from his/her father and the 12 allele from his/her mother—is also pq. Therefore, the probability of having an 8, 12 genotype at the TPOX locus is pq + pq or 2pq. Using the probabilities provided, the chance of two Caucasians having the same 8, 12 genotype is calculated as 2 x 0.535 x 0.041 = 0.044, or 4.4%.

STR alleles are the basis of the U.S. Federal Bureau of Investigation (FBI)'s Combined DNA Index System (CODIS) (see Figure 6.24). CODIS uses 20 different loci to create a profile of an individual. The profile has 40 numbers, which correspond to the number of repeats on both chromosomes of the 20 different loci.

PCR in Paternity Testing

One of the first applications of RFLP analysis was paternity testing. This type of testing was expensive and time-consuming. With the invention of PCR, paternity testing using STR analysis is now inexpensive, and the results are available within hours. Paternity testing compares the DNA profile of a child with the profiles of the potential parents. Human cells are diploid, having two copies of each chromosome (with the exception of the Y chromosome in males). Chromosomes are inherited from parents, one copy from the mother and one copy from the father. The number of STR repeats in the child must match the number of repeats on at least one of the father's chromosomes (see Figure 6.25). Otherwise, the male individual is not the child's biological father.

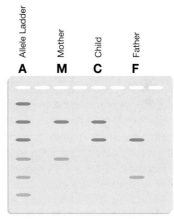

Figure 6.25. **Paternity test.** In this simple example of a paternity test using one STR locus, the child shares one STR allele with the mother and one STR allele with the father.

PCR in Human Migration

Understanding how humans evolved and migrated across the world has been assisted by studies using PCR and DNA sequencing. For example, mitochondrial DNA (mtDNA) sequences of modern humans have been amplified by PCR and used to trace the history of mtDNA mutations back to a maternal ancestor—Mitochondrial Eve—from whom all humans are derived. Similar strategies have been used on the Y chromosome to trace back the male line.

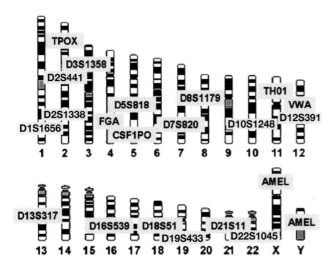

Figure 6.24. **CODIS loci.** The Combined DNA Index System (CODIS) uses 20 STR loci across the human genome and the amelogenin (AMEL) locus to determine an individual's gender.

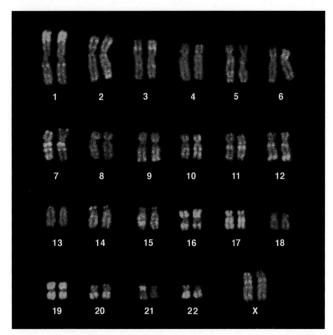

Figure 6.26. **Alu sequences in the human genome.** A karyotype showing human metaphase chromosomes stained with a fluorescent probe to the Alu sequence (green) and counterstained for DNA (red). Photo credit: Bolzer et al. (2005).

Scientists investigating the origin of the human species are sequencing the Neanderthal genome and have obtained sequencing data suggesting that humans and Neanderthals interbred. Interestingly, 1–4% of Neanderthal DNA is present in modern humans but not in certain African human populations, indicating that the interbreeding occurred after humans migrated away from Africa.

Human migration can also be tracked using differences in the occurrence of repetitive elements in human genomes. Short interspersed nuclear elements (SINEs) have been randomly inserted into our genomes over millions of years. The elements are derived from retrotransposons—mobile sequences of DNA that are also called jumping genes. One such repetitive element is called the **Alu** sequence. This is a DNA sequence approximately 300 bp long that is repeated, one copy at a time, almost one million times within the human genome (see Figure 6.26). The Alu name comes from the Alu I restriction enzyme recognition site that is found in this sequence. Some occurrences of the Alu sequence, such as those found in the PV92 region of chromosome 16, are so recent that they are polymorphic among populations, which means that some humans have them and some do not. Alu elements have been used to trace human migration and disprove paternity.

PCR in Wildlife Conservation

Protecting endangered species from poachers and preventing illegal trafficking of plants and animals is of international importance, and PCR has become a tool in wildlife conservation. For example, STR analysis has been used to type sturgeon eggs from Russia, distinguish whale and dolphin meat, and identify teak and mahogany woods illegally removed from rain forests. It has also been used to identify elephant remains, ivory carvings, rhinoceros horns (see Figure 6.27), and bear gall bladders. The limiting factor in using PCR in conservation efforts is obtaining funding to support the research. Every animal or plant species has a unique genome, and human STR sequences cannot be used for other organisms. Research must be conducted to find the optimal sequences needed to identify each species. Funding must be found for that research and for the field tests that follow.

Figure 6.27. **Dehorned rhinoceros.** Rhinoceroses are dehorned by conservationists to protect them from poachers, who may otherwise kill them for their horns.

Chapter 6 Essay Questions

1. Discuss research on human origins and their migration patterns using Alu elements.

2. Discuss why some countries choose to use GM crops over conventional crops. Include statistics on GMO production in various countries.

3. Review the current status of personal sequencing services such as 23andMe.

Additional Resources

Resources on forensics analysis:

Butler JM (2009). Fundamentals of Forensic DNA Typing (Burlington: Academic Press).

Neme L (2009). Animal Investigators: How the World's First Wildlife Forensics Lab Is Solving Crimes and Saving Endangered Species (New York: Scribner).

CODIS, or Combined DNA Index System, Federal Bureau of Investigation. fbi.gov/services/laboratory/biometric-analysis/codis. Accessed Jan 23, 2018.

DNA Evidence: Basics of Analyzing, National Institute of Justice. nij.gov/topics/forensics/evidence/dna/basics/pages/analyzing. aspx. Accessed May 8, 2018.

Resources on human origins:

Wells S (2007). Deep Ancestry: Inside the Genographic Project (Washington: National Geographic Society).

Green R et al. (2006). Analysis of one million base pairs of Neanderthal DNA. Nature 444, 330–336.

Original article for Figure 6.26:

Bolzer A et al. (2005). Three-dimensional maps of all chromosomes in human male fibroblast nuclei and prometaphase rosettes. PLoS Biol 3, 5.

Activity 6.A STR PCR Analysis

Overview
Short tandem repeat (STR) analysis is widely used in forensics. STRs are short repetitive sequences that occur across the genome at different loci. The number of repeats at each locus varies among individuals, with each individual having two different repeats (alleles) at each locus, one for each chromosome. These alleles can be amplified by PCR. Primers are designed to bind to the region on either side of the STR and to amplify the region from each chromosome, generating two PCR products. The length of each PCR product is compared to that of a DNA standard called an allele ladder. Each band of the allele ladder represents a specific number of repeats found at that locus that can be directly compared with the PCR products, allowing fast analysis of the gel.

In this activity, you will play the role of a forensic investigator and amplify DNA representing a fictional STR locus—the BXP007 locus. Template DNA samples from a crime scene and four suspects are provided. You will amplify PCR products from the template DNA using a PCR master mix and then run the PCR products on a 3% agarose TAE gel. The profile of the DNA from the crime scene and the four suspects will be determined using the allele ladder. The data will then be used to determine whether DNA from any of the suspects matches that found at the crime scene. If there is a match in the DNA profile, the suspect(s) cannot be excluded from the investigation.

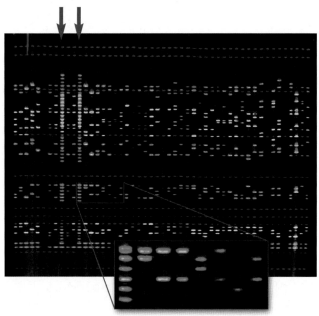

Figure 6.28. **Gel of STR analysis of multiple loci using multiplex PCR.** Each vertical lane (except the allele ladders shown by arrows) is the result of a single PCR reaction from an individual. The multiple bands are PCR products generated by multiple sets of primers. (The primers have been fluorescently labeled with different colors to help differentiate PCR products from different loci.) The magnified section shows an allele ladder on the first lane and PCR products from seven individuals at one locus. Image courtesy of Linda Strausbaugh, University of Connecticut-Storrs.

Note: This activity does not use human DNA. Actual STR analysis cannot be conducted on agarose gels because the PCR products are too short to resolve. Actual STR analysis requires specialized electrophoresis and analysis equipment (see Figure 6.28). This activity simulates STR analysis. Activity 6.C provides an opportunity to conduct PCR on human DNA—your own!

Tips and Notes
- PCR is extremely sensitive; therefore, it is very important not to contaminate any of the reagents used in the reaction

- Always use aerosol-barrier pipet tips when setting up PCR reactions to prevent the transfer of aerosols from the pipet to the sample

- Keep the master mix and reaction tubes on ice until they are ready to be placed into the thermal cycler

- Mix PCR reagents thoroughly and ensure that the reactions are at the bottom of the tube. If necessary, use a microcentrifuge with adaptors for 0.2 ml PCR tubes to pull the liquid to the bottom of the tube. Pulse-spin tubes for 5–10 seconds

Safety Reminders: Appropriate personal protective equipment (PPE) should be worn at all times, and attention should be given when using electricity in the presence of liquids. Safety measures built into the equipment should prevent accidental exposure; however, should any leakage occur, notify your instructor before proceeding. If ethidium bromide, fluorescent stains, or UV light is used, special precautions are required.

Research Question
- Does the DNA profile of any of the suspects match the DNA profile of the crime scene sample?

Objectives
- Set up five PCR reactions with DNA from the crime scene and from the four suspects
- Load and run PCR products on a 3% agarose TAE gel
- Determine the DNA profiles of the suspects and crime scene DNA using the allele ladder

Skills to Master
Refer to Laboratory Skills Assessment Rubric (Appendix E) and Laboratory Notebook Rubric (Appendix F) for more details.
- Record laboratory notebook entries
- Set up PCR reactions
- Use a thermal cycler
- Perform agarose gel electrophoresis (Activity 4.C)
- Analyze an agarose gel (Activity 4.D)

Activity 6.A STR PCR Analysis

Student Workstation Materials

Part 1: Setting Up PCR Reactions

Items	Quantity
Thermal cycler (shared)	1
Microcentrifuge or mini centrifuge with adaptors for PCR tubes (optional) (shared)	1
2–20 µl adjustable-volume micropipet and aerosol-barrier tips	1
PCR tubes	5
PCR tube rack	1
Container of ice	1
Laboratory marking pen	1
Waste container	1
Master mix (MMP) (on ice)	120 µl
Crime Scene DNA (on ice)	25 µl
Suspect A DNA (on ice)	25 µl
Suspect B DNA (on ice)	25 µl
Suspect C DNA (on ice)	25 µl
Suspect D DNA (on ice)	25 µl

Part 2: Running the Gel

Items	Quantity
Horizontal gel electrophoresis chamber	1
Power supply (shared)	1
Digital imaging system (optional) (shared)	1
Microcentrifuge or mini centrifuge with adaptors for PCR tubes (optional) (shared)	1
Rocking platform (optional) (shared)	1
UV light source (optional) (shared)	1
2–20 µl adjustable-volume micropipet and aerosol-barrier tips	1
Microcentrifuge tube rack	1
PCR tube rack	1
Gel staining tray	1
Laboratory marking pen	1
PCR reactions from part 1	5
5x Orange G sample loading buffer (SLB)	60 µl
or UView™ 6x loading dye and stain (UView)	50 µl
Allele ladder (AL)	25 µl
1x TAE electrophoresis buffer*	300 ml
1x Fast Blast DNA stain**	100 ml
3% agarose TAE gel***	1

* If necessary, make 1x TAE electrophoresis buffer (see part 1 of Activity 2.D).
** Alternative staining methods may be used. See Appendix B for details.
***If necessary, prepare a 3% agarose TAE gel according to the protocol in Activity 4.B.
   This may require extra 1x TAE electrophoresis buffer.

Prelab Focus Questions

1. Explain why STR analysis is the method of choice for forensics investigators.
2. Why are aerosol-barrier tips used when pipetting PCR reagents?
3. Why are allele ladders used instead of DNA size standards to analyze STRs?

Activity 6.A STR PCR Analysis

Activity Protocol

Part 1: Setting Up PCR Reactions

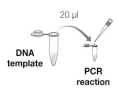

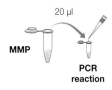

1. Label five PCR tubes **CS**, **A**, **B**, **C**, and **D**, and include your group name or initials on the tubes. Place the tubes into a PCR tube rack.

2. Pipet 20 µl of the appropriate template DNA into the corresponding PCR tube.

    Note: Use a fresh tip for each DNA sample.

3. Using a fresh tip each time, transfer 20 µl of the blue master mix (**MMP**) into each PCR tube containing template DNA. Pipet up and down to mix the master mix and template DNA. Cap each tube after adding master mix and place on ice.

    Note: If necessary, pulse-spin the PCR tubes for 5–10 sec in a microcentrifuge or mini centrifuge with adaptors for PCR tubes to pull contents to the bottom of the tube.

4. When instructed to do so, place the PCR tubes in the thermal cycler.

5. If necessary, program the thermal cycler with the following program:

Initial denature:	94°C for 2 min
35 cycles of:	94°C for 30 sec
	52°C for 30 sec
	72°C for 1 min
Final extension:	72°C for 10 min
Hold:	15°C for ∞

6. When all samples are in the cycler, start the PCR.

    Note: The reactions will take 2–3 hr to complete. Since there is no need to wait for the cycler to finish, the reactions can be left in the cycler overnight.

7. Store samples at 4°C in the refrigerator until they are ready to be run on the gel.

8. If necessary, prepare 1x TAE (refer to part 1 of Activity 2.D) and a 3% agarose TAE gel (refer to Activity 4.B) for the next part of the activity.

Part 2: Running the Gel

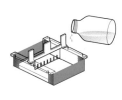

1. Retrieve the five PCR reactions from the refrigerator. Pulse-spin the PCR tubes for 5–10 sec in a microcentrifuge or mini centrifuge with adaptors for PCR tubes to pull the liquid to the bottom of the tube.

2. Using a fresh tip for each sample, pipet 10 µl of Orange G sample loading buffer (**SLB**) or 8 µl of UView 6x loading dye and stain (**UView**) into each tube. Pipet up and down to mix.

Activity 6.A STR PCR Analysis

3. Place the 3% agarose TAE gel into the electrophoresis chamber. Fill the electrophoresis chamber with sufficient 1x TAE buffer to cover the gel by approximately 2 mm.

4. Check that the wells of the agarose gel are near the black (–) electrode or cathode and that the bottom edge of the gel is near the red (+) electrode or anode.

5. Using a fresh tip for each sample, load 20 µl if using Orange G sample loading buffer or 10 µl if using UView 6x loading dye and stain of each sample into the gel in the following order:

   Lane 1 **AL** (Allele ladder)
   Lane 2 **CS** (Crime scene)
   Lane 3 **A** (Suspect A)
   Lane 4 **B** (Suspect B)
   Lane 5 **C** (Suspect C)
   Lane 6 **D** (Suspect D)

   The order in which the samples are loaded in the gel should be recorded in your laboratory notebook.

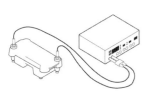

6. Place the lid on the electrophoresis chamber. Connect the electrical leads to the power supply, red to red and black to black.

7. Turn on the power and run the gel at 100 V for 30 min.

   Note: If using the fast gel running buffer (0.25x TAE), run the gel at 200 V for 20 min (see Appendix A).

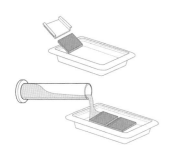

8. When the electrophoresis run is complete, turn off the power and remove the lid from the chamber. Carefully remove the gel and tray from the electrophoresis chamber. Be careful—the gel is very slippery. If using UView 6x loading dye and stain or other fluorescent stains, use a UV light source to visualize DNA bands. Otherwise, slide the gel into the staining tray and carefully pour a sufficient volume of 1x Fast Blast DNA stain to submerge the gel.

   Note: For alternative DNA staining options, see Appendix B.

9. Stain the gel overnight to visualize the DNA bands and then record the results. (Optional) A rocking platform may be used to enhance staining.

Activity 6.A STR PCR Analysis

Results Analysis

Determine the genotype for each DNA sample at the BXP007 locus by comparing the banding pattern of the DNA fragments on the agarose gel with the allele ladder. The allele ladder contains sizes of all the alleles known to occur at this locus; allele names are indicative of the number of repeats at each allele (see Figure 6.29). There are eight possible alleles: 1, 2, 3, 4, 5, 7, 10, and 15. In the example in Figure 6.29, the crime scene DNA (CS) has a 5–2 genotype; the genotype of suspect A is 7–4. Record the genotypes for each DNA sample from your results. Then determine whether any of the suspects matches the DNA found at the crime scene. Suspects whose DNA match the crime scene DNA cannot be excluded from the investigation.

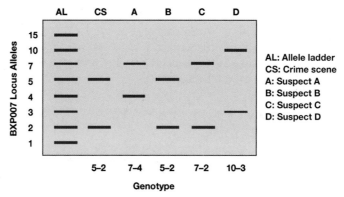

Figure 6.29. **Example of genotypes of various DNA samples.** The allele ladder should be used to determine the genotypes of the test samples. For example, CS has a band with 5 repeats and another with 2 repeats, therefore its genotype is 5–2 for the BXP007 locus.

Postlab Focus Questions

1. What are the genotypes of the crime scene and suspect DNA samples? What can you conclude from the results?

Self Assessment

Assess whether you have mastered the skills of this activity using the Laboratory Skills Assessment Rubric in Appendix E.

Assess your experimental write-up in your laboratory notebook using the Laboratory Notebook Rubric in Appendix F.

Activity 6.B GMO Detection by PCR

Overview
Many crops have been genetically modified to improve yields. On July 29, 2016, President Barack Obama signed into law the National Bioengineered Food Disclosure Standard (Public Law No. 114-216) which, in part, directs USDA to establish a national standard to disclose certain food products or ingredients that are "bioengineered." The activity in this section will allow you to experimentally determine whether foods purchased from the grocery store contain genetically modified (GM) crops.

Two methods are currently used to identify genetically modified crops: enzyme-linked immunosorbent assay (ELISA) and PCR. ELISA is an antibody-based test that identifies the recombinant proteins expressed by the crops. ELISA can test only fresh produce and is specific to the genetic modification of the crop. For example, an ELISA that is designed to check for the presence of Bacillus thuringiensis (Bt) proteins can detect only Bt GM corn, not herbicide-tolerant GM corn. ELISA is inexpensive and can be performed in the field with little expertise. PCR is performed in a laboratory and is more expensive than ELISA; however, PCR can identify GM content in highly processed foods because DNA is much more stable than proteins. Unlike ELISA, in which a single crop is tested for its GM modification, GM PCR tests can detect GM content from multiple GM crops. This is possible because agricultural scientists use only a small number of regulatory DNA sequences (promoter and terminator sequences) to control the expression of the inserted genes. Therefore, approximately 85% of GM plants have common promoter and terminator sequences; the 35S promoter from cauliflower mosaic virus (CaMV) and the nopaline synthase (NOS) terminator from *A. tumefaciens*. These sequences are detected in this activity. The percentage of GM plant material in food can also be determined using real-time PCR; this can be explored in an extension to this activity.

In this activity, food containing corn, soy, or another suspected GM plant will be ground with a mortar and pestle. gDNA will be extracted from the food at 95°C using InstaGene matrix. InstaGene matrix binds (chelates) divalent ions such as magnesium and prevents them from acting as cofactors for DNA degrading enzymes (DNases). PCR reactions will then be set up using the extracted gDNA and primers specific for genetic modifications. Once amplified, the PCR products will be run on agarose gels.

It is important to have controls for various steps in the activity to 1) check that gDNA was successfully extracted from the food, 2) confirm that the PCR reaction is not contaminated, and 3) validate that the thermal cycler and reagents are working as expected (see Figure 6.30). A set of primers that can amplify any plant gDNA (genetically modified or not) is used as a control to verify that gDNA is successfully extracted. Plasmid DNA that contains the target sequences of both the GM-specific primers and the plant primers is used as a control to verify that the reagents and the thermal cycler are

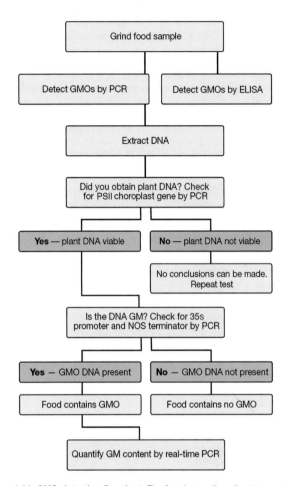

Figure 6.30. **GMO detection flowchart.** The flowchart outlines the steps used in determining whether foods have been genetically modified.

working as expected. Last, gDNA is extracted from a non-GM control food and used to verify that a positive result for genetic modification is not a result of contamination.

Tips and Notes
- The most common food products that contain genetically modified crops are processed corn snacks and soy food products, especially those that use soy as a filler, such as inexpensive ground beef products

- The mortar and pestle used for grinding foods is a potential source of PCR contamination and should be washed with a bottle brush and soap, wiped down with 10% bleach, and rinsed with tap water followed by distilled water before each use

- Using too much plant material can overwhelm the InstaGene matrix, leaving excess magnesium that can be used as a cofactor for DNases. Add only the specified amount of plant material to the InstaGene matrix. Use a transfer pipet instead

Activity 6.B GMO Detection by PCR

of a micropipet to transfer the ground food to the InstaGene matrix since micropipet tips are easily clogged by pieces of ground food

- PCR is extremely sensitive; therefore, it is very important not to contaminate any of the reagents used in the reaction

- Always use aerosol-barrier tips when setting up PCR reactions to prevent transferring aerosols from the pipet to the sample

- Keep the master mix and reaction tubes on ice until they are ready to be placed into the thermal cycler

- Mix PCR reagents thoroughly and ensure the reactions are at the bottom of the tube. If necessary, use a microcentrifuge with adaptors for 0.2 ml PCR tubes to pull the liquid to the bottom of the tube. Pulse-spin tubes for 5–10 seconds

Safety Reminders: Appropriate PPE should be worn at all times, especially when using 10% bleach, which is hazardous and can mark clothing. When using the centrifuge, it is important to balance each tube with an identically weighted tube in the opposite position and to ensure that the inner lid of the centrifuge is closed prior to spinning. Attention should be given when using electricity in the presence of liquids. Safety measures built into the equipment should prevent accidental exposure; however, should any leakage occur, notify your instructor before proceeding. If ethidium bromide, fluorescent stains, or UV light is used, special precautions are required.

Research Question

- Does the food tested contain genetically modified plant material?

Objectives

- Grind one test food and one non-GM control food using a mortar and pestle
- Extract gDNA with InstaGene matrix
- Set up six PCR reactions with test food gDNA, gDNA from non-GM control food, and GM-positive plasmid DNA using GMO primers (red) and plant primers (green) for each sample
- Load and run PCR products on a 3% agarose TAE gel
- Determine whether the test food has been genetically modified

Skills to Master

Refer to Laboratory Skills Assessment Rubric (Appendix E) and Laboratory Notebook Rubric (Appendix F) for more details.

- Record laboratory notebook entries
- Use a mortar and pestle
- Perform DNA extractions using InstaGene matrix
- Set up PCR reactions (Activity 6.A)
- Use a thermal cycler (Activity 6.A)
- Perform agarose gel electrophoresis (Activity 4.C)
- Analyze results on an agarose gel (Activity 4.D)

Activity 6.B GMO Detection by PCR

Student Workstation Materials

Part 1: Extracting Template DNA

Items	Quantity
Vortexer (optional) (shared)	1
Water bath or dry bath at 95°C (shared)	1
Microcentrifuge or mini centrifuge (shared)	1
Balance (shared)	1
Weigh boats	2
Pipet pump or filler (optional)	1
10 ml serological pipets or 10–25 ml graduated cylinder	1–2
1 ml sterile graduated transfer pipets	1–2
Microcentrifuge tube rack	1
Floating tube rack (optional)	1
Mortars and pestles	1–2
Laboratory marking pen	1
Waste container	1
Bottle brushes and laboratory soap (shared)	1
Distilled water (dH$_2$O)	25 ml
500 µl of InstaGene matrix in screwcap tubes	2
10% bleach in a squirt bottle (shared)	1
Test food sample	>2 g
Certified non-GM control food (optional)*	>2 g

* May have been prepared by instructor.

Part 2: Setting Up PCR Reactions

Items	Quantity
Thermal cycler (shared)	1
Microcentrifuge or mini centrifuge with adaptors for PCR tubes (shared)	1
2–20 µl adjustable-volume micropipet and aerosol-barrier tips	1
Microcentrifuge tube rack	1
PCR tubes	6
PCR tube rack	1
Container of ice	1
Laboratory marking pen	1
Plant master mix (PMM (green liquid)) (on ice)	65 µl
GMO master mix (GMM (red liquid)) (on ice)	65 µl
GMO positive control DNA	25 µl
gDNA from test food (extracted in part 1)	1
gDNA from non-GM control food (extracted in part 1 or by instructor)	1

Part 3: Running the Gel

Items	Quantity
Horizontal gel electrophoresis chamber	1
Power supply (shared)	1
Digital imaging system (optional) (shared)	1
Microcentrifuge or mini centrifuge with adapters for PCR tubes (optional) (shared)	1
Rocking platform (optional) (shared)	1
UV light source (optional) (shared)	1
2–20 µl adjustable-volume micropipet and aerosol-barrier tips	1
Microcentrifuge tube rack	1
PCR tube rack	1
Gel staining tray	1
Laboratory marking pen	1
PCR reactions from part 2	6
5x Orange G sample loading buffer (SLB)	70 µl
or UView 6x loading dye and stain (UView)	60 µl
PCR MW standard (MWS)	25 µl
1x TAE electrophoresis buffer*	300 ml
1x Fast Blast DNA stain**	100 ml
3% agarose TAE gel***	1

*   If necessary, make 1x TAE electrophoresis buffer (see part 1 of Activity 2.D).
**   Alternative staining methods may be used. See Appendix B for details.
*** If necessary, make a 3% agarose TAE gel according to the protocol in Activity 4.B. This may require extra 1x TAE electrophoresis buffer.

Prelab Focus Questions

1. Describe at least one procedure step that is important for reducing the chance of contamination of the PCR reaction and yielding false results.
2. Why is it important to weigh the samples before adding them to the mortar?
3. If the plant primers do not generate a PCR product when amplifying the test food, can you trust the data obtained using the GMO primers to amplify the test food?
4. If the GMO primers generate a PCR product with the non-GM control food, can you trust the data obtained using the GMO primers to amplify the test food?

Activity 6.B GMO Detection by PCR

Part 1: Extracting Template DNA

1. Label one screwcap tube containing InstaGene matrix **Non-GM** and the other **Test**.

2. Weigh 0.5–2 g of non-GM control food and put it into the mortar. Record the mass.

3. Add 5 ml of dH$_2$O for every gram of food. To calculate the volume of water needed, multiply the mass (in grams) of the food by 5, and add that many milliliters of water.

   Mass of food = _____ g x 5 = _____ ml

4. Grind the food with the pestle for at least 2 min to form a slurry.

5. Add another 5 ml of dH$_2$O for every gram of food. Mix or grind further with the pestle until the slurry is smooth enough to pipet.

   50 µl

6. Use the graduated transfer pipet to transfer 50 µl of ground slurry to the screwcap tube labeled **Non-GM**. Recap the tube and shake or vortex to mix. This is the non-GM template DNA.

7. Wash the mortar and pestle with soap, wipe them with 10% bleach, rinse them well with tap water, and do a final rinse with dH$_2$O.

8. Repeat steps 2–5 with the test food to prepare the test food sample. Use the graduated transfer pipet to transfer 50 µl of the test food slurry to the screwcap tube labeled **Test**. Recap the tube and shake or vortex to mix. This is the test template DNA.

9. Incubate the **Non-GM** and **Test** screwcap tubes at 95°C for 5 min.

10. Place the tubes in a centrifuge in a balanced configuration and centrifuge for 5 min at maximum speed.

    Note: If using a mini centrifuge that can reach only 2,000 x g, centrifuge for 10 min.

11. Proceed directly to part 2 or store the gDNA in the screwcap tubes at 4°C for up to 1 month. **Do not freeze the samples.**

Activity 6.B GMO Detection by PCR

Part 2: Setting Up PCR Reactions

1. Centrifuge the screwcap tubes containing the DNA templates at maximum speed for 2 min.

2. Label six PCR tubes 1 through 6, and include your initials. Cap the tubes to prevent contamination and place them in the PCR tube rack on ice. Each number corresponds to the following tube contents:

Tube Number	Master Mix	Template DNA
1	20 µl of PMM (green)	20 µl of **Non-GM** DNA
2	20 µl of GMM (red)	20 µl of **Non-GM** DNA
3	20 µl of PMM (green)	20 µl of **Test** DNA
4	20 µl of GMM (red)	20 µl of **Test** DNA
5	20 µl of PMM (green)	20 µl of GM-positive plasmid DNA
6	20 µl of GMM (red)	20 µl of GM-positive plasmid DNA

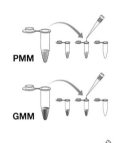

PMM

GMM

3. Using a fresh tip each time, pipet 20 µl of the indicated master mix to each PCR tube. Recap the tubes immediately after adding the reagent.

Supernatant

4. Using a fresh tip each time, pipet 20 µl of the indicated template DNA to each PCR tube. Mix the PCR reactions by pipetting gently up and down, and then recap the tubes. The tubes containing PMM should be green, while the tubes containing GMM should be red. Store the PCR tubes on ice.

   Note: Avoid disturbing or transferring the pellet of InstaGene matrix, which is at the bottom of the screwcap tubes that contain the template DNA. InstaGene matrix will inhibit the PCR reaction. If the InstaGene matrix is disturbed while attempting to transfer the template DNA, recentrifuge the screwcap tubes to pellet the beads.

   Note: Pulse-spin PCR tubes for 5–10 sec in a microcentrifuge or mini centrifuge with adaptors for PCR tubes to pull contents to the bottom of the tube.

5. When instructed to do so, place the PCR reactions in the thermal cycler.

6. If necessary, program the thermal cycler with the following program:

Initial denature:	94°C for 2 min
40 cycles of:	94°C for 1 min
	59°C for 1 min
	72°C for 2 min
Final extension:	72°C for 10 min
Hold:	15°C for ∞

7. When all samples are in the cycler, start the PCR.

   Note: The reactions will take 3–4 hr to complete. Since there is no need to wait for the cycler to finish, the reactions can be left in the cycler overnight.

8. Store samples at 4°C in the refrigerator until they are ready to be run on the gel.

Activity 6.B GMO Detection by PCR

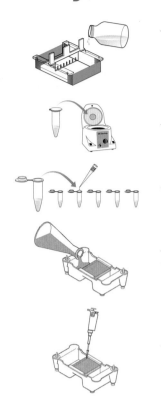

9. If necessary, prepare 1x TAE (refer to part 1 of Activity 2.D) and a 3% agarose TAE gel (refer to Activity 4.B) for the next part of the activity.

Part 3: Running the Gel

1. Retrieve the six PCR reactions from the refrigerator. Pulse-spin the PCR tubes for 5–10 sec in a microcentrifuge or mini centrifuge with PCR tube adaptors to pull the liquid to the bottom of the tube.

2. Using a fresh tip for each sample, pipet 10 µl of Orange G sample loading buffer (**SLB**) or 8 µl of UView 6x loading dye and stain (**UView**) into each PCR tube. Pipet up and down to mix.

3. Place a 3% agarose TAE gel into the electrophoresis chamber. Fill the electrophoresis chamber with sufficient 1x TAE buffer to cover the gel by approximately 2 mm.

4. Check that the wells of the agarose gel are near the black (–) electrode or cathode and that the bottom edge of the gel is near the red (+) electrode or anode.

5. Using a fresh tip for each sample, load 20 µl of each if Orange G sample loading buffer is used or 10 µl if UView 6x loading dye and stain is used into the gel in the following order:

Lane 1	Tube **1**
Lane 2	Tube **2**
Lane 3	Tube **3**
Lane 4	Tube **4**
Lane 5	Tube **5**
Lane 6	Tube **6**
Lane 7	PCR MW standard (**MWS**)

    The order in which the samples are loaded in the gel should be recorded in your laboratory notebook.

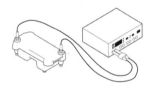

6. Place the lid on the electrophoresis chamber. Connect the electrical leads to the power supply, red to red and black to black.

7. Turn on the power and run the gel at 100 V for 30 min.

    Note: If using the fast gel running buffer (0.25x TAE), run the gel at 200 V for 20 min (see Appendix A).

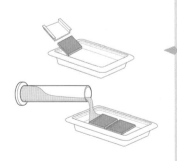

8. When the electrophoresis run is complete, turn off the power and remove the lid from the chamber. Carefully remove the gel and tray from the electrophoresis chamber. Be careful—the gel is very slippery. If using UView 6x loading dye and stain or other fluorescent stains, use a UV light source to visualize DNA bands. Otherwise, slide the gel into the staining tray and carefully pour a sufficient volume of 1x Fast Blast DNA stain to submerge the gel.

    Note: For alternative DNA staining options, see Appendix B.

9. Stain the gel overnight to visualize the DNA bands and then record the results. (Optional) A rocking platform may be used to enhance staining.

Activity 6.B GMO Detection by PCR

Results Analysis

The results of the electrophoresis should be recorded in your laboratory notebook. (See Figure 6.31 for the sizes of the bands in the PCR MW standard.) The plant primers used in this activity amplify the photosystem II (*PSII*) gene and generate a PCR product that is 455 bp. The GMO primers produce a PCR product that is approximately 200 bp. Summarize your results in your laboratory notebook. Comment on the result in each lane of the gel and generate conclusions based on your results. If your results are inconclusive, state how you came to this conclusion. If your results indicate the presence or absence of GM plants in the test food, state the evidence that supports your finding. The flowchart shown in Figure 6.30 can be used to determine whether the test food is positive for GM plants.

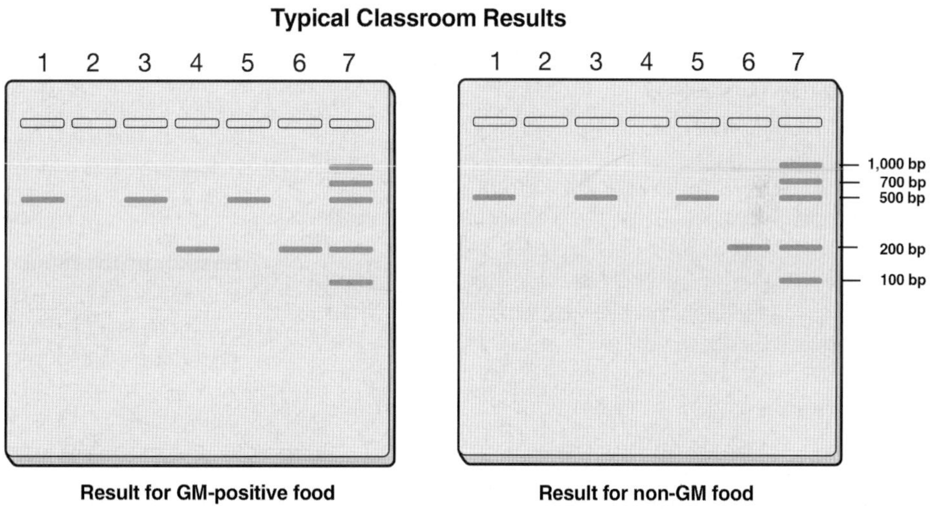

Typical Classroom Results

Result for GM-positive food

Result for non-GM food

Figure 6.31. **Agarose gels with PCR products from a GM-positive food and a non-GM control food.** In both gels, lanes 1 and 2 are PCR products from the non-GM control food, lanes 3 and 4 are PCR products from the test food sample, lanes 5 and 6 are PCR products from the GM-positive plasmid DNA, and lane 7 is the PCR MW standard. In each gel, the plant master mix (PMM) is used to amplify the samples loaded in lanes 1, 3, and 5, and the GMO master mix (GMM) is used to amplify the samples loaded in lanes 2, 4, and 6. The presence of a 455 bp band with PMM indicates that the food is positive for plant DNA and confirms that gDNA was successfully extracted from the food sample. The presence of a 220 bp band with GMM indicates that the food is positive for genetic modification.

Postlab Focus Questions

1. Did the PCR reagents and thermal cycler function as expected? What evidence do you have for this conclusion?
2. Was there any contamination of the PCR? How do you know?
3. Did the template DNA from the test food amplify with plant primers? Can you make conclusions regarding genetic modifications of the test food based on this result?
4. Was the test food genetically modified? What evidence do you have to support your conclusions?

Self Assessment

Assess whether you have mastered the skills of this activity using the Laboratory Skills Assessment Rubric in Appendix E.

Assess your experimental write-up in your laboratory notebook using the Laboratory Notebook Rubric in Appendix F.

Activity 6.C Detection of the Human PV92 Alu Insertion

Overview

Alu elements are short DNA sequences that occur only in the genomes of primates (including humans). They were inserted through the action of jumping genes, also called retrotransposons, and the elements are no longer mobile. Some Alu insertions, such as the Alu insertion in the PV92 region of human chromosome 16, are found only in humans and not in other species. The PV92 Alu element is believed to have been inserted relatively recently into the human genome because it is not found in all humans. The presence or absence of such elements can be used to estimate relatedness among populations. Data generated from scientific investigations into Alu insertions have helped shape current theories on human evolutionary history and the migration patterns of ancient human populations.

In this activity, the presence of the PV92 Alu repeat will be determined to estimate the frequency of an insert in a population—in this case, your class population. The data will provide a measure of molecular genetic variation. The presence or absence of the PV92 allele has no known correlation to disease or familial relationships. The PV92 Alu insert is approximately 300 bp long. If the Alu sequence is present following amplification of the PV92 region by PCR, the PCR product will be 941 bp. If the Alu sequence is absent, the PCR product will be 641 bp.

In this activity, template DNA will be extracted from cheek cells or hair follicles using InstaGene matrix. InstaGene matrix binds (chelates) divalent ions such as magnesium and prevents them from acting as cofactors for DNases (enzymes that degrade DNA). PCR reactions will be set up using the extracted template DNA. Following PCR, the PCR products will be run on 1% agarose TAE gels and genotypes will be determined from the gel results.

Tips and Notes

- When extracting cheek cells, it is important to gently chew the inside of your mouth to remove epithelial cells while swishing with the saline solution

- A small pellet of cells should be visible once the samples have been centrifuged. If not, add more mouth rinse and repeat the centrifugation

- When the samples are incubated at 56°C, shake them vigorously after 5 min to break up masses of cells. Shaking the samples will ensure that cells break open when heated to 95°C

- PCR is extremely sensitive; therefore, it is very important not to contaminate any of the reagents used in the reaction

- Always use aerosol barrier tips when setting up PCR reactions to prevent the transfer of aerosols from the pipet to the sample

- Keep the master mix and reaction tubes on ice until they are ready to be placed into the thermal cycler

- Mix PCR reagents thoroughly and ensure that the reactions are at the bottom of the tube. If necessary, use a microcentrifuge with adaptors to hold 0.2 ml PCR tubes to pull the liquid to the bottom of the tube. Pulse-spin tubes for 5–10 seconds

Safety Reminders: Handle only your own saliva sample and not those of others. Appropriate PPE should be worn at all times. When using the centrifuge, it is important to balance each tube with an identically weighted tube in the opposite position and to ensure that the inner lid of the centrifuge is closed prior to spinning. Attention should be given when using electricity in the presence of liquids. Safety measures built into the equipment should prevent accidental exposure; however, should any leakage occur, notify your instructor before proceeding. If ethidium bromide, fluorescent dyes, or UV light is used, special precautions are required.

Research Questions

- What is the genotype for the PV92 Alu insertion on chromosome 16 of the samples in your team?
- What is the frequency of the PV92 Alu element in the class?

Objectives

- Collect a cheek cell or hair follicle sample
- Extract gDNA with InstaGene matrix
- Set up a PCR reaction
- Load and run PCR products on a 1% agarose TAE gel
- Determine the genotype of DNA samples
- Calculate the frequency of PV92 Alu element in the class population

Skills to Master

Refer to Laboratory Skills Assessment Rubric (Appendix E) and Laboratory Notebook Rubric (Appendix F)for more details.
- Record laboratory notebook entries
- Perform DNA extractions using InstaGene matrix (Activity 6.B)
- Set up PCR reactions (Activity 6.A)
- Use a thermal cycler (Activity 6.A)
- Perform agarose gel electrophoresis (Activity 4.C)
- Analyze an agarose gel (Activity 4.D)
- Calculate allele frequency

Activity 6.C Detection of the Human PV92 Alu Insertion

Student Workstation Materials

Your instructor will determine whether the template DNA will be extracted from cheek cells or hair follicles. Use the appropriate inventory depending on the DNA extraction method.

Part 1A: Extracting Template DNA from Cheek Cells

Items	Quantity
Vortexer (optional) (shared)	1
Water bath or dry bath at 56°C (shared)	1
Water bath or dry bath at 95°C (shared)	1
Microcentrifuge or mini centrifuge (shared)	1
2–20 µl adjustable-volume micropipet and aerosol-barrier tips	1
100–1,000 µl adjustable-volume micropipet and aerosol-barrier tips (or one 1 ml sterile graduated transfer pipet per student)	1
1.5 ml microcentrifuge tubes	4
Microcentrifuge tube rack	1
Floating tube rack (optional)	1
Paper towels or tissues	4
Laboratory marking pen	1
Waste container	1
200 µl of InstaGene matrix in screwcap tubes	4
Cups or tubes with 10 ml of 0.9% saline solution	4

Part 1B: Extracting Template DNA from Hair Follicles

Items	Quantity
Vortexer (optional) (shared)	1
Water bath or dry bath at 56°C (shared)	1
Water bath or dry bath at 95°C (shared)	1
Microcentrifuge or mini centrifuge (shared)	1
Microcentrifuge tube rack	1
Floating tube rack (optional)	1
Tweezers or forceps	1
Razor blades or scissors	1–4
Laboratory marking pen	1
Waste container	1
200 µl of InstaGene matrix plus protease in screwcap tubes	4

Part 2: Setting Up PCR Reactions

Items	Quantity
Thermal cycler (shared)	1
Microcentrifuge or mini centrifuge with adaptors for PCR tubes (shared)	1
2–20 µl adjustable-volume micropipet and aerosol-barrier tips	1
Microcentrifuge tube rack	1
PCR tubes	4
PCR tube rack	1
Container of ice	1
Laboratory marking pen	1
Master mix (MM (yellow liquid)) (on ice)	95 µl
gDNA samples extracted in part 1	4

Part 3: Running the Gel

Items	Quantity
Horizontal gel electrophoresis chamber	1
Power supply (shared)	1
Digital imaging system (optional) (shared)	1
Microcentrifuge or mini centrifuge with adaptors for PCR tubes (optional) (shared)	1
Rocking platform (optional) (shared)	1
2–20 µl adjustable-volume micropipet and tips	1
Microcentrifuge tube rack	1
PCR tube rack	1
Gel staining tray	1
Agarose gel support film (optional)	1
Laboratory marking pen	1
PCR reactions from part 2	4
Completed +/+ control PCR reaction (shared)	1
Completed –/– control PCR reaction (shared)	1
Completed +/– control PCR reaction (shared)	1
Sample loading buffer (SLB)	80 µl
or UView 6x loading dye and stain (UView)	70 µl
EZ Load molecular mass ruler (MMR)	12 µl
1x TAE electrophoresis buffer*	300 ml
1x Fast Blast DNA stain**	100 ml
1% agarose TAE gel***	1

* If necessary, make 1x TAE electrophoresis buffer (see part 1 of Activity 2.D).
** Alternative staining methods may be used. See Appendix B for details.
*** If necessary, make a 1% agarose TAE gel according to the protocol in Activity 4.B. This may require extra 1x TAE electrophoresis buffer.

Prelab Focus Questions

1. Which cellular structures must be broken down to release the DNA from an animal cell?
2. Why do you need to shake or vortex InstaGene matrix extractions halfway through the 56°C incubation?
3. What are the possible genotypes at the PV92 Alu locus for any given person?

Activity 6.C Detection of the Human PV92 Alu Insertion

Activity Protocol

Part 1A: Extracting Template DNA from Cheek Cells

Note: Each person in the group should prepare one DNA sample using the following instructions. Each person should only handle his/her own saliva sample.

1. Label one 1.5 ml microcentrifuge tube and one screwcap tube containing 200 µl of InstaGene matrix with your initials.

2. Sip 10 ml of 0.9% saline solution and rinse your mouth vigorously for 30 sec. Gently chew the insides of your cheeks while rinsing to release cells. Expel the saline back into the cup or tube.

3. Pipet 1 ml of mouth rinse into the 1.5 ml microcentrifuge tube (not the screwcap tube containing InstaGene matrix).

4. Place the tubes into the microcentrifuge in a balanced configuration. Place the hinges of the tubes facing outward so that the pellet will be easy to locate after spinning. Centrifuge at maximum speed for 2 min.

   Note: If using a mini centrifuge that can reach only 2,000 x g, centrifuge for 5 min.

5. Locate the cell pellet. The pellet should be white and be the size of a match head. If the pellet is smaller than a match head, remove the supernatant, add another 1 ml of mouth rinse to the same tube, and repeat the centrifugation.

6. After pelleting the cells, pour off the saline solution. Be careful not to lose the pellet. Blot the tube on a paper towel. A small amount of saline (<50 µl) should remain in the bottom of the tube.

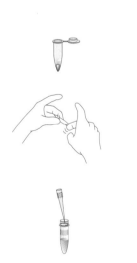

7. Resuspend the pellet by vortexing or flicking the tube so that no clumps of cells remain.

8. Using a 2–20 µl micropipet set to 20 µl, transfer all of the resuspended cells to the screwcap tube containing InstaGene matrix. This may require multiple transfers.

9. Screw the cap tightly on the tube. Vortex or shake the tube vigorously to mix the tube contents.

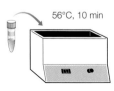

56°C, 10 min

10. Incubate the screwcap tubes at 56°C for 10 minutes in a water bath. After 5 min of incubation, vortex or shake the tubes vigorously, and then place them back in the 56°C water bath for the remaining 5 min.

Activity 6.C Detection of the Human PV92 Alu Insertion

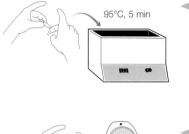

11. Remove the tubes from the 56°C waterbath or dry bath and vortex or shake. Then incubate the tubes at 95°C for 5 min. After the incubation is over, vortex or shake the tubes.

12. Place the tubes into the microcentrifuge in a balanced configuration. Centrifuge at top speed for 5 min. The supernatant contains the DNA template.

    Note: If using a mini centrifuge that can reach only 2,000 x g, centrifuge for 10 min.

13. Proceed directly to part 2 or store the gDNA in the screwcap tubes at 4°C for up to 1 month. **Do not freeze the samples.**

Part 1B: Extracting Template DNA from Hair Follicles

Note: Each person in the group should prepare one DNA sample using the following instructions.

1. Label one screwcap tube containing 200 μl of InstaGene matrix plus protease with your initials.

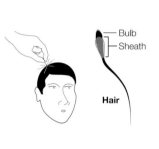

2. Using tweezers or forceps, collect 2 hairs from your head, arm, or leg. Select hairs that have either a noticeable sheath (a coating of epithelial cells near the base of the hair) or a good size root (the bulb shaped base of the hair). Using scissors or a razor blade, trim the hair, leaving the last 2 cm of the base of the hair. Place the two trimmed hairs into the screwcap tube containing InstaGene matrix plus protease.

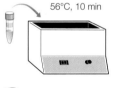

3. Incubate the screwcap tubes at 56°C for 10 min in a water bath. After 5 min of incubation, vortex or shake the tubes vigorously and then place them back in the 56°C water bath or dry bath for the remaining 5 min.

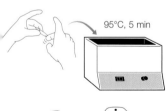

4. Remove the tubes from the 56°C water bath or dry bath and then vortex or shake them. Then incubate the tubes at 95°C for 5 min, and then vortex or shake them again.

5. Place the tubes into the microcentrifuge in a balanced configuration. Centrifuge at maximum speed for 5 min. The supernatant contains the DNA template.

    Note: If using a mini centrifuge that can reach only 2,000 x g, centrifuge for 10 min.

6. Proceed directly to part 2, or store the gDNA in the screwcap tubes at 4°C for up to 1 month. **Do not freeze the samples.**

Activity 6.C Detection of the Human PV92 Alu Insertion

Part 2: Setting Up PCR Reactions

Note: Each person in the group should prepare one PCR reaction using the following instructions.

1. Centrifuge the screwcap tubes containing the DNA templates at maximum speed for 2 min.

2. Label a PCR tube with your initials and place it in the PCR tube rack on ice.

Supernatant

3. Pipet 20 µl of DNA template from the screwcap tubes into the PCR tube.

   Note: Avoid disturbing or transferring the the pellet of InstaGene matrix, which is at the bottom of the screwcap tubes that contain template DNA. InstaGene matrix will inhibit the PCR reaction. If the InstaGene matrix is disturbed while attempting to transfer the DNA template, recentrifuge the screwcap tubes to pellet the beads.

MM

4. Pipet 20 µl of the master mix into the PCR tube. Mix by pipetting up and down 2–3 times. Cap the PCR tube tightly. The mixture should be yellow.

   Note: Pulse-spin the PCR tubes for 5–10 sec in a microcentrifuge or mini centrifuge with adaptors for PCR tubes to pull the contents to the bottom of the tube.

5. When instructed to do so, place the PCR tubes in the thermal cycler. Control reactions prepared by the instructor should also be placed in the thermal cycler at this point.

6. If necessary, program the thermal cycler with the following program:

Initial denature:	94°C for 2 min
40 cycles of:	94°C for 1 min
	60°C for 1 min
	72°C for 2 min
Final extension:	72°C for 10 min
Hold:	15°C for ∞

7. When all samples are in the cycler, start the PCR.

   Note: The reactions will take 3–4 hr to complete. Since there is no need to wait for the cycler to finish, the reactions can be left in the cycler overnight.

8. Store samples at 4°C in the refrigerator until they are ready to be run on the gel.

9. If necessary, prepare 1x TAE (refer to part 1 of Activity 2.D) and a 1% agarose TAE gel (refer to Activity 4.B) for the next part of the activity.

Activity 6.C Detection of the Human PV92 Alu Insertion

Part 3: Running the Gel

1. Retrieve the PCR reactions from the thermal cycler or refrigerator. Pulse-spin the PCR tubes for 5–10 sec in a microcentrifuge or mini centrifuge with adaptors for PCR tubes to pull the liquid to the bottom of the tube.

2. Using a fresh tip for each sample, pipet 10 µl of sample loading buffer (**SLB**) or 8 ul of UView 6x loading dye and stain (**UView**) into each PCR tube. Pipet up and down to mix.

3. Place a 1% agarose TAE gel into the electrophoresis chamber. Fill the electrophoresis chamber with sufficient 1x TAE buffer to cover the gel by approximately 2 mm.

4. Check that the wells of the agarose gel are near the black negative electrode or cathode and that the bottom edge of the gel is near the red positive electrode or anode.

5. Using a fresh tip for each sample, load the indicated volume of each sample into the gel in the following order:

Lane 1	EZ Load molecular mass ruler (**MMR**)	10 µl
Lane 2	+/+ control	10 µl
Lane 3	–/– control	10 µl
Lane 4	+/– control	10 µl
Lane 5	Student sample	20 µl*
Lane 6	Student sample	20 µl*
Lane 7	Student sample	20 µl*
Lane 8	Student sample	20 µl*

* If using UView 6x loading dye and stain, load 10 ul of each sample.

The order in which samples are loaded into the gel should be recorded in your laboratory notebook.

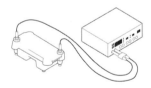

6. Place the lid on the electrophoresis chamber. Connect the electrical leads to the power supply, red to red and black to black.

7. Turn on the power and run the gel at 100 V for 30 min.

    Note: If using the fast gel running buffer (0.25x TAE), run the gel at 200 V for 20 min (see Appendix A).

8. When the electrophoresis run is complete, turn off the power and remove the lid from the chamber. Carefully remove the gel and tray from the electrophoresis chamber. Be careful—the gel is very slippery. If using UView 6x loading dye and stain or other fluorescent stain, use a UV light source to visualize DNA bands. Otherwise, slide the gel into the staining tray and carefully pour a sufficient volume of 1x Fast Blast DNA stain to submerge the gel.

    Note: For alternative DNA staining options, see Appendix B.

9. Stain the gel overnight to visualize the DNA bands and then record the results. (Optional) A rocking platform may be used to enhance staining.

10. (Optional) Trim away any unloaded lanes and air dry on gel support film. Tape the dried gel into your laboratory notebook.

Activity 6.C Detection of the Human PV92 Alu Insertion

Results Analysis

Interpretation of Results

The PCR primers used for the reaction bracket the region where the Alu element resides within the PV92 region of chromosome 16. The primers amplify a 641 bp PCR product if the Alu element is absent. Since the Alu element is approximately 300 bp, the PCR product will be 941 bp if the Alu element is present. Each individual has two copies of chromosome 16, and the PV92 Alu element may be present or absent on each of these chromosomes. If an individual is homozygous positive for the Alu insertion, he/she will have the PV92 Alu element on both copies of chromosome 16; in this case, the primers will amplify a 941 bp product from each chromosome, resulting in a single 941 bp band on the agarose gel. If an individual is homozygous negative for the Alu insertion, the PV92 Alu element will be absent from both chromosomes and two 641 bp PCR products will be amplified, resulting in a single 641 bp band. If an individual is heterozygous for the Alu insertion, the Alu element will be present on only one of the chromosomes, and both a 941 bp PCR product and a 641 bp PCR product will be amplified, resulting in two bands on the agarose gel (see Figure 6.32).

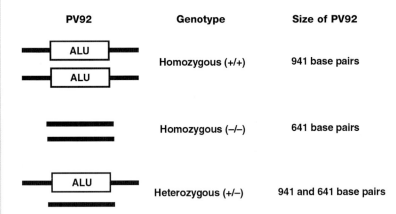

Figure 6.32. **PV92 genotypes.** The presence or absence of the PV92 Alu insert changes the length of the PCR product by 300 bp. Homozygous individuals will have two alleles that are the same, so a single PCR product will be observed on the gel. Two PCR products will be amplified in heterozygous individuals, resulting in two bands on the agarose gel.

After staining the gel, record the number of bands for each sample and estimate the size of each band using the size standard (see Figure 6.33 for the band sizes in the standard). The 941 bp band will be much fainter than the 641 bp band (see lane 7 of Figure 6.33) because it is easier to amplify short PCR products than long ones.

Use the size standard and the bands from the control reactions in the gel to determine the genotype of the student samples. Record your conclusions.

Allelic Frequency Calculation

Once the genotypes of all students are collected, the frequency of the PV92 Alu allele in the class population can be calculated. First determine the total number of possible alleles in the class population. Since each person has two alleles — one on each chromosome 16 — multiply the total number of people in the class that have a confirmed genotype by 2. For example, if there are 10 students in the class, there will be 20 possible alleles.

Second, determine the total number of (+) alleles (that is, the total number of PV92 alleles that contain the Alu insert) in the class population. Each homozygous positive student (+/+) has two (+) alleles, each homozygous negative (–/–) student has no (+) alleles, and each heterozygous student

Activity 6.C Detection of the Human PV92 Alu Insertion

has one (+) allele. For example, if your class has 3 (+/+) students, 5 (+/−) students, and 2 (−/−) students, the number of (+) alleles in the class population would be (3 x 2) + (5 x 1) + (2 x 0) = 11.

Third, calculate the frequency of the (+) allele by dividing the number of (+) alleles in the class by the total number of possible alleles in the class population. In this example, the frequency would be 11/20 or 0.55.

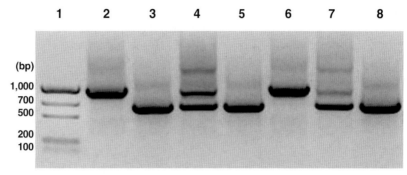

Figure 6.33. **Example of PV92 PCR results.** Lane 1, EZ Load molecular mass ruler; lane 2, homozygous (+/+) control; lane 3, homozygous (−/−) control; lane 4, heterozygous (+/−) control; lane 5, homozygous (−/−) student sample; lane 6, homozygous (+/+) student sample; lane 7, heterozygous (+/−) student sample; lane 8, homozygous (−/−) student sample.

Self Assessment

Assess whether you have mastered the skills of this activity using the Laboratory Skills Assessment Rubric in Appendix E.

Assess your experimental write-up in your laboratory notebook using the Laboratory Notebook Rubric in Appendix F.

Postlab Focus Questions

1. What are the genotypes for the PV92 Alu insertion on chromosome 16 of the samples in your team?
2. What is the frequency of the PV92 Alu (+) allele in your class?

Activity 6.D Fish DNA Barcoding

Overview

What Is DNA Barcoding? Have you ever ordered a California roll at a sushi counter and wondered exactly what sort of seafood made up the "imitation crab" in your meal? Or have you ever been to a seafood restaurant in New England and ordered scrod and wondered what fish you were getting? Once a piece of seafood has been processed and filleted, it can be difficult to tell what species the fish is. An explosion of quick and inexpensive methods to isolate, purify, amplify, and sequence DNA has brought new methods to help identify different species, whether they are fish sold at the market or newly discovered species. Using DNA-based technologies, a multinational alliance of scientists is now cataloging life using a DNA barcoding system in order to accelerate the discovery of new species and develop powerful new tools to monitor and preserve Earth's vanishing biodiversity. In much the same way that a UPC (universal product code) barcode can differentiate a carton of milk from a bag of carrots when they are scanned into the cash register at a grocery store, DNA sequences can be used to uniquely identify different species. This is the basis of DNA barcoding.

Once a DNA barcode is generated, in order to perform a barcode search it is neccessary to have a searchable database that links the DNA barcodes generated by researchers from known and highly characterized biological specimens to the specimens' formal names and other important information (including data that allow the origin and current location of the source specimen to be easliy tracked and verified by other researchers if neccessary).

Since DNA barcodes from unknown samples will be compared against those in the database, it is critical that a) the database contains sequence data of the highest quality possible, and b) that the identity given to the species from which the reference DNA barcode is generated is assigned correctly. For example, if the code in the grocery store cash register database for button mushrooms, which tend to sell for $2.00/lb, was accidently switched with the code for morel mushrooms, which tend to sell for $20.00/lb, many people would be very unhappy when their button mushrooms suddenly cost ten times more than expected!

For this reason, scientists developed a highly regulated database called the Barcode of Life Data Systems (BOLD) reference library, which stores all the high-quality reference DNA barcode records. Through the BOLD Identification System (BOLD-IDS), a query (unknown or unverified) barcode sequence obtained from an unknown tissue sample of food product can be compared against reference barcode sequences contained in the BOLD reference library to determine the identity of the unknown specimen. The BOLD Systems website can be found at boldsystems.org.

In this activity you will extract DNA from fish tissue obtained from a local grocery store, restaurant, fishing trip, or other source. You will set up a PCR reaction for each fish DNA extract you generated as well as two control PCR reactions. One control PCR reaction will be the positive control, which will use pCOI plasmid DNA as your target sequence. The other reaction will be a negative control, which will use water instead of target DNA. Gel electrophoresis will allow you to determine the success of your PCR reactions by visualizing the size of your amplified DNA. The size of the band that corresponds to your successfully amplified COI gene PCR product should be approximately 650 bp. You may also notice an additional band less than 100 bp in size. This band corresponds to unincorporated primers from your PCR reaction, which can stick to each other in what is known as a primer-dimer.

The UView™ 6x loading dye and stain you will add to each of your samples contains a fluorescent compound that binds to DNA. During gel electophoresis it will comigrate with your DNA and allow your DNA to be visualized with UV light. In the DNA sequencing stage, you will be submitting your PCR products to a sequencing facility to be purified and sequenced. Once you obtain these sequences, you will use bioinformatics tools to analyze the quality of your data, assemble a consensus sequence from your high-quality sequencing results, and search the Barcode of Life Database (BOLD) to determine the closest match for your fish sample.

Tips and Notes

- PCR involves amplication of DNA and therefore it is critical to use proper technique to avoid any cross contamination between fish samples during DNA extraction. Do not recycle cutting implements, pipet tips, or containers. If using gloves, change gloves in between the handling of different fish samples

Safety Reminders: Eating, drinking, smoking, and applying cosmetics are not permitted in the work area. Wearing protective eyewear and gloves is strongly recommended. Wash your hands with soap before and after this exercise. If any solution gets into your eyes, flush with water for 15 minutes. Lab coats or other protective clothing should be worn to avoid staining clothes. Place all used razor blades into a sharps container. Special precautions are required when UV light is used.

Research Question

- What is the identity of a fish sample?

Activity 6.D Fish DNA Barcoding

Objectives

- Estimate or weigh out the proper amount of fish tissue and mince it as finely as possible
- Extract DNA from fish tissue with buffers and a spin column
- Set up a PCR reaction for each fish DNA extract, a positive control using pCOI plasmid DNA as the target sequence, and a negative control using water instead of target DNA
- Load and run PCR products on a 1% agarose TAE gel
- Send PCR products to a sequencing facility to be purified and sequenced
- Use bioinformatics tools to analyze the quality of the data, and assemble a consensus sequence from the high-quality sequencing results
- Search the Barcode of Life Database (BOLD) to determine the closest match for the fish sample

Skills to Master

Refer to Laboratory Skills Assessment Rubric (Appendix E) and Laboratory Notebook Rubric (Appendix F) for more details.

- Record laboratory notebook entries
- Perform DNA extractions
- Set up PCR reactions (Activity 6.A)
- Perform agarose gel electrophoresis (Activity 4.C)
- Analyze an agarose gel (Activity 4.D)
- Perform bioinformatics analysis

Student Workstation Materials

Part 1A: Preparing fish samples

Items	Quantity
Fish samples each in a separate weigh boat	2
Razor blades, plastic knives, or other cutting implements	2
(**Note:** it is critical to use one implement per fish sample)	
Empty 2 ml microcentrifuge tubes with caps	2
100–1,000 adjustable-volume micropipet and tips	1
Marking pen	1
Disposable gloves (optional)	4–8

Part 1B: Extracting DNA from fish samples

Items	Quantity
Resuspension solution (R)	500 µl
Lysis solution (Lys)	600 µl
Neutralization solution (N)	600 µl
Matrix	600 µl
Wash solution (W)	2 ml
Distilled water	300 µl
Spin columns	2
Empty 2 ml microcentrifuge tubes without caps	4
100–1,000 adjustable-volume micropipet and tips	1
Marking pen	1
Water bath or dry bath set to 55°C (shared)	1
Microcentrifuge (shared)	1–2

Part 2A: PCR Amplification of DNA

Items	Quantity
Ice bath	1
Fish DNA samples from Part 1B	2
(+) sample	10 µl
(–) sample	10 µl
COI master mix (CMM)	150 µl
PCR tubes	4
2–20 µl adjustable-volume micropipet	1
20–200 µl adjustable-volume micropipet	1
2–20 µl pipet tips, aerosol barrier	1 rack
20–200 µl pipet tips, aerosol barrier	1 rack
Marking pen	1

Part 2B:
Preparing PCR samples for electrophoresis and sequencing

Items	Quantity
Ice bath	1
PCR reactions from Part 2A	4
2 mL microcentrifuge tubes	7
2–20 µl adjustable-volume micropipet	1
20–200 µl adjustable-volume micropipet	1
2–20 µl pipet tips, aerosol barrier	1 rack
20–200 µl pipet tips, aerosol barrier	1 rack
Marking pen	1

Part 3: Running the Gel

Items	Quantity
Electrophoresis samples labeled E from Part 2	4
Molecular weight ruler	25 µl
UView 6x loading dye and stain (UView)	15 µl
Sterile water	40 µl
2–20 µl adjustable-volume micropipets	1
2–20 µl pipet tips, aerosol barrier	1 rack
1% agarose TAE gel *	1
1% or 0.25% TAE electrophoresis buffer	250 ml
Gel electrophoresis chamber	1
Power supply (may be shared by multiple groups)	1
Marking pen	1
UV transilluminator or imaging system	1

Part 4A: Sequencing

Items	Quantity
SEQ samples from Part 2	3
Marking pen	1
Parafilm (1 piece per microcentrifuge tube)	1

Part 4B: Bioinformatics

Materials needed but not provided	Quantity
Forward and reverse sequencing files for samples	varies
Computer with internet access	1

Activity 6.D Fish DNA Barcoding

Prelab Focus Questions

1. What parts of the cell must be broken down to extract DNA?

2. It is important to keep track of the location of the DNA at each stage of purification. For the following steps of the protocol, state whether the DNA is in the pellet, in the supernatant, bound to the column, or in the flowthrough:

   a. After centrifuging down the neutralized fish tissue lysate (pellet or supernatant).

   b. After centrifuging the supernatant through the column (column or flowthrough).

   c. After centrifuging the wash solution through the column (column or flowthrough).

   d. After centrifuging the elution solution through the column (column or flowthrough).

3. How do researchers target the portion of DNA to be amplified during PCR?

4. Do you expect the pCOI plasmid to generate a PCR product? What about the negative control? Why or why not?

5. What do think the results should look like for each sample in the gel?

6. Why is it necessary to have PCR products purified before sequencing? (Hint: Think about how sequencing works and what might interfere with that if left over from PCR).

7. What primers will be used to sequence your PCR product?

8. What steps of the process may impact the quality of the sequence that will be generated?

Activity 6.D Fish DNA Barcoding

Part 1A: Preparing fish samples

Initials   Initials

1. Label one capped 2 ml microcentrifuge tube for each of your fish samples (that is, "**1**" for fish sample 1, "**2**" for fish sample 2, etc.). Also label with your initials.

2. Cut a piece of fish muscle up to 100 mg in mass, approximately the size of an eraser on a pencil, from your first fish sample. Place the piece in a new weigh boat and slice it with a razor blade or cutting implement until finely minced. Transfer the sample into the appropriately labeled microcentrifuge tube.

3. Properly discard the razor blade or cutting implement. If wearing gloves, change gloves before handling the next piece of fish. If not, wash hands thoroughly.

4. Using a new razor blade or cutting implement, cut a piece of fish muscle up to 100 mg in mass, approximately the size of an eraser on a pencil, from your second fish sample. Place the piece in a new weigh boat and slice it with a razor blade until finely minced. Transfer the sample into the appropriately labeled microcentrifuge tube. Properly discard the razor blade or cutting implement.

Part 1B: Extracting DNA from fish samples

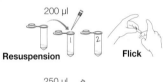

200 µl

Resuspension   **Flick**

1. Add 200 µl of resuspension solution (**R**) to your two microcentrifuge tubes containing minced fish and flick the tubes several times to ensure full submersion of the fish sample in the resuspension solution.

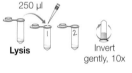

250 µl

Lysis   Invert gently, 10x

2. Add 250 µl of lysis solution (**Lys**) to each tube and mix gently by inverting tubes 10 times to mix contents. Do not vortex! Vortexing may shear genomic DNA, which can inhibit PCR amplification.

55°C, 10 min

Water bath

3. Incubate samples at 55°C for 10 min. The samples do not need to be shaken during incubation.

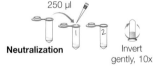

250 µl

Neutralization   Invert gently, 10x

4. Add 250 µl of neutralization solution (**N**) to each microcentrifuge tube and mix gently by inverting tubes 10 times to mix contents (do not vortex). A visible cloudy precipitate may form.

12,000–14,000 x g, 5 min

5. Centrifuge the tubes for 5 min at top speed (12,000–14,000 x g) in the microcentrifuge. A compact pellet will form along the side of the tube. The supernatant contains the DNA. If there are a lot of particulates remaining in the supernatant after centrifugation, centrifuge the tubes for 5 additional min.

6. Snap (do not twist!) the bottoms off of the spin columns and insert each column into a capless 2 ml microcentrifuge tube.

Activity 6.D Fish DNA Barcoding

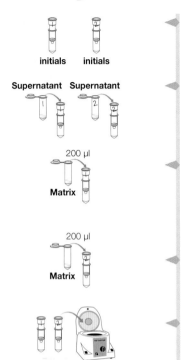

initials   initials

Supernatant   Supernatant

200 µl

Matrix

200 µl

Matrix

12,000–14,000 x g, 30 sec

7. Label one spin column **1** for Fish 1 and a second spin column **2** for Fish 2. Also label the columns with your initials.

8. Transfer the entire supernatant (500–550 µl) of each fish sample from step 5 into the appropriately labeled spin column. Try not to get any of the particulates into the spin column because they will clog the column and prevent you from continuing.

9. Thoroughly mix the contents of the tube labeled **Matrix** by vortexing or repeatedly shaking and inverting the tube to make sure particulates are completely resuspended before use.

10. Add 200 µl of thoroughly resuspended matrix to the first column containing fish extract and pipet up and down to mix.

11. Using a new pipet tip, add 200 µl of thoroughly resuspended **Matrix** to the second column containing fish extract and pipet up and down to mix.

12. Centrifuge the columns for 30 sec at full speed.

    Note: Take care to spin the column for only 30 sec. Drying the matrix completely at this point will result in loss of DNA.

13. Remove the spin column from the 2 ml microcentrifuge tube, discard the flowthrough at the bottom of the 2 ml tube, and replace the spin column in the same tube. Add 500 µl of wash solution (**W**) and centrifuge for 30 sec.

    Note: Take care to spin the column for only 30 sec. Drying the matrix completely at this point will result in loss of DNA.

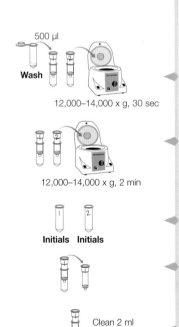

500 µl

Wash

12,000–14,000 x g, 30 sec

12,000–14,000 x g, 2 min

Initials   Initials

Clean 2 ml
tube with "1"

Initials

14. Repeat step 13 to wash samples again.

15. Remove the spin column from the 2 ml microcentrifuge tube, discard the flowthrough at the bottom of the 2 ml tube, and replace the spin column in the same tube. Centrifuge columns for a full 2 min to remove residual traces of ethanol and dry out the matrix.

16. Label two clean 2 ml capless microcentrifuge tubes with your fish sample number and your initials.

17. When your 2 min spin is completed, remove the spin columns and discard the 2 ml microcentrifuge wash tubes.

18. Place the spin column for each sample into the correctly labeled new capless 2 ml microcentrifuge tube from step 16.

Activity 6.D Fish DNA Barcoding

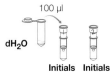

dH₂O

Initials Initials

initials initials

1 2 (+) (−)

CMM

1

19. Using a fresh pipet tip for each sample, add 100 µl of distilled water to each spin column, being careful not to touch the resin. Recover the DNA by centrifuging for 1 min at full speed.

20. Label two clean 2 ml microcentrifuge tubes (with caps) **Fish 1** and **Fish 2** and your initials.

21. Transfer the recovered DNA into the appropriately labeled 2 ml microcentrifuge tube with caps and store the DNA at 4°C until you are ready to proceed.

Part 2A: PCR Amplification of DNA

1. Label four PCR tubes with your initials and the sample name (**1** for fish sample 1, **2** for fish sample 2, (**+**) for the PCR positive control DNA, (**−**) for the PCR negative control). Keep the tubes on ice for the remaining steps.

2. Using a fresh aerosol filter pipet tip each time, add 35 µl of COI master mix (**CMM**) to each PCR tube, capping each tube immediately after the addition of liquid.

3. Using a fresh aerosol filter tip for each tube, add 5 µl of the appropriate DNA sample directly into the **CMM** liquid in each corresponding PCR tube and pipet up and down to mix. Recap each tube immediately after adding DNA.

Tube Name	Master Mix DNA
1	35 µl CMM, 5 µl fish sample 1
2	35 µl CMM, 5 µl fish sample 2
(+)	35 µl CMM, 5 µl (+) sample
(−)	35 µl CMM, 5 µl (−) sample

4. If necessary, program the thermal cycler with the following program:

Initial denature:	94 °C for 2 min
40 cycles of:	94 °C for 1 min
	60 °C for 1 min
	72 °C for 2 min
Final extension:	72 °C for 10 min
Hold:	15 °C for ∞

5. When all samples are in the cycler, start the PCR.

   Note: The reactions will take 3–4 hr to complete. Since there is no need to wait for the cycler to finish, the reactions can be left in the cycler overnight.

6. Store samples at 4°C in the refrigerator for Part 2B.

Activity 6.D Fish DNA Barcoding

Part 2B: Preparing PCR Samples for Electrophoresis and Sequencing

1, E   2, E   (+), E   (–), E
Initials Initials Initials Initials

1. Label four 2 ml microcentrifuge tubes with both your initials and **E**. E stands for electrophoresis. Now label one of these tubes **1**, one tube **2**, one tube (**+**), and one tube (**–**).

1, PCR
Product   5 µl

1, E
Initials

2. Transfer 5 µl from each PCR reaction into the 2 ml microcentrifuge tube corresponding to that sample.

1,   2,   (+),
SEQ   SEQ   SEQ
Initials Initials Initials

3. Label three 2 ml microcentrifuge tubes with both your initials and **SEQ**. SEQ stands for sequencing. Now label one of these tubes **1**, one tube **2**, one tube (**+**). You will not be sequencing your negative control sample.

4. Using a fresh aerosol filter tip for each tube, transfer 30 µl from each PCR reaction into the 2 ml microcentrifuge tube corresponding to that sample.

1, PCR   30 µl
Product

1, SEQ
Initials

5. Store your samples at 4°C until you are ready to proceed with electrophoresis and sequencing.

Part 3: Gel Electrophoresis

Sterile   5 µl
water

1, E 2, E (+), E (–), E

1. Add 5 µl sterile water to each PCR product sample (four samples). Use a new pipet tip for each tube.

UView   2 µl

1, E 2, E (+), E (–), E

2. Add 2 µl of UView 6x loading dye and stain (**UView**) to each sample, using a new pipet tip each time. Mix samples well and pulse-spin.

3. Place a 1% agarose TAE gel into the electrophoresis chamber. Fill the electrophoresis chamber with sufficient 1x or 0.25x TAE buffer to cover the gel by approximately 2 mm.

4. Check that the wells of the agarose gel are near the black negative electrode or cathode and that the bottom edge of the gel is near the red positive electrode or anode.

5. Load the agarose gel in the following lane order and volume, using a new pipet tip each time.

Lane	Sample
1	Empty
2	Empty
3	20 µl PCR molecular weight ruler (**MWR**)
4	12 µl (**+**) **E**
5	12 µl (**–**) **E**
6	12 µl **1 E**
7	12 µl **2 E**
8	Empty

Activity 6.D Fish DNA Barcoding

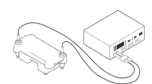

6. Ask your instructor whether the electrophoresis buffer your electrophoresis unit contains is 0.25x TAE or 1x TAE.

   If your buffer is 0.25x TAE, run the gel at 200 V for 20 min.

   If your buffer is 1x TAE, run the gel at 100 V for 30 min.

7. Visualize the gel on a UV transilluminator or other imaging system. No gel staining is required as the loading dye contains a fluorescent compound that will allow visualization of DNA with UV light.

Part 4A: Sequencing

1. Thoroughly cover your capped **1 SEQ**, **2 SEQ**, and **(+) SEQ** tubes with Parafilm to prevent leakage during transport.

2. Record the sample names on your tubes and make sure these match the names your instructor is submitting to the sequencing facility. This is the only way you can identify the correct sequencing data file for each sample.

3. Give your samples to your instructor for sequencing.

Part 4B: Bioinformatics

Download the Fish DNA Barcoding Kit Bioinformatics Guide (bulletin 6398) at bio-rad.com/fishbarcoding for instructions on how to carry out the bioinformatics portion of the activity.

Self Assessment

Assess whether you have mastered the skills of this activity using the Laboratory Skills Assessment Rubric in Appendix E.

Assess your experimental write-up in your laboratory notebook using the Laboratory Notebook Rubric in Appendix F.

Postlab Focus Questions

1. Where are the sequencing primer sequences on your PCR product? How did they get there?
2. What two aspects of primer design have been used in this activity to ensure successful DNA amplification from a wide variety of fish samples?

Chapter 6 Extension **Activities**

1. Investigate additional foods or crops for genetic modifications or look for occurrences of gene transfer of the genetically modified DNA into non-GM plants in the vicinity of GM crops.
2. Design and test additional primers to specific plant genetic modifications such as those in Bt or Roundup Ready crops.
3. Investigate the PV92 Alu allelic frequency of a specific population, such as your school, and determine whether it is in Hardy-Weinberg equilibrium. Use the bioinformatics database at the Dolan DNA Learning Center to compare class frequencies to other groups around the world. (See Chromosome 16: PV92 PCR Informatics Kit instruction manual, Bio-Rad bulletin 4110052 for more information.)
4. Perform plant DNA extraction and nested PCR using Bio-Rad's Nucleic Acid Extraction module in combination with the GAPDH PCR module.
5. Explore real-time PCR using Bio-Rad's Crime Scene Investigator PCR Basics™ Real-Time PCR Starter kit.
6. Determine the proportion of GM material in GM-positive food using Bio-Rad's GMO Investigator™ Real-Time PCR Starter kit.

Protein Structure & Analysis

Summary

Proteins perform most of the work, transmit most of the signals, and form many of the structures required for life. Scientists have learned how to engineer and manufacture proteins for many uses, including drugs like insulin and growth hormone, as industrial enzymes in paper and detergent manufacturing, and as tools in life science research.

A protein is a chain of amino acids (a polypeptide) that is folded into a specific, three-dimensional (3-D) shape that is necessary for the protein to perform its function. The sequence of the amino acid chain is based on the DNA sequence of the gene that encodes the protein. In addition proteins have different chemical properties based on their amino acid composition, such as net charge, hydrophobicity, and size.

Proteins must be quantified, analyzed, and purified to be used in biotechnology. The amount of protein in a sample is quantified using colorimetric assays. Proteins are analyzed using techniques such as polyacrylamide gel electrophoresis (PAGE), in which proteins are separated based on their size. Proteins are purified from a mixture using chromatography, which separates proteins based on physical and chemical properties. Bioinformatics, the computer analysis of genes and proteins, is also an important tool in investigating proteins. Because protein production is expensive, precautions are taken throughout the production process to catch and solve problems and ensure consistent results. The activities of this chapter introduce you to skills required for protein analysis.

7.1 Protein Synthesis

Each protein is encoded by a gene. Genes serve as templates for the synthesis of messenger RNA (mRNA), which in turn is used as the template for the synthesis of a polypeptide chain of amino acids, which is then folded to make the functional protein. The synthesis of mRNA occurs through a process called transcription, and the synthesis of protein occurs through a process called translation. Though the fundamental aspects of transcription and translation are universal among living organisms, there are some key differences in how these processes occur and are regulated in bacteria and eukaryotes.

Bacterial Transcription

In bacteria transcription is regulated by proteins that bind to a region upstream of a gene called a promoter, and transcription is initiated by the RNA polymerase binding to the promoter.

RNA polymerase is an enzyme that generates the mRNA molecule, and RNA polymerase activity is very similar to that of DNA polymerase in DNA replication (see Figure 6.6), except that it produces an RNA molecule instead of DNA (see Figure 7.1). RNA polymerase unwinds the DNA, then travels down one strand of the DNA, adding ribonucleotide molecules to a growing a chain of mRNA that is complementary to the DNA sequence. Since the complementary sequence is RNA, uracil is used in place of thymine. The process continues until the RNA polymerase reaches a termination sequence at the end of the gene, at which point the enzyme and the mRNA are released from the DNA.

Unlike DNA replication, multiple mRNA molecules can be transcribed at the same time. In bacteria, mRNA can immediately be translated into a polypeptide; in eukaryotes, mRNA is made in the nucleus and must be exported into the cytoplasm for translation (see Figure 7.2).

Eukaryotic Transcription

Eukaryotic genes are more complex than bacterial genes and the complexity affects how they are transcribed. Eukaryotic genes are composed of multiple units called **exons** that are separated by non-coding DNA sequences called **introns** (see Figure 7.3). When a eukaryotic gene is transcribed, the exon and intron sequences are present in the initial mRNA, called pre-mRNA. Pre-mRNA is then processed to remove the introns and join the exons in a process called RNA splicing. Eukaryotic mRNA also is capped with a modified guanine base added to the 5' end, and a long sequence of adenosines called a poly(A) tail is added to the 3' end (see Figure 7.3). These additions are thought to protect the mRNA from degradation. The final, processed mRNA is called mature mRNA. Transcription and mRNA processing are carried out in the nucleus and translation occurs on the rough endoplasmic reticulum (RER) in the cell cytoplasm.

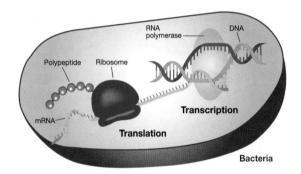

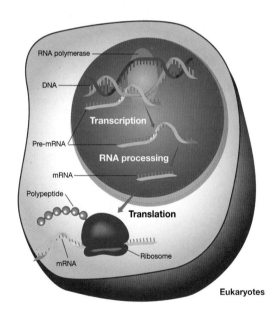

Figure 7.2. **Bacterial and eukaryotic protein synthesis**. Transcription and translation occur concurrently in bacteria since there is no nucleus. Eukaryotes transcribe and process mRNA in the nucleus, then export the mRNA for translation.

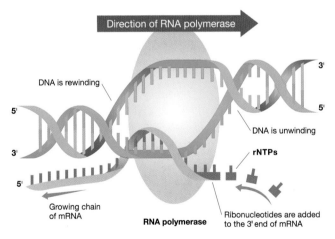

Figure 7.1. **Transcription.** RNA polymerase reads the 3' to 5' DNA strand and transcribes a 5' to 3' mRNA.

Translation

Translation is the conversion of the information encoded in mRNA into a sequence of amino acids. Translation is based on a set of rules called the **genetic code**, which maps units of three nucleotides from the mRNA to a corresponding amino acid. Each three-nucleotide unit is called a codon and there are 64 possible codons encoding specific amino acids (or signals to stop translation). There are only 20 standard amino acids, so multiple codons encode the same amino acid (see Figure 7.5).

In most cases, the first codon of an mRNA (the start codon) codes for the amino acid methionine, whose sequence is AUG. There are also three codons for which there are no corresponding amino acids, and these are called stop codons. The genetic code is shared among organisms like bacteria, plants, and humans.

Specialized RNA molecules called **transfer RNAs (tRNAs)** read and translate the genetic code; that is, they connect the nucleotide sequence on the mRNA to the amino acid sequence. tRNAs are RNA molecules that have been folded into 3-D structures. At one end of the tRNA is a three-base sequence, or anticodon, that binds a matching codon on the mRNA. At the other end is an amino acid (see Figure 7.4). There is a matching tRNA for each codon (except for the three stop codons).

Translation consists of three steps: initiation, elongation, and termination (see Figure 7.4). It occurs within a large complex called a ribosome, which is made up of protein and **ribosomal RNA (rRNA)**. The ribosome contains three sites that host the tRNAs: peptidyl sites (P sites), aminoacyl sites (A sites), and exit sites (E sites). tRNAs move through these sites as the mRNA is translated. The ribosome also contains enzymes that form the peptide bonds between amino acids, creating the polypeptide chain.

Ribosomes contain a small and a large subunit that bind the mRNA at different times. To initiate translation, the small subunit binds the mRNA, and the first tRNA (carrying methionine) binds to the complementary sequence AUG on the mRNA (see Figure 7.4). Next, the large subunit binds and the tRNA that matches the next codon on the mRNA binds. The first tRNA is in the P site and the second is in the A site. A **peptide bond** forms between the amino acids of the tRNAs and the ribosome moves three bases along the mRNA, placing the first tRNA molecule in the E site, from which it is released from the ribosome. A third tRNA moves into the A site and the next peptide bond is formed. This addition of new amino acids to the growing polypeptide chain is called elongation (see Figure 7.4).

Elongation extends the polypeptide chain until a stop codon is reached. At that point, a release factor binds, the ribosomal subunits and polypeptide are released, and the process is terminated. The mRNA is repeatedly translated until it is degraded, so multiple polypeptide chains are generated. The ribosomes and tRNAs are also recycled.

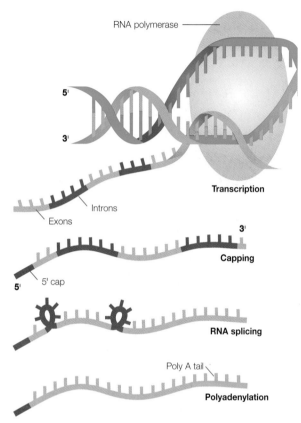

Figure 7.3. **Eukaryotic RNA processing.** Following transcription, RNA processing takes place inside the nucleus of eukaryotic cells. Introns are removed and exons are spliced together, a 5' guanine cap and a 3' poly(A) tail are added, and the mRNA is transported out of the nucleus through nuclear pores before being translated in the cytoplasm.

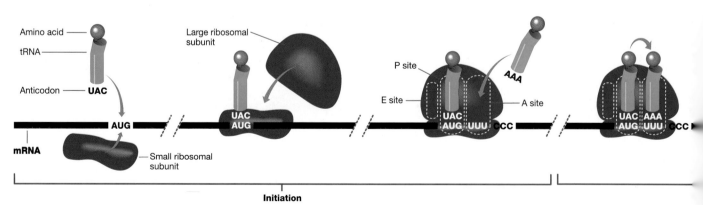

Figure 7.4. **Translation.** An mRNA is read by a ribosome to make a polypeptide. Translation occurs in three phases: initiation, elongation, and termination.

Posttranslational Modifications

The availability of proteins is regulated by the rates of transcription and translation; however, the activity of proteins is also regulated by modifications made after translation by a process called posttranslational modification. Posttranslational modifications are much more common in eukaryotes than bacteria. Posttranslational modifications include additions of carbohydrate molecules (glycosylation) and phosphate groups (phosphorylation). For example, glycosylation facilitates interactions between cells and their environment, as when glycosylated proteins in the plasma membrane of white blood cells help the cells infiltrate sites of infection.

7.2 Protein Structure

Proteins are 3-D biological molecules that do the work and form the structure of cells and organisms. Protein functionality depends on the molecule's size, shape, and chemical properties.

Amino Acids

Amino acids are the building blocks of proteins and the combinations of amino acids that make up a protein give the protein a distinctive set of chemical properties. The protein's biological activity results from chemical interactions between it and other molecules in the cell or organism.

Amino acids are hydrocarbons that have a central carbon attached to an amino group (NH_2), a carboxyl group (COOH), and a side chain (R group) that varies among different amino acids (see Figure 7.6). There are 20 standard amino acids in protein polypeptide chains. Each is a different size and the R group gives each amino acid its unique properties. The smallest amino acid, glycine, has a single hydrogen atom as its R group, while the largest standard amino acid, tryptophan, has aromatic rings in its R group.

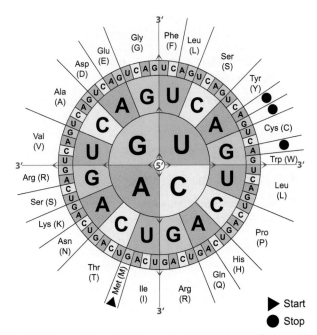

Figure 7.5. **The genetic code**. Each sequence of three bases of mRNA is a codon that encodes an amino acid or stop signal. To identify the amino acid encoded by a codon, begin in the center circle and find the quadrant that matches the first base of the codon (1st base). Staying within that quadrant, move out from the center ring by ring selecting the second and third bases in the codon without ever crossing a radial line. Then read the amino acid in the outside ring. **Note:** When translating a coding DNA sequence, substitute T for U in the chart. (See Figure 7.7 for the abbreviations for amino acids, for example, Phe stands for phenylalanine.)

Figure 7.6. **Amino acid structure**. An amino acid has an amino (NH_2) functional group at one end and a carboxyl (COOH) functional group at the other end. The side chain (R group) varies among amino acids.

Amino acids are classified by their R groups. They can be positively charged (basic), negatively charged (acidic), polar, or nonpolar. The chemical interactions of the R groups with each other and the environment affect the 3-D structure of the protein (see Figure 7.7).

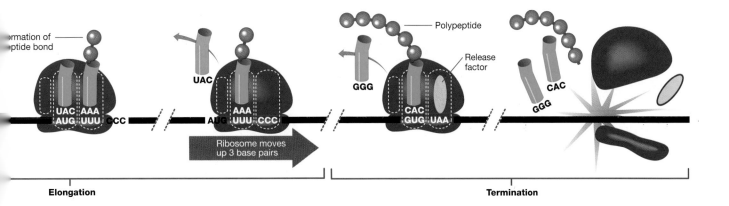

Elongation

Termination

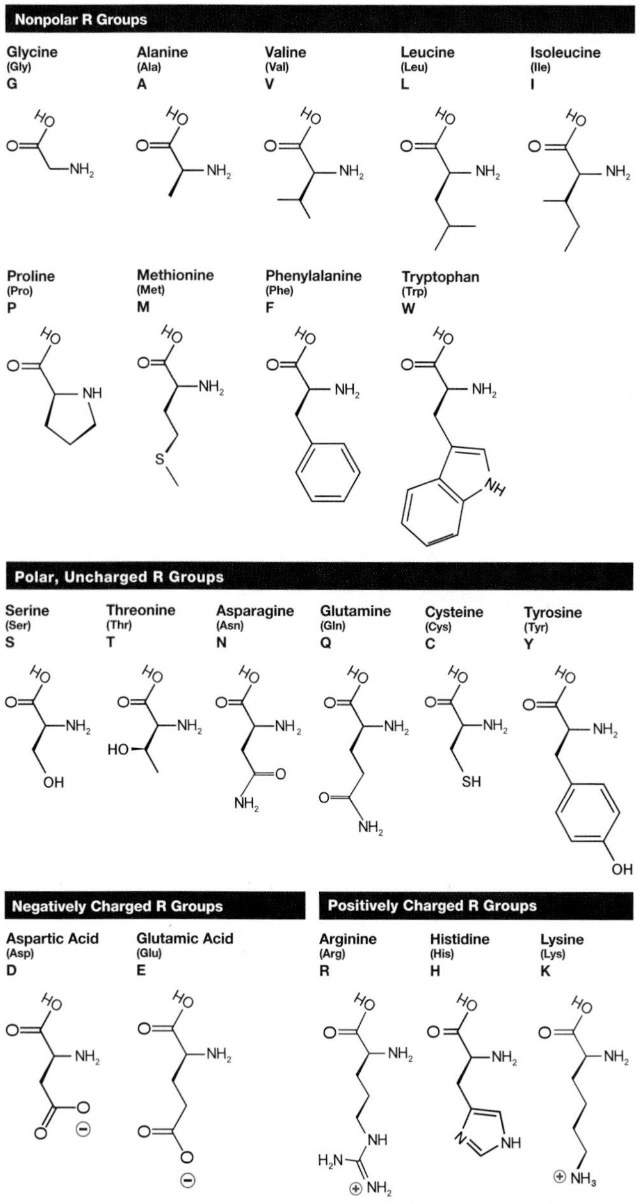

Nonpolar R Groups

Glycine (Gly) **G** | **Alanine** (Ala) **A** | **Valine** (Val) **V** | **Leucine** (Leu) **L** | **Isoleucine** (Ile) **I**

Proline (Pro) **P** | **Methionine** (Met) **M** | **Phenylalanine** (Phe) **F** | **Tryptophan** (Trp) **W**

Polar, Uncharged R Groups

Serine (Ser) **S** | **Threonine** (Thr) **T** | **Asparagine** (Asn) **N** | **Glutamine** (Gln) **Q** | **Cysteine** (Cys) **C** | **Tyrosine** (Tyr) **Y**

Negatively Charged R Groups

Aspartic Acid (Asp) **D** | **Glutamic Acid** (Glu) **E**

Positively Charged R Groups

Arginine (Arg) **R** | **Histidine** (His) **H** | **Lysine** (Lys) **K**

Figure 7.7. **Amino acids**. Amino acids have properties that are based on their R groups. They can be charged or uncharged, aromatic, polar or nonpolar.

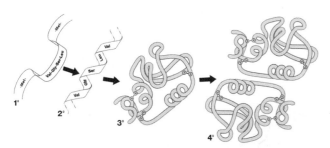

Figure 7.8. **The four levels of protein structure**. The primary structure of a protein is the sequence of amino acids. The secondary structure is generated by hydrogen bonds among the amino acids that form alpha helices and beta pleated sheets. The tertiary structure forms from interactions of the amino acid side chains that generate the 3-D structure. The quaternary structure results from multiple tertiary structures folding together to form a functional protein.

Protein Folding

Proteins have four levels of structure (see Figure 7.8):

- The **primary structure** is the amino acid sequence in the polypeptide chain. Remember that during translation, amino acids are joined together by peptide bonds that are formed by enzymes within the ribosome. The peptide bond forms between the carboxyl group of one amino acid and the amino group of another, releasing one molecule of water (see Figure 7.9). The peptide bonds form the backbone of a protein and the R groups hang from this backbone. A string of amino acids is called a polypeptide chain

- The **secondary structure** is generated when hydrogen bonds between amino acids in close proximity to each other form two regular structures, alpha helices and beta pleated sheets. Alpha helices are tight coils and beta pleated sheets are strands of 3–10 amino acids that lie side by side (see Figure 7.10)

- The **tertiary structure** is the 3-D shape of the polypeptide chain when it is folded. Tertiary structure is determined by interactions of the amino acid side chains. For example, hydrophobic side chains may interact in the interior of globular proteins, preventing exposure of the hydrophobic residues to the aqueous environment

- The **quaternary structure** reflects interactions among multiple polypeptide chains (see Figure 7.8); not all proteins have quaternary structure. Each polypeptide is called a protein subunit and multiple subunits join together to form the protein. Hemoglobin, the protein that carries oxygen in blood, has four subunits

If a protein is not folded properly (misfolded), it will not function correctly. In nature, such misfolding can cause disease. In Alzheimer's disease, for example, large quantities of misfolded proteins accumulate in the brain. In the biotechnology industry, manufacturing recombinant proteins that fold correctly can be very difficult and may require experimentation with multiple culture systems.

7.3 Proteins in Biology

Proteins are critical to life and perform many functions, including:

- Support (for example, the cellular cytoskeleton)

- Metabolism (for example, enzymes and hormones)

- Protection (for example, antibodies)

- Communication (for example, signal transduction proteins that act as messengers between and within cells)

- Regulation (for example, transcription factors)

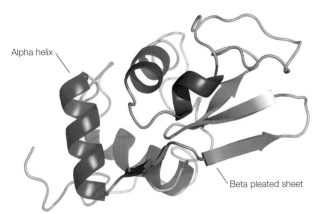

Figure 7.9. **Peptide bond**. Amino acids are joined by peptide bonds that form between the carboxyl and amino groups of adjacent amino acids. The condensation reaction releases one molecule of water.

Figure 7.10. **3-D structure of lysozyme**. Lysozyme is an enzyme found in animal secretions, such as mucus and tears, that breaks down bacterial cell walls, helping to protect animals from infection. Lysozyme will be used in Activity 7.C. The alpha helices are colored blue and the beta pleated sheets are colored red. In this representation of protein structure, the direction of the polypeptide chain is indicated in the beta pleated sheets by arrows. Protein structure derived by Diamond R. (1974) J Mol Biol. 82, 371–391, PDB ID # 4LYZ. Image courtesy of Matthew Betts.

Alpha helix

Beta pleated sheet

Protein solubility, the ability of proteins to dissolve in water, is vital to protein function. Some proteins need to be soluble (for example, enzymes in saliva), while others need to be insoluble (for example, keratin in hair and nails). Proteins are divided into two major groups: fibrous proteins are almost all insoluble and shaped like fibers, and globular proteins are typically soluble and roughly spherical.

Fibrous Proteins

Fibrous proteins usually form structures in a cell or in organisms. They are the proteins that are visible in tissue because they are insoluble. Keratin, myosin, and collagen are examples of fibrous proteins (see Table 7.1). Myosin and actin are the major proteins in muscle and make up the thick and thin filaments that slide past each other to contract the muscle.

Table 7.1. **Examples of fibrous proteins**.

Protein	Structures	Function
Keratin	Skin, fur, and nails	Covering and protection
Myosin	Thick filaments of muscle fibers	Muscle contraction
Collagen	Skin, ligaments, tendons, and extracellular matrix	Support

Globular Proteins

Whereas fibrous proteins form structures, globular proteins perform most of the signaling functions and mediate most of the chemical changes in cells and organisms. Globular proteins include antibodies, transcription factors, signal transduction proteins, some hormones, and enzymes (see Table 7.2).

Table 7.2. **Classes of globular proteins**.

Protein Class	Function
Antibodies	Recognize pathogens
Transcription factors	Regulate gene expression
Signal transduction proteins	Transmit signals from outside to inside cells and within cells
Hormones	Regulate physiological activities
Enzymes	Catalyze chemical and biochemical reactions

Enzymes

Enzymes help direct chemical reactions by making them occur faster than they would normally. They can alter proteins, nucleic acids, carbohydrates, lipids, metabolites, and other chemicals in and around cells. The molecules that are altered by enzymes are called **substrates** and the molecules resulting from the reactions are enzyme **products**. Enzymes are biological catalysts; that is, they change the rate of reactions without being consumed by the reaction. They change the rate of reaction by decreasing the activation energy necessary to start the reaction.

A substrate binds a pocket on the enzyme called the active site, which, according to the induced fit hypothesis, is thought to change shape slightly as the substrate binds (see Figure 7.11) and help to stabilize chemical intermediates as the substrate is transformed into product, lowering the activation energy.

The activity of an enzyme can be affected by many factors, including heat and pH. Heat can increase rates of reaction up to a point because substrates will collide with the enzyme more frequently. Changes to the environmental pH will affect the shape and charge of the amino acids in the active site of the enzyme, which can affect its functionality; each enzyme has an optimal pH. Too much heat or the wrong pH can denature the enzyme.

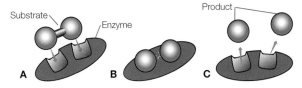

Figure 7.11. **Enzyme and substrate interaction**. Substrate binds to the active site of an enzyme. This induces the enzyme shape to change slightly, catalyzing a reaction. As the products exit the active site, the enzyme reverts to its original shape.

Editing Genes to Cure Disease

For people with cystic fibrosis (CF), a debilitating genetic disease affecting the lungs, pancreas, and other organs, the hope of living a normal life has in the past been a matter of science fiction. As a genetic disease, CF is caused by mutations in a person's genomic DNA, which traditionally has been impossible to fix. However, with the emergence of CRISPR as a therapeutic tool, for those who suffer with CF and a multitude of other genetic diseases, there is now potential for curative therapies.

In a 2013 study, scientists demonstrated that they could use CRISPR to develop therapies that target the cells and tissues affected by CF that cause difficulty breathing and frequent infections, among other symptoms. What if, however, mutations that cause CF could be fixed in embryos, before the disease has had a chance to take hold? What if mutated genes could be corrected in the reproductive cells (sperm and eggs) of parents to prevent transmission of the disease in the first place?

Though the ethical debates surrounding genetic engineering and gene editing are not new, the fact that CRISPR can be more accurate and even easier to perform than other technologies has reignited concerns about potential misuse. One of the main areas of concern surrounding CRISPR is its use in reproductive cells, in germline therapies. Using CRISPR to prevent transmission of hereditary disease could stop suffering before it starts, but might there be unintended consequences to a population? If CRISPR could be used to edit out disease, could it also be used for genetic enhancement, to change eye color, intelligence, athletic ability, or height? What would the implications be of creating so-called "designer babies"? Even if such designer changes could be deemed "safe," the therapies would likely be expensive, putting them out of reach of many. Who should decide which genetic changes are truly necessary or beneficial, and for whom?

Bioethicists around the world generally agree that human germline editing should not be attempted at this time. As of 2014, 40 countries discouraged or banned research on germline editing because of ethical and safety concerns. There is also an international effort, launched in December 2015 and led by the U.S., U.K., and China, to harmonize regulation of the application of genome editing technologies. Despite these bans, should studies into making germline therapies safe and effective be allowed to continue? What kind of oversight is needed?

7.4 Proteins in Biotechnology

Proteins are vital tools in biotechnology. Enzymes can perform industrial tasks more efficiently than just using chemicals and heat, which saves time and money. Many healthcare drugs such as hormones and antibodies are made from proteins. Scientists are also improving the functionality of natural proteins using recombinant DNA technology so that they perform their tasks more efficiently. Proteins are essential for life science research; for example, enzymes such as restriction enzymes are the foundation of molecular biology, and PCR is dependent on DNA polymerases.

Proteins in Industry

The most common proteins utilized in industry are enzymes. Enzymes are used to perform industrial processes, and large biotechnology companies such as Novozymes and Amano Enzyme specialize in manufacturing enzymes for industrial purposes, a $2 billion industry. Some of the enzymes manufactured are recombinant and some are purified from natural sources. Table 7.3 lists industrial uses for some enzymes.

Table 7.3. **Industrial uses for enzymes**.

Type of Enzyme	Substrate	Industry	Product/Use
Lipases	Fats and oils	Soap and detergent	Laundry detergent
Proteases	Proteins	Soap and detergent	Laundry detergent
		Tannery	Leather
		Food supplements	Digestive aids
Amylases	Starch	Soap and detergent	Laundry detergent
		Food	High fructose corn syrup
		Textile	Destarch fabric
		Brewing	Release sugars
		Waste management	Treat water
Cellulases	Cellulose	Bioenergy	Convert plant material to glucose
		Brewing	Digest barley

Researchers are always looking for new enzymes that can improve the performance of existing processes. For example, more efficient cellulases are being sought for the bioenergy industry, where they are used to release the solar energy stored in the cellulose of plants to produce alternatives to fossil fuel. More efficient cellulases can improve the efficiency of the process used for energy production and, in turn, lower the cost of fuel.

Synthetic biology is a discipline that solves biological problems in part by engineering the proteins of entire metabolic pathways to create synthetic organisms that can generate novel products or perform useful functions. For example, in the biofuel industry, scientists are developing organisms that can process plant-based feedstock for cellulosic ethanol versus engineering single enzymes to be added to a reaction. Synthetic biologists are also generating efficient plastic-consuming bacteria for bioremediation.

Proteins in Healthcare

Diseases can be caused by protein deficiencies or by defective proteins, and proteins manufactured by the healthcare industry can be used to treat some of these diseases. For example, recombinant human growth hormone is used to treat growth hormone deficiency (once called pituitary dwarfism) and manufactured insulin is used to treat diabetes.

Recombinant proteins are used as drugs to treat medical conditions. For example, some stroke victims are given recombinant tissue plasminogen activator (tPA), which helps to dissolve blood clots in the brain and can greatly increase survival rates. In another example, collagen sponges made of purified collagen from animals like pigs are routinely used for dressing severe burns and ulcers. The protein provides a porous structure for the patient's own cells to enter in order to speed healing (see Figure 7.12).

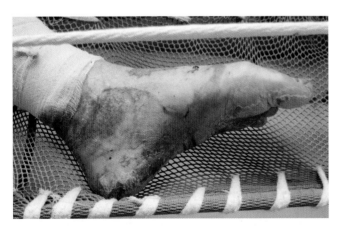

Figure 7.12. **Skin graft over a burn**. Skin grafted over a severe burn joins with existing skin as it heals. Purified collagen assists the development of artificial skin where skin grafts are not possible.

Some medical conditions are diagnosed using recombinant proteins. For example, survivors of thyroid cancer are treated with recombinant thyroid stimulating hormone to detect any remaining cancer cells. The proteins used most frequently in disease diagnosis are antibodies. Antibodies can specifically identify proteins associated with medical conditions. Antibodies and their role in healthcare are discussed in Chapter 8.

Proteins as Tools in Life Science Research

Life science researchers use many different proteins as tools in their experiments. As discussed in previous chapters, purified enzymes from many different sources are used in molecular biology research, for example, restriction enzymes and ligases to construct plasmids and DNA polymerases for PCR.

Proteins are also used in biochemical research. For example, antibodies identify specific molecules, as will be discussed in Chapter 8. Visible proteins, such as green fluorescent protein, can be fused to other proteins so that scientists can track the location of the protein in a cell or organism. In a similar fashion to DNA size standards, proteins of known size are used as size standards to help scientists determine the sizes of unknown proteins during protein electrophoresis (see Figure 7.13).

250 kD

150 kD

100 kD
75 kD

50 kD

37 kD

25 kD
20 kD

15 kD

10 kD

Figure 7.13. **A protein size standard**. Bio-Rad's Precision Plus Protein™ Kaleidoscope™ protein size standard contains ten highly purified recombinant proteins that have been color-coded to help identification on the gel.

Scott Chilton, PhD
Marketing Manager, Maravai Life Sciences

Scott manages the marketing team at Maravai Life Sciences in San Diego, CA, where he creates marketing strategies and product development plans so that the company can provide products that help customers with their scientific research. As part of this work, he leads a team to answer a variety of marketing questions such as: *What products should we produce? How are customer needs changing? What is the best way to advertise our products?* He works on many projects across the company and relies on both his science background and his strategic decision-making abilities.

Scott's interest in science began in a high school biotechnology course and continued to grow throughout his undergraduate studies at the Massachusetts Institute of Technology (MIT) in Cambridge, MA. While at MIT, he took opportunities to gain laboratory experience, including summer internships and research positions at Bio-Rad Laboratories in Hercules, CA and various academic laboratories. Scott graduated with a BS in biology, and then went on to earn his PhD in biochemistry at Harvard University in Cambridge, MA, where he studied the proteins involved in bacterial uptake of DNA from the environment.

Image courtesy of Scott Chilton

After graduate school, Scott held roles at Bio-Rad Laboratories where his science background helped to develop new products and communicate their values to customers. He then moved on to become a Product Brand Manager at Natera, Inc., a provider of genetic tests for reproductive health. There he oversaw the marketing of a noninvasive prenatal screen for aneuploidy. He helped develop marketing and sales plans to communicate the benefits of this test to large numbers of doctors. At Maravai Life Sciences, he incorporates all this experience to lead marketing efforts across several partner companies.

Scott's advice to people starting out in biotechnology is to not restrict yourself by developing only lab skills. Outside interests, activities, and hobbies can help develop other skills useful to your career. Biotechnology companies need people who not only understand the science behind their products, but who can also think critically about problems and develop and communicate solutions. Professional and social networks are also important, as they can provide valuable work and development opportunities. Finally, have some fun. If you are not having fun, it might be time to look at other opportunities.

7.5 Methods of Protein Analysis

There are a number of standard techniques that scientists use to analyze proteins. These include protein quantitation and the purification and separation of proteins using chromatography and polyacrylamide gel electrophoresis (PAGE).

Protein Quantitation

Scientists often need to know how much protein is present in a sample so it can be used appropriately. To measure protein concentration, proteins in solution are added to chemically reactive dyes that change color in the presence of proteins. The degree of color change is directly proportional to the amount of protein in the sample (this is based on the Beer-Lambert law, which states that when a solute absorbs light of a particular wavelength, the absorbance is directly proportional to the concentration of the solute in solution). Assays that measure changes in color are called colorimetric assays.

To measure the color change, an instrument called a spectrophotometer is used (see Figure 7.14). Spectrophotometers measure how much of a specific wavelength of light is absorbed by a material. They do this by shining a narrow beam of light of a set wavelength through the sample and measuring the amount of light that passes through the sample using a detector. The sample is placed in a specialized transparent vessel called a cuvette that has a specific light path length (usually 1 cm).

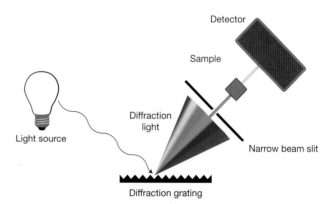

Figure 7.14. **The components of a spectrophotometer**. White or UV light from a light source is directed toward a diffraction grating or prism that separates the light into the different wavelengths, or colors, of light. The diffraction grating is rotated to select a particular wavelength. When a sample is present, a portion of the selected wavelength of light is absorbed by the sample. The detector measures the amount of light that is not absorbed by the sample.

The concentration of protein is determined by first generating a standard curve of the absorbance of known protein concentrations and then using the standard curve to determine the concentration of the unknown samples by comparing their absorbance values. There are multiple colorimetric tests available for determining protein concentration. These include the biuret test, the Lowry assay, and the Bradford assay. Each assay has a different level of sensitivity and compatibility with chemicals in protein solutions.

Biuret Test

The **biuret test** is the oldest method for determining protein concentration and was first used in 1833 by Ferdinand Rose. In the test, a color change is caused by copper ions in alkaline solution, forming a violet-colored complex in the presence of peptide bonds. The violet solution absorbs light at 540 nm. This test is not often used to quantify proteins because it is not very accurate or sensitive. It is, however, a useful qualitative way to detect the presence or absence of proteins.

Lowry Assay

The **Lowry assay** also uses copper ions to cause a color change and was developed by Oliver Lowry in 1951. His paper describing the development of the assay was the most-cited paper of all time. It has been cited over 300,000 times, which indicates how important determining protein concentration is to researchers. The Lowry assay measures how proteins that have been treated with copper interact with a reagent called Folin-Ciocalteu reagent. The interaction causes a color change that absorbs light at 750 nm (see Figure 7.15).

Figure 7.15. **Bio-Rad's *DC*™ protein assay.** Based on the Lowry assay, this assay has been modified to allow greater compatibility with the detergents used to prepare protein samples. The tube on the left is the control (no protein) and the tubes to the right have increasing amounts of protein.

Bradford Assay

The **Bradford assay** was first described by Marion Bradford in 1976 and uses a dye called Coomassie Brilliant Blue G-250, which interacts with the R groups of specific amino acids, especially arginine. The dye is used in a solution called the Bradford reagent and in the absence of protein is a reddish-brown color. In this state it has a maximum absorption at 470 nm. When the dye interacts with protein, it changes to a stable form that is blue (see Figure 7.16) and its peak absorption shifts from 470 nm to 595 nm. The amount of blue color is proportional to the amount of protein in the sample; the more protein, the more intense the blue color. The Bradford assay uses a quicker and simpler protocol than the Lowry assay but offers a similar level of sensitivity.

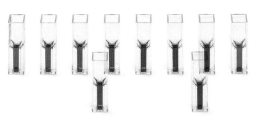

Figure 7.16. **Bradford assay.** The cuvette on the left is a control with no protein, and the cuvettes to the right contain increasing amounts of protein.

Protein Properties Used in Analysis

Each protein has a unique amino acid composition that gives it unique properties. Electrophoresis and chromatography are techniques to separate proteins by exploiting differences in their physical properties, such as size and charge.

Protein Size

The component base pairs of DNA are roughly the same size, which allows the size of DNA molecules to be measured as length in base pairs. With proteins however, the sizes of component amino acids differ depending on the sizes of their R groups. For this reason, protein size is measured as its mass; the unit of measure for proteins is the dalton (Da). Each Da is one-twelfth the mass of a carbon atom, which is approximately the mass of a proton. Most proteins are more than 100 amino acids long and are measured in kilodaltons (kD). Proteins vary greatly in size; for example, titin in muscle is 3 million Da, while thymosin is 5,000 Da (see Table 7.4), and this makes size a useful property for protein separation. It should be noted that scientists commonly use the term relative molecular weight (MW), rather than molecular mass (MM), to describe the mass of a protein.

Table 7.4. **Molecular masses of muscle proteins.**

Protein	MM (in kD)	Function
Titin	3,000	Centers myosin in sarcomere
Dystrophin	427	Anchors to plasma membrane
Filamin	280	Crosslinks actin filaments into a gel
Spectrin	240	Attaches filaments to plasma membrane
Myosin heavy chain	210	Slides actin filaments
Nebulin	107	Regulates actin assembly
α-actinin	100	Bundles actin filaments
Gelsolin	90	Fragments actin filaments
Fimbrin	68	Bundles actin filaments
Actin	42	Forms filaments
Tropomyosin	35	Strengthens actin filaments
Troponin (T)	30	Regulates contraction
Myosin light chains	15–25	Slides actin filaments
Troponin (I)	19	Regulates contraction
Troponin (C)	17	Regulates contraction
Thymosin	5	Sequesters actin monomers

Protein Charge

Proteins can also be separated by their charge. The amino acids that make up a protein have different charges and the net charge of a protein depends on the pH of its environment (see Chapter 2 for a review of pH). The **isoelectric point** (**pI** or **IEP**) of a protein is the pH at which the protein carries no net charge. Each protein has a specific pI. If a protein is placed in a pH lower than its pI, the protein will have a net positive charge; if the pH is above the pI, the protein will have a net negative charge.

Protein Mass Spectrometry

Protein mass spectrometry is a highly sensitive method for determining the size (mass) of a protein. Because it can be used with pure samples as well as complex mixtures, it has many applications in biotechnology research and manufacturing.

With purified samples, protein mass spectrometry is one of the best methods for precisely determining the mass of a protein and its amino sequence. It can also be used to study the protein's 3-D structure, identify any posttranslational modifications, such as glycosylation, and determine its interactions with other proteins. Mass spectrometers produce detailed data not available through other methods, but they are rather large and intricate instruments that require specialized training. For this reason, biotechnology labs either maintain shared mass spectrometry facilities or send samples to outside facilities for analysis.

Protein mass spectrometry can also be used to identify and quantitate the proteins in complex samples. It is a primary method for identifying the proteins present in a particular cell type, organelle, or membrane. As such, it is commonly used in biotechnology research aimed at discovering proteins associated with diseases, with the goal of using those proteins as markers in tests for those diseases.

With mass spectrometry, proteins can be analyzed intact; however, scientists often first digest, or break proteins into fragments before analysis. This technique is known as protein mass fingerprinting (PMF) (see Figure 7.17). In PMF, proteins are broken into fragments using enzymes called proteases. **Proteases** break proteins at predictable places within each peptide sequence. For example, the protease trypsin cuts at the carboxyl side of the amino acids lysine or arginine, except when either is followed by proline. The resulting peptides can then be analyzed by a mass spectrometer, which produces a spectrum showing the mass of each fragment. The data are compared to a database containing known protein sequences, or even predicted translations of DNA sequences. Computer programs can be used to predict peptide fragments based on the amino acid sequence of a protein and generate a theoretical mass spectrum. An experimentally observed mass spectrum of an unknown protein can be compared to theoretical mass spectra of known proteins to find a match.

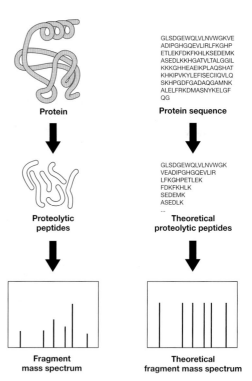

Figure 7.17. **Protein mass fingerprinting (PMF)**. In PMF, proteins are digested into fragments using proteases and then analyzed by mass spectrometry. Those data are compared with theoretical data to find the best match.

Protein Electrophoresis

Organisms contain proteins of many different sizes and charges, and these proteins can be separated and visualized using electrophoresis. Proteins are, on average, much smaller than DNA fragments, so a much tighter matrix is required for their separation. For this reason, polyacrylamide is used to separate proteins rather than agarose, which is commonly used to separate DNA, and the process is called **polyacrylamide gel electrophoresis** (**PAGE**).

The chemical structure of proteins is more complex than that of DNA, which makes protein separation more complex than separating DNA. In contrast to DNA, which is uniformly negatively charged, each protein has a different net charge. This means that in an electrical field, some proteins move toward the anode and some move toward the cathode. In addition, proteins have different shapes based on their tertiary and quarternary structures, which also affects their migration through a gel matrix.

To separate proteins based on their size, scientists treat proteins with a detergent, sodium dodecyl sulfate (SDS), which coats the proteins and gives them an overall negative charge (see Figures 7.18 and 7.19). The degree of negative charge is proportional to the molecular

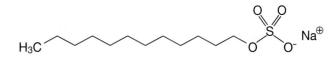

Figure 7.18. **The chemical structure of sodium dodecyl sulfate (SDS)**.

mass of the protein; that is, proteins treated with SDS have a similar mass-to-charge ratio. Since SDS is a detergent, it performs two additional functions: it breaks open cell membranes to release proteins (cell lysis) and partially denatures the proteins. The 3-D structure of proteins is fully disrupted by heating samples to 95°C. In addition, sometimes samples are treated with reducing agents such as dithiothreitol or β-mercaptoethanol, which break disulfide bonds between cysteine amino acids. Once treated with SDS and heat, proteins are linear, negatively charged chains and migrate through the gel based on their size rather than their original net charge. PAGE performed using SDS in this manner is called SDS-PAGE.

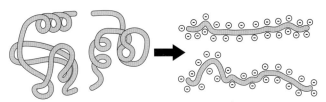

Figure 7.19. **The effect of SDS on proteins**. SDS partially denatures proteins and coats them in a negative charge that is proportional to their mass.

Polyacrylamide Gels

The polyacrylamide gels used in PAGE are very thin (approximately 1 mm) and are cast between two transparent plates. Polyacrylamide gels are used in vertical electrophoresis chambers (see Figure 7.20). In these chambers, the electrical current passes through the gel from an electrode in a reservoir of buffer at the top of the gel to an electrode in a reservoir at the bottom. Polyacrylamide gels are usually run at 200 volts for 30 minutes to 1 hour, or until the dye front from the loading dye has reached the bottom of the gel. Sometimes gels, for example TGX™ gels, can be run at 300 volts for 20 minutes when they are formatted for these conditions. The same power supply used for horizontal gel electrophoresis is used for SDS-PAGE.

Figure 7.20. **Mini-PROTEAN® Tetra cell**. This vertical electrophoresis chamber runs up to four gels.

Polyacrylamide gels are made by polymerizing acrylamide in the presence of bis-acrylamide, tetramethylethylenediamine (TEMED), and ammonium persulfate in a chemical reaction. Unlike agarose gels, polyacrylamide gels cannot be remelted. Acrylamide in its monomeric form can cause cancer and deaden nerves but is safe once polymerized (for example, polyacrylamide is used in some diapers to

absorb moisture). Precast (ready-made) commercial PAGE gels are frequently used to avoid the hazards of polyacrylamide gel preparation.

The percentage of acrylamide in a gel depends on the size of the protein being investigated. Small proteins are best resolved on high-percentage gels (such as 15%), while large proteins are best resolved on low-percentage gels (7.5%). If a wide range of protein sizes is being investigated in a sample, a gradient gel (with a gradient of acrylamide percentages) is used (see Figure 7.21).

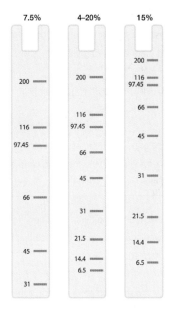

Figure 7.21. **Protein migration in different percentage polyacrylamide gels**. The numbers next to each band indicate the molecular mass of a protein in kD. In a 7.5% gel, proteins of large molecular mass are resolved, while in a high-percentage gel, proteins of low molecular mass are resolved. In a 4–20% gradient gel, large and small proteins are separated in the same gel.

Electrophoresis Running Buffer

Most commonly, Tris/glycine/SDS (TGS) running buffer is used for SDS-PAGE. Tris is a buffer that helps maintain the pH of the running buffer, SDS keeps the proteins negatively charged and denatured, and glycine ions help resolve the proteins into tight bands. TGS is made or purchased as a 10x solution and must be diluted to 1x before use.

Discontinuous Buffer Systems

A discontinuous buffer system is used in SDS-PAGE. Protein samples are loaded into vertical wells in an SDS-PAGE gel (Figure 7.23). Proteins at the bottom of the well enter the gel before proteins at the top of the well, which can theoretically lead to protein bands that are fuzzy and unresolved. This problem is remedied through use of a discontinuous buffer system.

In a discontinuous buffer system, the gel has two sections: a stacking gel made with a very low percentage of acrylamide (4–5%) and a resolving gel made with a higher percentage of acrylamide (see Figure 7.23). Due to differences in pH and molarity between the

How To...

Set Up a Vertical Electrophoresis System

The Mini-PROTEAN Tetra system is a vertical electrophoresis system from Bio-Rad. Each company's vertical electrophoresis equipment is set up differently, so refer to the instruction manual for the specific equipment to be used.

1. Prepare a Mini-PROTEAN TGX gel. Pull off the green plastic strip at the bottom of the gel and remove the green comb. Notice the gel has two plates, a short plate and a long plate.

2. Remove the electrode assembly from the tank and open the green clamps.

3. Place the gel into the electrode assembly that has the banana plug jacks with the short plate facing inward (see Figure 7.22). Place a second gel on the opposite side of the electrode assembly. (**Note:** If only one gel is run, use a buffer dam to replace the second gel, inserting it so that the notch faces inward.)

4. Push both gels toward each other, making sure they are flush against the notched green gaskets and that the short plates sit just below the notches of the gasket. Lock the green clamps over the gels.

5. Lower the electrode assembly into the tank on the side of the tank with the plastic tabs (see Figure 7.22). Make sure to match the red banana plug to the red oval on the tank.

6. Fill the inner chamber of the electrode assembly with 1x TGS electrophoresis buffer so the short plates of the gels are covered, and add buffer to the outer tank up to the 2 Gels line. The gels are now ready to be loaded. (**Note:** If using Bio-Rad's rapid electrophoresis protocol, 300 V for 18 minutes, be sure to add buffer to the 4 gels mark regardless of the number of gels being run.)

Note: This procedure describes how to set up Bio-Rad's Mini-PROTEAN Tetra system using a single electrode assembly. Refer to the instruction manual of the Mini-PROTEAN Tetra cell if running four gels using the companion running module.

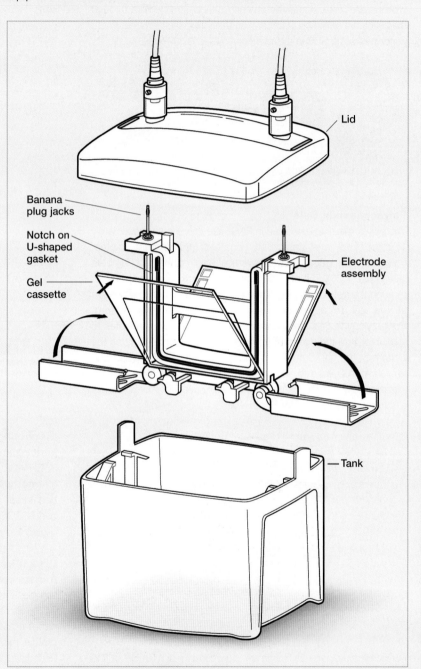

Figure 7.22. **Assembling the Mini-PROTEAN Tetra cell**.

running buffer and the stacking gel, the proteins in the samples compress into very narrow bands by the time they reach the top of the resolving gel so all of the proteins enter the resolving gel at the same time. The stacking of the proteins at the intersection of the two gels can be observed by watching the migration of colored protein size standards and the dye front of the gel.

Gradient gels do not have separate stacking and resolving gels. Instead, the percentage of acrylamide increases continuously down the gel, for example from 4 to 20%. Gradient gels provide a wider range than single-percentage gels and allow separation of small and large proteins in the same gel (see Figure 7.21).

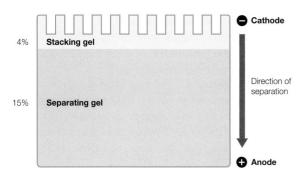

Figure 7.23. **A discontinuous polyacrylamide gel**. A discontinuous polyacrylamide gel has two sections: a low-percentage stacking gel and a high-percentage separating gel.

Visualizing Proteins in Polyacrylamide Gels

Following electrophoresis, protein band patterns can be visualized and subjected to qualitative and quantitative analysis. Since most proteins cannot be seen in a gel with the naked eye, protein visualization can be achieved through the use of total or specific protein stains or a more recent development, stain-free gel technology.

There are many different protein stains made from dyes or metals that have different levels of sensitivity. The most commonly used stain is Coomassie Brilliant Blue stain, which is made from dyes that bind to specific amino acid side chains and stain proteins blue. Gels are stained with Coomassie stain and then destained to remove dye that has not bound to proteins. Protein gels stained with Coomassie stain have distinct blue bands against a colorless background. Other protein stains include silver stain, which is very sensitive, and fluorescent stains that require special imaging equipment.

Stain-free gels are used in the same way that SDS-PAGE gels are used except they contain reagents that modify the tryptophan amino acids when the gel is exposed to UV light. This causes the protein bands to fluoresce and be visible without the need to stain, which can save time.

Other Types of PAGE

Polyacrylamide has uses other than in SDS-PAGE. Acrylamide gels can be used to separate proteins and other molecules based on physical properties other than size.

Native PAGE

Proteins can also be separated based on their net charge-to-mass ratio rather than their mass alone (as in SDS-PAGE). Samples are prepared without SDS and are run on gels in the absence of SDS. This is called native polyacrylamide gel electrophoresis. Native PAGE is useful because proteins in the samples are not denatured. They may still be associated with other molecules required for functionality. In addition, proteins can be recovered from native gels and retain their activity.

Isoelectric Focusing

Isoelectric focusing (IEF) separates proteins based on their isoelectric point (pI), the point at which they have no overall charge. Some IEF uses thin strips made of polyacrylamide that have an immobilized pH gradient (IPG) from one end to the other; one strip is used per sample. Samples are loaded onto the IPG strips in an IEF apparatus and a current is applied to the strip. Proteins move along the strip until their net charge is zero. Isoelectric focusing is also the first step in two-dimensional (2–D) protein electrophoresis.

Two-Dimensional PAGE

Two-dimensional PAGE separates proteins by their pI and their size in two different dimensions. First, proteins in a sample are separated by isoelectric point using IEF. Next, the IPG strip is applied to the top of a specialized SDS-PAGE gel with a single well the size of the IPG strip and the proteins migrate from the strip into the SDS-PAGE gel and are separated by size. This creates one gel per sample and each protein within the sample is represented by a spot on the gel (see Figure 7.24). Samples are compared to each other using software that looks for differences in the pattern of spots. Proteins are identified by cutting the protein from the gel and using downstream assays such as mass spectrometry to identify them.

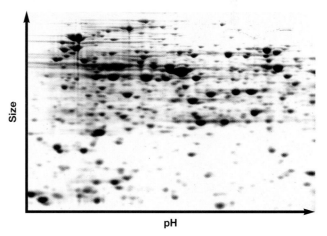

Figure 7.24. **2-D PAGE**. E. coli proteins were loaded onto an IPG strip and separated by their pI by IEF (horizontal dimension). The strip was then applied to an SDS-PAGE system and the proteins were separated by their molecular masses (vertical dimension). Each spot on the gel represents a different protein.

Nucleic Acid Separation

Polyacrylamide gels can also be used to separate small fragments of DNA. In this case, Tris/boric acid/EDTA (TBE) buffer is used for the gel and electrophoresis buffer. Polyacrylamide gels should be used only for DNA fragments of less than 1 kb. Special polyacrylamide gels called sequencing gels, can resolve DNA down to a single base, though their use has been largely replaced by automated sequencing instruments. Automated sequencers use capillary electrophoresis, which is a polyacrylamide gel in a long and extremely narrow tube. Samples separate through the tube by size and are recorded as they pass a detector.

Protein Chromatography

Chromatography is commonly used by scientists to separate biological molecules such as proteins from complex mixtures. For example, if a recombinant protein is produced in bacterial cells, chromatography is frequently used to purify the recombinant protein from the bacterial proteins, lipids, and carbohydrates.

Chromatography involves a mobile phase (a liquid) flowing over a stationary phase (a solid). In column chromatography, the stationary phase is typically a **resin** packed in a cylindrical glass or plastic column (see Figure 7.25) and the mobile phase is a buffer solution. Chromatography resins are solid materials composed of small beads; the specific molecular characteristics of the beads vary based on the type of chromatography being performed.

Mixtures of molecules are separated based on each molecule's degree of affinity for the mobile and stationary phases. For example, if molecule A has a stronger affinity for the stationary phase than molecule B, then A will migrate through the chromatography column more slowly than B.

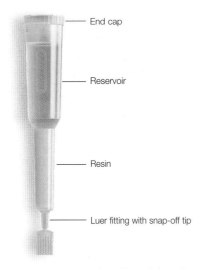

Figure 7.25. **A prepacked chromatography column**. The resin in a column is called a bed and the amount of resin in a column is called the bed volume. The area above the bed that holds the sample, wash, or solvent is called the reservoir.

The sample solution containing the protein of interest in a buffer is placed on top of the resin in the chromatography column and flows into the column. As the sample solution passes through the beads, some molecules will be retained on the column and others will pass right through the column. The solution flows out of the bottom of the column and is collected in a series of tubes; these **fractions** each contain a set volume and are analyzed for the presence of the protein of interest. In some types of chromatography, the column is washed with one type of solution to remove proteins that are not tightly bound to the resin and eluted with another solution that releases the protein of interest from the column.

Size Exclusion Chromatography

Size exclusion chromatography (**SEC**) separates molecules based on their size. The beads in SEC resins are porous, meaning that they have holes in them (see Figure 7.26). Small molecules enter the beads through the pores but larger molecules are too big to enter the pores. As a result, small molecules move slowly through the column, while the large molecules flow quickly through the column (see Figure 7.27). Proteins are separated into fractions based on the rate at which they flow through the column.

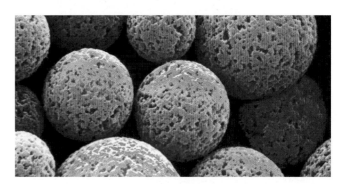

Figure 7.26. **Size exclusion chromatography resin**. An electron micrograph showing individual beads in an SEC resin. Note the pores on the surface of the beads.

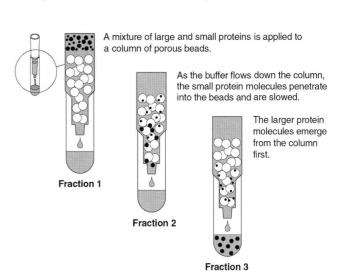

A mixture of large and small proteins is applied to a column of porous beads.

As the buffer flows down the column, the small protein molecules penetrate into the beads and are slowed.

The larger protein molecules emerge from the column first.

Fraction 1

Fraction 2

Fraction 3

Figure 7.27. **Size exclusion chromatography**. SEC separates large molecules from small molecules by retarding small molecules in the pores of the beads of the chromatography resin, while large molecules pass around the beads.

Hydrophobic Interaction Chromatography

Hydrophobic interaction chromatography (HIC) uses hydrophobic ("water-fearing") resins to separate proteins based on their hydrophobicity, the degree to which they are seemingly repelled by water. A protein's hydrophobicity depends on the ratio of hydrophobic and hydrophilic ("water-loving") amino acids as well as the concentration of salt in its environment. Proteins are added to the HIC column in a very high salt environment. The hydrophobic regions bind to the hydrophobic resin to reduce their exposure to the high salt (highly ionic) solution conditions. Proteins are eluted from the column with gradually decreasing levels of salt and are collected in fractions. The more hydrophobic the protein, the lower the salt concentration required to release the protein (see Figure 7.28). In the absence of salt, the most hydrophobic proteins elute from the column.

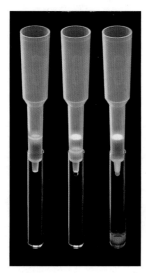

Figure 7.28. **Green fluorescent protein (GFP) on a HIC column**. GFP is a hydrophobic protein and can be separated from a mixture using HIC chromatography. Left, GFP in a mixed sample is added to the column and binds to the resin at the top of the column in the presence of high salt buffer, while nonhydrophobic proteins are eluted. Middle, GFP migrates through the column in the absence of salt. Right, GFP is eluted from the column and collected in a fraction.

Ion Exchange Chromatography

Ion exchange chromatography separates molecules based on their net charges. A cation exchange resin is negatively charged and binds molecules that are positively charged (cations). An anion exchange resin is positively charged and binds negatively charged molecules (anions). Samples are loaded in low salt buffers. If samples carry a charge that is opposite that of the resin, they bind, but if they are of the same charge as the resin, they flow through. The binding is reversible, and proteins are eluted into fractions using buffers with gradually increasing concentrations of salt.

Biotech In The Real World

Vaccines in Bananas

Vaccines have been revolutionary in preventing infectious diseases; nevertheless, millions of infants die every year without immunization, especially in remote and impoverished parts of the world. Many vaccines are expensive and require refrigeration, and some cultures resist vaccination, especially by injection. As one solution, researchers are trying to engineer fruits and vegetables to produce vaccines. If this effort is successful, it could overcome all three of the primary obstacles to widespread immunization in the developing world: the vaccines could be produced cheaply, they would not require refrigerated storage, and they would provide immunity after the vaccine is eaten, avoiding injections.

Initial efforts used potatoes, because potatoes are edible and are relatively easy to transfect. In 1998, a small clinical trial tested a vaccine expressed in potatoes against a toxin from diarrhea-causing bacterium. In this trial, 10 of the 11 individuals who ate raw chunks of vaccine-producing potato developed an immune response to the toxin. Potatoes engineered with a vaccine against the hepatitis B virus produced similar results in small clinical trials in 2005. Since raw potatoes are not pleasant to eat, researchers shifted to producing vaccines in more edible plants, such as tomatoes and bananas. Bananas, in particular, are appealing to most children and can be grown in many of the regions of the world affected by low vaccination rates.

Genetically engineered bananas generated the desired immune response in mice, but even after over 20 years of research there are no commercially available engineered fruit-based vaccines. First, a technical problem associated with this research is increasing the amount of antigen proteins in the recombinant vaccines to levels high enough to provoke an immune response while also ensuring that the proteins are glycosylated properly. Second, it is difficult to control the amount made and ingested (the dosage). Third, producing vaccines against diseases found only in developing countries is not profitable for pharmaceutical companies, making it a low priority. Additionally, many countries have strict regulations regarding genetically engineered foods and stringent purity standards for vaccines, both of which hamper distribution. Lessons learned from this research, though, have raised the possibility of using plants as factories for vaccines. Tobacco is being used to quickly produce vaccines against diseases like Ebola, malaria, and even the flu. Perhaps other edible plants might produce other vaccines, plants that might be dried to create plant powders within which the dosage and purity could be better controlled. In many countries, these could also be a culturally acceptable alternative to injections.

Affinity Chromatography

Affinity chromatography separates molecules based on their affinity for a specific binding partner, the ligand. Affinity chromatography is often used to purify a single protein, such as a manufactured protein, from a mixture (see Figure 7.29). The beads of the resin are coated with ligands and the protein of interest binds specifically to the ligand on the resin while other proteins in the sample flow through and are washed away.

Ligands can be antibodies, cofactors, enzymes, or enzyme substrates. Alternatively, recombinant proteins can be tagged with peptides that specifically bind to a resin. For example, a short sequence of histidines (called a His-tag) specifically binds nickel ions (Ni^{2+}) on a resin.

Proteins are eluted by adding another molecule that binds to the ligand more strongly than the molecule of interest or by reducing the pH to alter the protein structure and reduce its affinity for the column resin.

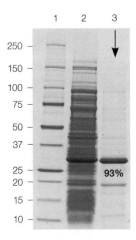

Figure 7.29. **SDS-PAGE gel of a purified affinity-tagged protein**. A His-tagged protein was purified on a Ni^{2+}-charged affinity chromatography resin, Profinity™ IMAC resin. The original sample and the fraction containing the purified protein were then electrophoresed. Lane 1, protein size standards; lane 2, initial mixture of proteins loaded onto the column; lane 3, eluted purified protein that is 93% pure.

Analyzing the Results of Chromatography

The results of chromatography can be analyzed in multiple ways: the proteins in each fraction can be analyzed by SDS-PAGE (see Figure 7.29), assays can be performed on each fraction (for example, in an enzyme assay if an enzyme is being purified), the protein concentration in each fraction can be determined using a spectrophotometer or a protein quantitation assay, or specialized chromatography instruments such as the NGC™ Chromatography system can monitor the conductivity and absorbance of the sample as it elutes from the column.

Scaling Chromatography

Chromatography can be performed on a small scale in a laboratory using columns with diameters of <1 cm. Small-scale chromatography can be performed manually or using laboratory instruments that control the process.

When chromatography is used in manufacturing facilities to produce industrial enzymes or biological drugs, it is performed on a much larger scale (process scale), for example, using columns of 1 meter in diameter or larger (see Figure 7.30). The principles of the chromatographic separation are the same at all scales but special care needs to be taken when processes are scaled up. Process-scale chromatography is highly controlled and automated to ensure the chromatography is performed as efficiently as possible.

Figure 7.30. **Process-scale chromatography columns**. These columns have a diameter of 1 meter. Columns like these are used by pharmaceutical companies to purify biological drugs.

Protein Analysis Using Bioinformatics

Advances in biotechnology and computing have led to the generation of vast quantities of raw data, including protein sequences. Bioinformatics allows multiple sequences to be compared, which enables researchers to determine how closely related one sequence is to another, or which areas in a sequence have been conserved across species. This provides clues to which amino acids are essential for protein function.

The amino acid sequences of proteins can be aligned with each other using tools such as Clustal Omega. For example, alignment of the sequences of the same protein from different species can reveal sections in which the amino acid sequence is the same in each protein, suggesting that these areas are vital for protein function (see Figure 7.31). Clustal Omega can also generate phylogenetic trees showing how proteins are related to each other (see Figure 7.32), or how a protein or protein function evolved. Analysis of this kind is accomplished in seconds by the computer but would take weeks to do manually.

187: Yeast	FVMGVNEEKYTSD-LKIVSNASCTTNCLAPLAKVINDAFGIEEGLMTTVHSLTATQKTVD
261: Human	FVMGVNENDYNPGSMNIVSNASCTTNCLAPLAKVIHERFGIVEGLMTTVHSYTATQKTVD
293: Mouse	FVMGVNEKDYNPGSMTIVSNASCTTNCLAPLAKVIHENFGIVEGLMTTVHSYTATQKTVD
285: Rat	LVMGVNEKDYNPGSMTVVSNASCTTNCLAPLAKVIHERFGIVEGLMTTVHAYTATQKTVD
195: Nematode worm	YVVGVNHEKYDASNDHVVSNASCTTNCLAPLAKVINDNFGIIEGLMTTVHAVTATQKTVD
186: Fruit fly	FVCGVNLDAYKPD-MKVVSNASCTTNCLAPLAKVINDNFEIVEGLMTTVHATTATQKTVD

Figure 7.31. **Alignment of GAPDH protein sequences using Clustal Omega**. Section of a Clustal Omega sequence alignment. Amino acids with similar properties are the same color. Clustal Omega alignments are investigated further in Activity 7.F. Data are available from enzyme classification number EC 1.2.1.12 on the ExPASy Proteomics server of the UniProt Knowledgebase (uniprot.org, accessed February 23, 2018). Sievers et al. (2011).

The tertiary and quarternary structures of proteins and their interactions with binding partners are predicted using a subdiscipline of bioinformatics called protein modeling. Protein modeling methods help explain how proteins function; for example, how small-molecule drugs might inhibit disease-causing proteins. The 3-D structures of proteins are determined by analyzing the diffraction patterns produced when X-rays strike crystallized proteins, a field called X-ray crystallography. Nuclear magnetic resonance (NMR) is also used to determine protein structures. Three-dimensional models of proteins are stored in the Protein Data Bank (PDB) and can be viewed, rotated, and colorized to help scientists understand more about proteins and how they function.

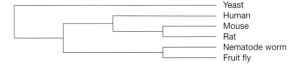

Figure 7.32. **Phylogenetic tree**. Clustal Omega alignment of GAPDH proteins from different organisms. Sequences that are most similar are clustered together and are connected by shorter lines. Data are available from enzyme classification number EC 1.2.1.12 on the ExPASy Proteomics server of the UniProt Knowledgebase (uniprot.org, accessed February 23, 2018). Sievers et al. (2011).

7.6 Protein Production in Industry

Protein Production

Protein products have many applications in industry. There are also several strategies to manufacture them. Though proteins can be extracted from the organ tissues that produce them in nature, it is often too difficult, impractical, or expensive to do so. For example, insulin must be extracted from pancreas tissue, and the milk coagulant chymosin is isolated from the fourth stomach of unweaned calves. The cost and difficulty of these extraction methods are too high to meet the needs of diabetics and cheesemakers. To get around these limitations, biotechnology companies develop and use recombinant DNA techniques to clone the genes for the desired proteins and express large amounts of them in another cell type. With this approach, they may also use recombinant DNA technology to study and improve the functionality of the protein.

Today, insulin is produced using recombinant DNA technology. The human insulin gene is cloned and inserted into a plasmid which is transformed into either *Escherichia coli* (*E. coli*) or *Saccharomyces cerevisiae* (baker's yeast). The transformed cells are grown on a small scale in the laboratory to confirm that they can produce the human insulin protein. At this stage, scientists use many of the protein analysis techniques described in the chapter to determine how much insulin is produced and whether it meets the desired specifications, for example, proper function and posttranslational modifications.

Next, the growth of the recombinant cells is scaled up for production. The bacteria or yeast are grown in sequentially larger vessels, and eventually in large fermentation tanks called bioreactors. Some bioreactor tanks can hold up to 20,000 liters of cells and growth media. During these stages, scientists carefully monitor the growth conditions to ensure proper protein production and prevent contamination by other bacteria or fungi (see Chapter 3). The bacteria produce more and more of the recombinant insulin protein as they grow. Finally, the protein is extracted and purified by filtration and chromatography.

Though bacteria grow quickly, they do not produce a lot of protein, and growing bacterial cell cultures in bioreactors can be expensive. In addition, proteins may have structural features, including posttranslational modifications that bacterial cells cannot create. For example, some proteins have sugar molecules attached to them, and they do not function properly unless those sugars are present in exactly the right place. For these types of proteins, companies must use yeast, plant, or animal cell cultures, or even whole plants or animals (see Protein-Based Drug Production in Chapter 1) to produce more, or more active proteins.

Quality Control in Protein Production

Protein production is expensive. Precautions are taken throughout the production process, or in-process, to ensure that the product is meeting specifications and to minimize waste. Biotechnology companies design process checkpoints, which are critical steps when specific criteria are measured and assessed. If a process fails to meet criteria, adjustments can be made or the production run is stopped. These checkpoints also provide added assurance to regulatory agencies like the U.S. Food and Drug Administration (FDA) that the process is being conducted in a controlled way. Examples of in-process checkpoints include using a spectrophotometer to measure the quantity of cells in a bioreactor or monitoring the pH of a buffer as it flows through a chromatography column. The decision of which tests to include usually depends on the risks involved. For example, the risk of contamination in liquid culture media is high. A quality control technician might test samples taken at critical points in the process to confirm that no contaminants were introduced during production.

Quality control (QC) testing of finished products may include assays such as mass spectrometry, gel electrophoresis, western blotting, or other immunoassays to confirm purity and identity. Other methods such as spectrophotometry and protein assays may be used to determine the quantity of product. Additional functional tests are important to ensure that the product performs properly and is safe. Throughout the production process QC personnel must document the results of each test performed. Before beginning the production of a new batch of product, manufacturing personnel receive a batch record, which is assigned a unique identifying number by the quality assurance (QA) department. The batch record includes instructions and places to record relevant information during the production. The same batch number is included on all test samples taken during the production run. QC personnel follow test records to perform each assay and again include the unique batch number on the test record. In this way, all documents used for a particular batch are linked by the unique batch number. After the production run is complete QA reviews all documentation to determine whether the product was made correctly and documented properly. If the final product meets all required specifications, it is released for sale.

Chapter 7 Essay Questions

1. Discuss the principles behind sample preparation in SDS-PAGE.

2. Research affinity chromatography and discuss some advantages and disadvantages of different types of affinity tags on recombinant proteins.

3. Discuss why it is important to use various organisms for protein production.

4. If a mushroom was found to have cellulase activity, discuss experiments that could be conducted to identify, characterize, and isolate the cellulase enzyme(s) to investigate whether it is a viable enzyme for biofuel production.

Additional Resources

Biochemistry textbooks:

Nelson DL and Cox MM (2017). Lehninger Principles of Biochemistry (New York: W.H. Freeman).

Stryer L et al. (2015). Biochemistry (New York: W.H. Freeman).

Background information on polyacrylamide electrophoresis:

Guide to Polyacrylamide Gel Electrophoresis and Detection. Bio-Rad Bulletin 6040.

Molecular Weight Determination by SDS-PAGE. Bio-Rad Bulletin 3133.

Bioinformatics resources:

The National Institutes of Health National Center for Biotechnology Information hosts bioinfomatics databases such as GenBank and has a wealth of information on bioinformatics. ncbi.nlm.nih.gov. Accessed June 29, 2018.

Sayers EW et al. (2010). Database resources of the National Center for Biotechnology Information. Nucleic Acids Res. 38: D5–16.

RCSB Protein Data Bank. rcsb.org. Accessed February 23, 2018.

Berman H et al. (2003). Announcing the worldwide Protein Data Bank. Nature Structural Biology, 10, 980.

Website with educational bioinformatics activities:

Digital World Biology. digitalworldbiology.com. Accessed June 26, 2018.

Activity 7.A Protein Quantitation Using the Bradford Assay

Overview

Protein quantitation is one of the most commonly performed procedures in a biotechnology laboratory. In this activity, the protein concentration of milk will be measured using the Bradford assay. In the presence of proteins, the Bradford reagent changes color from reddish-brown to blue and its peak absorption of light changes from 470 nm to 595 nm. The color change is visible and the amount of blue is proportional to the amount of protein. The color change can be quantified using a spectrophotometer set to read the absorbance of light at 595 nm in wavelength. The concentration of protein can then be determined by constructing a standard curve using the absorbance of samples with known concentration. Using the standard curve, the protein concentration in the test sample can be determined.

Tips and Notes

- It is important to thoroughly mix samples with the Bradford reagent. Samples can be pipetted up and down in the cuvette using a 100–1,000 µl micropipet to mix. Alternatively, the top of the cuvette can be sealed with Parafilm and the cuvette inverted five times

- Samples must be incubated at room temperature for at least 5 min after mixing with the Bradford reagent

- Orient the cuvette correctly in the spectrophotometer so that light passes through the sample in a 1 cm path length. Ensure that the cuvettes are very clean. Do not handle the cuvette where the light passes through since fingerprints can interfere with the reading. Wipe off fingerprints with a laboratory tissue

- If a spectrophotometer is not available, visually compare the samples to the protein standards

- If a colorimeter is available instead of a spectrophotometer, measure the samples at 635 nm, which is the wavelength closest to 595 nm

- When using 4 ml cuvettes that fit a Spectronic 20 (also known as a Spec 20) spectrophotometer, change the volumes in the procedure to dispense 2.5 ml of Bradford reagent to each cuvette and add 50 µl of protein sample

Safety Reminder: Follow all laboratory safety procedures. Read and be familiar with the SDSs for the Bradford reagent. Wear appropriate PPE. Bradford reagent can stain skin and clothes.

Research Question

- How much protein is present in milk?

Objectives

- Perform a Bradford assay on two milk samples
- Perform a Bradford assay on a series of standards of known protein concentration
- Use a spectrophotometer to measure the absorbance of Bradford assay samples
- Generate a standard curve of protein concentration against the absorbance at 595 nm
- Determine the protein concentration in the milk samples

Skills to Master

Refer to Laboratory Skills Assessment Rubric (Appendix E) and Laboratory Notebook Rubric (Appendix F) for more details.
- Record laboratory notebook entries
- Perform a Bradford assay
- Use a spectrophotometer (Activity 5.D)
- Generate a standard curve using absorbance values
- Determine experimental values using a standard curve

Student Workstation Materials

Items	Quantity
Spectrophotometer (shared)	1
Computers with graphing software (optional)	1–4
2–20 µl adjustable-volume micropipet and tips	1
20–200 µl adjustable-volume micropipet and tips	1
100–1,000 µl adjustable-volume micropipet and tips, or 1 ml graduated transfer pipet	1
Microcentrifuge tubes	2
Microcentrifuge tube rack	1
1.5 ml standard cuvettes*	10
Graph paper	4
Parafilm (approx. 10 cm strip) (optional)	1
Laboratory marking pen	1
Waste container	1
1x phosphate buffered saline (PBS)	250 µl
Bradford reagent	12 ml
Protein standards (0.125, 0.250, 0.500, 0.750, 1.000, 1.500, and 2.000 mg/ml)	25 µl of each
Milk sample A	5 µl
Milk sample B	5 µl

* When using cuvettes larger than 1.5 ml, increase the volume of protein samples and Bradford reagent proportionally (see Tips and Notes).

Prelab Focus Questions

1. Calculate how much 10 mg/ml bovine gamma-globulin is needed to prepare 50 µl of 0.5 mg/ml bovine gamma-globulin.
2. What is the color of the Bradford reagent before and after it reacts with protein?

Activity 7.A Protein Quantitation Using the Bradford Assay

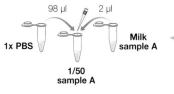

Protocol

1. Label one empty microcentrifuge tube **1/50 sample A** and label another tube **1/50 sample B**.

2. Pipet 2 µl of milk sample A and 98 µl of 1x PBS into the microcentrifuge tube labeled **1/50 sample A**. Mix well by pipetting or vortexing.

3. Repeat step 2 for milk **sample B**.

4. Label two cuvettes, one **Sample A** and one **Sample B**.

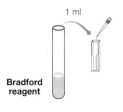

5. Pipet 20 µl of the 1/50 diluted milk samples into the corresponding cuvettes.

6. Label eight cuvettes for the protein standards as follows:
    Blank
    0.125
    0.250
    0.500
    0.750
    1.000
    1.500
    2.000

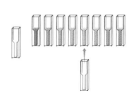

7. Pipet 20 µl of 1x PBS into the cuvette labeled **Blank**.

8. Pipet 20 µl of each protein standard into the corresponding cuvette.

9. Add 1 ml of the 1x Bradford reagent to all ten cuvettes. Mix completely, either by pipetting up and down with a 100–1,000 µl micropipet (using a fresh tip for each sample) or by covering each cuvette with a small piece of Parafilm and inverting the cuvette five times. Incubate cuvettes at room temperature for 5 min.

10. After 5 min, visually compare the cuvettes containing the milk samples to the cuvettes containing the protein standards. Determine the standard that most closely matches the color of each milk sample. Estimate the protein concentration of the milk based on the visual comparison.

11. Set the spectrophotometer to measure absorbance at a wavelength of 595 nm. This step is not necessary if your spectrophotometer measures the absorbance for multiple wavelengths at once.

12. Insert the cuvette labeled **Blank** into the spectrophotometer. Use the blank to set the spectrophotometer to zero absorbance (100% transmittance).

Activity 7.A Protein Quantitation Using the Bradford Assay

Sample A

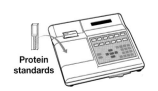

Protein standards

13. Remove the cuvette labeled **Blank** and insert **Sample A** into the spectrophotometer. Read the absorbance. Repeat with **Sample B** and record the values.

14. If instructed, read the absorbance of the seven protein standards and record the absorbance values.

Results Analysis

Estimate the protein concentration of each milk sample by visually comparing them with the protein standards. Record your estimates in a data table. Remember that each milk sample was diluted 50 times; therefore, the protein concentration of the original milk sample is obtained by multiplying the protein concentration of the 1/50 sample by 50.

Generate a standard curve of the known concentrations of the protein standards and their absorbance at 595 nm either manually or using graphing software (see options 1 and 2, below). Use the standard curve or the equation of the line to determine the protein concentration of the milk samples in mg/ml. Alternatively, generate a standard curve using the spectrophotometer if the instrument has the capability (see option 3, on next page). Record the data.

Compare the protein concentration you calculated from the standard curve with those you estimated visually. How good were your estimates?

Compare the calculated protein concentrations to those provided in the nutritional information on the carton of milk. You will need to convert the concentration on the milk carton to mg/ml.

Option 1: Estimating Protein Concentration Using Graph Paper
Plot the absorbance (in absorbance units (AU)) on the y-axis and the protein concentration (in mg/ml) on the x-axis on regular graph paper. Add a linear line of best fit (or trend line) through the data points. Find the absorbance of sample A on the y-axis and read across to the line of best fit. Then read down to find the protein concentration. Repeat for sample B.

Alternatively, determine the equation of the line ($y = mx + b$), where m is the slope of the line and b is the point where the line intercepts the y-axis. Use this equation to calculate the protein concentration of the milk samples.

Option 2: Estimating Protein Concentration Using Graphing Software
The protein concentration can be estimated using graphing software such as Microsoft Excel or Google Sheets. When working with graphing software, open a new spreadsheet and enter the absorbance in column A and the protein concentrations in column B. Using the graph function, highlight both columns and graph the values using a scatter plot. Add a linear trend line (line of best fit). Use the option to display the equation of the line ($y = mx + b$), where m is the slope of the line and b is the point where the line intercepts the y-axis. Select the equation of the line to calculate the protein concentration of the milk samples from their absorbances. The software will also provide the coefficient of determination r^2, which is a measure of how well the regression line fits the data points. The closer the r^2 value is to 1, the better the fit to the standard curve.

Activity 7.A Protein Quantitation Using the Bradford Assay

Option 3: Estimating Protein Concentration Using a Spectrophotometer
A standard curve can be generated by some spectrophotometers and used to automatically calculate protein concentration.

Postlab Focus Questions

1. How much protein is present in milk?
2. How do your visual estimates of protein concentration compare to the protein concentrations determined by the spectrophotometer?
3. How do the protein concentrations determined using the Bradford assay compare to those printed on the milk carton?
4. Why might it be important or useful to know the protein concentrations in milk or other liquids or foods?

Self Assessment

Assess whether you have mastered the skills of this activity using the Laboratory Skills Assessment Rubric in Appendix E.

Assess your experimental write-up in your laboratory notebook using the Laboratory Notebook Rubric in Appendix F.

Activity 7.B Size Exclusion Chromatography

Overview

Size exclusion chromatography separates mixtures of molecules based on their mass. Size exclusion chromatography resin is comprised of microscopic porous beads that temporarily trap small molecules and slow their progress through the resin, while larger molecules pass around, or are excluded from, the beads (see Figure 7.26). The smaller the molecules, the slower they move through the beads. The resin used in this activity fractionates (or separates) molecules that are 60,000 Da or smaller, which enter pores in the beads and pass through the column more slowly. Molecules that are larger than 60,000 Da pass around the beads and are excluded from the column. The upper size limit for molecules that cannot penetrate the pores of the beads is referred to as the exclusion limit of a column.

The resin in the column is called the column bed. The sample to be separated is dissolved in a buffer, also called a solvent. The sample is pipetted gently onto the column bed. The molecules in the sample enter the top of the column bed, filter through and around the porous beads, and ultimately pass (or elute) through a small opening at the bottom of the column. For this process to be completed, additional column buffer is pipetted onto the column bed after the sample has entered the bed to facilitate the molecules' movement through the resin. The eluate (or flowthrough) is collected dropwise into collection tubes in fractions, which are sequentially ordered. A specific number of drops is collected into each tube, or fraction. The larger molecules, which pass quickly through the column, will be in the early fractions. The smaller molecules, which penetrate the pores of the resin and become temporarily trapped, will be collected in the later fractions.

In this activity, a mixture containing hemoglobin and vitamin B12 will be separated. Hemoglobin is brown, while vitamin B12 is pink. The purpose of the activity is to determine whether these molecules can be separated using size exclusion chromatography and which molecule is larger than the other.

Tips and Notes

- When removing the end of the column, avoid twisting the tab. Instead, snap off the end with a sharp 90° bend. This will generate a larger opening and will increase the sample flow through the column

- Insert a small wedge of paper between the column and the collection tube to prevent the formation of an airtight seal between the column and tube, which could increase air pressure and slow the flow of sample through the column

- Insert the pipet as far into the column as possible without touching the resin and pipet slowly and gently to avoid any disturbance of the column bed

Safety Reminders: Follow all laboratory safety procedures. Read and be familiar with all SDSs for this activity and wear appropriate PPE.

Research Questions

- Can hemoglobin and vitamin B12 be separated using size exclusion chromatography?
- Which molecule is larger, hemoglobin or vitamin B12?

Objectives

- Prepare a size exclusion chromatography column
- Apply a mixture of hemoglobin and vitamin B12 to a size exclusion column
- Collect fractions from the column

Skill to Master

Refer to Laboratory Skills Rubric (Appendix E) and Laboratory Notebook Rubric (Appendix F) for more details.
- Record laboratory notebook entries
- Perform size exclusion chromatography

Student Workstation Materials

Items	Quantity
1 ml graduated transfer pipet*	1
Collection tubes	12
Column end cap	1
Test tube rack or microcentrifuge tube rack	1
Laboratory marking pen	1
Waste container	1
Column buffer	4 ml
Protein mixture	50 µl
Size exclusion column	1

* Adjustable-volume micropipets (20–200 µl and 100–1,000 µl) and pipet tips can be used instead.

Prelab Focus Questions

1. Do smaller or larger molecules exit a size exclusion column first? Explain your reasoning.
2. Describe what is meant by the exclusion limit of a size exclusion chromatography resin.
3. With regard to a chromatography experiment, what is a fraction?

Activity 7.B Size Exclusion Chromatography

1. Label 10 collection tubes sequentially from **1** to **10**. Label the final two tubes **Waste** and **Column buffer**.

2. Remove the cap and snap off the end of the sizing column, then place the column into a collection tube. Wait for the storage buffer to drain into the **Waste** tube. Cap the bottom of the column with a column end cap as soon as the column stops dripping.

   Note: When all of the storage buffer enters the column, the surface of the resin appears grainy.

3. Place the column into the collection tube labeled **1**.

4. Uncap the column. Gently apply one drop (or 50 µl) of the protein mix onto the top of the column bed. Place the pipet into the column close to but not touching the column bed to minimize disturbance of the resin.

   Note: Size separation works best when the column bed is undisturbed.

5. Gently pipet 250 µl of column buffer onto the column bed. Place the pipet close to the column bed and let it drip down the side of the column wall to minimize disruption of the resin. Collect drops into tube **1**.

6. Add another 250 µl of column buffer to the top of the column. Place the pipet close to the column bed and let it drip down the side of the column wall. Continue to collect drops into tube **1**.

7. Transfer the column to tube **2** and add 3 ml of column buffer to the top of the column matrix. Collect 5 drops of buffer in tube **2**.

Activity 7.B Size Exclusion Chromatography

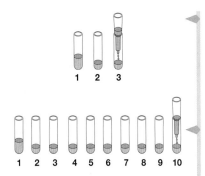

8. When 5 drops have been collected in tube **2**, transfer the column to tube **3**. When 5 drops have been collected into tube **3**, lift the column off and transfer it to the next tube. Continue to collect 5 drops of buffer into each collection tube through tube **9**.

9. Collect 10 drops in tube **10**. Cap the column. Observe the tubes. Record your results, including the color and its intensity in each fraction.

If necessary, seal the collection tubes with Parafilm and store them at 4°C.

Results Analysis

Hemoglobin is brown and vitamin B12 is pink. Assess the column fractions in the collection tubes and record the color and the intensity of the color of the eluate in each tube. Determine whether the molecules have been successfully separated and determine which molecule is larger.

Postlab Focus Questions

1. Were hemoglobin and vitamin B12 separated? What is your evidence?
2. Which column fractions contained hemoglobin and which contained vitamin B12?
3. Which is larger, hemoglobin or vitamin B12? What is your evidence?
4. Think about the exclusion limit of the column used in this activity. What can you conclude about the size of the molecules separated by this column?

Self Assessment

Assess whether you have mastered the skills of this activity using the Laboratory Skills Assessment Rubric in Appendix E.

Assess your experimental write-up in your laboratory notebook using the Laboratory Notebook Rubric in Appendix F.

Activity 7.C GFP Purification by Hydrophobic Interaction Chromatography

Overview

Many recombinant proteins, including those used as pharmaceutical drugs, are produced by expressing proteins from cloned genes in bacteria. The bacteria are then broken open (lysed) and the recombinant protein is purified using chromatography. In this activity, recombinant green fluorescent protein (GFP) will be purified from bacteria that have been transformed with the pGLO™ plasmid that was used in Activity 5.B; GFP will subsequently be separated from the thousands of endogenous proteins present in *E. coli*. Purifying GFP is a great way to learn about chromatography because unlike most proteins that are invisible to the naked eye, GFP can be visualized with UV light. Therefore, it is easy to track the progress of GFP through the chromatography experiment.

Chromatography is a powerful technique for separating proteins in a complex mixture. Hydrophobic interaction chromatography (HIC) separates molecules based on their hydrophobicity. Hydrophobic means water fearing; therefore, hydrophobic molecules do not mix well with water. Hydrophobic substances naturally bind to other hydrophobic substances. Some amino acids are hydrophobic leading to hydrophobic regions on the surfaces of proteins. Depending on their amino acid sequence, some proteins are more hydrophobic than others, including GFP. A HIC column is filled (or packed) with hydrophobic resin. In a high salt buffer, hydrophobic proteins bind to the resin. The more hydrophobic the protein, the more tightly it binds to the resin. When the salt concentration is lowered, the hydrophobic proteins begin to release from the resin. Highly hydrophobic proteins elute only in buffer with a very low salt concentration, while less hydrophobic proteins elute in higher salt concentrations; this phenomenon is used to separate more hydrophobic proteins from less hydrophobic proteins.

In this activity, *E. coli* transformed with pGLO plasmid will be cultured in LB/amp/ara broth that induces GFP expression. The bacteria will be isolated from the broth and broken open (lysed) using an enzyme called lysozyme, followed by a freeze-thaw step during which ice crystal formation helps to break open the cell membranes. Cellular debris will be centrifuged into a pellet and the supernatant will be applied to an HIC column in a high-salt buffer. Hydrophobic proteins in the supernatant will bind to the HIC resin. The salt concentration of the column buffer will be gradually lowered so that proteins with lower hydrophobicity will be released from the column and collected in fractions. Finally, an elution buffer that contains no salt will be added so that the most hydrophobic proteins will be released and collected in the final fraction.

Tips and Notes

- Make sure that the bacteria transformed with pGLO plasmid (from Activity 5.B) are viable. Approximately 24–72 hr ahead of the laboratory, restreak the transformed bacteria from stab cultures, glycerol stocks, or agar plates (that are less than 2 weeks old) onto fresh LB/amp/ara agar plates

- Once the liquid cultures have been centrifuged, it is very important to resuspend the pellet thoroughly in TE buffer to ensure that the cells are separated from one another. This enables the lysozyme and the freeze-thawing to properly lyse the cells

- Track the location of the GFP through the experiment using a UV light. When the bacteria are intact, the GFP should be inside the bacteria. Before lysis, GFP should first be observed in the bacterial colony, then in the liquid culture, and finally in the cell pellet. Once the bacteria are lysed, GFP should be in the supernatant

- When removing the end of the column, avoid twisting the tab. Instead, snap off the end with a sharp 90° bend to generate a larger opening and increase the sample flow through the column

- Insert a small wedge of paper between the column and the collection tube to prevent the formation of an airtight seal between the column and tube, which could increase air pressure and slow the flow of the sample through the column

- Insert the pipet as far into the column as possible without touching the resin, and pipet slowly and gently to avoid any disturbance of the column bed

- To collect concentrated GFP, use an extra collection tube or microcentrifuge tube. Observe the progress of GFP through the column. As soon as GFP starts to elute, transfer the column to the fresh collection tube. Once GFP has been eluted, transfer the column back to tube **3**

Safety Reminders: Follow all laboratory safety procedures. Inform your instructor if you are allergic to antibiotics and do not handle ampicillin. Use aseptic technique when handling bacteria and dispose of microbial waste properly. Read and be familiar with all SDSs for this activity. Wear appropriate PPE. Special precautions are required when using UV lights.

Research Question

- Can GFP be separated from a mixture of proteins using HIC?

Activities Protein Structure and Analysis

Activity 7.C GFP Purification by Hydrophobic Interaction Chromatography

Objectives

- Use aseptic technique to inoculate LB/amp/ara broth with bacteria transformed with the pGLO plasmid
- Grow liquid cultures overnight
- Centrifuge the bacterial culture and lyse bacteria
- Centrifuge the bacterial cell lysate to isolate the supernatant
- Apply the supernatant to a HIC column
- Apply buffers with different salt concentrations to the column
- Collect fractions

Skills to Master

Refer to Laboratory Skills Assessment Rubric (Appendix E) and Laboratory Notebook Rubric (Appendix F) for more details.

- Record laboratory notebook entries
- Aseptically inoculate a liquid culture from a plate (Activity 3.C)
- Make a bacterial lysate from a liquid culture
- Perform HIC
- Use a microcentrifuge (Activity 5.C)

Student Workstation Materials

Part 1: Inoculation of Cell Cultures

Items	Quantity
Incubator (shared)	1
Shaking incubator, shaking water bath, or incubator with mini-roller (shared)	1
Pipet pump or filler*	1
5 ml sterile serological pipet*	1
14 ml culture tubes	2
Test tube rack (optional)	1
Bunsen burner**	1
Inoculation loop**	1–4
UV light	1
Parafilm (shared)	1 roll
Laboratory marking pen	1
Microbial waste container	1
LB/amp/ara broth	4 ml
LB/amp agar plate (optional)	1
LB/amp/ara agar plate (optional)	1
pGLO transformed *E. coli* (from agar plate, stab culture, glycerol stock, or from activity 5.B)	1

* Sterile 1 ml graduated transfer pipets or 100–1,000 µl adjustable-volume micropipets with sterile tips can be used instead to transfer 2 ml of LB/amp/ara broth.
** If metal inoculation loops are available, one loop is sufficient and must be sterilized with a Bunsen burner. Alternatively, four disposable plastic loops can be used.

Part 2: Bacterial Lysis

Items	Quantity
Microcentrifuge or mini centrifuge (shared)	1
20–200 µl adjustable-volume micropipet and tips, or 1 ml graduated transfer pipet	1
100–1,000 µl adjustable-volume micropipet and tips, or 1 ml graduated transfer pipet	1
2.0 ml microcentrifuge tube	1
Microcentrifuge tube rack	1
UV light	1
Laboratory marking pen	1
Microbial waste container	1
Overnight cultures from part 1	2
TE buffer	300 µl
Lysozyme	50 µl

Part 3: Preparation of Bacterial Lysate and HIC Column

Items	Quantity
Microcentrifuge or mini centrifuge (shared)	1
100–1,000 µl adjustable-volume micropipet and tips, or 1 ml graduated transfer pipet	1
2.0 ml microcentrifuge tube	1
Microcentrifuge tube rack	1
Collection tube	1
Column end cap	1
UV light	1
Laboratory marking pen	1
Microbial waste container	1
Bacterial lysate from part 2	1
Binding buffer (4M $(NH_4)_2SO_4$)	300 µl
Equilibration buffer (2M $(NH_4)_2SO_4$)	3 ml
HIC column	1

Part 4: Purification of GFP by HIC

Items	Quantity
100–1,000 µl adjustable-volume micropipet and tips, or 1 ml graduated transfer pipet	1
Microcentrifuge tube rack	1
Collection tubes	3
Column end cap	1
UV light	1
Laboratory marking pen	1
Microbial waste container	1
Supernatant from part 3	1
Wash buffer (1.3 M $(NH_4)_2SO_4$)	300 µl
TE buffer	800 µl
Equilibrated HIC column	1

Prelab Focus Questions

1. Why are ampicillin and arabinose added to the LB broth when growing the liquid cultures?
2. What would you see if a green pGLO colony from the LB/amp/ara agar plate was streaked onto LB/amp agar, incubated at 37°C overnight, and viewed under UV light?
3. What would you see if a white pGLO colony from the LB/amp agar plate was streaked onto LB/amp/ara agar, incubated at 37°C overnight, and viewed under UV light?
4. Why is lysozyme added before freezing the cells?

Activity 7.C GFP Purification by Hydrophobic Interaction Chromatography

Part 1: Inoculation of Cell Cultures

1. If necessary, prepare one LB/amp agar plate, one LB/amp/ara agar plate, and 50 ml of LB/amp/ara broth using the instructions in Activity 3.A.

2. If plates with pGLO-transformed bacterial colonies are not provided by your instructor, use aseptic technique to streak one LB/amp agar plate and one LB/amp/ara agar plate with pGLO-transformed *E. coli* (see Activity 3.C, part 1B, step 8).

3. Incubate the plates overnight at 37°C. If the plates are not used the next day, seal them with Parafilm and store them at 4°C for up to 2 weeks.

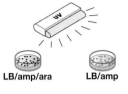

LB/amp/ara   LB/amp

4. Examine the plates using a UV light. Identify several single green colonies (ones not touching other colonies) on the LB/amp/ara agar plate. Identify several single white colonies on the LB/amp agar plate.

5. Label one 14 ml culture tube **+** and one **–**, and add your initials and date. Use aseptic technique to add 2 ml of LB/amp/ara broth to each tube (your instructor may have already done this for you).

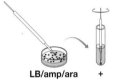

LB/amp/ara   **+**

6. Flame the loop and let it cool down. Use the loop to scrape a single green colony from the LB/amp/ara agar plate. Transfer the loop into the culture tube labeled **+** and swirl it in the broth to disperse the bacteria. Close the tube.

    Note: A sterile plastic inoculation loop can also be used to pick the colonies but it should not be flamed.

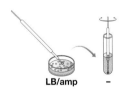

LB/amp   **–**

7. Repeat step 6 to transfer a white colony from the LB/amp agar plate to the LB/amp/ara broth in the culture tube labeled **–**.

8. Place the culture tubes into a shaking incubator, shaking water bath, or incubator with tube roller. Grow the culture overnight at 32°C. Predict whether GFP will be expressed in the cultures.

    Note: If an incubator is not available, incubate the samples at room temperature for 48 hr with shaking or rolling. This protocol is not recommended if the samples cannot be shaken or rolled while incubating at room temperature. After the overnight growth, the cultures can be stored at 4°C for up to 1 week.

Activity 7.C GFP Purification by Hydrophobic Interaction Chromatography

Part 2: Bacterial Lysis

1. The overnight cultures should appear cloudy. Examine the cultures with a UV light and record whether the cultures glow green. Determine whether your predictions were correct. Discard the tube labeled − as microbial waste.

   Note: GFP is best observed with the UV light in dark or low-light conditions.

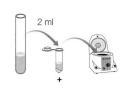

2. Label one 2.0 ml microcentrifuge tube **+** and add your initials. Transfer 2 ml of the liquid culture from the culture tube labeled **+** into the microcentrifuge tube labeled **+**. Centrifuge the microcentrifuge tube for 5 min at maximum speed. Ensure that the microcentrifuge is balanced with another tube that contains approximately the same volume.

   Note: Make sure to accommodate your classmates' tubes to ensure economic use of the microcentrifuge.

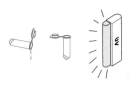

3. Observe the sample under UV light and record your observations. Pour out and discard the supernatant.

4. Resuspend the bacterial pellet in 250 µl of TE buffer and rapidly pipet up and down to completely resuspend the bacteria.

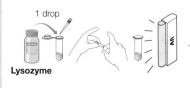

5. Add 50 µl, or 1 drop, of lysozyme to the resuspended bacteria to initiate enzymatic digestion of the bacterial cell wall. Mix the contents gently by flicking the tube five times or by pipetting up and down. Observe the tube under the UV light and record your observations.

6. Place the microcentrifuge tube in the freezer until the next laboratory period. The freezing causes the bacteria to rupture completely.

Part 3: Preparation of Bacterial Lysate and HIC Column

1. Thaw the microcentrifuge tube containing the bacterial lysate using the warmth of your hand. Centrifuge the tube for 10 min at maximum speed to pellet the insoluble bacterial debris. Ensure that the microcentrifuge is balanced with another tube that contains approximately the same volume.

   Note: Make sure to accommodate your classmates' tubes to ensure economic use of the microcentrifuge.

2. While the tube is spinning, prepare the prefilled HIC column. Label the top of the column with your initials. Remove the cap and snap off the bottom. Drain the storage buffer from the column into a collection tube; this will take 3–5 min.

Activity 7.C GFP Purification by Hydrophobic Interaction Chromatography

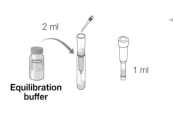

Equilibration buffer

2 ml

1 ml

3. Equilibrate the column by dripping 2 ml of equilibration buffer down the wall of the column. Equilibration buffer has the same salt concentration as the sample that will be applied to the column. Drain the buffer to the 1 ml mark on the column.

4. Recap the top of the column and add a column end cap to the bottom of the column.

   Note: Store the column at room temperature until ready to perform the chromatography.

5. After the 10 min centrifugation of the cell lysate, examine the microcentrifuge tube with a UV light. Record your observations.

250 μl

+ +

6. Label a clean microcentrifuge tube **+** and add your initials. Transfer 250 μl of the supernatant into the new microcentrifuge tube, making sure to avoid disturbing the pellet.

250 μl

Binding buffer

7. Transfer 250 μl of binding buffer to the supernatant and mix by pipetting up and down. (Binding buffer is a very high-salt buffer.)

   Note: Once binding buffer is added to the supernatant, the sample can be stored in the refrigerator until ready to perform the chromatography.

Part 4: Purifying GFP by HIC

1. Label three collection tubes **1**, **2**, and **3** and place them in a rack. Remove the caps from the top and bottom of the column and place the column in collection tube **1**. Allow the equilibration buffer to drain until it reaches the surface of the column bed.

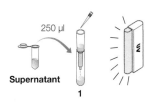

250 μl

Supernatant

1

2. Carefully and gently load 250 μl of the supernatant in binding buffer (from step 7 in part 3) onto the top of the column. Place the pipet close to the column bed and let it drip down the side of the column wall. Examine with UV light. Record your observations. When the column stops dripping, transfer it to collection tube **2**.

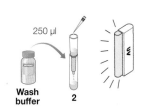

250 μl

Wash buffer

2

3. Add 250 μl of wash buffer to the column. Let the entire volume flow into the column and collect the eluate in collection tube **2**. (Wash buffer is a medium-salt solution.) Examine with UV light and record your observations. When the column stops dripping, transfer it to collection tube **3**.

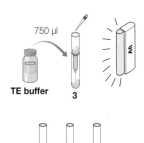

750 μl

TE buffer

3

4. Add 750 μl of TE buffer to the column. Let the entire volume flow into the column and collect the eluate in collection tube **3**. (TE buffer contains no salt.) Examine with UV light. Record your observations.

1 2 3

5. Examine all three collection tubes and record your observations.

Activity 7.C GFP Purification by Hydrophobic Interaction Chromatography

Results Analysis

Record the progress of GFP throughout the experiment, from the initial bacterial colony to the collection tubes. Identify and record the fraction in which the GFP was collected.

Postlab Focus Questions

1. Based on your results, explain the roles or functions of these buffers.
   Hint: How does the name of the buffer relate to its function?

   a. Binding buffer
   b. Wash buffer
   c. TE (or elution) buffer

2. Was GFP separated from bacterial proteins? What evidence do you have to support your answer?

Self Assessment

Assess whether you have mastered the skills of this activity using the Laboratory Skills Assessment Rubric in Appendix E.

Assess your experimental write-up in your laboratory notebook using the Laboratory Notebook Rubric in Appendix F.

Activity 7.D SDS-PAGE of Fish Muscle

Overview

Polyacrylamide gel electrophoresis (PAGE) is a fundamental tool in biotechnology. It is usually used to separate proteins but it can also be used to separate small fragments of DNA or RNA. SDS-PAGE uses sodium dodecyl sulfate (SDS) to coat proteins and change the natural net charge of the proteins to a negative charge that is relative to their size. This allows electrophoresis to be used to separate proteins based on their size.

In this activity, polyacrylamide gel electrophoresis will be used to compare the proteins from muscle cells in various fish species (see Figure 7.33). The differences can be used to infer evolutionary relationships among the species using a phylogenetic tree or cladogram. First, pieces of muscle from different fish species will be mixed with Laemmli sample buffer, which denatures the proteins. The Laemmli sample buffer also coats the proteins in SDS, which makes them negatively charged and ready to be electrophoresed. The protein extracts will be heated and then electrophoresed in a polyacrylamide gel. Once the proteins in the gel are stained, proteins of the same mass will be identified on the gel and used to construct a cladogram to infer relationships among the fish.

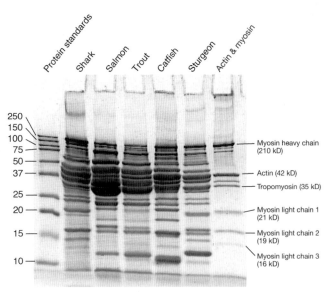

Figure 7.33. **SDS-PAGE of five fish samples**. Five fish samples (shark, salmon, trout, catfish, and sturgeon) were electrophoresed on a precast polyacrylamide gel. Lane 1, Precision Plus Protein™ Kaleidoscope™ standards; lane 2, shark; lane 3, salmon; lane 4, trout; lane 5, catfish; lane 6, sturgeon; and lane 7, actin and myosin standard.

Tips and Notes

- Fresh or frozen raw muscle tissue can be used in this activity. Muscle tissue can be frozen for a few months. It is easier to cut frozen tissue

- Tissue samples to be added to the Laemmli sample buffer should be small (approximately a 2 mm cube). Increasing the amount of tissue will cause the gel to be overloaded, which will affect gel analysis

- Although fish is recommended for this activity, clams, scallops, octopus, shrimp, snails, lean chicken, beef, pork, lamb, and other similar protein sources can also be used

- Polyacrylamide gels should not be touched with bare hands or fingers for safety reasons and to prevent deposition of proteins from skin onto the gel, which may show up when staining or imaging the gel

Safety Reminders: Follow all laboratory safety procedures. Read and be familiar with all SDSs for this activity. Wear appropriate PPE (including gloves) when handling polyacrylamide gels and protein stain. Protein stain can stain hands and clothing. Attention should be given when using electricity in the presence of liquids. Safety measures in the equipment should prevent accidental exposures; however, should any leakage occur, notify your instructor before proceeding. Inform your instructor if you are allergic to any fish or seafood, and do not handle fish samples. If UV light is used, special precautions are required.

Research Questions

- What are the sizes of fish muscle proteins?
- Can protein profiles determine the relatedness among organisms?

Objectives

- Extract protein from five muscle samples
- Electrophorese muscle proteins using SDS-PAGE
- Stain and/or image gels
- Generate a standard curve
- Determine the masses of proteins in the gel
- Generate a cladogram using a shared character matrix

Skills to Master

Refer to Laboratory Skills Assessment Rubric (Appendix E) and Laboratory Notebook Rubric (Appendix F) for more details.
- Record laboratory notebook entries
- Extract proteins using Laemmli sample buffer
- Perform SDS-polyacrylamide gel electrophoresis (SDS-PAGE)
- Use a power supply (Activity 4.C)
- Stain polyacrylamide gels
- Process stain-free polyacrylamide gels (optional)
- Generate a standard curve using a protein size standard
- Determine protein molecular masses using a standard curve
- Generate a character matrix and a cladogram

Activity 7.D SDS-PAGE of Fish Muscle

Student Workstation Materials

Part 1: Laemmli Extraction of Proteins

Items	Quantity
Water bath or dry bath set to 95°C (shared)	1
100–1,000 µl adjustable-volume micropipet and tips, or 1 ml graduated transfer pipet	1
1.5 ml microcentrifuge tubes	5
1.5 ml screwcap microcentrifuge tubes	5
Microcentrifuge rack	1
Floating tube rack (optional)	1
Razor blade or scissors	1
Laboratory marking pen	1
Waste container	1
Laemmli sample buffer (LSB)	1.5 ml
Fish muscle samples	5 species

Part 2: Electrophoresis of Samples

Items	Quantity
Mini-PROTEAN Tetra cell electrophoresis module (shared)	1
UV light source (optional) (shared)	1
Sample loading guide (optional) (shared)	1
Buffer dam (optional)*	1
Power supply (shared)	1
Water bath or dry bath set to 95°C (shared)	1
Rocking platform (optional) (shared)	1
2–20 µl adjustable-volume micropipet	1
Prot/Elec™ pipet tips	7 tips
Gel staining tray	1
Gel opening key (shared)	1
Laboratory marking pen	1
Waste container	1
Fish protein extracts from part 1	5 species
Actin and myosin standard	12 µl
Precision Plus Protein Kaleidoscope standards (Std)	6 µl
1x Tris/glycine/SDS (TGS) running buffer	500 ml
Bio-Safe™ Coomassie stain (shared)	50 ml
4–20% 10-well Mini-PROTEAN TGX gel or 4-20% 10-well Mini-PROTEAN TGX Stain-Free Protein Gel **	1

*   Buffer dam is required if 1 gel is run in the Mini-PROTEAN Tetra cell.
** Alternative visualization of proteins may be used by using a 4-20% 10-well Mini-PROTEAN TGX Stain-Free Protein Gel. See Appendix C for details.

Part 3: Gel Drying and Analysis

Items	Quantity
Digital imaging system (optional) (shared)	1
Computers with graphing software (optional)	1–4
Ruler	1
Semilog graph paper	4

1. State two reasons why SDS is used in Laemmli sample buffer.
2. Why are polyacrylamide gels used instead of agarose gels to separate and analyze proteins?
3. Why are proteins measured in daltons instead of the number of amino acids?

Activity 7.D SDS-PAGE of Fish Muscle

Part 1: Laemmli Extraction of Proteins

1. Label five 1.5 ml microcentrifuge tubes and five 1.5 ml screwcap tubes with the names of the five fish samples and add your initials so that there are two labeled tubes for each fish.

2. Add 250 µl of Laemmli sample buffer to each microcentrifuge tube (but not to the screwcap tubes).

3. Cut a piece (approximately a 2 mm cube) from each fish sample and transfer it into the corresponding microcentrifuge tube. Close the lids.

4. Flick the microcentrifuge tubes 15 times to agitate the fish in the sample buffer and then incubate the tubes for 5 min at room temperature.

5. Carefully transfer the buffer from each microcentrifuge tube into the corresponding screwcap microcentrifuge tube by pouring. **Do not transfer the fish.**

6. Heat the protein extracts in the screwcap tubes for 5 min at 95°C. Proceed to part 2 immediately or store protein extracts at –20°C for up to 6 months.

Part 2: Electrophoresis of Samples

1. If your vertical electrophoresis system is different from the one shown here, follow your instructor's directions for setting it up.

2. Prepare the Mini-PROTEAN TGX gel by pulling the tape off of the bottom of the gel and removing the comb.

3. Clamp the gel into the electrode assembly with the short plate facing inward. Clamp another gel or a buffer dam on the oppose side of the electrode assembly. Refer to How To… Set Up a Vertical Electrophoresis System on page 246 for more information.

4. Place the electrode assembly into the gel tank, making sure to match the red electrode with the red side of the tank and the black electrode with the black side of the tank.

Activity 7.D SDS-PAGE of Fish Muscle

5. Fill the inner chamber of the electrode assembly completely with 1x TGS. Pour 1x TGS to the proper fill line of the outside chamber. If a sample loading guide is used, place it into the inner chamber.

6. Heat the fish samples and actin and myosin standard to 95°C for 2–5 min. (Do not heat the Precision Plus Protein Kaleidoscope standards.)

7. Load your gel according to the table:

Lane	Volume	Sample
1	empty	empty
2	empty	empty
3	5 µl	Precision Plus Protein Kaleidoscope standards (Stds)
4	10 µl	fish sample 1
5	10 µl	fish sample 2
6	10 µl	fish sample 3
7	10 µl	fish sample 4
8	10 µl	fish sample 5
9	10 µl	actin and myosin standard (AM)
10	empty	empty

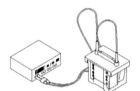

8. Place the lid on the electrophoresis chamber, making sure to first remove the sample loading guide if it was used. Connect the electrical leads into the power supply in the following orientation: red to red and black to black.

9. Turn on the power and run the gel at 300 V for 18 min. As the power is switched on, look for the release of bubbles from the electrode in the inner chamber. Watch the progression of the blue tracking dye in the Laemmli sample buffer and the separation of the prestained standards down the gel as the electrophoresis progresses.

   Note: Some gel formulations can be run at a higher voltage, such as 300 V for 18 min. Check manufacturer guidelines.

10. When the electrophoresis run is complete, turn off the power and remove the lid from the chamber. Remove the electrode assembly and pour the buffer from the inner chamber back into the tank. Remove the gel cassette from the electrode assembly.

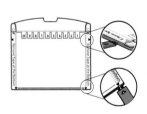

11. Remove the gel from the gel cassette using the opening key. Line up the arrows on the opening lever with the four arrows on the cassette to open the cassette.

    Note: For alternative visualization of proteins using stain-free gels, see Appendix C.

12. Very gently, pick up the gel and transfer into a gel staining tray. Handle the gels from the bottom, not from the top.

13. Rinse the gels in tap water in the gel staining tray. If time allows, wash the gels three times for 5 min.

14. Pour off the water and add 50 ml of Bio-Safe Coomassie stain. Stain two gels per staining tray. For best results, stain gels for at least 1 hr with gentle shaking.

Activity 7.D SDS-PAGE of Fish Muscle

15. Discard the stain and rinse the gels two times. Destain the gels in a large volume of water overnight, changing the water at least once. Blue-stained bands will be visible on a clear gel after destaining.

16. Scan, photocopy, or image the gels while they are still wet. **Note:** For digital imaging, gel images are best recorded while they are still wet.

Results Analysis

Record the banding pattern on the gel by taking a photo or imaging using a gel documentation system.

Part 1: Determine the Molecular Masses of Fish Proteins

Step 1: Label the gel and estimate protein sizes.

   a. Label the gel image with the sizes of the Precision Plus Protein Kaleidoscope standards (see Figures 7.13 and 7.34).

   b. Using Figure 7.33 as a guide, label the proteins of the actin and myosin standard.

   c. Using the protein standard as a guide, roughly estimate the sizes of actin, tropomyosin, and the myosin subunits of the actin and myosin standard.

   d. Compare your estimates with the actual sizes of the proteins given in Table 7.4.

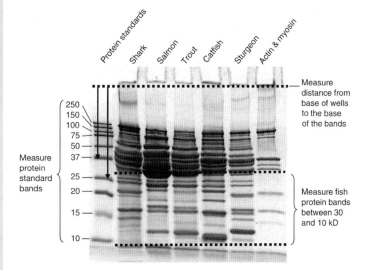

Figure 7.34. **SDS-PAGE of fish muscle proteins**. Five fish samples (shark, salmon, trout, catfish, and sturgeon) were electrophoresed on a precast polyacrylamide gel. Lane 1, Precision Plus Protein Kaleidoscope standards; lane 2, shark; lane 3, salmon; lane 4, trout; lane 5, catfish; lane 6, sturgeon; lane 7, actin and myosin standard. Analysis should be done on samples below actin and above the 10 kD standard as indicated.

Activity 7.D SDS-PAGE of Fish Muscle

Step 2: Create a standard curve.

To determine the molecular masses of the proteins, create a standard curve by plotting the known molecular masses of the proteins in the Precision Plus Protein Kaleidoscope standards against the distance they have migrated down the gel from the bottom of the well.

   a. Determine the distance traveled by each protein in the Precision Plus Protein Kaleidoscope standards by measuring the distance from the bottom of the well to the bottom of each band (see Figure 7.34).

   b. Record your measurements in a table in your laboratory notebook.

   c. Generate a standard curve using either semilog graph paper provided by your instructor or graphing software such as Microsoft Excel or Google Sheets. (Refer to Activity 4.D for directions on constructing a standard curve.)

Step 3: Measure the migration of fish proteins and estimate the protein size.

   a. Identify the actin band in each fish sample and measure the distance traveled by each band below actin and above the 10 kD standard in each sample (see Figure 7.34) and record these measurements in a table.

   b. Use the standard curve to determine the size of the proteins. Note: This smaller range of proteins is chosen to reduce analysis time; if desired, determine the molecular mass of the larger protein bands for each fish as well.

If available, digital gel imaging software (software that analyzes the gel automatically and provides band sizes without the need to measure migration of the proteins) can also be used for this activity. However, it is important to understand how the software calculates the sizes of the proteins and to acquire the skills needed to determine the sizes of unknown proteins using a standard curve.

Part 2: Generate a Cladogram

Step 1: Prepare data for analysis.

The banding patterns, or protein profiles, of the fish can be visually compared to one another and similarities among samples can often be distinguished. For example, salmon and trout have very similar profiles (see Figure 7.34). Based on this information, an assumption can be made that these fish share characteristics and are more closely related to each other than they are to shark.

To do a formal analysis of the fish samples that were analyzed in this activity, make a table similar to Table 7.5 to determine the fish that share specific bands. Count the number of bands for each fish.

Activity 7.D SDS-PAGE of Fish Muscle

Table 7.5. **Summary of various proteins that are present in fish species.**

Distance Migrated, mm	Protein Molecular Mass, kD	Shark	Salmon	Trout	Catfish	Sturgeon
25	32.5	X				
26	31.5		X	X	X	X
26.5	31.0	X				
27.5	30.0		X	X	X	X
28.5	29.1					
29	28.6	X	X	X	X	
30	27.6			X		X
30.5	27.1					X
32	25.6		X	X	X	
33	24.7					X
34.5	23.2		X	X		
35.5	22.2					X
36	21.7	X				
36.5	21.2	X	X	X	X	
37	20.7					X
37.5	20.2		X	X		
38	19.7				X	
38.5	19.3				X	
39	18.8	X				X
39.5	18.3					X
40.5	17.3		X	X		
41	16.8				X	
41.5	16.3					
42	15.8		X	X		X
43	14.8					
44	13.9	X				X
45	12.9		X	X		
46	11.9				X	
46.5	11.4			X		
47	10.9					X
47.5	10.4				X	
51.5	6.5			X		
52	6.0	X				
	COUNT	**8**	**10**	**13**	**10**	**12**

Step 2: Generate a matrix showing shared proteins.

   a. Generate a matrix similar to Table 7.6 to record the number of shared bands. Common protein bands are shared characteristics and shared characteristics are what evolutionary biologists use to determine relatedness among organisms. Two species that share more characteristics with each other than with a third species are considered to be more closely related to each other than to the third species.

   b. Compare each fish protein profile and record the number of shared bands in a matrix similar to Table 7.6. For example, only two bands from shark are shared with any other fish in Table 7.6. Salmon has ten bands in common with trout but only five bands in common with catfish and three with sturgeon.

Activity 7.D SDS-PAGE of Fish Muscle

Table 7.6. **Shared characteristic matrix for five fish species**. Bands common among fish species were counted and recorded. **Note:** Since the lower left part of the table is a repetition of the top right section, it can be left blank.

	Shark	**Salmon**	**Trout**	**Catfish**	**Sturgeon**
Shark	8	2	2	2	2
Salmon		10	10	5	3
Trout			13	5	4
Catfish				10	2
Sturgeon					12

Step 3: Construct cladograms.

Use the data gathered on the banding patterns of the fish to construct a cladogram.

a. Draw the trunk of the cladogram and find the fish that has the least number of bands in common with the other fish (in Table 7.6, shark has only two bands in common with the other fish).

b. Draw a branch off the trunk near the bottom and label it for fish (in this case, shark) (see Figure 7.35A).

c. Identify the two fish with the most number of bands in common (in Table 7.6, trout and salmon have ten bands in common). Draw a branch toward the top of the trunk, and label the two ends of the branch for the two fish (in this case, trout and salmon) (see Figure 7.35B). It does not matter which fish is on which branch.

d. Determine the fish that has the next most in common (in Table 7.6, catfish has five bands in common with salmon and trout) and add a branch below the two fish that have the most in common (see Figure 7.35C).

e. Repeat this procedure to place the final fish (see Figure 7.35D).

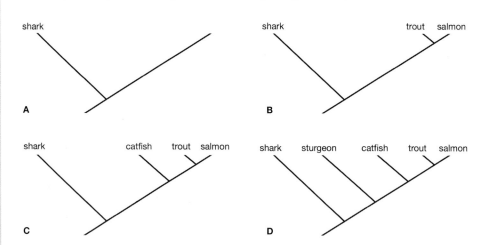

Figure 7.35. **Construction of a cladogram**.

Activity 7.D SDS-PAGE of Fish Muscle

Postlab Focus Questions

1. Which samples on your gel share the most bands in common?
2. Which samples share the fewest bands in common?
3. What new information did you learn from the cladogram that was not evident from the gel?
4. Research the evolutionary relationships among the fish species you investigated. Do your results support using protein profiles to determine the relatedness among organisms?

Self Assessment

Assess whether you have mastered the skills of this activity using the Laboratory Skills Assessment Rubric in Appendix E.

Assess your experimental write-up in your laboratory notebook using the Laboratory Notebook Rubric in Appendix F.

Activity 7.E Biofuel Enzyme Assay

Overview

The international research community is currently searching for effective cellulase enzymes to make cellulosic ethanol production economically and technically viable. The target raw materials for cellulosic ethanol production are byproducts of agriculture, such as corn leaves and stalks, or fast-growing, high-yield plant material, such as switch grass or poplar trees. The energy in these materials is bound up in cellulose and extracting that energy mechanically and chemically is inefficient and expensive. Naturally occurring cellulases — enzymes that hydrolyze cellulose — are being investigated to improve the process of cellulosic ethanol production. In this activity, the enzymatic activity of cellobiase will be investigated. Cellobiases are a subfamily of cellulases that digest cellobiose. Cellobiose is a breakdown product of cellulose enzymatic digestion and is a disaccharide composed of two glucose molecules bound by a 1,4 β-glucoside linkage. Cellobiases act at the last step in cellulose digestion to break cellobiose into two glucose molecules. The glucose can then be fermented to produce cellulosic ethanol.

In this experiment, the activity of cellobiase will be investigated. Enzyme activity is usually determined by measuring the rate of product formation (glucose) or the rate of substrate depletion (cellobiose). Although cellobiose is the natural substrate of cellobiase, there is no simple method to quantitatively detect the production of glucose or the disappearance of the cellobiose. To detect enzyme activity, p-nitrophenyl glucopyranoside is used as the substrate. p-Nitrophenyl glucopyranoside is a glucose molecule bound to a colorimetric substrate, p-nitrophenol. If cleaved, it turns yellow when placed in a basic solution (see Figure 7.36). The intensity of the yellow color is proportional to the amount of product and can be estimated visually by using a spectrophotometer at a wavelength of 410 nm.

In part 1 of the activity, the rate of product formation in the presence of enzyme will be investigated and compared to the rate of product formation in the absence of enzyme. This part of the activity provides a baseline for the subsequent parts.

In part 2 of the activity, the effect of temperature on enzyme activity is investigated. Remember that optimal enzyme activity is dependent on the temperature of the reaction (for example, many restriction enzymes have optimal performance at 37°C, while Taq DNA polymerase works best at 72°C); therefore, investigating enzyme activity involves finding the best temperature for the reaction.

In part 3 of the activity, the presence and activity of naturally occurring cellobiases in mushrooms will be investigated. Mushrooms are decomposers and should contain cellulases that help them decompose plants and wood.

Tips and Notes

- When using larger volumes of sample in larger cuvettes for some spectrophotometers, proportionally increase all the volumes in the procedure

- This activity generates a lot of data that must be analyzed. Remember to properly record data in labeled tables as they are generated. Make sure to describe the results that were analyzed, the experimental conditions, and the part of the procedure that was used to generate the data. Organizing data is a very important skill to master in laboratory research

- If a time point is missed (for example, if the 2 min time point is actually taken at 3.5 min), record the actual time the time point is taken in your laboratory notebook (3.5 min) and use that information in your analyses. Although the experiment is not conducted exactly as planned or exactly as the procedure dictates, the data are still valid and can be used to generate conclusions. On the other hand, if you write in your lab notes any time point other than the one actually taken, the data will be invalid and it will be difficult to generate conclusions. For example, if an incorrect time point is used, the rate curve for the enzyme may not be linear

- For part 3 of the activity, use only mushrooms from the grocery store. Do not collect mushrooms from the field. Many mushrooms are poisonous

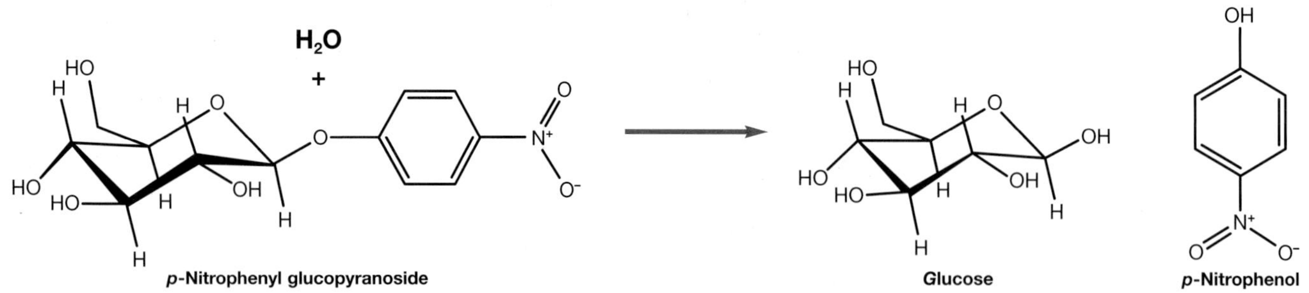

Figure 7.36. **The breakdown of p-nitrophenyl glucopyranoside by cellobiase**. The cleavage of p-nitrophenyl glucopyranoside results in glucose and p-nitrophenol as products. This reaction occurs at an extremely slow rate in the absence of cellobiases; however, it is considerably accelerated in the presence of cellobiases.

Activity 7.E Biofuel Enzyme Assay

Safety Reminders: Follow all laboratory safety procedures. Read and be familiar with all SDSs for this activity. Stop solution is a strong base and can cause burns; therefore, extra precaution should be taken when working with chemicals. All laboratory safety protocols should be followed and appropriate PPE must be worn for this activity. Since mushrooms collected from the field can be poisonous, it is recommended that **only** mushrooms obtained from the grocery store be used for this activity.

Research Questions

- What is the rate of reaction of cellobiase?
- What effect does temperature have on cellobiase function?
- Do mushrooms have cellobiase enzymes?

Objectives

- Add an enzyme to a substrate and stop the reaction at a series of time points. Analyze the amount of product formed and compare the rate of product formation in the presence and absence of enzyme
- Add an enzyme to a substrate at three different temperatures, analyze the amount of product formed, and compare the rate of product formation at different temperatures
- Extract proteins from mushrooms, test mushroom extracts for cellobiase activity, and determine the rate of product formation for the mushroom enzymes

Skills to Master

Refer to Laboratory Skills Assessment Rubric (Appendix E) and Laboratory Notebook Rubric (Appendix F) for more details.

- Record laboratory notebook entries
- Perform a serial dilution
- Use a thermometer
- Use a spectrophotometer (Activity 5.D)
- Perform time course experiments
- Organize experimental data
- Generate a standard curve using absorbance values (Activity 7.A)
- Determine experimental values using a standard curve (Activity 7.A)
- Determine the initial rate of an enzyme reaction

Student Workstation Materials

Part 1: Determining the Reaction Rate in the Presence or Absence of an Enzyme

Note: Do not discard the substrate, enzyme, or stop solution. These reagents will be used in multiple parts of this activity.

Items	Quantity
Spectrophotometer (shared)	1
Computers with graphing software (optional)	1–4
100–1,000 µl adjustable-volume micropipet with tips, or 1 ml graduated transfer pipets*	1
Beaker with water to rinse transfer pipets (optional)	1
15 ml conical tubes	2
Test tube rack	1
1.5 ml semimicro cuvettes**	12
Cuvette rack (optional)	1
Stopwatch or timer	1
Laboratory tape (optional) (shared)	1 roll
Parafilm (approx. 10 cm strip) (optional)	1
Laboratory marking pen	1
Waste container	1
Distilled water (dH$_2$O)	1.5 ml
Assay buffer	1 ml
1.5 mM substrate	10 ml
Enzyme	2.5 ml
Stop solution	12 ml
200 nmol/ml *p*-nitrophenol***	1 ml

* If 100–1,000 µl micropipets are not available, at least four transfer pipets are required.

** When using cuvettes larger than 1.5 ml, increase the volumes of samples and reagents proportionally (see Tips and Notes).

*** If your instructor has already prepared the serial dilution of standards (steps 1–3 of part 1), you will have 1 ml of each standard solution (S1–S5) at your workstation.

Activity 7.E Biofuel Enzyme Assay

Part 2: Determining the Effect of Temperature on the Reaction Rate

Items	Quantity
Spectrophotometer (shared)	1
Computers with graphing software (optional)	1–4
Water bath or dry bath at 37°C (shared)	1
100–1,000 µl adjustable-volume micropipet with tips, or 1 ml graduated transfer pipets*	1
Thermometer	1
Beakers to hold 37°C water (optional if dry bath is on bench) and water to rinse transfer pipets (optional)	2
1.5 ml microcentrifuge tubes	6
Microcentrifuge tube rack	1
Floating tube rack	1
1.5 ml semimicro cuvettes**	3
Stopwatch or timer	1
Parafilm (approx. 10 cm strip) (optional)	1
Container of ice	1
Laboratory marking pen	1
Waste container	1
Standards (S1–S5)	5 cuvettes
Enzyme	1 ml
1.5 mM substrate	6.5 ml
Stop solution	6 ml

* If 100–1,000 µl micropipets are not available, at least four transfer pipets are required.
** When using cuvettes larger than 1.5 ml, increase the volumes of samples and reagents proportionally (see Tips and Notes).

Part 3: Testing the Cellobiase Activity of Mushroom Extracts

Items	Quantity
Spectrophotometer (shared)	1
Computers with graphing software (optional)	1–4
Microcentrifuge or mini centrifuge (shared)*	1
Balance (shared)	1
Weigh boat	1
100–1,000 µl adjustable-volume micropipet with tips, or 1 ml graduated transfer pipets**	1
Beaker with water to rinse transfer pipets (optional)	1
15 ml conical tube	1
Test tube rack	1
1.5 ml microcentrifuge tube	1
1.5 ml semimicro cuvettes***	6
Mortar and pestle	1
Stopwatch or timer	1
Laboratory marking pen	1
Waste container	1
1.5 mM substrate	3.25 ml
Extraction buffer	5 ml
Stop solution	3.25 ml
Mushroom	1

* If a microcentrifuge is not available, the mushroom extract can be filtered using filter paper, cheese cloth, or a strainer.
** If 100–1,000 µl micropipets are not available, at least four transfer pipets are required.
*** When using cuvettes larger than 1.5 ml, increase the volumes of samples and reagents proportionally (see Tips and Notes).

Prelab Focus Questions

1. Why is improving biofuel production efficiency important?
2. How does an enzyme speed up a reaction?
3. What is the natural substrate of cellobiase?
4. What is the natural product of a reaction catalyzed by cellobiase?

Activity 7.E Biofuel Enzyme Assay

Part 1: Determining the Reaction Rate in the Presence or Absence of an Enzyme

Note: If your instructor already prepared the standards in advance, skip to step 4.

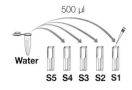

1. (Optional) Prepare a twofold serial dilution of p-nitrophenol standards. Label five cuvettes **S1–S5**. Add 500 µl of dH$_2$O to the **S1** cuvette. Add 250 µl of dH$_2$O to the **S2, S3,** and **S4** cuvettes.

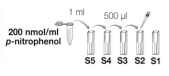

2. (Optional) Add 1 ml of 200 nmol/ml p-nitrophenol to the cuvette labeled **S5**. Add 500 µl from the **S5** cuvette to the **S4** cuvette and mix. Add 500 µl from the **S4** cuvette to the **S3** cuvette and mix. Add 500 µl from the **S3** cuvette to the **S2** cuvette and mix. Remove 500 µl from the **S2** cuvette and discard. Do not add p-nitrophenol to the **S1** cuvette.

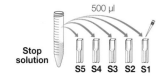

3. (Optional) Add 500 µl of stop solution to each cuvette labeled **S1–S5** and mix. These are your standards. Calculate and record the amount of p-nitrophenol in each cuvette.

    Note: Maintain these cuvettes throughout the course of the activity and do not discard them until instructed to do so. Seal the cuvettes with Parafilm and store them at 4°C upon the completion of each activity.

4. Label five cuvettes **1 min, 2 min, 4 min, 6 min,** and **8 min**.

5. Label the two remaining cuvettes **Start** and **End**.

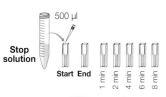

6. Pipet 500 µl of stop solution into each labeled cuvette.

7. Label one empty 15 ml conical tube **Enzyme reaction** and the other **Control**.

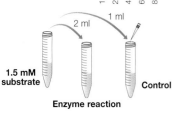

8. Pipet 2 ml of 1.5 mM substrate into the 15 ml conical tube labeled **Enzyme reaction**. Pipet 1 ml of 1.5 mM substrate into the conical tube labeled **Control**.

    Note: Read and understand the following steps fully before proceeding. These steps are time sensitive. Use a fresh pipet tip for each addition.

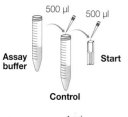

9. Pipet 500 µl of assay buffer into the 15 ml conical tube labeled **Control** and gently mix by pipetting up and down. Remove 500 µl of this mixture and add it to the cuvette labeled **Start**.

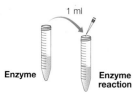

10. Pipet 1 ml of enzyme into the 15 ml conical tube labeled **Enzyme reaction** and gently mix with the pipet. **START YOUR TIMER.**

Activity 7.E Biofuel Enzyme Assay

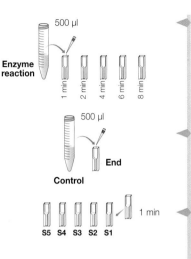

Enzyme reaction

500 µl

1 min · 2 min · 4 min · 6 min · 8 min

11. At 1 min, 2 min, 4 min, 6 min, and 8 min, remove 500 µl of solution from the **Enzyme reaction** tube, and add it to the appropriately labeled cuvette containing the stop solution.

500 µl

End

Control

12. After all the enzyme samples have been collected, remove 500 µl of solution from the **Control** reaction tube and add it to the cuvette labeled **End**.

S5 S4 S3 S2 S1   1 min

13. Visually compare the seven experimental cuvettes with the standards. Determine which standard most closely matches the color of each experimental cuvette and estimate the amount of p-nitrophenol using the information in Table 7.7.

Table 7.7. **p-Nitrophenol standards**.

Standard	Amount of p-Nitrophenol, nmol
S1	0
S2	12.5
S3	25
S4	50
S5	100

14. Set the spectrophotometer to measure absorbance at a wavelength of 410 nm.

S1

15. Insert the standard cuvette labeled **S1** into the spectrophotometer. Use **S1** (the blank) to set the spectrophotometer to zero absorbance (100% transmittance).

16. Remove the blank and replace with the cuvette labeled **S2**. Read and record the absorbance value. Repeat with samples in the cuvettes labeled **S3** through **S5** and record the absorbance values.

1 min

17. Read the absorbance of the seven experimental cuvettes and record the absorbance values.

18. Do not discard your stock solutions of the enzyme, 1.5 mM substrate, stop solution, assay buffer, and standards since they will be used for the next two parts of the activity. Rinse out the two 15 ml conical reaction tubes and cuvettes with lots of water and save them for later activities.

Part 2: Determining the Effect of Temperature on the Reaction Rate

Label up
here

1. Label three cuvettes **Ice**, **RT**, and **37°C**.

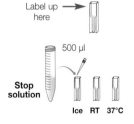

500 µl

Stop solution

Ice   RT   37°C

2. Pipet 500 µl of stop solution into each cuvette.

Activity 7.E Biofuel Enzyme Assay

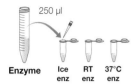

3. Label three 1.5 ml microcentrifuge tubes **Ice enz**, **RT enz**, and **37°C enz**. Pipet 250 µl of enzyme into each microcentrifuge tube.

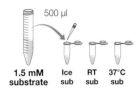

4. Label three 1.5 ml microcentrifuge tubes **Ice sub**, **RT sub**, and **37°C sub**. Pipet 500 µl of 1.5 mM substrate into each microcentrifuge tube.

5. Obtain a beaker of 37°C water, or have a 37°C dry bath on your bench.

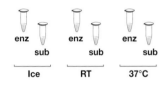

6. Place the tubes labeled **Ice enz** and **Ice sub** in the ice container. Place the tubes labeled **37°C enz** and **37°C sub** in warm water at 37°C on your bench. Leave the tubes labeled **RT enz** and **RT sub** on the laboratory bench. Allow the tubes to equilibrate to their respective temperatures for at least 5 min.

   Note: Using a thermometer, check the temperature of the room, the ice, and the 37°C water. Record these values and use them in your data analysis.

   Note: Read and understand the following steps fully before proceeding. These steps are time sensitive.

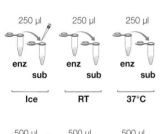

7. Have a stopwatch ready. Pipet the 250 µl of enzyme from the tube labeled **Ice enz** into the tube labeled **Ice sub**, and place the tube containing your enzyme and substrate mix back on ice. Add 250 µl of the **RT enz** to the **RT sub**, and place that tube back on the bench. Add 250 µl of the **37°C enz** to the **37°C sub**, and put that tube back into the 37°C water bath. **START YOUR TIMER.**

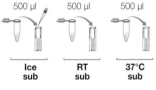

8. After 2 min, use a fresh tip for each temperature reaction to transfer 500 µl from each reaction tube to the appropriately labeled cuvette containing the stop solution. Perform this step in the same order the reactions were mixed: first **Ice**, then **RT**, and finally **37°C**.

9. Visually compare the three experimental cuvettes with the five standards. Determine which standard most closely matches the color of each experimental cuvette and estimate the amount of *p*-nitrophenol using the information in Table 7.7.

10. Set the spectrophotometer to measure absorbance at a wavelength of 410 nm.

11. Read the absorbance of the three experimental cuvettes using standard cuvette **S1** as a blank and record the absorbance values.

12. Save your stock solutions of the enzyme, 1.5 mM substrate, stop solution, and assay buffer since they will be used for the next parts of the activity. Rinse out your cuvettes with lots of water and save them for later activities.

Activity 7.E Biofuel Enzyme Assay

Part 3: Testing the Cellobiase Activity of Mushroom Extracts

1. Weigh out approximately 1 g of mushroom and put it into a mortar. Record the type of mushroom and its mass in your notebook.

Extraction buffer

2. Add 2 ml of extraction buffer for every gram of mushroom into the mortar.

   To calculate the amount of extraction buffer you need, multiply the mass (in g) of the mushroom by 2 and add that many ml of extraction buffer.

   Mass of mushroom _____ g x 2 = _____ ml

3. Using a pestle, grind your mushroom in extraction buffer to produce a slurry. Scoop the slurry into a 1.5 ml microcentrifuge tube. Label the tube with your initials.

Mushroom extract

4. Place the tube in the microcentrifuge and balance it with another tube that contains approximately the same volume. Centrifuge the tubes for 2 min (or 5 min if using a mini centrifuge) at top speed to pellet the mushroom debris.

   Note: Make sure to accommodate your classmates' tubes to ensure economic use of the microcentrifuge.

Label up → here

5. Label six cuvettes **Blank**, **1 min**, **2 min**, **4 min**, **6 min**, and **8 min**. Pipet 500 µl of stop solution into each cuvette.

6. Label a 15 ml conical tube with the name of the mushroom and add 3 ml of 1.5 mM substrate to the tube.

   Note: Read and understand the following steps fully before proceeding. These steps are time sensitive.

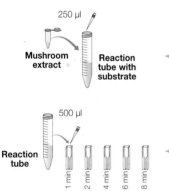

250 µl

Mushroom extract   **Reaction tube with substrate**

7. Pipet 250 µl of mushroom supernatant into the reaction tube containing the substrate. Take care to avoid the pellet of mushroom debris. **START YOUR TIMER.**

500 µl

Reaction tube

1 min   2 min   4 min   6 min   8 min

8. At 1 min, 2 min, 4 min, 6 min, and 8 min, remove 500 µl of solution from the reaction tube, and add it to the appropriately labeled cuvette containing the stop solution.

9. Add 500 µl of extraction buffer to the cuvette labeled **Blank** and add one drop of mushroom extract. The mushroom extract is added to the blank to exclude the color of the mushroom extract from the optical density reading.

10. Set the spectrophotometer to read absorbance at a wavelength of 410 nm.

Activity 7.E Biofuel Enzyme Assay

Blank

1 min

11. Insert the cuvette labeled **Blank** into the spectrophotometer. Use the blank to set the spectrophotometer to zero absorbance (100% transmittance).

12. Read the absorbance of the experimental samples and record the absorbance values.

Results Analysis

Part 1: Determine the Initial Reaction Rate in the Presence or Absence of an Enzyme

Step 1: Determine the amount of p-nitrophenol generated in each time point.
First, estimate the amount of p-nitrophenol in the reaction samples.

a. Compare the color of the reaction samples to the colors of the standards and decide to which standard each is most similar. Hold the cuvettes against a white background to analyze. Use Table 7.7 to estimate the amount of p-nitrophenol in each sample.

b. Record your estimates in your laboratory notebook using a table similar to Table 7.8.

Table 7.8. **Comparison of reaction cuvettes to standard cuvettes**.

Cuvette	Experimental Conditions	Standard Most Similar (S1–S5)	Amount of p-Nitrophenol, nmol
Start (0 min)	No enzyme		
End (8 min)	No enzyme		
1 min	Enzyme		
2 min	Enzyme		
4 min	Enzyme		
6 min	Enzyme		
8 min	Enzyme		

Next, to make an accurate determination of the amount of p-nitrophenol generated at each time point, create a standard curve by plotting the known amount of p-nitrophenol in the standards against their absorbance at 410 nm.

c. Use graph paper or graphing software such as Microsoft Excel or Google Sheets to generate a standard curve and determine the equation of the line. (See Activity 7.A for more information on generating standard curves based on absorbance and note that in Activity 7.A absorbance is plotted against the protein concentration, while in this activity absorbance is plotted against the amount of p-nitrophenol.)

d. Use the standard curve or the equation of the line to accurately determine the amount of p-nitrophenol in each of the experimental samples.

e. Record the results using a table similar to Table 7.9.

Activity 7.E Biofuel Enzyme Assay

Table 7.9. **Amount of *p*-nitrophenol calculated from standard curve**.

Cuvette	Experimental Conditions	Absorbance at 410 nm	Amount of *p*-Nitrophenol, nmol Calculated from the Standard Curve
Start (0 min)	No enzyme		
End (8 min)	No enzyme		
1 min	Enzyme		
2 min	Enzyme		
4 min	Enzyme		
6 min	Enzyme		
8 min	Enzyme		

Step 2: Determine the rate of reaction in the presence and absence of an enzyme.
Determine the initial rate of reaction in the presence and absence of enzyme and plot the amount of product (*p*-nitrophenol) against time. Note that the slope of the line in the linear region of the graph gives the initial rate of reaction. Later in the reaction, if the substrate becomes scarce as it is used up, the reaction slows down (see the 10 min time point in Figure 7.37).

a. Plot the data for the rate of product formation in the presence of enzyme (five data points) on a graph similar to the one in Figure 7.37.

b. Determine the slope of the line, which is the change in y divided by the change in x for the reaction in the presence of enzyme. The slope is measured in nmol/min. For example, in Figure 7.37:

The change in y = 100 nmol − 12.5 nmol = 87.5 nmol

The change in x = 8 min − 1 min = 7 min

The initial rate of product formation is 87.5 nmol/7 min = 12.5 nmol/min

c. Using the same method, determine the slope of the line for the reaction without enzyme (two data points).

d. Draw conclusions about the effect of an enzyme on the rate of reaction.

Activity 7.E Biofuel Enzyme Assay

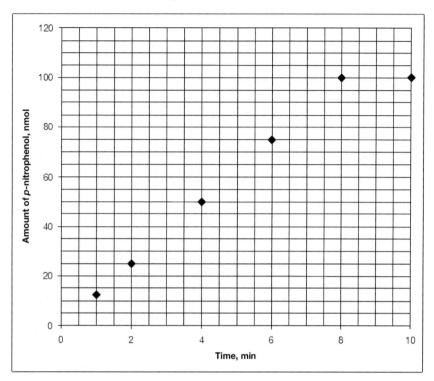

Figure 7.37. **Rate curve for an enzyme reaction**. The amount of product formed over time is plotted. The reaction rate from 1 min to 8 min is linear; however, the reaction slows between 8 min and 10 min as the substrate is depleted.

Part 2: Determine the Effect of Temperature on the Reaction Rate

Step 1: Determine the amount of p-*nitrophenol generated in each reaction.*
Analyze the raw data using the information provided in step 1 of part 1.

    a. Use the standard curve or visually inspect the samples against the standards to determine the amount of *p*-nitrophenol formed (in nmol) at each temperature.

    b. Record your data in a data table.

Step 2: Determine the initial rates of product formation at each temperature.
Calculate the initial rate of product formation for the enzyme at each temperature by following the steps below. Record the data in your laboratory notebook in a table similar to Table 7.10.

    a. Enter the amount of *p*-nitrophenol produced at the 2 min time point for each reaction temperature in the table. Since data for 0 min were not collected, assume that the amount of *p*-nitrophenol at 0 min is 0 nmol.

    b. Calculate the change in product for the reaction at 0°C by subtracting the amount of product at 0 min from the amount of product at 2 min. Enter results in a table similar to Table 7.10. Repeat for the data from the room temperature and 37°C reactions.

       Change in product (nmol) = product at 2 min (nmol) – product at 0 min (nmol)

    c. Calculate the change in time for the reaction at 0°C by subtracting the start time from the end time. Enter the results in your table. Repeat for the data from the room temperature and 37°C reactions.

       Change in time (min) = end time (min) – start time (min)

Activity 7.E Biofuel Enzyme Assay

d. Calculate the initial results of product formation for the reaction at 0°C by dividing the change in product by the change in time. Enter results in your table. Repeat for the data from the room temperature and 37°C reactions.

$$\text{Inital rate of product formation (nmol/min)} = \frac{\text{Change in product (nmol)}}{\text{Change in time (min)}}$$

For example, if the 37°C sample generated 22 nmol of p-nitrophenol after 2 min:

The change in product would be 22 nmol – 0 nmol = 22 nmol

The change in time would be 2 min – 0 min = 2 min

The initial rate of product formation would be 22 nmol/2 min = 11 nmol/min

Table 7.10. **Production of p-nitrophenol at different temperatures**.

Temperature	Product, nmol 0 min	Product, nmol 2 min	Change in Product, nmol	Change in Time, min	Initial Rate nmol/min,
0°C	0			2	
22°C*	0			2	
37°C	0			2	

* Input room temperature as measured during the experiment.

Step 3: Plot the effect of temperature on the rate of enzymatic reaction.

a. Generate a graph showing the effect of temperature on the rate of reaction by plotting the rate of p-nitrophenol produced (nmol/min) on the y-axis and reaction temperature (°C) on the x-axis.

b. Draw conclusions about the effect of temperature on the rate of reaction.

c. Predict what might happen at 60°C and 100°C. If you have an opportunity, test your predictions.

Part 3: Test the Cellobiase Activity of Mushroom Extracts

a. Determine the amount of p-nitrophenol produced for each time point.

b. Plot the data on a chart similar to the one in Figure 7.37.

c. Calculate the initial rate of reaction for the mushroom enzyme.

d. Draw conclusions and include considerations of what factors are not controlled in this experiment.

Activity 7.E Biofuel Enzyme Assay

Self Assessment

Assess whether you have mastered the skills of this activity using the Laboratory Skills Assessment Rubric in Appendix E.

Assess your experimental write-up in your laboratory notebook using the Laboratory Notebook Rubric in Appendix F.

Activity 7.F Exploring Bioinformatics with GFP

Overview

Bioinformatics is the use of information technology and computer science to perform biological research. Bioinformaticians create and use databases, algorithms, and computational models. Most laboratory scientists use tools and databases created by bioinformaticians to conduct their own research. In this activity, you will investigate GFP using several bioinformatics databases and tools that are commonly used by scientists.

GenBank is a database of DNA and protein sequences. Each sequence in GenBank was submitted by a researcher and the information about the researcher and the sequence can be found in the GenBank display. Each submission is identified by an accession number. In part 1 of this activity, the accession number for the pGLO plasmid will be used to search GenBank for the pGLO plasmid sequence. **Note:** The pGLO plasmid was originally called pBAD-GFPuv.

Once scientists obtain a DNA sequence, they can determine if there are any genes in the sequence. Software can search DNA sequences for the presence of putative genes by looking for certain patterns. For example, genes have a start codon (usually ATG) that encodes the first amino acid of the protein and a stop codon that signals the ribosome to release from the mRNA. These codons have to be in frame with each other in a pattern of triplets called an open reading frame (ORF). It would be difficult and laborious to look for ORFs manually but software does it in seconds. In part 2 of this activity, NEBcutter® will be used to generate a restriction site map of the pGLO plasmid and to identify ORFs on the plasmid. (NEBcutter was introduced in Activity 4.A.) Since not all ORFs code for actual proteins, the Basic Local Alignment Search Tool (BLAST) will be used to determine if the identified ORFs encode proteins. The putative protein sequence for each ORF will be queried against the protein database in GenBank using BLAST to find the protein sequences that most closely match the ORF. The GenBank submissions for the close matches can then be examined and used to determine what protein, if any, the ORF encodes.

The Protein Data Bank (PDB) is a database that hosts 3-D protein structures in a similar way to GenBank's hosting of DNA and protein sequences. Structures in the PDB are identified by a PDB identification number (PDB ID) that is similar to an accession number in GenBank. The PDB page for the protein also links to a molecular viewing program called Jmol that allows the protein structure to be rotated in 3-D space. The PDB ID for GFP is provided, and in part 3 of the activity; it will be used to search the PDB database for GFP and then view the GFP protein structure.

The same protein in different organisms often has a similar secondary or tertiary structure but a different primary structure (that is, a different amino acid and DNA sequence). Multiple protein or DNA sequences can be aligned to provide information on their similarity. This can provide information about protein conservation among species and about the relatedness of different species. In part 4 of the activity, an alignment program called Clustal Omega will be used to align the GFP sequences from three different jellyfish species and the GFP sequence from the pGLO plasmid (which encodes a mutated GFP, called the cycle 3 mutant). Clustal Omega will also generate a phylogenetic tree based on the protein sequences to identify proteins that are most similar to each other.

Tips and Notes

- Open multiple browser windows to keep the previous results open while moving on to the next step

- If the data are not in a printable form (for example, 3-D protein structure views of GFP using Jmol), take screen shots (screen captures) of the information generated and copy the image into a printable file such as a Microsoft Word or Google document

- The tools used in this activity are always available online; therefore, they can be explored further on your own time at home or in the library

Research Questions

- In which open reading frame on the pGLO plasmid is the *GFP* gene found?
- Does GFP have more beta pleated sheets or alpha helices?
- Which jellyfish species have the most similar GFP?

Objectives

- Find the GenBank submission for the pGLO plasmid
- Use NEBcutter to generate a restriction site map and to identify ORFs on the pGLO plasmid
- Use BLAST to identify the proteins, if any, encoded by the ORFs on the pGLO plasmid
- Find a PDB submission of GFP
- Use Jmol to view and manipulate the 3-D protein structure of GFP
- Use Clustal Omega to align GFP from different jellyfish species and create a phylogram

Activity 7.F Exploring Bioinformatics with GFP

Skills to Master

Refer to Laboratory Skills Assessment Rubric (Appendix E) and Laboratory Notebook Rubric (Appendix F) for more details.

- Record laboratory notebook entries
- Search GenBank for DNA sequences using accession numbers
- Use NEBcutter software to predict restriction sites (Activity 4.A)
- Use NEBcutter to identify ORFs
- Use BLAST to identify proteins from amino acid sequence
- Use PDB identifiers to retrieve and view 3-D protein structures in Jmol
- Create a phylogram using Clustal Omega
- Record detailed software instructions into a laboratory notebook (Activity 4.A)

Student Workstation Materials

Items	Quantity
Computers with Internet connection	1–4
Printer (shared)	1

Prelab Focus Questions

1. Which U.S. agency operates GenBank?
2. What is an accession number?
3. What are the two main types of abbreviations for amino acids? Provide both abbreviations for alanine, tryptophan, and isoleucine.
4. What is an open reading frame (ORF)?
5. What information is provided by a phylogenetic tree?

Activity 7.F Exploring Bioinformatics with GFP

> ### Protocol
>
> **Part 1: Using the pGLO Accession Number to Find the GenBank Submission**
>
> 1. Go to the National Center for Biotechnology Information (NCBI) website (ncbi.nlm.nih.gov).
>
> 2. In the search window, type **U62637**, the GenBank accession number for the pGLO plasmid, which was originally named PBAD-GFPuv.
>
> 3. Click the link **Cloning vector PBAD-GFPuv, complete sequence** to load the GenBank display of the pGLO plasmid sequence.
>
> 4. The top of the display gives the reference information for the sequence. (The original name of the pGLO plasmid was pBAD-GFPuv.)
>
>   a. Identify the authors of reference 1 of the sequence.
>
>   b. Identify the number of bases in the sequence.
>
> 5. Scroll down the page. Look at the **FEATURES** section.
>
>   a. How many coding sequences (CDS) or proteins are encoded by this sequence?
>
> 6. Scroll down the page. Look at the **ORIGIN** section.
>
>   a. What are the first five bases of the plasmid sequence?
>
> **Part 2: Using NEBcutter to Determine the Restriction Site Map and ORFs of pGLO Plasmid**
>
> 1. Go to the New England BioLabs website (neb.com) and locate NEBcutter under the NEB tools **Tools of Resources** dropdown menu.
>
> 2. If the GenBank sequence window has been closed or this is a new lesson period, repeat steps 1–3 of part 1 to open up a browser window containing the GenBank submission page for the pGLO plasmid.
>
> 3. Scroll back to the top of the GenBank display page. Under the title **Cloning Vector, PBADGFPuv, complete sequence**, select the link **FASTA**. FASTA format is a way to write DNA or protein sequences that is recognized by many biological database search tools (see Figure 7.38). FASTA format starts with a > sign, followed by a file name; the sequence starts on the next line.
>
>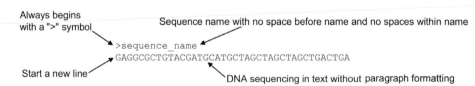
>
> Figure 7.38. **FASTA format**.
>
> 4. Select and copy the information from the > sign to the end of the sequence, including the file name of the sequence.

Activity 7.F Exploring Bioinformatics with GFP

5. Paste the sequence into the large box in the NEBcutter window. Click the **Circular** radio button at **The sequence is** section, leave the other settings on their defaults, and click **Submit**.

6. Once submitted, the window will refresh and show an image of the plasmid with restriction sites and open reading frames (ORFs) identified. Click on one of the larger ORFs.

7. A new page with the amino acid sequence will load. To identify the protein (if any) that the ORF encodes, BLAST uses the amino acid sequence to search a database containing millions of protein sequences. BLAST will look for sequences that match the ORF (query sequence). Click the **BLAST this sequence at NCBI** button.

8. The BLAST web page at NCBI automatically opens with the different sections for data entry already filled in. For **Enter Query Sequence**, the sequence in the box should match the protein sequence of the ORF in the NEBcutter window; **Non-redundant protein sequences (nr)** should be selected in **Choose Search Set**; and **Program Selection** should have **blastp (protein-protein BLAST)** selected.

9. Click the **BLAST** button at the bottom of the page. BLAST results can take awhile to load and the page may refresh multiple times.

10. Once the results are loaded, scroll down the page to the table entitled **Sequences producing significant alignments**, under the multicolored chart. The sequences at the top of the table most closely match the query sequence. Click the accession number for the top result, which will open the GenBank submission for the sequence.

11. Record the accession number and the definition of the sequence in a table. (Do not print BLAST results since they can run for hundreds of pages.) Refer to the pGLO plasmid map (Figure 5.20) and identify which gene (if any) this ORF encodes. You may need to check multiple accession numbers if the protein name is not obvious; remember that some ORFs will not encode proteins.

12. Go back to the NEBcutter circular sequence display. Use BLAST to identify the remaining ORFs. Record the data.

13. From the NEBcutter circular sequence display, identify two restriction enzymes that could cut the GFP sequence from the plasmid.

Part 3: Using Jmol to View 3-D Protein Structure of GFP

1. Go to the RCSB Protein Data Bank website (rcsb.org). This website hosts a database of 3D protein structures.

2. Enter **3i19**, the PDB ID for GFP, in the search field at the top of the page. Click **Search**.

3. Locate the image of the protein. Under the image, click the **Structure** link or the **3D View** tab.

   Note: A new 3D version of the protein structure will load that can be rotated.

4. Hold the cursor over the image, and click and hold. Then move the cursor to rotate the structure. Observe the protein from all sides.

   a. What type of secondary structure makes up the structure that looks like a barrel?

   b. Are there alpha helices present? Where?

Activity 7.F Exploring Bioinformatics with GFP

5. The options in the box to the right of the 3D structure in the **Structure View** tab allow it to be viewed in different ways. What other information can you learn about GFP by adjusting these views?

Part 4: Using Clustal Omega to Compare GFP from Different Jellyfish Species

GFP was originally isolated from a jellyfish species called *Aequorea victoria*. It has since been found in other jellyfish species. A sequence alignment tool called Clustal Omega will be used to compare the differences between the GFP protein from various jellyfish species and GFPuv protein (the mutated GFP from the pGLO plasmid).

1. Go to the NCBI website (ncbi.nlm.nih.gov).

2. Open a blank text document with a word processing program such as Microsoft Word.

    Note: These are protein sequences and use a single letter amino acid code (see Figure 7.7).

3. Find the following GFP protein sequences by searching with the accession numbers:

    AAC53663: GFPuv (cloning vector pBAD-GFPuv is also known as pGLO)
    AAA27722: green fluorescent protein (*Aequorea victoria*)
    AAN41637: green fluorescent protein (*Aequorea coerulescens*)
    AAK02062: green fluorescent protein (*Aequorea macrodactyla*)

4. For each sequence, select the FASTA sequence (by following part 2, steps 3 and 4). Copy and paste the sequence with the > sign and sequence name into a text file. Paste the four FASTA sequences sequentially after one another. Start each FASTA sequence on a new line. Save the text file.

    Note: Ensure that your word processing program does not insert paragraph marks (¶) or spaces at the end of each line of the FASTA formatted text. If it does, delete them so there are no interruptions in the sequence. Clustal Omega will not process sequences with interruptions.

5. Open the Clustal Omega website (ebi.ac.uk/Tools/msa/clustalo/).

6. Select and copy all the FASTA sequences from the text file and paste them into the Clustal Omega window under **Sequences in any format**.

7. Select output format as **Clustal with numbers**. Leave other settings as default and click **Submit**.

Table 7.11. **Color assignments of amino acids in Clustal Omega.**

Color	Amino Acids	Amino Acid Properties
Red	AVFPMILW	Small (small and hydrophobic)
Blue	DE	Acidic
Magenta	RK	Basic
Green	STYHCNGQ	Hydroxyl, sulfhydryl, amine, and G
Gray	Others	Unusual amino/imino acids

Activity 7.F Exploring Bioinformatics with GFP

8. View the aligned sequences. Use the accession numbers to identify the sequences.

    a. What symbols are used below the alignment when all the amino acids match?

    b. What symbols are used when the amino acids do not match?

    c. Using the **Show Colors** option (see Table 7.11 for an explanation of the colors) and looking at the structures of the amino acids in Figure 7.7, can you deduce what the different symbols mean when the amino acids do not match?

    d. The GFP from *A. victoria* was mutated to enhance the fluorescence of the protein. The mutated form of the *GFP* gene was cloned into the pGLO plasmid (pBAD-GFPuv). Identify the amino acids that were changed in the pGLO GFP sequence.

9. Click **Send to Simple Phylogeny**.

10. Go straight to **STEP 2 — Set your Phylogeny options** and in the **CLUSTERING METHOD** dropdown menu, select **UPGMA**. Leave all other settings as default. Click **Submit**.

11. Look at the phylogram and toggle between the **Cladogram** and **Real** radio buttons.

    a. Which GFP proteins are more alike?

    b. Which GFP proteins are most different?

    c. What can you conclude from the phylogram?

Results Analysis

Record each step of the investigation in a step-by-step manner so that the laboratory notebook serves as a reference for using the tools and databases in the future. Print or summarize the results of each part and paste printouts into your laboratory notebook, including accession numbers where appropriate. Draw conclusions based on the data that have been generated. The questions in the protocol should assist with conclusions.

Postlab Focus Questions
1. In which ORFs on the pGLO plasmid are the GFP and beta-lactamase genes found?
2. Does GFP have more beta pleated sheets or alpha helices?
3. Which jellyfish species has GFP most similar to GFP from *Aequorea victoria*? Can you make conclusions on the relatedness of these jellyfish species based on this result?

Self Assessment

Assess whether you have mastered the skills of this activity using the Laboratory Skills Assessment Rubric in Appendix E.

Assess your experimental write-up in your laboratory notebook using the Laboratory Notebook Rubric in Appendix F.

Chapter 7 Extension **Activities**

1. Compare the protein concentration of different sources of milk or other protein-rich solutions, such as tears, protein shakes, and cooking broth.
2. Use a combination of chromatography methods to sequentially purify a mixture of proteins and confirm the results using SDS-PAGE.
3. Separate the proteins in the GFP purification into additional fractions by using a graduated series of wash buffers from 1.3 M to 0 M of ammonium sulfate. Run each fraction on an SDS-PAGE gel to determine the level of purity of the GFP.
4. Test the effects of temperature and pH on cellobiase activity in mushroom extracts.
5. Investigate the conservation of a protein of interest among organisms and create phylogenetic trees using NCBI databases, BLAST, and Clustal Omega.

Immunological Applications

Summary

Immunology is the study of the immune system, the body's system of defense against disease. Antibodies produced by the immune system are the workhorses of the immune response. These proteins are made in response to specific pathogens: each antibody recognizes only a single, small area on the surface of a single molecule. This specificity makes antibodies useful as tools in biotechnology.

There are many applications for antibodies in the research lab, in the hospital, at the veterinary clinic, in the field — just about anywhere assays are performed. Over-the-counter test kits, such as pregnancy tests, use antibodies, as do laboratory techniques such as fluorescence microscopy. Antibodies are also used as drugs to treat cancer and autoimmune diseases or to prevent organ rejection after a transplant. The number of applications involving antibodies will only increase as new technologies are developed. This chapter provides background information about immunology and various immunoassays, such as ELISA and western blotting. The activities in this chapter introduce you to skills necessary to use antibodies in laboratory assays.

8.1 The Immune System

The body protects itself from infection using a variety of strategies, including physical and chemical barriers, antibodies that circulate in the blood, and immune cells that attack foreign substances and invading microorganisms. **Immunity** is the resistance of an organism to infection or disease, and an immune response is triggered by the presence of something foreign to the body. There are two types of immunity: innate immunity and acquired immunity.

Innate immunity refers to the natural immunological defenses that protect against pathogens and are present at birth. These defenses include immune system components such as circulating macrophages and natural killer cells that do not change with exposure to pathogens and do not have much specificity for particular pathogens.

Acquired, or adaptive, immunity is a response to a specific foreign substance. Some immune cells can adapt so that they are able to

"remember" and recognize specific invaders. This helps prepare the body for future attacks. Though individuals (except those who are immune-compromised) are born with the ability to respond to these substances, the system must be activated by an initial contact with the invader. The initial contact, or **immunization**, begins a cascade of events that allows the body to mount a specific response upon subsequent exposure to the invader, hence the term acquired immunity.

Factors capable of eliciting an immune response are called **antigens**. Antigens can be microorganisms (such as viruses or bacteria), microbial products (such as toxins produced by some bacteria or the protein components of microbes), foreign proteins, DNA and RNA molecules, drugs, and other chemicals. Antigens prompt antibody production from white blood cells called B cells. Each antibody recognizes and binds to a single shape on an antigen called an **epitope**, in essence flagging it, which helps the immune cells recognize and attack foreign invaders. Everyone (except those who are immune-compromised) has circulating antibodies and immune cells that collectively recognize a huge number of antigenic substances.

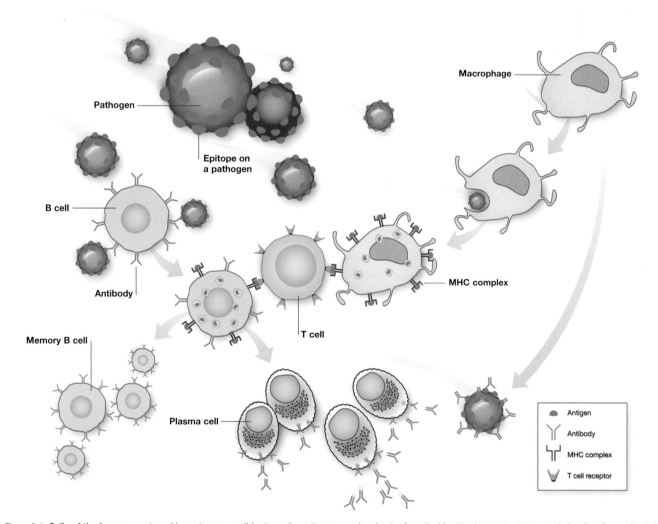

Figure 8.1. **Cells of the immune system.** Macrophages engulf foreign cells, pathogens, and molecules from the blood and present antigens on their cell surfaces with their major histocompatibility complexes (MHC). The MHCs are recognized by T cells, which attract more immune cells to the site of infection. Like macrophages, B cells present antigens on their surface to attract T cells. T cells kill whole cells that are infected by a virus to prevent the spread of infection. B cells also proliferate in response to T cells, forming both memory B cells that are part of the secondary immune response and plasma cells that secrete antibodies. Secreted antibodies bind to pathogens like flags, making it easier for other immune cells to find and destroy them.

Cells of the Immune System

Macrophages, B cells, and T cells are the soldiers of the acquired immune response (see Figure 8.1) and each cell type has a different function. Macrophages serve two primary functions: they remove foreign cells and molecules from the blood by phagocytosis (innate immunity) and they process antigens and present them on their cell surfaces (acquired immunity). Macrophages present antigenic epitopes on their cell surfaces to be recognized by T cells using the major histocompatibility complex (MHC). The T cells draw more immune cells to the site of infection, causing inflammation. Both B cells and T cells are lymphocytes (white blood cells) and each recognizes a single specific epitope. (B cells mature in the bone marrow and T cells mature in the thymus, which helps with remembering their names.) B cells produce antibodies. The number of different circulating antibodies in the human body has been estimated to be between 10^6 and 10^{11} so there is usually an antibody ready for any antigen. The huge number and diversity of antibodies are possible because B cells have the ability to rearrange their DNA to make different antibody genes. Like macrophages, B cells present antigenic epitopes on their surface to attract T cells. T cells have two main functions: they stimulate the proliferation of B cells that have bound to an antigen and they kill whole cells that are infected by a virus to prevent the virus from infecting other cells.

Antibodies

Antibodies are extremely specific, meaning that they recognize and bind to only one epitope or antigen. This makes them very useful as tools in science and medicine.

Antibodies are proteins that are also called **immunoglobulins (Ig)**. They are produced by B cells and can either remain attached to the B cells or become free floating in the bloodstream and other body fluids. There are five types of antibodies in mammals (IgA, IgD, IgE, IgG, and IgM) and each is unique in its structure and function:

- IgA is a first line of defense against invading microorganisms; it is found in external secretions such as tears, saliva, milk, and in mucosal secretions of the respiratory, genital, and intestinal tracts
- IgD initiates B cell activation and activates other immune cells to produce antimicrobials in the respiratory immune system
- IgE is a primary component in allergic reactions; it binds allergens and stimulates histamine release
- IgM is found in serum and is responsible for the primary immune response
- IgG (see Figure 8.2) is the most abundant immunoglobulin in internal body fluids, constituting about 15% of total serum protein in adults. It provides most of the immune response against invading pathogens

IgG is the antibody type that has been most utilized in biotechnology. It comprises two heavy and two light chains, each with beta sheet structure, which are held together by disulfide bonds to make a distinctive Y shape (see Figures 8.2 and 8.3). All

IgG molecules share an overall similarity except in their antigen binding sites, which are at each end of the heavy and light chains.

The two antigen binding sites on an antibody are identical and unique to the antibody so they allow each antibody to bind two antigens simultaneously. The antigen binding region (called the Fab region for fragment, antigen binding) is different for each antibody. It is the variability at this region that allows for the generation of millions of different antibodies. The remainder of the molecule (in black and grey in Figure 8.2) is called the constant region (or Fc for fragment, constant) and is the same for all antibodies within an isotype.

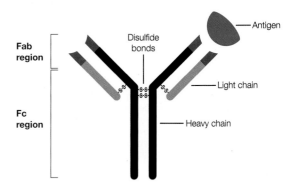

Figure 8.2. **Antibody structure**. Schematic of IgG antibody structure showing the two heavy and light chains. The red Fab region is variable among antibodies and contains the antigen binding site. The grey and black constant FC regions are the same in each IgG antibody.

An epitope is a part of an antigen that is recognized by the immune system. Each antibody recognizes a single epitope, and multiple antibodies may recognize and bind to different epitopes on a single antigen. For example, a HIV virus particle (virion) has many potential epitopes that may be recognized by many different antibodies. One antibody may recognize the amino terminus of p24, a HIV protein, and another may recognize the carboxy terminus of p24.

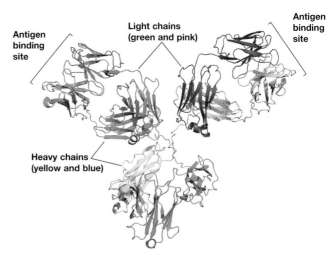

Figure 8.3. **Structure of an IgG antibody**. The structure was determined by X-ray crystallography. The two heavy chains are colored blue and yellow, and the light chains are green and pink. Protein structure derived by Harris L. et al (1998), J Mol Biol 275, 861–872, PDB ID #1IGY. Image courtesy of Matthew Betts.

Immune Response

When exposed to an antigen, either accidentally (for example, by natural exposure) or intentionally (for example, by vaccination), an individual mounts a primary immune response. Within 1–2 weeks, there is a rise in antibody production directed against the antigen (this is called seroconversion), which is dominated by the IgM class of antibodies. IgM production is usually followed by production of IgG. After that, antibody levels decrease (see Figure 8.4).

The second time the individual is exposed to the antigen, whether weeks or years after immunization, the immune response is more pronounced and much more rapid due to the immunity that has been acquired by the individual. In this secondary response, IgM is produced in detectable amounts in a matter of days, followed by a large production of IgG (see Figure 8.4). IgG is generated in much greater quantities and persists in the blood for a much longer time than in the primary response. Other classes of immunoglobulin may also be produced and antibody production may continue for months or even years.

Acquired immunity is the basis for the series of vaccinations that we undergo as we grow up. In the 1790s, long before there was an understanding of how the immune system functioned, it was discovered that inoculation with pus from a cowpox lesion prevented infection with smallpox, a disease related to cowpox. Children in the U.S. are now routinely vaccinated against diseases such as measles, mumps, rubella (German measles), diphtheria, tetanus, pertussis (whooping cough), polio, *Haemophilus influenzae* type b, hepatitis B, rotavirus, varicella (chicken pox), hepatitis A, and pneumococcal disease.

8.2 Antibodies as Tools

Scientists have taken advantage of the high specificity that antibodies show for their epitopes, making antibodies essential tools in many areas of science and medicine. Their utility depends on the ability of researchers to generate antibodies that are specific for molecules of interest. Antibodies can be manufactured in the laboratory, both in vivo and in vitro. In fact, in vivo techniques have been in use for more than 100 years. Though most antibodies are produced by traditional methods using animals or animal-derived cells, techniques for making antibodies using recombinant DNA technology are becoming more common. There are two major types of produced antibodies: polyclonal antibodies and monoclonal antibodies.

Polyclonal Antibodies

Polyclonal antibodies are generated by immunizing an animal by injecting it with an antigen (see Figure 8.5). After the animal has generated an immune response, some of its blood is collected. The solid components of the blood, such as cells, are then removed, leaving the liquid fraction, called serum, which contains antibodies. Ordinarily, mammals such as rabbits, goats, or donkeys are used, though antibodies are also produced in chickens.

The immunization schedule depends on the immune response (see Figure 8.4) but usually has an initial immunization followed by one or more booster immunizations to maximize the levels of antibodies against the molecule of interest prior to harvest. Since the injected antigen has many different epitopes that provoke immune responses, the isolated serum will contain a mixture of antibodies that recognize the different epitopes on the antigen. The antibodies are called polyclonal because they are from many (poly) B cell clones (clonal) in the animal's blood.

Rate of Seroconversion

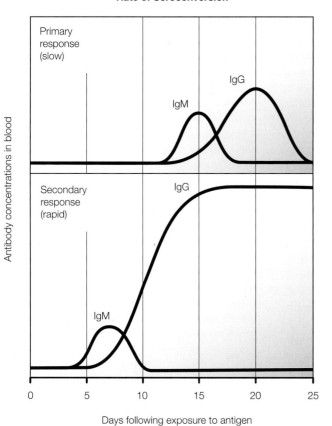

Figure 8.4. **Seroconversion after immunization**. The quantity of circulating antibody in the blood, called the antibody titer, can be measured. After initial exposure to the antigen, the titer of IgM in the blood rises, followed by the IgG titer. After another exposure, even months or years after the first exposure, the IgG titer rises rapidly.

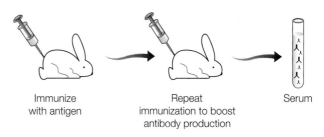

Figure 8.5. **Production of polyclonal antibodies**.

Using Immunoassays
to Save the Panda

The habitat of giant pandas once ranged across most of China and into the neighboring countries of Myanmar and Northern Vietnam. Today, however, the majority of giant pandas live in isolated or fragmented groups nestled high in the mountains of four provinces in China. Destruction of their habitat through farming, deforestation, and urban development along with climate change and poaching have all contributed to this decline. As a conservation measure in 1984, the giant panda was listed as an endangered species under the United States Endangered Species Act.

One way conservationists attempt to save species from extinction is through breeding programs. They study the reproductive cycles of an animal like the giant panda and intervene to promote successful mating, pregnancy, birth, and survival of offspring. With pandas, this is a race against the biological clock, because female pandas are fertile for only 1–3 days a year. Zoos and other conservation centers increase the chances of successful panda pregnancies and births by (1) ensuring that male and female pandas have access to one another during critical breeding days, and (2) using artificial insemination to increase the chance of fertilization. A key to success is knowing when a female panda is ready to conceive.

Dozens of hormones and enzymes are required to support ovulation in female mammals such as the giant panda, and cyclical changes in the levels of these hormones indicate when the animals are ovulating and ready to conceive. Since the female giant panda experiences these hormone changes only once per year, determining the precise timing is critical for reproductive success, especially in captivity and when using artificial insemination. Researchers use enzyme-linked immunosorbent assays (ELISAs) to determine the presence and amounts of hormones in panda urine samples. ELISAs are also used to confirm successful fertilization (pregnancy).

For the giant panda, conservation efforts have been successful enough that in 2016 the International Union for Conservation of Nature (IUCN) changed the panda's status from endangered to vulnerable. This reclassification is a sign of success in preserving the panda species, but it does not mean that conservation efforts are complete. The panda may still return to endangered status due to threats to its habitat and reproductive success. Learning more about the reproductive cycle of pandas will be critical to conserving the species.

For example, to prepare antibodies that bind **myosin**, a motor protein found in muscle cells, purified myosin from a chicken is injected into a rabbit. Two weeks later, the injection is repeated and a week after that the antibody level is checked. If the level of antibodies is adequate, blood containing the antibodies is drawn from the rabbit and the cells of the blood are removed, leaving the serum (which is called antiserum for chicken myosin since it contains antibodies against chicken myosin). In common terminology, this antiserum would be called rabbit anti-chicken myosin antiserum, and it can be used directly or the antibodies could be purified from it. Polyclonal antiserum has the advantage of being simple and inexpensive to produce; the disadvantage is that no two batches, even those made in the same animal, are exactly the same.

Monoclonal Antibodies

Polyclonal antibodies are too variable for many antibody applications, such as diagnostic testing. In these cases, a single antibody type from a single B cell is preferable. B cells producing single antibodies can be isolated from the spleens of immunized animals and grown in cell culture. But as discussed in Chapter 3, these primary cells die after a few weeks in the laboratory, which limits production of the antibody to quantities lower than those generally needed for research and commercial applications. Cells grown from a single cell are called clones and antibodies from B cell clones are called **monoclonal antibodies**.

In the 1970s, a technique was developed for producing large amounts monoclonal antibodies, earning a Nobel Prize for its inventors, Georges Köhler and César Milstein. Using the technique, B cells from immunized mice are fused with myeloma (immortal tumor) cells (see Figure 8.6). These hybrid cells, called hybridomas, grow indefinitely in cell culture and so overcome the limitations of primary B cell cultures. Each hybridoma cell line is monoclonal and produces antibodies against a single antigenic epitope. The monoclonal antibodies generated by hybridomas can be collected and purified from the cell culture growth medium with almost no batch-to-batch variability. The ability to produce monoclonal antibodies that are consistent from batch to batch is crucial for diagnostic and pharmacological uses.

Humanized Monoclonal Antibodies for Medical Use

The specificity of antibodies for their epitopes makes them ideal candidates for medical therapies. For example, an antibody that recognizes a tumor antigen can be attached to a chemotherapeutic or radioactive drug and used to deliver the drug specifically to targeted tumor cells, sparing the patient many of the side effects of conventional chemotherapy or radiation treatment (see Figure 8.7).

Traditional monoclonal antibodies made in mice are problematic for medical use because they may elicit an immune response in the human body that results in their destruction. Nevertheless, several mouse monoclonal antibodies are currently approved for human clinical use. For example, Muromonab-CD3 (trade name Orthoclone

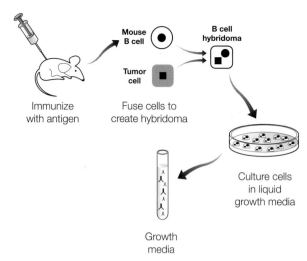

Figure 8.6. **Production of monoclonal antibodies**.

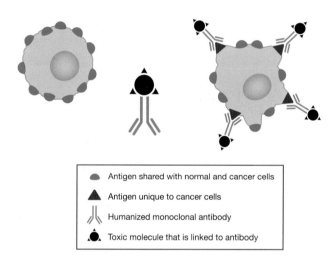

Figure 8.7. **Toxic antibodies as drugs**. Antibodies that recognize epitopes on cancer cells can be linked to toxic or radioactive molecules, enabling targeted treatment of only the cells of interest.

OKT3) is used to prevent rejection after organ transplants. The antibody binds T cells and causes them to undergo apoptosis (cell death), thereby preventing T cell proliferation and transplant rejection. The drug is for short-term use and lasts less than 1 day in the body before it is degraded by the immune system. Another example is Rituxumab (trade name Rituxan), which is used to treat non-Hodgkin's lymphoma and rheumatoid arthritis among other uses. Rituxumab is a chimeric monoclonal antibody against CD20. It is found primarily on B cells and triggers their death. Similar to human IgG, it is present for several months after administration.

Recombinant DNA technology has yielded several new antibody production methods that avoid the problems encountered with mouse antibodies. Fusion proteins combine genes from different organisms to produce a protein product that contains sequences from both species. A **humanized antibody** is a human antibody that has been engineered to contain some of the antigen binding

region (Fab) of a mouse monoclonal antibody. While the mouse protein sequences provide the specificity for the antibody, the human portion is familiar to the body and does not evoke an immune response. Trastuzumab (trade name Herceptin) (see Figure 8.8) is an example of a humanized monoclonal antibody that is used to treat cancer (see Chapter 1, page 7).

Mice have also been genetically engineered to produce human monoclonal antibodies. The XenoMouse is a transgenic system in which the genes for human antibodies have been cloned into a line of mice so that when the mice are immunized, they produce human antibodies rather than mouse antibodies from their B cells. The XenoMouse has been used to produce Panitumumab (trade name Vectibix), a human monoclonal antibody used to treat colorectal cancer.

Another method that is used to generate highly specific human recombinant monoclonal antibodies is Human Combinatorial Antibody Libraries (HuCAL). HuCAL is a compilation of more than 45 billion synthetically designed functional human antibody genes cloned into a phage library. An in vitro selection technology can be used to select genes from the library that encode for antibodies that bind to almost any antigen.

When used as medicine, antibodies are treated the same as other drugs; they are subjected to the same rigors of U.S. Food and Drug Administration (FDA) testing and clinical trials as other medicines (see Chapter 1).

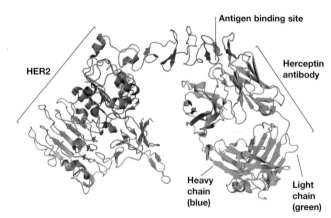

Figure 8.8. **Herceptin binding to HER2**. One Fab region of the Herceptin antibody is shown (compare to Figure 8.3). The green structure is the Fab light chain and the blue structure is the Fab heavy chain. The pink protein is the antigen HER2, a membrane protein. Note the antigen binding site of the antibody interacting with the HER2 antigen. The antibody binds a region of HER2 that is adjacent to the plasma membrane. Binding in this region blocks HER2 signaling, stopping the uncontrolled cell proliferation mediated by HER2. Protein structure derived by Cho H. et al (2003), Nature 421, 756–760, PDB ID #1N8Z. Image courtesy of Matthew Betts.

8.3 Immunoassays

Immunoassays are tests that exploit the specificity with which antibodies bind molecules to detect or measure those molecules in biological samples.

Over-the-counter pregnancy tests are immunoassays that use antibodies to detect the presence of human chorionic gonadotropin (hCG), a hormone that appears in the blood and urine soon after a woman becomes pregnant (see Figure 8.9). These are qualitative immunoassays, determining the presence or absence of the hormone but not the amount of hormone.

Other immunoassays are quantitative and can be used to determine the amount of a particular substance. For example, the biotechnology company Agdia, Inc. sells kits for testing GM crops (see Figure 8.10), including a test that quantifies the amount of

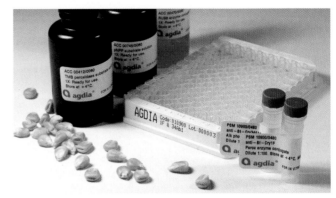

Figure 8.10. **Agdia immunoassay kit**. Many kits on the market can be used to detect and/or quantify transgenic proteins in agricultural products. Photo courtesy of Agdia, Inc.

transgenic Cry1 protein in Bt corn. Controls containing known quantities of the transgenic protein are assayed at the same time as the experimental sample. By comparing the results of the controls and experimental samples, the amount of protein in the experimental samples can be determined.

Immunoassays are available in many different formats and the common components among the assays are antibodies. Different types of immunoassays are discussed later in this chapter and include:

- Dipstick tests, such as the pregnancy test in Figure 8.9 are the simplest to use

- Enzyme immunoassays (EIA) include enzyme-linked immunosorbent assays (ELISAs) and other similar tests that are performed in microplates (see Figure 8.11) or tubes

- Ouchterlony or double diffusion assays that are easy to perform and use agarose gels

- Western blots that are performed on samples that have been separated by SDS-PAGE and blotted onto membranes

- Immunohistological or immunocytological assays that are performed on tissue or cell samples on microscope slides

- High-throughput assays that include multiplex bead assays, which utilize microplates and microscopic beads, and fluorescence-activated cell sorting that is performed on whole cells

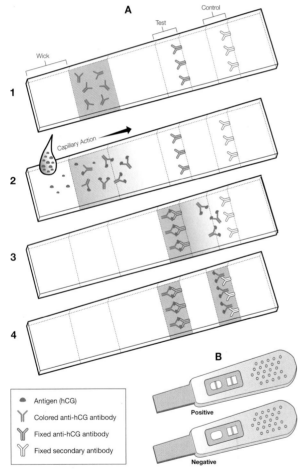

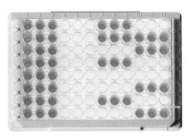

Figure 8.11. **Microplate used for ELISA**. Microplates, also called microtiter plates or enzyme immunoassay (EIA) plates, are made from polystyrene and can be used to assay up to 96 samples.

Figure 8.9. **Pregnancy test**. Over-the-counter pregnancy test kits are immunoassays that detect human chorionic gonadotropin (hCG). **A**, graphic of pregnancy test. The wick area of the dipstick is next to an area coated with anti-hCG antibody labeled with a pink compound (step 1). The strip is dipped in urine. If hCG is present, it will bind to the pink antibody and the pink hCG-antibody complex will migrate up the strip by capillary action (step 2). When the pink complex reaches the first test zone, a narrow strip containing an unlabeled, fixed anti-hCG antibody, the complex will bind and concentrate there, making a pink stripe (step 3). The dipsticks have a built-in control zone containing an unlabeled fixed secondary antibody that binds any pink antibodies (present in both positive and negative results) in the second stripe (step 4). Thus, every valid test will give a second pink stripe, but only a positive pregnancy test will give two pink stripes. **B**, test strip with positive and negative results.

Select an Antibody

Table 8.1. **Example of antibody product information for two different antibodies.** Note that there is usually more detailed background information on downloadable product data sheets available for antibodies.

Example of primary antibody information in a product catalog	α-Actin (1A4) Epitope: N-terminus Applications: WB, IF, E, C, IHC-F Target species: human Cross Reactivity: pig, rabbit, rat, mouse, sheep Isotype: mouse IgG2a	β-Actin (ACTB) Epitope: N-terminus Applications: WB, IHC-F Target species: human Cross Reactivity: rat, mouse Isotype: rabbit IgG
Interpretation	Mouse monoclonal antibody called 1A4 that recognizes an epitope on the N-terminus of α-actin in humans with cross-reactivity in pigs, rabbits, rats, mice, and sheep, and has been shown to work for western blot, immunofluoresence, ELISA, and immunohistochemistry in paraffin and frozen samples.	Rabbit polyclonal antibody called ACTB that recognizes β-actin in humans with cross reactivity in mice and rats, and has been shown to work for western blot and immunohistochemistry of frozen samples.
Type of secondary antibody required	Anti-mouse secondary	Anti-rabbit secondary

Immunoassays require antibodies, and a number of factors need to be considered when choosing an antibody for a particular application.

1. Choose a primary antibody. The primary antibody must bind to the antigen of interest, and the selection of antibody will depend on the application. Antibodies prepared against native proteins rarely recognize the denatured form of the proteins and vice versa. In experiments in which the target molecules are examined in cells or tissue samples, such as fluorescence microscopy, the proteins are in native conformation and the primary antibody must be prepared against the native protein. In western blotting, however, proteins are normally reduced and denatured, so the primary antibody should be prepared against denatured protein. Information provided with the antibody will usually state its suitability for a particular application (see Table 8.1).

2. Decide whether to use a monoclonal or polyclonal primary antibody. Monoclonal antibodies are usually more expensive than polyclonals but provide a higher level of specificity because they bind to single epitopes on the molecule of interest. Polyclonal antibodies, on the other hand, can bind to many different epitopes on the target, which can make the assay more sensitive but may also increase nonspecific background.

3. Choose a secondary antibody. The secondary antibody must recognize the primary antibody. First, determine the species in which the primary antibody was made. The secondary antibody must recognize antibodies from that species (for example, an anti-mouse secondary antibody for a mouse monoclonal primary antibody). Then select the species used to generate the secondary antibody. Companies offer secondary antibodies made in many animals, including rabbits, goats, and donkeys. The species matters only if the samples will be derived from any of these animals. In that case, choose a secondary antibody made in a different species.

4. Choose a label for the secondary antibody. The secondary antibody must be labeled so it can be detected and the label used will depend on the type of assay to be run. For example, for an ELISA or western blot, the label might be an enzyme, such as horseradish peroxidase, that oxidizes a colorimetric substrate. For electron microscopy, the label might be tiny gold particles that are electron dense. For immunofluorescence microscopy or FACS, the label might be a fluorescent compound, such as fluorescein or rhodamine.

Finally, be aware that even with careful selection, all antibodies must be empirically tested and may not be suitable for a particular assay. It is likely that a few different antibodies will need to be tested to find one that binds appropriately. In addition, for all applications the assay must be optimized to find the best assay conditions for the antibody.

Labeling and Detecting Antibodies

To be useful in immunoassays, antibodies must be made visible, for example, by covalently linking (conjugating) the antibodies to a chemical label. Examples of labels include:

- Fluorescent dyes that allow localization of antigens in cells or tissues using fluorescence microscopy

- Enzymes that oxidize chromogenic (color-producing) substrates, which produces visible color only where the enzyme-linked antibodies have bound

- Enzymes that cleave light-emitting substrates (chemiluminescent substrates) that can be detected using photographic film or digital imaging systems

An antibody that recognizes an antigen of interest is called a **primary antibody**, and primary antibodies can be detected directly by labeling them as described above. Labeling antibodies, however, is both time-consuming and expensive. Considering the sheer number of primary antibodies in use, this strategy is usually impractical.

Luckily, the properties of antibodies themselves solve this problem. It is possible to use an additional antibody (a **secondary antibody**) that recognizes all primary antibodies of one type (like IgG) derived from one species. The secondary antibody binds to the primary antibody, which in turn binds to the antigen. This is called indirect detection (see Figure 8.12).

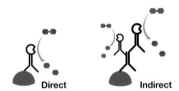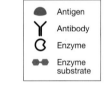

⬤	Antigen
Y	Antibody
Ϭ	Enzyme
⬤—⬤	Enzyme substrate

Direct   **Indirect**

Figure 8.12. **Direct and indirect detection using primary and secondary antibodies**. In direct detection (left), the primary antibody is labeled with an enzyme that cleaves or oxidizes a substrate. In indirect detection (right), a secondary antibody is labeled with an enzyme and multiple molecules of secondary antibody bind to each primary antibody. This results in more enzyme binding, which increases the amount of substrate cleaved or oxidized and amplifies the signal.

Secondary antibodies are prepared by injecting antibodies made in one species into another species. Antibodies from different species are different enough from each other that they are recognized as foreign proteins and provoke an immune response. For example, to make a secondary antibody that recognizes a mouse primary antibody, mouse antibodies are injected into another animal, for example, a goat. After the goat mounts an immune response, the goat serum contains antibodies that recognize and bind to mouse antibodies. This antiserum is called goat anti-mouse antiserum and can be used to detect antibodies generated in mice.

Secondary antibodies can be labeled using the strategies described above and are used to detect the primary antibodies in an immunoassay. Secondary antibodies also amplify the signal from primary antibodies. Because secondary antibodies are usually polyclonal, multiple secondary antibodies can bind to a single primary antibody since they recognize different epitopes

(see Figure 8.12). This amplifies the signal and means that less primary antibody is needed when a secondary antibody is used, reducing costs considerably.

Enzyme-Linked Immunosorbent Assay (ELISA)

The **enzyme-linked immunosorbent assay (ELISA)** uses antibodies to detect an antigen or antibody of interest by exploiting the tendency of proteins to **adsorb** (stick) to a plastic microplate (see Figure 8.11). Since the development of ELISA in the 1970s, it has become widely used in basic scientific research, medical diagnostics, forensics, agriculture, public health administration, and industrial quality control. There are several variations of ELISA but all involve the basic steps of immobilizing antigens or antibodies to plastic microplates and detecting them with one or more antibodies. ELISAs can detect either antigens or antibodies, and ELISAs can be either direct (using only a primary antibody) or indirect (using a secondary antibody to detect the primary antibody).

ELISA for Direct Antigen Detection

The simplest form of ELISA is the direct ELISA (see Figure 8.13). In a direct ELISA, only a primary antibody is used.

1. The sample is added to a microplate and the proteins and other molecules within the sample bind to the plastic in each well. Excess material is then washed off.

2. A primary antibody that has been labeled with an enzyme is added to the well, where it binds to its antigen if present. Unbound antibody is washed away.

3. A chromogenic (color producing) substrate is added to the wells. In wells that contain the antigen, the enzyme bound to the antibody will catalyze a chemical reaction and the solution in the well will change color.

Direct ELISA is not used as frequently as other types of ELISA, as it requires a primary antibody labeled with enzyme. In addition, direct ELISAs are less sensitive than ELISAs that employ a secondary antibody for signal amplification.

ELISA for Indirect Antigen Detection

Indirect ELISA (see Figure 8.14) takes full advantage of the ability of polyclonal secondary antibodies to amplify signals by binding to multiple epitopes on an antigen. In indirect ELISA, both primary and enzyme-linked secondary antibodies are used.

1. The sample is again added to the wells of a plastic microplate and excess material is washed off.

2. A primary antibody that is NOT labeled with an enzyme is added to the well, where it binds to its antigen if present. Unbound antibody is washed away.

3. Before a substrate is added, an enzyme-linked secondary antibody is added which binds to the primary antibody. Unbound antibody is washed away.

4. A chromogenic substrate is added and the enzyme bound to the antibody will catalyze a the color-change reaction in wells where the antigen is present.

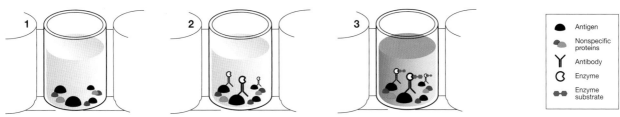

Figure 8.13. **Direct ELISA**. In the direct assay, an enzyme-linked primary antibody binds directly to the antigen. Wells are coated with sample (step 1). After washing wells to remove unbound sample, enzyme-linked antibody is added and binds to the antigen (step 2). After washing, colorimetric substrate is added (step 3). If antigen is present, the substrate reacts with the enzyme-linked antibody and the solution turns blue.

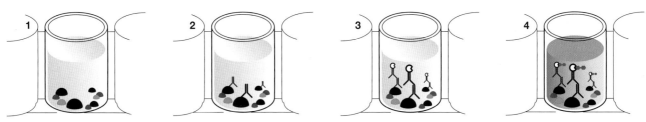

Figure 8.14. **Indirect ELISA**. In the indirect assay, purified antigens are added to the microplate wells and adsorb to the plastic wells (step 1). After washing wells to remove unbound sample, unlabeled primary antibody is added and binds to the antigen (step 2). After washing, enzyme-linked secondary antibody is added and binds to the primary antibody (step 3). After washing, colorimetric substrate is added (step 4). If antigen is present, the substrate reacts with the enzyme-linked antibody and the solution turns blue.

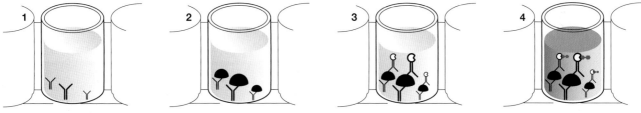

Figure 8.15. **Direct sandwich ELISA**. In a sandwich assay, a primary antibody is first bound to the plate (step 1), then a sample is added (step 2), which may or may not contain the antigen. A second primary antibody that is linked to an enzyme is then added (step 3). The second antibody will react with a chromogenic substrate only if antigen is present (step 4).

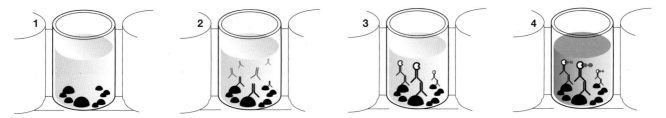

Figure 8.16. **Detecting serum antibodies in an ELISA**. Purified antigens are added to the microplate wells and adsorb to the plastic wells (step 1). A serum sample containing millions of different antibodies is added to the well and antibodies that recognize the antigen will bind (step 2). The wells are washed to remove unbound antibodies, then enzyme-linked antibodies are added to the well and bind to the serum antibodies (step 3). After washing the wells, colorimetric substrate is added (step 4). If serum antibodies bound to the antigen, the substrate reacts with the enzyme-linked antibody and the solution turns blue (step 4).

Sandwich ELISA for Antigen Detection

A variation of the antigen detection assay is the sandwich assay, in which the antigen is sandwiched between two layers of antibody (see Figure 8.15). The sandwich assay can be either direct or indirect.

In a sandwich ELISA, a primary antibody specific for the antigen is bound first to the microplate. After washing, the sample is added. If the antigen is present, it will bind to the antibody on the plate. After washing, a second primary antibody that binds to a different epitope on the same antigen is added. If the antigen is present, the second antibody will also bind. The second primary antibody may be labeled (direct sandwich assay) or it may be unlabeled, in which case an enzyme-linked secondary antibody is then added (indirect sandwich assay). After the enzyme-linked antibody binds, chromogenic substrate is added. Any wells that contain the antigen will change color. Sandwich assays are commonly used in commercial diagnostic kits because the extra antibody increases the stringency of the assay. The increased stringency is because in a sandwich ELISA, the antigen must be recognized by two different primary antibodies for the assay to give a positive result.

ELISA for Antibody Detection

ELISAs can be used to detect the presence of antibodies against a specific antigen in a patient's blood. B cells produce and release antibodies into the bloodstream in response to exposure to a disease antigen. For example, a commonly used screening test for HIV infection looks for anti-HIV antibodies in serum. Some serum antibodies are detectable as soon as two weeks after HIV infection.

In antibody detection assays, antigens are bound to the plate. Serum from the patient is added to the well and if antibodies to the antigen are present, they bind to the antigen on the plate. After washing, enzyme-linked secondary antibodies that bind human antibodies are added to the wells. Unbound secondary antibodies are washed away, and a chromogenic substrate is added to the wells. In wells with bound serum antibodies, the enzyme will be present and the solution will change color (see Figure 8.16).

Numerous commercially available antibody detection ELISA kits are available for use in both human and veterinary medicine. Not all detect antibodies to disease agents such as HIV. An example is Bio-Rad's ANA screening test assays for the presence of antinuclear antibodies (see Figure 8.17). Antinuclear antibodies are antibodies that bind material from cell nuclei, and their presence indicates the person may have an autoimmune disease such as systemic lupus erythematosus (lupus or SLE) or scleroderma.

Figure 8.17. **Antibody detection kit**. The Bio-Rad ANA screening test assays for antinuclear antibodies in blood from patients suspected of having autoimmune disease.

Instruments for ELISA

ELISA results can be quantified using a microplate reader (see Figure 8.18). A microplate reader uses similar principles to those used by a spectrophotometer to accurately measure the absorbance of the samples in wells of a microplate at set wavelengths of light. Each chromogenic substrate generates a product that absorbs at a specific wavelength.

There are also fully automated instruments that perform the ELISA from start to finish, including reading the results at the end, with minimal handling required. These instruments are popular in clinical laboratories that have high throughput.

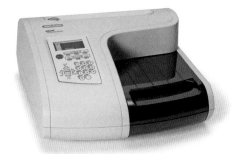

Figure 8.18. **A microplate reader**. Microplate readers like Bio-Rad's iMark™ microplate absorbance reader are able to measure the absorbance of multiple samples at once.

Ouchterlony Double Diffusion Assay

The **Ouchterlony**, or **double diffusion**, **assay**, developed in 1948 by Swedish scientist Örjan Ouchterlony, became popular because it was much easier to perform than other immunoassays of the day. One of the most common uses of the Ouchterlony assay was the determination of immunoglobulin type. It can also be used in forensics to determine if blood is of human or animal origin.

The Ouchterlony assay tests whether antisera or antibodies interact with antigens in a sample. It exploits the tendency of proteins to diffuse through a matrix. Antibodies and antigens are placed into holes in a slab of agarose gel; over time, they diffuse out of the holes into the agarose, and eventually the antibodies and antigens meet. If the antibodies bind to antigens in the sample, they form a large complex that precipitates and forms a visible white line (see Figure 8.19). Proteins form the large complex because multiple polyclonal antibodies bind to each antigen, and since each antibody binds two antigens, multiple antibodies and antigens become crosslinked together. If there is no reaction (in other words, if the antibodies do not recognize the antigen), then there is no visible precipitate.

Although fairly accurate, the Ouchterlony assay requires a large amount of antibody and antigen, so it has mostly been replaced by other immunoassays, such as ELISA.

Figure 8.19. **Ouchterlony double diffusion assay**. In this example, an antibody was placed in the center hole and antigens were placed in the four surrounding holes. The antibody recognized only the antigens in the holes at the top and right, which is shown by the visible precipitate between those sets of holes (see arrows).

Western Blotting

SDS-PAGE separates proteins based on their sizes (see Chapter 7). But the actual identities of the proteins can be inferred only if the protein's size is known, and the inference may be incorrect if a second protein of a similar size is in the sample. Therefore, SDS-PAGE is frequently followed by another type of immunoassay, western blotting, which can identify specific proteins in a sample and provide information about the protein's size and relative abundance in the sample.

In **western blotting**, proteins separated by SDS-PAGE are transferred to the surface of a membrane or **blot**, such as nitrocellulose membrane. The membrane is sturdier than a gel and it makes the proteins that were embedded in the gel more accessible to antibodies. This technique is similar to the Southern blot, and the name western blotting is a play on that blot's name.

Once the proteins have been transferred to the membrane, western blotting parallels an indirect ELISA in that an immobilized antigen on the membrane is detected with a primary antibody specific for the antigen (see Figure 8.20). The primary antibody is detected using an enzyme-linked secondary antibody. The position of the band on the membrane is highlighted by the reaction of a colorimetric or chemiluminescent substrate with the enzyme, which makes a colored band on the membrane or emits light in the position of the protein of interest (see Immunodetection below).

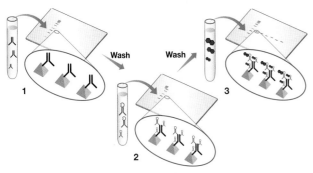

Figure 8.20. **Immunodetection in western blotting**. After the proteins have been transferred from the polyacrylamide gel to the membrane, the membrane is incubated with a primary antibody, which binds to antigen molecules immobilized on the membrane (step 1). Enzyme-linked secondary antibody is then added, which binds to primary antibody (step 2). In the final step, colorimetric enzyme substrate is added to the membrane and colored bands develop wherever antigen is detected (step 3).

Electroblotting

Proteins are transferred from the gel to the membrane (blotted) using a technique called **electroblotting**, which uses an electrical current to move the negatively charged proteins from the gel onto the membrane (see Figure 8.21). This is different from the capillary action used to draw molecules from the gel to the membrane in a traditional Southern blot. Electroblotting is more efficient in that more proteins are transferred more quickly than they would be by capillary action.

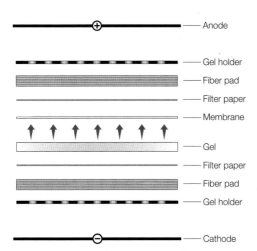

Figure 8.21. **Schematic of an electroblot**. The gel and membrane are layered between filter papers soaked in buffer and inserted into a gel holder cassette with fiber pads. The gel holder containing the gel is placed in buffer and a current is applied across the gel and membrane. The proteins move from the gel onto the surface of the membrane (see arrows).

In a western blot, the membrane is placed on top of the SDS-PAGE gel, and the membrane and gel are placed between two filter papers soaked in blotting buffer. This stack is sandwiched between fiber pads and enclosed in a gel holder cassette that squeezes the sandwich together. The entire cassette is submerged in blotting buffer between two electrodes.

Blotting buffer contains Tris, glycine, and alcohol. The Tris maintains the pH, the glycine ions conduct the current, and the alcohol helps the proteins bind to the membrane. An electrical current is applied across the whole stack and draws all the proteins from the gel toward the membrane, where they are trapped because the pores in the membrane are smaller than most of the proteins. The proteins blotted onto the surface of the membrane form a mirror image of the proteins separated in the gel.

Electroblotting can be performed vertically, with the blotting sandwich fully submerged in buffer (as described here) or can be performed using an apparatus called a semi-dry blotter, in which the blotting sandwich is soaked in buffer and placed horizontally between dry electrodes. Semi-dry blotters are able to perform electrblotting in as few as 3 minutes compared to tank electroblotting which can take several hours (see Figure 8.22).

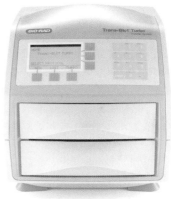

Figure 8.22. **Bio-Rad's Trans-Blot® Turbo™** is a semi-dry blotter. Each drawer can hold up to two mini gels.

Immunodetection

After the blot has been generated, the membrane is blocked using a protein-rich blocking solution, often made with dried milk. The proteins in the blocking solution bind to protein-binding sites on the membrane and to proteins (from the gel) that might bind nonspecifically to the antibodies used for detection. This blocking step is critical for preventing background staining and nonspecific detection of proteins.

The membrane is then incubated with a primary antibody, washed, incubated with a secondary antibody, and washed again. At each of these steps, the membrane and the solution it is in must be placed on a shaking platform to ensure even coverage of antibody and thorough washing.

Testing for HIV Status
at Home

According to the World Health Organization (WHO), in 2016 an estimated 37 million people worldwide were infected with the human immunodeficiency virus (HIV), with 1.8 million becoming newly infected that year alone. Treatments like antiretroviral therapy (ART) allow people with HIV to enjoy healthy lives, and this treatment can even prevent HIV transmission from infected pregnant women to their unborn children. However, in 2016, 40% of people infected with HIV around the world (over 14 million people) were unaware of their HIV status. In the U.S., the numbers were lower: 20% of those infected with HIV were unaware of their status, but they were responsible for up to 70% of the new infections each year. A key to treating people with HIV to both increase their longevity and prevent spread of the disease is providing access to testing.

HIV testing usually involves a visit to a doctor and ELISA tests to detect HIV antibodies in saliva, blood, or urine. But a visit to a doctor may be problematic. For some people, especially those in rural or undeveloped areas, access to doctors and testing may be difficult for logistical or financial reasons. For others, it may be inconvenient, too embarrassing, or, in some areas of the world, not anonymous or confidential. What if people could test themselves in the privacy of their homes?

In 2013, an in-home test for HIV became available in the U.S. The OraQuick HIV test is an oral swab test that detects HIV antibodies in saliva and produces results that can be read at home. This test makes it simpler and more private for people to learn their HIV status, but it has also raised concerns. If people refused to see a doctor for testing, would they follow up after an in-home test with a doctor for confirmatory testing, care, or support? How would new cases of HIV be reported if they did not follow up with a doctor? Would this test be a viable option for people in remote areas around the world?

OraQuick is 99.9% accurate at reading negative results and 91.7% accurate at identifying positive results. This means that almost 10% of people who are HIV positive may be incorrectly identified as HIV negative. In addition, infected people should not expect to see a positive result if they were infected less than 90 days before testing. What are the risks associated with a false negative result? Do the benefits of test accessibility (more people being tested) outweigh these risks?

The secondary antibody is usually linked to an enzyme, and a substrate for the enzyme is applied for detection. The enzyme can oxidize or cleave a colorimetric substrate similar to the substrate described in ELISA tests earlier, except that in a western blot the substrate forms a colored precipitate that adheres to the membrane where the antibody is bound, creating a visible band (See Figure 8.23).

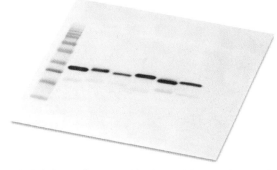

Figure 8.23. **Colorimetric western blot**. Myosin light chains from muscle samples of different animal species were detected using a primary antibody that recognizes myosin and visualized with a colorimetric substrate. The bands show the myosin proteins are of different sizes among the different species. The left lane contains a protein size standard.

Alternatively, the substrate may be chemiluminescent. These substrates emit light when cleaved by the enzyme, and the light is detected using photographic film or digital imaging equipment. Occasionally, the secondary antibody is linked to a fluorescent molecule and so no substrate is required; the membrane is visualized using digital imaging equipment that can detect the fluorescence. This option allows multiple proteins on the same blot to be detected concurrently using multiple antibodies, each linked to a different color of fluorescence (see Figure 8.24).

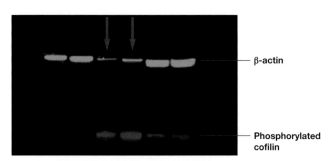

β-actin

Phosphorylated cofilin

Figure 8.24. **Multiplex fluorescent western blot**. β-actin and cofilin are cytoskeletal proteins. In this experiment, researchers studied the effect of reduced levels of β-actin on the phosphorylation of cofilin in human tissue culture cells. Antibodies labeled with green fluorescence were used to detect β-actin, and antibodies labeled with red fluorescence were used to detect phosphorylated cofilin on the same blot. The results show that when actin levels are reduced (see arrows), phosphorylation of cofilin protein increases.

Western Blotting in Diagnostic Testing

Western blots are also used in diagnostic testing. For example, if an ELISA is positive for HIV infection, the diagnosis must be confirmed by a second assay as a small percentage of ELISA results will be false positives. There are two types of commercially available western blot assays. The first type of kit detects the presence of proteins from the HIV virus. The patient's serum proteins are first analyzed by SDS-PAGE and secondly blotted onto a membrane that is then probed with anti-HIV antibodies. The second type of kit detects anti-HIV antibodies. HIV proteins (antigens) are provided on a strip of membrane. The membrane is used to detect anti-HIV antibodies in the patient's blood. Bio-Rad's GS HIV-1 western blot kit (Figure 8.25) is an example of the second type of kit.

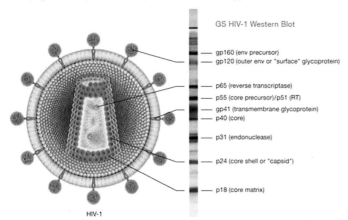

Figure 8.25. **HIV-1 structure and Bio-Rad GS HIV-1 western blot kit test strip.** HIV proteins are provided on a strip of membrane. The membrane strip is incubated in the patient's serum. If the patient's blood contains antibodies to HIV, the antibodies bind to the antigens on the test strip, which is then probed with an enzyme-linked anti-human antibody. Colored bands appearing after incubation in an enzyme substrate indicate positive results.

Using Antibodies to Study Cells and Tissues

Immunocytochemistry and **immunohistochemistry** are terms that refer to the use of antibodies for studying cells and tissues, respectively, under a microscope. With their specific binding capabilities, antibodies are excellent tools for examining cellular structures, following changes in cells or tissues, and visualizing processes that cannot be seen otherwise. The location of an antigen in a cell or tissue can be shown using an enzyme-linked antibody that deposits a colored enzyme product where it has bound. Cells or tissues are then counterstained with dyes to show the location of the antigen in relation to other cellular or tissue structures such as cell nuclei (see Figure 8.26).

Alternatively, the antibody can be labeled with a fluorescent molecule, in which case the technique is called immunofluorescence. Fluorescent molecules absorb light at one wavelength and emit light at a different wavelength. Immunofluorescence microscopy requires a special type of microscope, a fluorescence microscope, that has filters for exciting the fluorescent labels with specific wavelengths of light. As in a multiplex western blot, the use of different fluorescent labels and dyes allows multiple proteins to be detected at the same time (see Figure 8.27). The ability to locate multiple proteins in the same cell at the same time is invaluable in microscopy, since every cell and tissue sample is unique.

High-Throughput Immunoassays

Researchers probe the genomes and proteomes of organisms, discovering thousands of novel genes, proteins, and other biomolecules (analytes) that may or may not play roles in biological functions or in diseases and metabolic disorders. As more and more biomolecules are discovered, the need for screening and

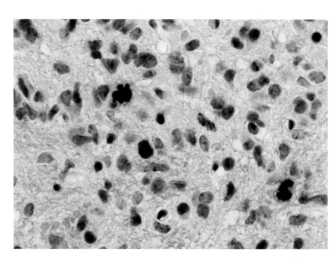

Figure 8.26. **Immunohistochemistry of a mouse brain tumor.** The primary antibody binds histone H3 protein, which is associated with DNA in the nucleus only when it is phosphorylated. The blue color is a DNA stain and indicates the nuclei of cells. The enzyme-linked secondary antibody produces a brown color when substrate is added to the sample. The cells that appear brown are positive for phospho-histone H3. Magnification 400x. Image courtesy of Joanna Phillips, University of California–San Francisco.

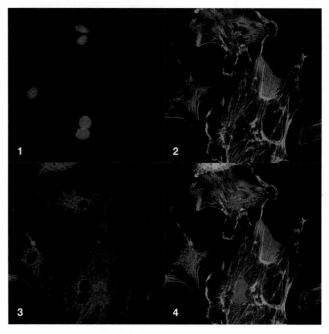

Figure 8.27. **Immunofluorescence microscopy.** All four panels show the same field of view. Panels 1, 2, and 3 are illuminated with different wavelengths of fluorescence, while panel 4 is a composite image of the previous three. In these cells, the primary antibody binds to microtubules and the secondary antibody is labeled with green fluorescence so the microtubules appear green in the micrograph. The nuclei are stained blue and mitochondria are stained red with other fluorescent dyes. Image courtesy of Kristi DeCourcy, Virginia Tech.

analyzing them all increases. Fortunately, advances in robotics, miniaturization, and computing, and the expansion of the field of bioinformatics have contributed to the development of **high-throughput immunoassays**, which can screen dozens to thousands of samples in only a matter of hours. High-throughput immunoassays wield the power of antibodies to screen tens to thousands of biomolecules, cells, or samples at one time.

CAR T Cell Therapy
— A new horizon in cancer treatment

In 2010, 5-year-old Emily Whitehead was diagnosed with acute lymphoblastic leukemia (ALL), and after two rounds of chemotherapy she had a relapse. Though doctors offered bone marrow transplantation, the Whiteheads chose instead to involve Emily in a study for a cutting edge therapy developed at the University of Pennsylvania. Known as CART-19, this treatment involved removing and genetically engineering Emily's own T cells and then placing them back into her body. T cells are white blood cells that circulate throughout the body, hunting and attacking foreign cells or substances. They do not normally attack cancer cells since they are not foreign, but Emily's T cells were reprogrammed to recognize and kill the tumor cells. Emily became the first child to receive CART-19 and is now a thriving teenager.

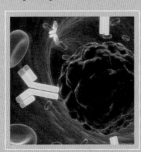

CART-19 is an example of a promising immunotherapy called chimeric antigen receptor CAR T cell therapy. Unlike current chemo and radiation therapies that kill many healthy cells in addition to rapidly dividing cancer cells, CAR T cell therapy only targets cancer cells. It involves genetically engineering a patient's own immune cells (T cells) to treat their cancer. It is perhaps the most personal of personalized medicine. Though these treatments show great promise, researchers caution that much remains to be discovered about immunotherapy. Unexpected side effects and whether the therapy can be used against solid tumors are among the questions being researched. As the work continues, though, it is likely immunotherapies will become even more effective, present fewer side effects, and treat other types of cancers.

Novartis, a pharmaceutical company based in Basel, Switzerland, commercialized a therapy based on the CART-19 technology, called tisagenlecleucel (Kymriah). This treatment, like Emily's, begins with removing white blood cells (T cells) from a patient's bloodstream and then engineering the cells to produce special receptors (CAR). These receptors allow the T cells to recognize the cancer cells as foreign. The CAR T cells are then grown in the laboratory and transfused back into the bloodstream of the patient, where they attack the cancer. Kymriah was approved by the FDA in August 2017 and is the first FDA-approved gene therapy.

Multiplex Bead Assays

Multiplex bead array assays use microscopic beads to provide quantitative measurements of large numbers of analytes, usually using an automated 96-well microplate format. As an example, Bio-Rad's Bio-Plex® system is a multiplex bead array system that can detect and quantify as many as 500 different antigens simultaneously, in a single assay. The system uses up to 500 sets of beads, each set with a unique fluorescent signal (or spectral signature, see Figure 8.28) and each conjugated to a different antibody.

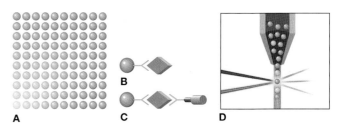

Figure 8.28. **Bio-Rad Bio-Plex system**. This high-throughput assay uses up to 500 bead sets, with each set possessing a unique spectral signature (**A**). The assay design is similar to the sandwich ELISA but the primary antibody is anchored to a bead rather than to an assay plate (**B**) and the second antibody is labeled with a fluorescent dye (**C**). The detector analyzes each bead individually (**D**).

The Bio-Plex assay resembles the sandwich ELISA (see Figure 8.15). First, fluorescently tagged beads with their bound antibodies are mixed with the sample solution and the antibodies bind to any antigen present. Then a second, fluorescently labeled antibody to the antigen is added to the sample. If there is antigen in the sample, it is bound and the bead has two fluorescent tags: one identifying the bead (and, therefore, the antibody) and the other from the second antibody bound to the antigen. If the antigen is not bound, the bead will be missing the second fluorescent tag.

After incubation, the beads are passed through a detection chamber where two lasers are directed at the sample. The first laser activates the unique fluorescent signal on the bead and identifies which of the 500 beads is being read. The second laser excites the fluorescent label on the second antibody, detecting whether the antigen has bound to the bead. As the beads are read one by one, the number of beads bound by antigen are counted and analzyed by a computer. The proportion of beads that have bound with antigen is measured and that number is used to determine the amount of antigen present in the sample.

Fluorescence-Activated Cell Sorting

Fluorescence-activated cell sorting (FACS) is a high-throughput assay that sorts cells into different populations based on the presence of antigens on their cell surfaces. In FACS, cells are incubated in a solution that contains different antibodies and each antibody type is labeled with a different fluorescent tag. Since the antibodies target cell surface molecules, living cells can be used in this assay because the antibodies do not have to cross the plasma membrane to reach their targets.

Once cells have bound to antibodies, they are sorted using an instrument called a fluorescence-activated cell sorter. The instrument separates the solution containing the cells into a stream of droplets, with one cell in each droplet. Each droplet proceeds single file past a laser and detection system. Similar to a bead in a multiplex assay, the cell in each droplet is targeted by a laser, which excites the fluorescent tag on the antibody. A detector records the results.

The next step is different from multiplex assays. Depending on which fluorescent tag is detected, the instrument gives the droplet a specific electrical charge. The droplet then passes through an electromagnetic field and is deflected into a specific tube depending on the charge (see Figure 8.29). The sorting may be as simple as separating one cell type tagged with one antibody from untagged cells, or it can be as complex as sorting cells labeled with up to 12 different fluorescently labeled antibodies. Current sorters can sort hundreds of thousands of cells in just minutes.

FACS is used extensively in medical diagnostics, particularly in the diagnosis and characterization of cancers. Cells of the immune system have thousands of different biomolecules on their surfaces, and the antigens on each cell surface can be used to characterize the cell type.

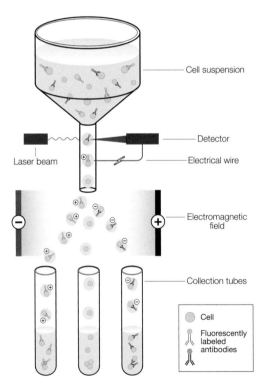

Figure 8.29. **Fluorescence-activated cell sorting (FACS)**. In FACS, cells are mixed with fluorescently labeled antibodies that bind cell surface antigens. The cells are separated into small droplets, one cell per droplet, and analyzed one at a time. Depending on the fluorescent signal, the cells are given specific electrical charges. When the cells pass through an electromagnetic field, they are deflected into different tubes depending on that charge.

Careers In Biotech

Sophy Wong, M.D.

Associate Clinical Professor of Medicine, University of California, San Francisco

Medical Director for HIV ACCESS and AIDS Education and Training Center

Photo credit: Young Whan Choi courtesy of Sophy Wong

Dr. Sophy Wong grew up exploring the natural world and communities around her. Curious by nature, she would take her time walking home from school, examining the plants, bugs, animals, buildings and people she saw along the way. She still loves to wander and stay curious, exploring ways to make a positive impact on the world, both professionally and personally.

Professionally, Sophy is a medical doctor and devotes much of her time to improving public health both locally and worldwide. As a freelance contractor she trains and advises health organizations in the San Francisco Bay Area that provide access to HIV and hepatitis testing, care, and prevention services, as well as general sexual healthcare. She cares for patients directly at Asian Health Services, a community health center in Oakland's Chinatown, and mentors clinical teams worldwide. In her own neighborhood, Sophy organizes and facilitates a meditation/mindfulness group for fellow social justice and service advocates.

Sophy is driven to make a difference in peoples' lives. She saw firsthand in Kenya, when HIV antiretroviral treatment (ART) became available, how timely treatment saves peoples' lives and improves their health. She believes the scientific tools needed to end the HIV and hepatitis C epidemics are available, but access to these tools around the world is still a hurdle. She dedicates her time to supporting organizations and cross-disciplinary groups of people that address the many economic, social, and geographical barriers that people face to access quality healthcare. For Sophy it's an exciting and rewarding endeavor with a great future.

For anyone interested in a career in biotechnology, Sophy advises to pursue your passions as much as you can. Doing what you love will not only give you a reason to wake up and be happy each day, it will also give you the motivation and energy to persist through the inevitable challenges in life. Using your skills and talents to do something positive in the world helps you stay connected with the rest of humanity and feel good about your work. Also, find and keep good mentors. Mentors guide us when we are not sure where we are going and provide advice for future challenges.

Chapter 8 Essay Questions

1. Describe the differences between monoclonal and polyclonal antibodies. Describe one application in which a monoclonal antibody would work best and one application best suited for a polyclonal antibody.

2. Antibodies are used to treat a number of human diseases. Report on a specific antibody drug, discussing how it was developed (for example, whether it is humanized) and how it works.

3. Compare the strategies of detecting GMOs by PCR and ELISA.

Additional Resources

Antibodies: A Laboratory Manual. Second Edition. Edited by EA Greenfield, ed. (Cold Spring Harbor, NY: Cold Spring Harbor Laboratory Press.)

Other resources:

UNAIDS. unaids.org. Accessed December 27, 2017.

World Health Organization. who.int/hiv/pub/en. Accessed December 27, 2017.

Activity 8.A Ouchterlony Double Immunodiffusion Assay

Overview

In this activity, the Ouchterlony double diffusion assay will be used to identify the origin of two animal blood samples. This assay is used to test whether antibodies recognize specific antigens. The antibodies and antigens are added to separate wells in an agarose gel and diffuse toward each other through the agarose matrix. When the molecules bind to each other, they form a large protein complex that precipitates out of solution, forming a visible white line in the gel at the site where the interaction occurs. If the molecules do not bind to each other, no complex is formed and no precipitate is visible. The presence of a white line indicates that a sample contains antigens that are recognized by the antibodies.

Figure 8.30. **An Ouchterlony assay**. An antibody was placed in the central hole and four serum samples were added to the outer holes. The serum samples in the top and right holes contained antigens that reacted with the antibody, creating a line of white precipitate. The samples in the bottom and left holes did not react.

In the following experiment, animal meat juices that contain blood proteins will be analyzed by an Ouchterlony assay. Antibodies that recognize chicken blood proteins will be placed in the center well and the surrounding wells will be filled with test blood samples. Since the antibodies are specific to chicken blood proteins, they will not recognize the blood of another animal. If the antibodies recognize the chicken blood proteins (the antigen), a white line will form between the two wells; however, if the antibodies do not recognize the antigen, no line will form. Two controls are used in the assay. The positive control is chicken serum that contains chicken blood proteins and the negative control contains buffer (with no blood proteins).

Tips and Notes

- Be sure to change tips between samples and reagents to avoid sample cross-contamination, which could give false results

Safety Reminder: Follow standard laboratory safety procedures. Wear appropriate PPE. Take proper precautions when handling raw meat and raw meat juices, especially chicken. Disinfect all surfaces that come into contact with raw meat or blood proteins. Wash hands thoroughly after any contact with raw meat or blood proteins. Read and be familiar with all SDSs for this activity.

Research Question

- Which of the test samples is from chicken?

Objectives

- Prepare the Ouchterlony assay plate
- Add the antibody and the test samples to the plate
- Interpret the results

Skills to Master

Refer to Laboratory Skills Assessment Rubric (Appendix E) and Laboratory Notebook Rubric (Appendix F) for more details.
- Record laboratory notebook entries
- Perform an Ouchterlony double immunodiffusion assay

Student Workstation Materials

Items	Quantity
2–20 µl adjustable-volume micropipet and tips	1
1 ml graduated transfer pipet	1
Razor blade or scissors	1
Laboratory marking pen	1
Waste container	1
Anti-chicken antibodies	20 µl
Positive control (chicken serum)	20 µl
Negative control (1x PBS)	20 µl
Test sample A	20 µl
Test sample B	20 µl
Ouchterlony assay plate with 1% agarose	1

Prelab Focus Questions

1. What is an antibody?
2. What is an antigen?
3. Can an antigen be bound by one or more polyclonal antibodies? Explain your answer.
4. How many antigens can bind to a single antibody? Where on the antibody do antigens bind?
5. Name three instances where antibodies are used in biotechnology.

Activity 8.A Ouchterlony Double Immunodiffusion Assay

Activity Protocol

1. Using a razor blade, cut off the end of the transfer pipet between the 250 µl and the 100 µl marks. Once cut, the pipet tip will be used to punch holes out of the agarose to prepare the assay plate.

2. Mark five dots on the bottom of the Ouchterlony assay plate in a crosswise pattern, with each dot approximately 1 cm away from the others. Squeeze the bulb of the transfer pipet and press the tip of the pipet into the agarose over the dot in the center of the plate. Release the bulb to suck out the small plug of agarose. This should leave a hole in the agarose.

3. Punch four more holes in the other positions as shown.

4. Label the bottom of the plate as shown.

5. Pipet 15 µl of the anti-chicken antibodies into the center hole.

6. Pipet 15 µl of the positive and negative controls and the test samples into the indicated holes.

7. Place the lid onto the petri plate. Incubate at room temperature for 24–72 hr. Keep the plate right side up.

   Note: Once a visible precipitate has formed, the plates can be sealed with Parafilm and stored at 4°C for up to 2 weeks.

Results Analysis

Examine the plates each day until a visible white line has formed in the area between the hole where the positive control was added and the hole where the anti-chicken antibodies were added. Sketch the results in your laboratory notebook and make conclusions based on how the test samples compared to the controls. Determine whether any of the test samples contain chicken blood proteins.

Postlab Focus Questions

1. Were any of the test samples from chicken? What evidence do you have to support your findings?
2. How could this assay be modified to test whether blood left at a crime scene originated from a human or a dog?

Self Assessment

Assess whether you have mastered the skills of this activity using the Laboratory Skills Assessment Rubric in Appendix E.

Assess your experimental write-up in your laboratory notebook using the Laboratory Notebook Rubric in Appendix F.

Activity 8.B Serum Antibody Detection by ELISA

Overview

In this activity, an enzyme-linked immunosorbent assay (ELISA) will be used to detect the presence or absence of an antibody. This type of diagnostic test is used to determine whether a person has been exposed to a disease-causing pathogen. Some diseases, such as HIV and Lyme disease, are very difficult to detect directly because the disease-causing pathogens may be present at levels that are too low to detect. Alternatively, disease-causing pathogens may be hidden inside cells and may not be present in bodily fluids that are easily accessible, such as blood, saliva, and urine. In diseases where the pathogen is difficult to detect, it is often easier to look for the body's reaction to the pathogen instead.

When a person is exposed to a pathogen, the body mounts an immune response that includes making antibodies specific for the pathogen. These antibodies are constantly produced at a low level well after initial exposure to the pathogen in case it is encountered again in the future. An ELISA is used to test whether a person's blood contains antibodies that recognize antigens from the disease-causing pathogen. If a person has antibodies to the pathogen, it is likely he/she has been exposed to that pathogen.

In this activity, you will play the role of a clinical scientist and perform an ELISA to detect the presence of antibodies in a serum sample taken from a patient. Serum is blood that has had the solid components (such as cells) removed.

This assay uses real antibodies and antigens; however, the test samples are simulated serum and are not of human origin. The antibodies and antigens for this assay are from chickens, rabbits, and goats. The enzyme used in this assay is horseradish peroxidase (HRP). The substrate is tetramethylbenzidine (TMB), a colorless solution that turns blue when it is oxidized by HRP in the presence of hydrogen peroxide.

This ELISA detects the presence or absence of an antibody; however, it is not designed to estimate the amount of antibody present. A test that gives a yes/no answer is called a qualitative test. In Activity 8.C, an ELISA is used to measure the amount of a substance. When a test is used to measure an amount or quantity, it is called a quantitative test.

Tips and Notes

- Make every attempt to not cross-contaminate samples; therefore, do not splash antigens or antibodies into adjacent wells and do change tips between samples

- When washing the wells, first turn the microplate strip upside down gently, let the liquid absorb into a paper towel, and discard the towel; then tap the strip upside down on a fresh paper towel to remove any liquid that is adhering to the wells

Safety Reminders: Follow all laboratory safety protocols and wear appropriate PPE for this activity. Read and be familiar with all SDSs for this activity. The antibodies and antigens used in this assay are not of human origin; they are from chickens, rabbits, and goats.

Research Questions

- Have the patients been exposed to the pathogen?
- Do the patient serum samples have antibodies to the antigen?

Objectives

- Coat the wells with antigen
- Add patient serum samples to the wells
- Add an enzyme-linked antibody to the wells to detect serum antibodies
- Add a substrate to the wells
- Determine if the serum samples are positive or negative for disease exposure

Skills to Master

Refer to Laboratory Skills Assessment Rubric (Appendix E) and Laboratory Notebook Rubric (Appendix F) for more details.
- Record laboratory notebook entries
- Perform a qualitative ELISA

Student Workstation Materials

Items	Quantity
20–200 µl adjustable-volume micropipet (or 50 µl fixed-volume micropipet) and tips*	1
Transfer pipet	1
Microcentrifuge tube rack	1
12-well microplate strips	2
Paper towels	stack
Laboratory marking pen	1
Waste container	1
Wash buffer	70–80 ml
Serum samples (250 µl of each)	4
Antigen (AG)	1.5 ml
Positive control serum (+)	500 µl
Negative control serum (−)	500 µl
Enzyme-linked antibody (ELA)	1.5 ml
Substrate (SUB)	1.5 ml

* 1 ml graduated transfer pipets can be used instead of a micropipet and the 25 µl mark can be used to measure 50 µl by filling to this line twice. (see Figure 2.29). A fresh pipet should be used for each new reagent. Approximately 10 transfer pipets are needed per team.

Prelab Focus Questions

1. Name three cells in the immune system that help protect us from disease.
2. Why are enzymes used in this immunoassay?
3. What kind of samples would make good positive and negative controls for this assay? Why?

Activity 8.B Serum Antibody Detection by ELISA

Protocol

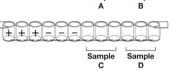

Sample A | Sample B

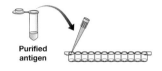

Sample C | Sample D

1. Label two 12-well microplate strips. On each strip, label the first three wells with a **+** for the positive controls and the next three wells with a **–** for the negative controls. Label the remaining wells to identify the serum samples being tested (three wells each).

Purified antigen

2. Use a fresh pipet tip to transfer 50 µl of purified antigen (AG) into each well of the two 12-well microplate strips.

3. Incubate the samples at room temperature for 5 min so that the antigen can bind to the plastic wells.

4. Wash protocol:

   a. Tip each microplate strip upside down onto a short stack of paper towels and gently tap the strip a few times to drain the wells. Make sure to avoid splashing sample back into wells.

   b. Discard the wet paper towels.

   c. Use a transfer pipet to fill each well with wash buffer, taking care not to spill the buffer over into neighboring wells.

   Note: Use the same transfer pipet for all washing steps.

   d. Tip each microplate strip upside down onto a small stack of paper towels and tap the strip a few times to drain the wells.

   e. Discard the wet paper towels.

5. Repeat step 4.

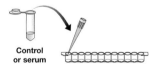

Control or serum

6. Use a fresh pipet tip to transfer 50 µl of the positive control (+) into the three **+** wells on each microplate strip.

7. Use a fresh pipet tip to transfer 50 µl of the negative control (–) into the three **–** wells on each microplate strip.

8. Use a fresh pipet tip to transfer 50 µl of the first serum sample into the corresponding three wells. Repeat this step for the remaining serum samples using a fresh pipet tip for each serum sample.

Activity 8.B Serum Antibody Detection by ELISA

9. Incubate the samples at room temperature for 5 min so that the serum antibodies can bind to the antigens.

Wash 2x

10. Repeat step 4 twice to wash the unbound controls or serum antibodies out of the wells.

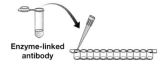

Enzyme-linked antibody

11. Use a fresh pipet tip to transfer 50 µl of enzyme-linked antibody (ELA) into each well of the microplate strips.

12. Incubate the samples at room temperature for 5 min so that the enzyme-linked antibodies can bind to the serum antibodies.

Wash 3x

13. Repeat step 4 three times to wash the unbound enzyme-linked antibody out of the wells.

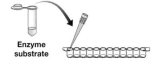

Enzyme substrate

14. Use a fresh pipet tip to transfer 50 µl of enzyme substrate (SUB) into each well of the microplate strips.

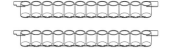

15. Incubate the samples at room temperature for 5 min. Observe and record the results into your notebook.

Results Analysis

Record detailed observations about the color of the positive and negative controls in each well in addition to the test sample wells. Although the color will continue to develop, the results should be recorded within 5–10 min.

Note: The substrate is light sensitive; therefore, light energy will catalyze the oxidation of the substrate over time and it will turn pale blue in the negative wells.

Postlab Focus Questions
1. Why were the wells washed after every step?
2. What could be the reason for a positive test if the patient did not have the disease?
3. Which samples were positive and which were negative? What evidence do you have?
4. Why were the tests performed in triplicate?

Self Assessment

Assess whether you have mastered the skills of this activity using the Laboratory Skills Assessment Rubric in Appendix E.

Assess your experimental write-up in your laboratory notebook using the Laboratory Notebook Rubric in Appendix F.

Activity 8.C Quantitative ELISA

Overview

An enzyme-linked immunosorbent assay (ELISA) can detect the presence of an antibody in a patient's serum using a purified antigen. This type of assay is performed in Activity 8.B. Alternatively, an ELISA can use a purified antibody to detect the presence of a specific antigen. This assay can be used in a medical test to test directly for the presence of a pathogen, in diagnostic tests (such as field tests for genetically modified foods) to track adulteration of non–genetically modified (non-GM) crops with GM products, or in laboratory tests to determine the presence of an antigen in a sample.

ELISAs can also measure quantities of antigens or antibodies. In this activity, an ELISA will be used to measure the amount of an antigen present in a sample. This type of quantitative test is used in the clinic to assay the levels of a pathogen or other proteins in a patient. For example, an ELISA may be used for an initial diagnosis if high levels of a protein indicate a medical condition. The technique may also be used to check how well a patient is responding to a treatment. It is also used in research to compare antigen levels among samples.

The amount of antigen is determined by comparing the intensity of the blue color of the test samples with a standard curve produced from a dilution series of samples with known amounts of antigen. The intensity of the color can be estimated visually or quantitated using a microplate reader. A microplate reader uses similar principles to a spectrophotometer to accurately measure the absorbance of the samples at set wavelengths of light. The blue color is detected using a 655 nm filter.

The antigen in this assay is chicken gamma globulin (a type of chicken antibody), while the primary antibody is a polyclonal rabbit anti-chicken antibody. The enzyme-linked secondary antibody is a polyclonal goat anti-rabbit antibody linked to horseradish peroxidase (HRP). The substrate is tetramethylbenzidine (TMB), a colorless solution that turns blue when it is oxidized by HRP in the presence of hydrogen peroxide. TMB turns yellow under acidic conditions.

Tips and Notes

- Make every attempt to not cross-contaminate samples. Avoid splashing antigens or antibodies into adjacent wells and change tips between samples. When washing the wells, first turn over the microplate strip gently and let the liquid absorb into a paper towel. Then tap the strip on a fresh paper towel to remove any liquid that is adhering to the wells

- Ensure that all serial dilutions are thoroughly mixed before preparing the next dilution .

- The enzymatic reaction can be stopped to provide more accurate results by adding 50 µl of 0.18 M sulfuric acid to the wells. The acid should be added to the wells in the same order as the substrate. Since acid turns the product yellow, the absorbance of the sample should be read in a microplate reader using a 450 nm filter

Safety Reminders: Follow all laboratory safety procedures. Wear appropriate PPE. Since acid can cause burns, take appropriate precautions when using acid. Read and be familiar with all SDSs for this activity. The antibodies and antigens used in this assay are from chickens, rabbits, and goats.

Research Question

- How much antigen is present in the test sample?

Objectives

- Serially dilute the antigen in the microplate wells to prepare the standards
- Perform an ELISA on the standards and the test samples to detect the antigen
- Visually compare the test samples to the standards and estimate the antigen concentration
- Read the standards and the samples using a microplate reader
- Generate a standard curve and determine the amount of antigen in the test samples

Skills to Master

Refer to Laboratory Skills Assessment Rubric (Appendix E) and Laboratory Notebook Rubric (Appendix F) for more details.

- Record laboratory notebook entries
- Perform a serial dilution (Activity 7.E)
- Perform a quantitative ELISA
- Use a microplate reader
- Generate a standard curve using absorbance values (Activity 7.A)
- Determine experimental values using a standard curve (Activity 7.A)

Activity 8.C Quantitative ELISA

Student Workstation Materials

Items	Quantity
Microplate reader (optional) (shared)	1
Computers with graphing software	
(only if a microplate reader is available) (optional)	1–4
20–200 µl adjustable-volume micropipet	
(or 50 µl fixed-volume micropipet) and tips	1
Transfer pipet	1
Microcentrifuge tube rack	1
12-well microplate strips	2
Paper towels	stack
Graph paper (only if a microplate reader is available)	4
Laboratory marking pen	1
Waste container	1
Wash buffer	70–80 ml
1x phosphate buffered saline (PBS)	1 ml
Test samples (250 µl of each)	2
Control antigen (1,000 ng/ml)	250 µl
Positive control sample (+)	250 µl
Negative control sample (−)	250 µl
Primary antibody (PA)	1.5 ml
Enzyme-linked antibody (ELA)	1.5 ml
Substrate (SUB)	1.5 ml
0.18 M sulfuric acid (optional)	1.5 ml

Prelab Focus Questions

1. Give two instances when it would be important to know the concentration of an antigen in a given solution.

2. Assume a stock solution of antigen has a concentration of 2 mg/ml. If this stock solution is serially diluted by 50% ten times, what will the final concentration of antigen be in ng/ml?

Activity 8.C Quantitative ELISA

Protocol

1. In the first 12-well microplate strip, label the outside wall of each well sequentially from **1–12**. These wells will contain the standards.

2. In the second 12-well microplate strip, label the first three wells with a **+** for the positive controls and the next three wells with a **–** for the negative controls. Label the last two sets of three wells for the test samples.

3. In the first microplate strip, pipet 50 µl of 1x PBS into the wells of the microplate strip labeled **2** through **12**.

4. Add 100 µl of 1,000 ng/ml antigen to the first well of the first microplate strip labeled **1**.

5. Perform a serial dilution from well **1** through well **11** in the following manner:

   a. Using a fresh pipet tip, pipet 50 µl out of well **1** and add it to well **2**. Pipet up and down gently three times to mix the sample in well **2**.

   b. Using the same pipet tip, transfer 50 µl from well **2** to well **3** and mix as you did in step 5a.

   c. Repeat this transfer and mixing step, moving to the next well each time. **STOP** when you reach well **11** and discard 50 µl of solution from well **11** into a waste container.

6. In the second microplate strip, use a fresh pipet tip to pipet 50 µl of the positive control (**+**) into the three **+** wells.

7. Use a fresh pipet tip to transfer 50 µl of the negative control (**–**) into the three **–** wells.

8. Use a fresh pipet tip to transfer 50 µl of each of the test samples into the corresponding wells, making sure to change pipet tips between each test sample.

9. Incubate the samples for 5 min at room temperature so that the antigen can bind to the plastic wells.

10. Washing protocol:

    a. Tip the microplate strip upside down onto a short stack of paper towels and gently tap the strip a few times. Make sure to avoid splashing sample back into the wells.

    b. Discard the wet paper towels.

    c. Use a transfer pipet to fill each well with wash buffer, taking care not to spill over into neighboring wells.

    Note: Use the same transfer pipet for all washing steps.

    d. Tip the microplate strip upside down onto a short stack of paper towels and gently tap the strip a few times.

    e. Discard the wet paper towels.

Activity 8.C Quantitative ELISA

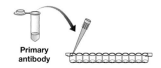

Primary antibody

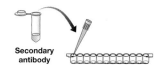

Secondary antibody

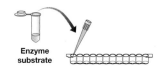

Enzyme substrate

11. Repeat step 10.

12. Use a fresh pipet tip to transfer 50 µl of primary antibody (**PA**) into each well of both microplate strips.

13. Incubate the samples at room temperature for 5 min so that the primary antibody can bind to the antigen.

14. Repeat step 10 **twice** to wash the unbound primary antibody out of the wells.

15. Pipet 50 µl of enzyme-linked antibody (**ELA**) into each well of both microplate strips.

16. Incubate the samples at room temperature for 5 min so that the enzyme-linked antibody can bind to the primary antibodies.

17. Repeat step 10 **three times** to wash the unbound enzyme-linked antibody out of the wells.

18. Use a fresh pipet tip to transfer 50 µl of enzyme substrate (**SUB**) from the brown tube into each well of both microplate strips.

19. Wait 5 min for the blue color to develop. (Optional) To stop the enzyme reaction, add 50 µl of 0.18 M sulfuric acid to each well in the same order the substrate was added. (In the presence of acid, the product will turn yellow.)

20. Visually compare the intensity of the color in the test wells to the intensity of the color in the standards. Identify the standard that most closely matches the test wells and estimate the concentration of antigen in the test samples.

21. If a microplate reader is available, insert the strips firmly into the microplate frame in the correct orientation. Place the frame (with the strips) into the microplate reader and close the lid. Read the absorbance of the samples using a 655 nm filter if the solutions in the wells are blue. Use a 450 nm filter if the solutions in the wells are yellow.

Results Analysis

In the first part of the analysis, estimate the concentrations of the test samples by visually comparing the intensity of the color in the experimental wells with the intensity of the color of the wells of the standards.

The concentrations of the antigen in the wells of the dilution series are:

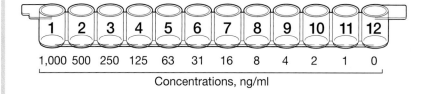

1	2	3	4	5	6	7	8	9	10	11	12
1,000	500	250	125	63	31	16	8	4	2	1	0

Concentrations, ng/ml

Activity 8.C Quantitative ELISA

If a microplate reader is used to measure the absorbance of each of the wells in the dilution series, generate a standard curve by plotting the concentration of the antigen in the standards (in ng/ml) against the absorbance in absorbance units (AU) using graph paper or graphing software.

Calculate the average absorbance and the standard deviation for all three wells of the positive control, negative control, and each test sample.

Last, use the standard curve to determine the average concentration of each test sample from the average absorbance values.

Postlab Focus Questions
1. What is the concentration of antigen in the unknown samples?
2. How did the data from the microplate reader compare to the visual estimation?

Self Assessment
Assess whether you have mastered the skills of this activity using the Laboratory Skills Assessment Rubric in Appendix E.

Assess your experimental write-up in your laboratory notebook using the Laboratory Notebook Rubric in Appendix F.

Activity 8.D Western Blotting

Overview

In this activity, you will use a western blot to identify a muscle protein, myosin light chain 1 (MLC1), from the full complement of proteins extracted from fish.

In a western blot, proteins are separated by SDS-PAGE and transferred to a membrane. The membrane is then probed with a primary antibody to a protein of interest, followed by incubation with an enzyme-linked secondary antibody. A colorimetric substrate for the enzyme is then used to identify where the protein of interest is located.

The beginning of this activity is similar to Activity 7.D. Proteins will be extracted from the muscle tissues of different fish species and separated by electrophoresis on a vertical polyacrylamide gel (see Activity 7.D). The proteins from the gel are then transferred or blotted to a solid membrane made of nitrocellulose. This transfer is usually performed by electroblotting, in which a stack of fiber pad, filter paper, gel, membrane, filter paper, and fiber pad (see Figure 8.31) is soaked in buffer and an electrical current is applied across the stack. During electroblotting, the current draws the negatively charged proteins from the gel onto the membrane. Proteins blotted onto the membrane form a mirror image of their position in the gel. Blotting deposits the proteins on the surface of the membrane and makes them more accessible to antibodies than they were when they were embedded in the gel. In addition, a membrane is easier to work with since it is less fragile and sturdier than a gel.

The membrane or blot is then incubated in blocking solution, which is a protein-rich solution. Blocking solution blocks protein-binding sites on the membrane that might otherwise nonspecifically bind the antibodies used for detection and result in background.

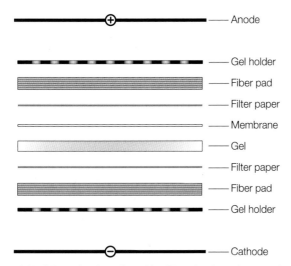

Figure 8.31. **Schematic of an electroblot**. The gel and membrane are layered between filter papers soaked in buffer and inserted into a gel holder cassette with fiber pads. The gel holder containing the gel is placed in buffer and a current is applied across the gel and membrane. The proteins move from the gel onto the surface of the membrane.

Blocking buffer is made from dried milk, which is an inexpensive source of proteins.

After blocking, the membrane is incubated with a primary antibody that recognizes MLC1. The primary antibody is then washed away and an enzyme-linked secondary antibody is added. Once the secondary antibody is washed away, the membrane is incubated with a substrate for the enzyme. Where the antibodies have bound, the substrate is catalyzed into a purple precipitate that binds to the membrane. The purple color corresponds to the band on the original gel that contained MLC1.

The primary antibody is a mouse monoclonal antibody generated against rabbit myofibrils. The secondary antibody is a goat anti-mouse antibody linked to horseradish peroxidase (HRP). The substrate is based on 4-chloro-1-naphthol (4CN), a colorless substrate that is oxidized and forms a purple precipitate in the presence of HRP.

Tips and Notes

- If available, protein extracts from Activity 7.D that have been stored at –20°C can be used. If protein extracts are available, it is not necessary to perform part 1 of the activity

- When setting up the western blot, it is vital that the membrane and filter papers are presoaked in blotting buffer and that the membrane is not allowed to dry out during the protocol

- Be sure there are no air bubbles between the gel and the membrane. The presence of bubbles will result in patches on the membrane where no proteins transferred

- Be sure to wear clean gloves at all times. Fingerprints contain proteins, which will be visible on the blot

- Ensure even distribution of the antibody by shaking the incubation tray when incubating the membrane with each antibody

Safety Reminders: Follow all laboratory safety procedures. Read and be familiar with all SDSs for this activity. Wear appropriate PPE (including gloves) when handling polyacrylamide gels and protein stain. Protein stain can stain hands and clothing. Attention should be given when using electricity in the presence of liquids. Safety measures in the equipment should prevent accidental exposures; however, should any leakage occur, notify your instructor before proceeding. Inform your instructor if you are allergic to any fish or seafood, and do not handle fish samples. If UV light is used, special precautions are requried.

Objectives

- Extract fish muscle proteins using Laemmli buffer
- Run samples on a polyacrylamide gel
- Blot proteins onto a nitrocellulose membrane
- Incubate the membrane with antibodies to MLC1
- Determine the size of MLC1

Activity 8.D Western Blotting

Research Questions
- Can myosin light chain 1 (MLC1) be detected in the fish tissue?
- What size is MLC1?

Skills to Master
Refer to Laboratory Skills Assessment Rubric (Appendix E) and Laboratory Notebook Rubric (Appendix F) for more details.
- Record laboratory notebook entries
- Extract proteins using Laemmli sample buffer (Activity 7.D)
- Perform SDS-PAGE (Activity 7.D)
- Use a power supply (Activity 4.C)
- Perform western blotting
- Generate a standard curve using a protein size standard (Activity 7.D)
- Determine protein molecular masses using a standard curve (Activity 7.D)

Student Workstation Materials

Part 1: Extracting Proteins

Items	Quantity
Water bath or dry bath at 95°C (shared)	1
100–1,000 µl adjustable-volume micropipet and tips, or 1 ml graduated transfer pipet	1
1.5 ml microcentrifuge tubes	5
1.5 ml screwcap microcentrifuge tubes	5
Microcentrifuge tube rack	1
Floating tube rack (optional)	1
Razor blade or scissors	1
Laboratory marking pen	1
Waste container	1
Laemmli sample buffer (LSB) with DTT	1.5 ml
Fish muscle samples	5 species

Part 2: Running the Gel

Items	Quantity
Mini-PROTEAN® Tetra cell electrophoresis module (shared)	1
UV light source (optional) (shared)	1
Sample loading guide (optional) (shared)	1
Buffer dam (optional)*	1
Power supply (shared)	1
Water bath or dry bath at 95°C (shared)	1
Rocking platform (optional) (shared)	1
2–20 µl adjustable-volume micropipet	1
Prot/Elec™ pipet tips	7 tips
Laboratory marking pen	1
Waste container	1
1x Tris/glycine/SDS (TGS) electrophoresis buffer	500 ml per box
Fish protein extracts from part 1	5 species
Actin and myosin standard	12 µl
Precision Plus Protein™ Kaleidoscope™ prestained standards (Std)	6 µl

4–20% 10-well Mini-PROTEAN® TGX™ gel or 4-20% 10-well Mini-PROTEAN TGX Stain-Free Protein Gel **	1

* A buffer dam is required if 1 gel is run in the Mini-PROTEAN Tetra cell.
** Alternative visualization of proteins may be used by using a 4-20% 10-well Mini-PROTEAN TGX Stain-Free Protein Gel. See Appendix C for details.

Part 3: Electroblotting

Items	Quantity
Mini-PROTEAN Tetra cell electrophoresis tank and lid (shared)	1
Mini Trans-Blot® inner module (shared)	1
Gel holder	1
Fiber pads	2
Frozen cooling unit (shared)	1
Power supply (shared)	1
Rocking platform (optional) (shared)	1
Ruler	1
Trays to prepare and assemble blots	2
Gel opening key (shared)	1
Roller	1
Soft pencil	1
Blotting buffer	1 L
Blocking solution (optional)	25 ml
Blotting paper	2
Nitrocellulose membrane	1

Part 4: Detecting Proteins

Items	Quantity
Computers with graphing software (optional)	1–4
Ruler	1
Incubation tray	1
Semilog graph paper	4
Plastic wrap (shared)	1 roll
Paper towels	2 sheets
Laboratory marking pen	1
Waste container	1
Distilled water (dH$_2$O)	100 ml
Blocking solution (if not used in part 3)	25 ml
Anti-MLC1 primary antibody	10 ml
Enzyme-linked secondary antibody	10 ml
Substrate	10 ml
Wash buffer	200 ml

Prelab Focus Questions
1. Why are polyacrylamide gels used instead of agarose gels to analyze proteins?
2. MLC1 is approximately 22 kD, myosin heavy chain is 200 kD, and actin is 42 kD. Which protein will migrate fastest through the gel? Why?
3. Why are proteins blotted from the gel onto the membrane?
4. State two pieces of information the western blot result provides about the protein of interest.

Activity 8.D Western Blotting

Protocol

Part 1: Extracting Proteins

1. Label five 1.5 ml microcentrifuge tubes and five 1.5 ml screwcap tubes with the names of the five fish samples so there are two labeled tubes for each fish.

2. Add 250 µl of Laemmli sample buffer to each microcentrifuge tube (but not to the screwcap tubes).

3. Cut a piece (approximately a 2 mm cube) from each fish sample and transfer it into the corresponding microcentrifuge tube. Close the lids.

4. Flick the microcentrifuge tubes 15 times to agitate the fish in the sample buffer and incubate the tubes for 5 min at room temperature.

5. Carefully transfer the buffer from each microcentrifuge tube into the corresponding screwcap tube by pouring. **Do not transfer the fish.**

6. Heat the protein extracts in the screwcap tubes for 5 min at 95°C. Proceed to part 2 immediately or store protein extracts at –20°C for up to 6 months.

Part 2: Running the Gel

1. If your vertical electrophoresis system is different from the one shown here, follow your instructor's directions for setting it up.

2. Prepare the Mini-PROTEAN TGX gel by pulling the tape off the bottom of the gel and removing the comb.

3. Clamp the gel into the electrode assembly with the short plate facing inward. Clamp another gel or a buffer dam on the oppose side of the electrode assembly. Refer to How To… Set Up a Vertical Electrophoresis System on page 246 for more information.

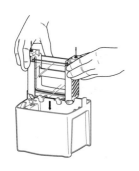

4. Place the electrode assembly into the gel tank, making sure to match the red electrode with the red side of the tank and the black electrode with the black side of the tank.

Activity 8.D Western Blotting

5. Fill the inner chamber of the electrode assembly completely with 1x TGS. Pour 1x TGS to the proper fill line of the outside chamber. If a sample loading guide is used, place it into the inner chamber.

6. Heat the fish samples and actin and myosin standard to 95°C for 2–5 min. (Do not heat the Precision Plus Protein standards.)

7. Load your gel according to the table:

Lane	Volume	Sample
1	empty	empty
2	empty	empty
3	5 µl	Precision Plus Protein Kaleidoscope prestained standards
4	5 µl	fish sample 1
5	5 µl	fish sample 2
6	5 µl	fish sample 3
7	5 µl	fish sample 4
8	5 µl	fish sample 5
9	5 µl	actin and myosin standard (AM)
10	empty	empty

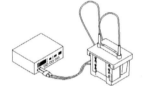

8. Place the lid on the electrophoresis chamber, making sure to first remove the sample loading guide if it was used. Connect the electrical leads to the power supply in the following orientation: red to red and black to black.

9. Turn on the power and run the gel at 200 V for 30 min. As the power is switched on, look for the release of bubbles from the electrode in the inner chamber. Watch the progression of the blue tracking dye in the Laemmli sample buffer and the separation of the prestained standards down the gel as the electrophoresis progresses.

   Note: Some gel formulations can be run at higher voltage, such as 300 V for 18 min. Check manufacturer guidelines.

10. When the electrophoresis run is complete, turn off the power and remove the lid from the chamber. Remove the electrode assembly and pour the buffer from the inner chamber back into the tank. Remove the gel cassette from the electrode assembly.

11. Examine the gel and record your observations in your laboratory notebook. The protein size standards should be visible in the gel. If this is not the case, inform your instructor.

12. Proceed to the next part. If necessary, store the gel in the gel cassette overnight at 4°C.

Part 3: Electroblotting

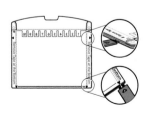

1. Remove the gel from the gel cassette using the opening key. Line up the arrows on the opening lever with the four arrows on the cassette to open the cassette.

2. Very gently pick up the gel and transfer it to a gel staining tray. Handle the gels from the bottom, not from the top.

Activity 8.D Western Blotting

3. Using a ruler, chop the top and bottom off the gel.

4. Equilibrate the gel in blotting buffer for 15 min on a rocking platform.

5. Soak the fiber pads thoroughly in blotting buffer.

6. Mark the nitrocellulose membrane with your initials using a pencil and wet the membrane in blotting buffer along with the blotting paper.

7. Make a blotting sandwich, being sure to remove air bubbles from between the layers:

    a. Obtain a container large enough to fit the plastic gel holder and add sufficient transfer buffer until the container is filled approximately 1 cm deep. Place the gel holder cassette in the container with the black side immersed in the buffer and the clear side outside of the buffer.

    b. Lay one wet fiber pad flat on the black plastic.

    c. Place a piece of wet blotting paper onto the fiber pad and roll out any air bubbles. Ensure that the buffer just covers the blotting paper. The liquid assists in squeezing air bubbles out of the sandwich.

    d. Place the gel squarely onto the blotting paper. Wet the roller and carefully roll over the gel to push out air bubbles.

    e. Carefully place the wet nitrocellulose membrane squarely onto the gel, making sure that the side with your initials is facing down. Move the membrane as little as possible once it has been placed on the gel since proteins begin to blot immediately. Roll out any air bubbles between the gel and membrane.

    f. Place a second sheet of wet blotting paper on top of the nitrocellulose membrane and roll out any air bubbles. Place the second wet fiber pad onto the blotting paper.

Activity 8.D Western Blotting

8. Fold the clear plastic side of the gel holder over the sandwich and clamp it to the black plastic side by sliding over the white clip. This tight fit will squeeze the sandwich together. Keep the gel holder partly submerged in blotting buffer.

9. Insert the gel holder into the Mini Trans-Blot inner module. Ensure that the black side of the gel holder is next to the black side of the Mini Trans-Blot module.

10. Place the inner module into the electrophoresis chamber. Add a frozen cooling unit and fill the chamber with blotting buffer to the level of the white clip on the gel holder.

    Note: The blotting buffer used to prepare the sandwich can be used in the tank.

11. Place the lid on the electrophoresis chamber. Connect the electrical leads to the power supply in the following orientation: red to red and black to black. Turn on the power and run the blot at 20 V for 2.5 hr.

    Note: Depending on your power supply, your instructor may request that the gels be blotted at 100 V for 30 min.

12. After blotting, proceed directly to the next part of the activity. Alternatively, store the blots in the tanks submerged in blotting buffer overnight at room temperature or dismantle the sandwiches and store the blots in blocking buffer overnight at 4°C.

13. After dismantling the blot, examine it and record your observations. The protein size standard should be visible on the membrane. If this is not the case, inform your instructor.

Part 4: Detecting Proteins

1. If the membrane is not blocked overnight, immerse it in 25 ml of blocking solution and incubate it for 15 min to 2 hr at room temperature on a rocking platform.

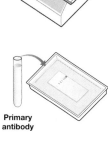

Primary antibody

2. Discard the blocking solution and incubate the membrane with 10 ml of primary antibody for 10–20 min on a rocking platform. (The rocking platform should be set fast enough to ensure constant coverage of the membrane.)

3. Rinse the membrane quickly in 50 ml of wash buffer and then discard the wash buffer.

Wash buffer

4. Add 50 ml of wash buffer to the membrane for 3 min on the rocking platform at a medium speed setting.

Activity 8.D Western Blotting

Secondary antibody

Wash buffer

Substrate

5. Discard the wash and incubate membrane with 10 ml of secondary antibody for 5–15 min on the rocking platform at a fast speed setting.

6. Rinse the membrane quickly in 50 ml of wash buffer and discard the wash buffer.

7. Add 50 ml of wash buffer and wash the membrane for 3 min on the rocking platform at a medium speed setting.

8. Discard the wash buffer and add 10 ml of substrate.

9. Incubate the membrane with the substrate for 10–30 min, either with manual shaking or on a rocking platform. Watch the color development.

10. Rinse the membrane twice with distilled water and blot dry with paper towel.

11. Air dry for 30–60 min and then cover in plastic wrap.

Results Analysis

Record the banding pattern on the blot by tracing, photocopying, or imaging using digital imaging equipment. Alternatively, tape the plastic-covered blot into your laboratory notebook. Label the copy of the blot with the sizes of the bands in the Precision Plus Protein Kaleidoscope standard (see Figure 7.34). Use the protein standard as a guide to roughly estimate the size of the MLC1 band.

Note: The anti-MLC1 monoclonal antibody recognizes MLC1 most specifically, but it also recognizes another closely related member of the MLC family, myosin light chain 2 (MLC2). Therefore, there may be two bands in some lanes of the blot. MLC1 is the larger band. Make conclusions about what proteins are present in each of the fish samples.

Determine the Molecular Weight of MLC1

To make an accurate determination of the molecular weight (MW) of MLC1 from each fish sample, create a standard curve by plotting the known molecular masses of the proteins in the Precision Plus Protein Kaleidoscope prestained standards against the distance the proteins traveled from the top of the blot.

• Estimate the top of the blot by drawing a horizontal line (parallel to the bands) approximately 2 mm above the largest (250 kD) protein of the protein size standard

Activity 8.D Western Blotting

- Measure the distance traveled by each protein in the protein size standard from the top of the blot to the bottom of each band

- Use either semilog graph paper or graphing software such as Microsoft Excel to generate a standard curve. (Refer to Activity 4.D for directions on constructing a standard curve)

Once the standard curve has been generated, measure the distance traveled by the MLC1 band for each fish. Record the measurements in a table and use the standard curve to determine the size of the proteins.

How did your rough estimate compare to the value generated from the standard curve? Is the MLC1 protein from the different fish the same size?

Self Assessment

Assess whether you have mastered the skills of this activity using the Laboratory Skills Assessment Rubric in Appendix E.

Assess your experimental write-up in your laboratory notebook using the Laboratory Notebook Rubric in Appendix F.

Postlab Focus Questions

1. How many bands were visualized for each sample?
2. What is the molecular weight of MLC1 for each fish?
3. If the molecular weights of MLC1 were different among fish species, suggest one reason for this observation.
4. How can you use western blotting to identify actin protein in insect samples?

Chapter 8 Extension **Activities**

1. Develop an Ouchterlony assay to test for blood from other animals.
2. Develop an ELISA test to identify chicken blood proteins.
3. Develop an ELISA test to quantify the amount of *Bacillus subtilis* in soil samples.
4. Use bioinformatics to investigate the differences in the sizes of the MLC1 protein in different species.
5. Develop a western blot for actin.

Research Projects

Summary

Scientific research involves developing and testing hypotheses on the way to developing theories. A hallmark of scientific research is that scientists use the findings of other scientists as a foundation for their own work, and each researcher contributes something novel to the scientific community. Student research should be conducted in the same way: projects should be founded on existing information or even the work of previous students and should enable you to apply your theoretical knowledge and technical skills to a real-world problem. This develops the skills in critical thinking and problem solving required for scientific research and allows you to practice techniques you have learned in class. This chapter provides recommendations for student research projects, including tips for writing a project proposal, performing background research, and designing experiments. It also reviews the various venues available for conducting research and presenting results as well as the different methods for communicating findings and funding projects. Finally, more than 100 biotechnology project ideas are provided to stimulate your imagination for your own research projects and general tips for different types of projects are presented.

9.1 What Is Research?

Research is driven by questions. How does this happen? What is this? What happens if I change something? How does this compare to something else? Experiments are then conducted in a controlled manner to answer these questions.

A **scientific fact** is an observation about the world that has been repeatedly confirmed. For example, saying that an organism has a particular enzyme is simply stating a fact. Facts do not attempt to explain an observation; they are the observation.

A **scientific hypothesis** is an educated guess or explanation of an observation that provides a basis for experimental testing. A hypothesis may incorporate scientific theories, previous data, or a novel idea. As an example, a hypothesis may be that a particular enzyme from a thermophilic organism found in a hot spring works more efficiently at high temperatures because this has been found to be true of other similar enzymes. This assumption can be tested by purifying the enzyme from the organism, performing an enzymatic reaction, and testing whether the enzyme converts its substrate faster when heated.

A **scientific theory** is an explanation of multiple related observations or events based on multiple hypotheses and supported many times by experimental results obtained by many independent researchers. For example, kinetic theory predicts that an increase in temperature increases the speed of molecular movement and so increases the chances of molecular collisions. This has been demonstrated countless times. In fact, the hypothesis about the enzyme in the previous paragraph is based on kinetic theory in addition to other observations.

A theory can predict events and occurrences. It has been established and advanced by many scientists and is generally accepted as true; however, it is still subject to testing and can be changed or modified if evidence is found that does not support it. In contrast to a hypothesis, which can be developed by a single individual and tested in a single experiment, a theory is supported by many individuals and experiments, and is usually developed over many years.

If a hypothesis cannot be experimentally tested, then it is outside the bounds of science. Science, as defined by the National Academy of Sciences' Committee on Revising Science and Creationism, is the "use of evidence to construct testable explanations and predictions of natural phenomena as well as the knowledge generated through this process."

Finally, a **scientific law** is based on repeated experimental observations and describes a natural phenomenon. It is a detailed description, usually involving math and equations, of how something happens. It does not, however, explain the phenomenon or why it happens. As an example, the ideal gas law, $PV = nRT$, describes the relationship between pressure (P), volume (V), moles (n), and

temperature (T) of a gas. It can and has been tested many, many times and provides a basis for predicting behaviors of gases and temperatures. Like a scientific theory, a scientific law can be changed if evidence appears to justify such a change.

Deductive and Inductive Research

Research is conducted in two ways: deductively and inductively (see Figure 9.1). Inductive and deductive research are complementary. For example, a hypothesis based on observations (inductive research) may then be tested deductively.

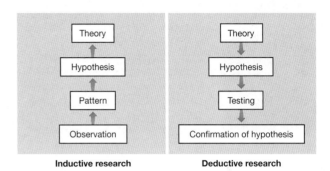

Figure 9.1. **Inductive and deductive research**.

Inductive research begins with observations. The researcher looks for patterns in data that may explain observations and develops a hypothesis based on these observations. These observations may then develop into a theory. For example, scientists may observe that a drug evokes an unexpected response in patients. They may then hypothesize that the drug causes this response because it upregulates a specific signaling pathway. This is inductive research; an observation resulting in a hypothesis.

Deductive research begins with a theory that leads to a hypothesis, which is then tested. For example, a scientist may hypothesize (based on previous experiments) that a molecule of a particular shape will block a signaling pathway. This may lead the scientist to test whether the molecule blocks the signaling pathway in a cell culture experiment. In this example, the scientist based a hypothesis on experimental data and then tested the hypothesis.

Conducting Research in the Real World

Many different factors drive research. In a biotechnology company, the goals of the company, the products it wants to develop, and its markets determine the types of research projects conducted. In academia, on the other hand, the driving force is often a curiosity about a natural phenomenon, a desire to understand more about a disease, or an interest in how some factor affects a system under investigation. These approaches complement each other to further developments that benefit the human race.

Research into a specific area may be performed by an individual, a group of scientists, or an entire research community (where each researcher may investigate a different aspect of a broad topic). Junior scientists usually work under the mentorship or supervision of more senior scientists who help define research questions, design the experiments, and analyze the data. Research is often

Sunny Choe, PhD

Senior Medical Scientist
Gilead Sciences, Inc.

Photo courtesy of Sunny Choe

Sunny is a Senior Medical Scientist at Gilead Sciences, Inc., a pharmaceutical company based in Foster City, California. Her role is to provide medical, clinical, and scientific information about Gilead's pharmaceutical products to help healthcare providers and their patients understand the options available to them. She also uses this expertise to assist clinical research studies, and she speaks at conferences, organizes and hosts Medical Affairs Advisory Boards, and trains speakers and sales staff.

Sunny had been interested in medicine since childhood, and this interest intensified after she took her first biology courses at Cornell University in Ithaca, New York. After earning a BA in biology in 1993, she worked as a research scientist in an HIV laboratory for several years as a way to help her choose whether to become a doctor or a researcher. Choosing research, she started graduate school in 1997 at Stanford University in Palo Alto, California, where she studied molecular pharmacology and virology.

After obtaining her doctorate in 2004, Sunny worked as a research scientist at ViroLogic, Inc. (subsequently Monogram Biosciences Inc.), studying HIV and HCV drug resistance. There, she developed assays that assess drug resistance to HIV and HCV antiretroviral therapies. It was during this time that while talking to a friend she learned about and became a Medical Scientist (also known as Medical Science Liaison, MSL). As an MSL, she educated healthcare providers about HIV and HCV drug resistance, resistance testing, and how to interpret assay results. She also collaborated on research studies and presented findings at scientific conferences. Sunny then went to work as an HIV/HCV MSL for Bristol-Myers Squibb and then to her current position at Gilead Sciences.

Sunny says that being a Medical Scientist satisfies both her interest in virus-mediated disease and her love of teaching. It is a profession that is in great demand. As specialty medicines become more common, healthcare providers face an increasing number of treatment options for their patients, but they have less time available to learn about them. The Medical Scientist provides important education about the options available to providers and their patients.

Sunny's advice to anyone starting a career in biotechnology is to not limit how and in what direction to take your career. Networking helps you meet people within your area of interest, but also can help you discover new career options. Find a good mentor to help guide you, and be creative — use internships, fellowships, volunteer activities, or even related jobs to gain experience in the field or area you wish to pursue. Though your path may not be direct, be flexible and you may find yourself in a profession you never knew existed.

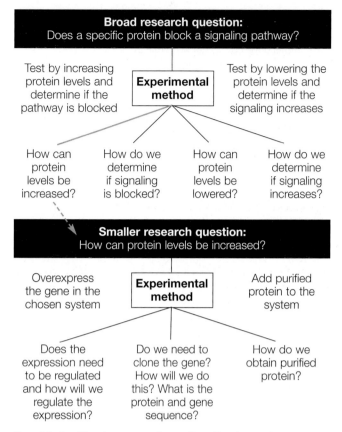

Figure 9.2. **Breaking down research questions**. Broad research questions are broken down into smaller, more manageable questions that can be addressed individually. Note that each smaller research question is a project in itself that could take the work of several scientists over many months to answer and can also be broken down into even smaller questions.

based on previous work conducted by the company or laboratory and is designed to find out more about an area in which the scientists already have expertise.

Most initial **research questions** are expansive and too broad to be tested by a single experiment or even by a single individual. Researchers can break down these broad questions into a series of smaller questions that can each be addressed by single experiments (see Figure 9.2). For example, a broad research question may be: does a particular protein block a signaling pathway? Different experimental approaches may be able to answer this question; for example, artificially increasing the level of the protein should block the pathway, while lowering the protein levels should increase signaling. In each case, a scientist needs to determine how to increase or decrease the protein levels, which raises additional questions that also need to be answered by additional experiments, and so on. By breaking down a broad research question into smaller questions and then answering the small questions, scientists lay foundations and then build on those foundations to address their original research question.

In academia, the head of the laboratory directs and coordinates the scientists in the laboratory to ensure that their individual research aligns with and makes progress toward an overarching research interest of the laboratory. Each scientist in the laboratory builds on the work of previous researchers to answer questions that contribute to the common laboratory goal. In a biotechnology company, the experimental work to resolve the small questions may be performed by different groups with expertise in particular areas and is directed by a lead scientist. For example, when developing a complex instrument for separating proteins, biochemists work on the protocol, mechanical engineers work on the pumps and moving parts, software engineers write the software, and electrical engineers design the boards and circuits.

Throughout any research project, communication between all the researchers and leads is critical. Clear verbal and written procedures must be provided and followed. In addition, team members must be informed of the project progress and roadblocks, so that experiments proceed smoothly and efficiently. If anything goes wrong or if mistakes are made, a researcher must feel comfortable coming forward to explain the situation. In the biotechnology industry this is critical to stop faulty but expensive production runs (or experiments) or to prevent harmful products from going to market. Similarly, students must keep teachers and other mentors aware of progress, problems, and even mistakes — they will help solve problems, address any potentially dangerous situations, and keep the research on track. It is a good idea to schedule regular (daily, weekly, or monthly) project review meetings with your research partners, teachers, and mentors.

Peer Review

Scientific progress relies on scientific research being conducted properly, meaning that appropriate controls are performed, valid conclusions are drawn, and the research is conducted in an ethical manner. Because only experts in the same field as the research being presented can judge its validity, these experts are called upon to review the validity and significance of the work. This is called peer review since the scientists' work is judged by their peers. The need to defend research against the criticisms of others strengthens the quality of science and maintains the credibility of scientific discovery.

Peer review can be informal or formal. Informal review can include comments and questions received during group meetings or at scientific meetings, at poster or oral presentations. Formal review is the official process where grants, research articles, and abstracts are reviewed prior to being funded or accepted for publication in a scientific journal. Reviewers question such things as was the research conducted significant? Were the methods used appropriate, or could something be done differently in a next set of experiments to obtain clearer or more robust results? Are the data analyzed correctly? Were proper positive and negative controls used? Are the conclusions valid, and what can be done to strengthen them?

Once peer review is conducted, a researcher must respond to the critiques received. Are the critiques valid? If so, what can be done to address them? If not, why not? All critiques require a response. Ultimately, it is the responsibility of all people involved in a project to evaluate their own and each other's work, to critically evaluate methods and results, and to incorporate critiques to improve upon the experiments they perform.

Clinical Trials and Data Fraud

"Publish or perish" is a common saying in research. It describes the real pressure researchers feel to publish their results — often, and in highly regarded journals — if they want to advance their careers. Sometimes, the drive to succeed can push researchers over an ethical line, and they modify or even create fraudulent data to move their research and careers forward. This can have many serious effects.

First, research fraud has a real human impact. If fraudulent data cover early stages of research, they lead to a huge waste of money later in subsequent bigger, more expensive studies involving people and clinical trials. A 2011 study showed that at least 6,500 patients received treatments in studies that were later retracted for fraud. Also, if fraudulent research is connected with later-stage clinical studies, patients may be treated with therapies that are ineffective or even dangerous.

A case in point is that of Paolo Macchiarini, an Italian surgeon who pioneered artificial trachea transplants. Macchiarini published research stating that, when seeded with a patient's own stem cells, the transplanted trachea developed into functional trachea. Unfortunately, the studies exaggerated the outcomes of his patients, and they falsely implied he had received ethical regulatory approval for his work, which he had not. In reality, all but one of his patients died of complications related to the transplants, and the one survivor had his transplant removed. Macchiarini was fired from the institute in which he worked, and he was found guilty of scientific misconduct in several papers. Despite this and the retraction of several of his papers, Macchiarini managed to secure a new position at a university in Russia as well as funding for similar research (also flawed) with esophageal transplants.

If papers are peer-reviewed, how does such fraud occur? Should some of the responsibility of publishing fraudulent studies rest with the publishers? Aside from duplicating the research to determine its validity, what can be done? Some journals are adopting statistical tests designed to detect fraudulent data in clinical studies. The tests look at how much the results deviate from what might normally be expected, with the idea that when scientists falsify data, they tend to make t too neat. Using these tests, investigators uncovered one of the largest cases of data fraud in history: Yoshitaka Fujii, an anesthesiologist studying a drug to prevent nausea and vomiting after surgery, is now the record holder for retractions, with 183. Should all journals use these techniques to uncover fraud during the review period for each paper? How can they be sure the tests are accurate? What would be the repercussions of false accusations? Are they worth the risks?

Sharing of Scientific Information

All the scientific advances in the world have been built on scientific discoveries of the past. The best science requires discussion and input from other scientists. Scientists constantly challenge and question each other's data, and this discourse results in high-quality research. Scientific discoveries are most reliable when different laboratories using different methods come to the same conclusions. Scientists share their discoveries with other scientists in many ways.

Peer-Reviewed Publications

The traditional way scientists formally share their work is by writing scientific papers and submitting them to scientific journals. The journal publisher then sends the papers to other respected scientists in the same field of study (peers) who review the work and send their opinions back to the publisher. Depending on the result of the review, the paper is accepted for publication or rejected if the work is not of sufficient quality. The scientist who submitted the paper may also be asked to perform some more work or provide additional data to meet the reviewers' criteria prior to publication.

Journals have different impacts on the scientific community depending on the stringency of their acceptance criteria. The scientific impact of a paper or journal is often quantified by counting the number of times it is cited (referenced) in future papers. *Nature*, a British journal first published in 1869, and *Science*, a publication of the American Association for the Advancement of Science (AAAS), are both interdisciplinary scientific journals that are considered the world's most prestigious scientific publications.

Conferences, Meetings, Seminars, and Posters

A less formal exchange of scientific ideas occurs through scientific meetings and seminars. Scientific meetings are conferences at which a group of scientists interested in a particular field of study congregate. Here scientists give talks or seminars to present their latest research. Junior scientists might present their work as posters during a poster session, standing by their posters to explain the data to other interested researchers. This format of scientific communication allows the audience to provide feedback to the researchers, both positive and negative, and allows scientists to network, find potential collaborators, and learn about new discoveries in their field prior to publication of a scientific paper.

9.2 Student Research Projects

For students, performing independent research is challenging and requires hard work, dedication, and commitment. An individual or a team is responsible for coming up with the ideas and converting those ideas into viable experiments. At times this can be difficult and frustrating, but working independently through a problem and finding a solution is extremely rewarding and is the best way to learn to be a scientist.

Student research projects can take on many different forms. An instructor may guide the project or the student may be entirely responsible for the project idea and its execution. Projects may involve the entire class, student teams, or be performed on an individual basis. The goals of the projects may also be different: the main goal could be to master laboratory skills or it could be to generate a winning entry in a science fair or poster competition. The type of project chosen will depend on the goal, the students' level of expertise, and the resources available.

Table 9.1. **U.S. and international high school research competitions**. Participation in national or international events usually requires first winning local or regional science fairs or competitions, or submitting a research paper to the competition. Details can be found on the website for the competition. If selected to attend an event, finalists are usually asked to prepare a poster and/or a short oral presentation on their work and to answer questions from judges. Note that competitions are subject to change; sites were accessed April 9, 2018.

Competition	Sponsor and/or Association	Entry Information
Intel International Science and Engineering Fair (student.societyforscience.org/intel-isef)	Intel	International science fair competition for all subjects Entrants qualified through regional science fairs
National FFA Agri-Science Fair (ffa.org/agrisciencefair)	Future Farmers of America, Cargill, Chevrolet, John Deere, and Syngenta	Agricultural science fair Entrants selected from local and regional agri-science fairs
Regeneron Science Talent Search (regeneron.com/science-talent-search)	Regeneron	Research competition Entrants qualified through submission of a research paper
National Junior Science and Humanities Symposia (jshs.org)	U.S. military	Research competition Entrants qualified through submission of a research paper to regional events
International Genetically Engineering Machine Competition (igem.org)	iGEM Foundation	International competition for application of synthetic biology Student teams with two academic instructors can enter to compete

Whole Class Projects

Different types of projects can be conducted by the whole class and the approach taken will depend on the goals for the research. For example:

- The class can conduct the same research in the same way. This approach lends power to the experimental results, which is important if the research generates novel data. If an experiment has been performed multiple times by different researchers with the same result, the results are much more likely to be valid than if the experiment were performed just once

- The class can approach the same research question but with each team using a different experimental approach. Teams can compete against each other, presenting their projects to the class and defending their experimental approaches and conclusions

- The class can investigate the same broad research question, with each team investigating a different aspect of the question. This can culminate in a research symposium where each team presents its findings to the class

Group Projects

Student teams may be assigned projects as part of a class assignment, be given a research area to explore, or be given the freedom to choose their own project. Working as part of a research team is common in the real world. Team work is rewarding since different areas of expertise are brought to bear on a problem. It also requires a collaborative approach and good communication skills to ensure that everyone on a team is included in the research.

When creating a research team, first find out about each others' strengths, weaknesses, and research interests. Then decide how to coordinate background research and whether you will conduct experiments together or divide the work to address more aspects of the research. Finally, decide when and how you will communicate the results of experiments to the entire team. Teams can achieve much more than individuals working alone but effort must be made to ensure that the team runs smoothly and that everyone has a voice.

Individual Projects

Science fairs and research competitions may require that a project be performed by an individual. Individual research projects can be challenging but they also offer a great sense of personal achievement. Individual projects often have the highest workload and take more time since the individual is responsible for every task. The initial plan, therefore, may need to be scaled back to fit the project into the time available.

Collaborating with Scientists

A project may be part of a larger collaboration with a researcher at a university or biotechnology company. As previously discussed, research is often broken down into different questions and the project team may be responsible for one section of the research (for example, purifying plasmids developed by the researcher and testing whether they express a protein of interest). Collaborative projects provide real-world experience and build connections with researchers that may become helpful in the future.

When seeking a collaboration with a researcher, you will likely have more success if the researcher has a personal connection to you or someone you know or to the school. For example, look for someone who has presented to the class, has a connection through your instructor, or is a friend of a parent. Often, researchers have ideas they would love to explore but do not have the time; so you may be able to step in and take on these projects. When contacting a scientist with or without a personal connection, be aware that they may be too busy to respond.

To increase the likelihood of a positive response, find out as much as possible about the scientist's research and write a concise, introductory email proposing the project idea. Specify the help needed (such as access to reagents or information) and demonstrate a desire to collaborate. Get help with the email from your instructor and be open to ideas proposed by the scientist, even if they differ from your original proposal.

When performing collaborative research, ensure that the experiments include the appropriate controls and that they are completed on time and with proper records in a laboratory notebook. Proper documentation is vital to having the collaborator trust the data generated. Be sure to always send the researcher a report on the project at its completion, successful or not, and thank them for their assistance.

Safety note: Minors should always involve their instructors when communicating with any individual about their work and ensure that their instructor is copied on all communications.

Venues for Conducting Research or Entering Competitions

While research for its own sake is satisfying, presenting research to the greater scientific community is extremely rewarding and can also generate recognition and financial awards.

High School Research

In high school, the primary opportunities for conducting and presenting research are science fairs, competitions (see Table 9.1), and internships, though informal collaborative opportunities may also exist.

Science fairs provide opportunities to present research and to compete for awards and scholarships. Science fairs and competitions can be at local, regional, state, national, and international levels. Depending on the affiliation of the event, winners of local and regional fairs can progress to state, national, and international events (see Figure 9.3). Science Buddies (see Figure 9.4) offers information about science fairs and provides tips for creating a winning science fair project. Be sure to visit the website if you plan to enter a science fair.

Make sure your project conforms to the rules specified by each science fair or research competition. Rules are more stringent in

Figure 9.3. **Science fair winners can progress through competition levels**.

more advanced stages of competition. In some cases, the project plan must be reviewed by a committee and guidelines on using live organisms and recombinant DNA must be followed. For an example of science fair rules, download the latest Intel International Science and Engineering Fair International Rules and Guidelines (student.societyforscience.org/intel-isef-forms).

Research can also be performed as part of an internship at a college, university, or local company. Internships provide an opportunity to work directly with a researcher as part of a team. Opportunities like this are rare and can be competitive. Some internships can be obtained through formal programs run by local or regional groups. Alternatively, informal opportunities can be found by networking with researchers and offering to help out.

Figure 9.4. **Science Buddies**. Science Buddies is a nonprofit, online resource that provides free science fair ideas, answers, and tools for students (sciencebuddies.org).

Undergraduate Research

In college, the primary opportunities for conducting and presenting research extend to include specific coursework, scientific conferences, and academic or corporate internships. You may have the opportunity to conduct independent research as part of a laboratory class. In this setting, you will be responsible for both the project idea and the experimental design. You may also be expected to present the results of your project at departmental or college-wide poster sessions. Posters are often presented alongside the research of graduate students and postdoctoral researchers, and the sessions are attended by all levels of scientists at the college. In addition, you may be able to submit a poster for presentation at a scientific meeting. Most scientific associations have an annual meeting, and many encourage students to present their research (see Table 9.2).

There are many opportunities at the undergraduate level to gain research experience through volunteering or internships in research laboratories. The best way to find volunteer opportunities is to walk the halls of your university, read the research posters on the walls, and then stop by and talk with the head of the

laboratory. Internships may be local or informal and performed at the college or university. Alternatively, there are many formal internship programs in the U.S. and elsewhere that may be at another university, institute, or government facility. See Table 9.3 for a selection of formal internship and research opportunities.

Many companies also offer both paid and unpaid internships that can provide exposure to the business environment as well as provide laboratory research experience. Company websites usually have information on internships in their careers/jobs section.

Experience gained from internships gives strength to a resumé upon graduation and can make a difference when applying for jobs. Not only does an internship help secure the technical skills required for the job but a potential employer will also appreciate that you chose to spend your summer breaks gaining experience.

Choosing a Research Project

As in the real world, a project may be assigned to you or you may have the opportunity to develop your own project idea. Even if a research topic is assigned, you will need to decide on the specific research question to answer and the experiments to perform.

Table 9.2. **Associations or conferences that encourage submission of student research**. A selection of associations and opportunities in the U.S. to present research. Websites were accessed April 9, 2018.

Association/Venue	Event
State Academies of Science	Annual meetings where student research competitions may be sponsored
American Association for the Advancement of Science (AAAS) (aaas.org)	Research presentation competition for undergraduates is held at the annual meeting in a major North American city each year
	Four regional divisions have similar competitions
American Institute for Biological Sciences (AIBS) (aibs.org)	Undergraduate research competition is held at the annual conference
Council on Undergraduate Education (CUR) (cur.org/)	Posters on the Hill conference on Capitol Hill in Washington, D.C. invites ~70 undergraduates to present their research and meet their political representatives
	National Conferences on Undergraduate Research is an annual event where students can present research in all disciplines
Butler University, Indianapolis, IN (urc.butler.edu)	Undergraduate research conference held on campus every year, typically hosting ~600 students from more than 30 midwestern colleges and universities

Designing a project from scratch can be intimidating since there are so many possibilities. Begin by considering the following questions:

- What techniques and subject areas have caught your interest as you have progressed through this course?

- What interests you outside of school?

- Do you have any hobbies that could be integrated into your project?

- Have you noticed a natural phenomenon that has piqued your curiosity and that could be the basis of a hypothesis?

- Is there a project performed by a previous student that can be extended or added to?

Later in this chapter, a list of project ideas is presented that may spark your imagination and give you some ideas for topic areas to investigate. Also, websites such as Science Buddies (sciencebuddies.org) can help with science fair projects and provide other online tools that can help narrow down a topic area.

Scoping a Research Project

Once you have a project idea, consider what is actually feasible given your skill set, timeline, and resource constraints. Ensure that you have the skills to address the project you choose. A well-executed, simple project will be more satisfying for you and more impressive to science fair or poster judges than a poorly executed, complex project. Consider the following questions:

- What skills do I/we have?

- What do I/we already know about the research area?

- Can I/we find a mentor?

- How much time can I/we commit to the project?

- What equipment and supplies are available?

- Do I/we have a budget to purchase more supplies?

- Does the project meet safety guidelines?

If the research area is totally new to you, then you will need to invest a lot of time to learn about the subject matter. Consider the

Table 9.3. **Formal undergraduate internship and research opportunities**. Programs are subject to change. Websites were accessed April 9, 2018.

Institute/Program	Program Name	Program Details
National Institutes of Health (NIH), Bethesda, MD (training.nih.gov/programs/ugsp)	The Undergraduate Scholarship Program	Scholarships for students from disadvantaged backgrounds combined with paid summer research training at the NIH during the summer as well as postgraduate employment.
National Institutes of Health (NIH), Bethesda, MD (training.nih.gov/programs/sip)	Summer Internship Program in Biomedical Research	Internships for high school and college students.
Cold Spring Harbor Laboratory, Cold Spring Harbor, NY (cshl.edu/education/undergraduate-research-program/)	The Undergraduate Research Program	Summer internships for students from around the world in many life science research fields.
National Aeronautics and Space Administration (NASA) (nasa.gov)	Various	Spring, summer, and fall opportunities that combine scientific research with professional hands-on engineering at NASA centers and with partners in the U.S.
Smithsonian Institution, National Museum of Natural History, Washington, D.C. (smithsonianofi.com/internship-opportunities/)	Various	Spring, summer, and fall opportunities that range from short-term part-time appointments to full-time year-long commitments.
U.S. Department of Energy (DOE) (energy.gov/student-programs-and-internships)	Various	Laboratory internships at the DOE national laboratories and technology centers.
EURO Scholars (euroscholars.com)	European Undergraduate Research Opportunities	A semester-long study-abroad program that combines research with coursework at a variety of European universities.
German Academic Exchange Service (daad.de/rise/en/)	Research Internships in Science and Engineering (RISE)	A summer opportunity to work on serious research with doctoral students at a German university or institute.
National Science Foundation (NSF) (nsf.gov/crssprgm/reu/)	Research Experiences for Undergraduates (REU)	An opportunity for a group of approximately ten undergraduates in institutions with an REU site to work on a specific research project with faculty and researchers.

Biotech In The Real World

Regeneron Science
Talent Search

Each year since 1942, the Society for Science & the Public (the Society), a non-profit organization based in Washington, D.C., has organized a national science talent search. This science competition has been sponsored in the past by industry giants Westinghouse and Intel Corporation, and is currently sponsored by Regeneron. It provides an opportunity for the country's best and brightest young scientists to conduct and present original research to nationally recognized professional scientists.

The Regeneron Science Talent Search is the nation's most prestigious science research competition for high school seniors. Any U.S. student who is attending his or her last year of secondary school may apply, and any independent, individual research is eligible. Students perform independent science research and write a research report that is similar to a journal article, explaining their experiments and conclusions. Entries are reviewed by three or more PhD scientists, mathematicians, or engineers in the subject area of the entry. The judges look for students who exhibit exceptional research skills, a commitment to academics and their extracurricular pursuits, innovative thinking, and promise as a scientist.

Chris Ayers, Science News, 2017. Used with permission

The judges name 300 students as Scholars, and each scholar and their school are awarded $2,000. From that select pool, 40 finalists win an all-expenses-paid trip to Washington, D.C., and the Regeneron Science Talent Institute, where they present their research to the public, meet with notable scientists, and compete for $1.8 million in awards. Scholars and finalists have come from public and private high schools across the country, and their research has covered areas as diverse as molecular and cell biology, bioengineering, genetics, mathematics, and thermodynamics. Each finalist receives a minimum $25,000 award, with a top award of $250,000. Alumni of this event hold more than 100 of the world's most coveted science and math honors, including the Nobel Prize and the National Medal of Science.

For more information, or to find out how you can enter the competition, go to student.societyforscience.org/regeneron-sts.

time available to do the research: multiple hours in one day will allow you to perform more complex experiments than having one hour twice a week. Projects are often limited by the amount of supplies available; this will also likely be a major constraint on your project. You may be able to raise funds to obtain supplies (see Funding Research Projects on page 343).

Enlist the help of your instructor in narrowing down your research project; however, do not expect your instructor or anyone else to develop your proposal or conduct your research. If your research question is too broad, try to break it down into smaller, more manageable research questions and choose one question to address (see Figure 9.2). For example, focus on optimizing a protocol that would be necessary for the broad research question.

Student: _____

Proposed Project Title: _____

Research Question/Hypothesis to Be Tested: _____

_____

Background/Introduction
- Give relevant background information
- State reasons why this question is interesting or important
- Provide citations of articles reviewed in the references section

Experimental Methods
Summarize the experiments being proposed and how they answer the research question
- Summarize the protocol for each type of experiment
- Detail the experimental controls to be used
- Give the timeline
- State the number of trials to be completed

Materials Required
- List the materials needed and their quantities

Data Analysis Plan
- State how data will be analyzed
- Include the types of statistical analyses to be used

Safety
- State any safety issues
- State how hazardous waste will be disposed of
- State what PPE is required

Expected Results
- Describe the expected results for the controls and experimental samples
- Discuss how the experimental plan might be modified depending on different results
- Discuss how the expected results would prove or disprove your hypothesis

Budget
- List the prices of materials required with quantities and source of materials

References
- Provide proper citations for all materials reviewed

Figure 9.5. **Sample project proposal**.

Planning a Project

Before performing a single experiment, develop a **project proposal**, one that incorporates the sections described in Figure 9.5. This proposal will help limit the scope of the project, ensuring that it fits into your timeline, and is feasible, considering your skills and the resources you have at your disposal. You may need to change your project from your initial idea as you develop the proposal and discover what is feasible.

Performing Background Research

Thorough background research is essential in a project not only for understanding the general research area but for identifying what is already known and not known about the actual research question. Since it is unlikely that a textbook will have information in sufficient depth for the needs of the project, you will need to use other resources, such as the Internet, libraries, peer-reviewed journal articles, and specialists in the research area.

Once you have collected the information, take time to evaluate it and prepare a summary. Determine which pieces of information pertain to your project and which are irrelevant. Write a summary of the information for the project proposal and a **citation** for each source of the data in the references section. It is important to ensure that the information used is from a reliable source and that facts can be verified.

Internet

When the Internet is used properly, it is a fantastic resource; however, great care must be taken to validate the information found online. In addition, it is important to avoid distractions when working on the Internet and focus research on the question at hand.

The best approach to using the Internet for research is to perform a search using scientific search engines (see Table 9.4) using key words from the research question or hypothesis. Then spend a set amount of time visiting the resulting websites to learn more about the subject area. Choose sites that are more likely to contain valid information, such as government or academic/university sites. Bookmark the sites that contain useful information, and take notes that include the site reference. Once the project is more defined, focus on specific questions and ensure that all facts and data can be verified. Keep a record of the references for all the information you gather.

Libraries

Libraries and librarians are great resources. Librarians can help you navigate library databases and catalogs and point you to useful references, including reference books. Reference books are valuable in that the information they contain is usually clearly stated and has been reviewed. University libraries are also a great source for peer-reviewed journal articles (see next section).

Peer-Reviewed Journal Articles

The gold standard for scientific data is peer-reviewed journal articles. Remember from the Sharing of Scientific Information section that scientists publish their work as research papers (articles) that have been reviewed by other scientists. Journal articles can be difficult to understand, so use them as a resource only once you have a good understanding of the field.

There are two types of peer-reviewed journal articles: primary research articles (these provide data, methods, etc. and represent original work) and review articles (these summarize current research questions and knowledge). Review articles are useful as initial resources when learning more about a specific subject. All journal articles, however, reference other related articles and so provide a great starting point for digging deeper into a subject area. See How To… Read a Journal Article.

Most scientific journals publish articles online. The **abstract** of an article is the summary of the research. Abstracts are freely available online, while only a subset of complete journal articles are free (open access). Open access articles can be found using the resources in Table 9.6.

Scientific journals may be available at the library of a local university. University librarians should be able to help you navigate the library system if you have all the information on the desired article (title, authors, journal, journal volume, and page numbers). Alternatively, contact a college student who may have full access to the article through his/her school library and ask him/her to share it with you. If those approaches don't work, you can also contact the authors of the paper for a copy of the article. These options take time, so try to not get hung up trying to obtain a paper that is not open access; there will usually be a free publication that contains sufficient information for your project.

Mentors

Mentors can really help with a project. Mentors do not need to be be experts in your chosen area, but they should be familiar with scientific research. They can help navigate through background research as well as guide experimental research. Finding a mentor

Table 9.4. **Resources to help find scientific and biological information**.

Resource	Website
VADLO Biological Search Engine	vadlo.com
Analytical Science Digital Library	asdlib.org
Wikipedia*	wikipedia.org
National Institutes of Health	nih.gov
Centers for Disease Control and Prevention	cdc.gov
Environmental Protection Agency	epa.gov
Howard Hughes Medical Institute	hhmi.org
United States Food and Drug Administration	fda.gov

* Wikipedia is a great starting point for information but contains many errors. Any facts or data from Wikipedia need to be verified, including links to the source of the information.
Note: Websites are subject to changes. Websites were accessed April 9, 2018.

Read a Journal Article

Reading journal articles or research papers is a learned skill. Initially, research papers may seem complicated and difficult, especially if you are not familiar with the topic. However, understanding their structure and organization can help. Research papers are organized in sections. Journals organize or combine the sections in different ways, but they follow a similar pattern. The major sections are described in Table 9.5 along with reading strategies to help you understand the research.

It can feel overwhelming to read a scientific paper from start to finish. Begin by reading the introduction, then the figures and results, and finally looking at the abstract, and discussion. But it's not always necessary to read a paper in the order it is presented. After reading through a few papers, find what reading strategy feels best for you.

Table 9.5. **Sections of a journal article**.

Section	Description	Strategies for Understanding While Reading
Title	Typically a short statement of the key point or topic of the paper.	• Write a summary of what the paper is about in your own words • Look up the parts of the title you do not understand
Authors	A list of people who contributed to the work and paper. The order usually shows the amount of work each author did; the first author did most of the work and organized the manuscript; the last author is usually the principal investigator of the lab.	• Determine who the authors are affiliated with. Are they at a university or college, industry, or a government institution? • Check footnotes for which author to contact with questions or to request more information
Abstract	A short summary of the paper. This is freely available for every paper and can help determine whether the article contains information of interest.	• Read to get an overview of the article and the research question • Read each sentence, one at a time, and check your understanding • Look up any words you do not understand • Try to summarize the abstract in a few sentences
Introduction	An overview of the research topic and experimental plan. Provides references to other papers in the same field.	• As you read, make sure you understand why the research is taking place • Summarize each paragraph in your own words
Materials and methods	Details on how the experiments were conducted. This information also helps you find materials and describes techniques that may be necessary for your experiments.	• Begin by skimming the section • Consider if there is enough information to be able to replicate the study. If necessary, look up methods that are not clearly described
Figures	Graphical representations of data illustrating the outcomes of experiments or diagrams that describe scientific models. Figures are further described in the text of the paper.	• Stop and study the figures • Before reading the caption try to understand the data and results presented in the figure • After reading the caption, try to write a sentence or two that summarizes the results shown in the figure and how it helps answer the experimental question
Results	The data generated by the research, presented in figures, tables, and graphs with a narrative to explain the research findings.	• As you read, ask yourself, what were the major findings? Did the data support the findings? Were the appropriate controls in place? • If there are complex statistics in the results, try to understand them, but don't focus on them • Refer back to the figures as you read to help clarify the text
Discussion	A summary of the results and explanations of the conclusions the authors have drawn from the data. In addition, the implications of these findings to the broader research community and recommendations for future research are presented.	• As you read, consider whether the conclusions are supported by the data. Do you agree with the conclusions based on the data presented? Are there other interpretations of the data?
References	A list of citations used in the paper that provide background or supporting information on a particular topic or technique.	• Use the references to look up additional information needed • The abstract of a paper is freely available online if you search for the title of the paper

Table 9.6. **Sources for open access journal articles**.

Resource	Information Provided
PubMed Central (ncbi.nlm.nih.gov/pmc/)	Free digital archive of life science journals provided by the National Institutes of Health.
Google Scholar (scholar.google.com)	Searches academic journals and provides a link to free versions of the articles, if they are available.
Microsoft Academic Research (academic.research.microsoft.com)	Searches academic journals and provides a link to free versions of the articles, if they are available.
Directory of Open Access Journals (doaj.org)	A directory of open access journals.
BioMed Central (biomedcentral.com)	A publisher of open access journals.
Public Library of Science "PLoS" (plos.org)	A publisher of open access journals.

Note: Websites are subject to changes. Websites were accessed April 9, 2018.

is similar to finding a collaborator (see Collaborating With Scientists), and a collaborator may act as a mentor. Always inform your instructor of all communication with mentors, and conduct any face-to-face meetings involving minors in the presence of a guardian.

Experimental Design

Every experiment should be performed in such a way that the researcher has confidence in the results and is sure that the experiment proceeded as expected. While this may sound simple, the interpretation of experiments is frequently difficult and each experiment requires time and money. The time taken early to properly plan experiments, therefore, is always worth the effort.

Try your best to truly understand the process you are investigating and the methods you are using. Use this knowledge to anticipate all possible outcomes, what each might mean, and the next steps to take should they occur. This will ensure your methods and approach are well thought out, and that all appropriate controls are included.

Remember to include peer review into your projects at multiple stages. Ask teachers, other students, and even scientists for advice regarding your project even before you start. Use their questions and critiques to fine tune your experiments before you run them. Present your results to them as well, and use their input, questions, and concerns to guide you as you refine your experiments, or as you write up your results. Anticipate the questions and concerns reviewers may have and address them ahead of time.

Methods

For each research question, you may have a choice of different methods or techniques you can use. Make sure you understand all of your options, their strengths and weaknesses, how each method differs in terms of the data it generates, and which method may be accessible to you in terms of cost, time, and expertise. Some techniques may call for expensive equipment that requires years of expertise to run, for example. In these cases, you may be able to partner with a facility that can run these analyses for you, or you may be able to find a cheaper, simpler alternative.

Whatever you choose as your method, determine whether you will need to design tests to demonstrate the assay or procedure is running as expected. Choosing appropriate controls is a first step (see Controls, below), but in some cases additional tests like instrument calibrations may be needed. In the biotechnology industry, **assay validation** is the process of proving your assay or process is consistently performing correctly and giving the expected outcome. Validated test methods are required for full compliance with good manufacturing practice (GMP) and good laboratory practice (GLP) regulations.

Controls

All experiments should have controls to confirm that the experiment worked as planned. Controls are samples or conditions that should give an expected result, either positive or negative, that provide confidence that experimental results are valid. A **positive control** is one that generates a positive result when used according to the experimental conditions; it allows you to trust any negative results you observe. A **negative control** generates a negative result and allows you to trust any positive results you observe. If a positive or negative control gives an unexpected result, the results of the test samples cannot be trusted. Take the time to consider what might go wrong in an experiment and which controls would best reveal these problems. Keep in mind that many experiments require multiple controls.

For example, in a PCR reaction:

- A positive control may contain a piece of DNA that is known to generate a PCR product. If it generates a band of the expected size, then you can be confident that the reaction ran as expected. If the positive control yields no PCR product, on the other hand, this indicates a problem with the reaction (for example, the thermal cycler was set incorrectly, a component of the master mix was left out, or the enzyme was expired)

- A negative control might contain no template DNA. It should not produce any band. Any product generated with the negative control may indicate contamination of some component of the PCR reaction

Trials and Sample Size

For experimental results to have statistical significance, the number of trials or samples used must be of sufficient quantity to make a valid conclusion. Since your project may be restricted by the availability of materials and laboratory time, plan your experiments carefully to make the most of what is available to you. For example, you may need to scale back the number of data points you collect to allow replication of the entire experiment. Also be sure to plan ahead for how the data will be analyzed.

Data Analysis

Laboratory experiments generate data that must be collected and analyzed to determine whether the research question has been answered and the hypothesis is supported. If the data are quantitative, mathematical and statistical analyses will be required, including the use of graphs and calculation of averages, standard deviations, and percent error. Statistical analyses help researchers analyze and compare data within and among experiments.

As discussed in the Experimental Design section, good scientific experiments and results are reproducible. An experiment must generate the same result more than once. As a rule of thumb, plan to repeat each experiment at least three times before having any confidence in the data.

Example of Data Analysis

The following example describes a hypothetical experiment examining whether the heat shock step is required for bacterial transformation. The hypothesis is that transformation efficiency will be reduced if the heat shock step is removed.

The experiment is performed by transforming bacteria with and without heat shock and calculating the transformation efficiencies in three independent experiments. The simulated results are shown in Table 9.7. The data are analyzed to find the average (mean) and standard deviation for each treatment (Table 9.8); data are then presented in a graph (see Figure 9.6). This example is designed to provide a reminder of how to perform statistical analysis and how to draw conclusions from the analysis.

Table 9.7. **Simulated data from bacterial transformation with and without heat shock**.

	Transformation Efficiency (CFU/µg DNA)	
	With Heat Shock	**Without Heat Shock**
Trial 1	6,000	4,000
Trial 2	5,400	1,800
Trial 3	3,500	1,200

The raw data in Table 9.7 indicate transformation efficiency is greatly reduced when there is no heat shock, which supports the hypothesis. However, the transformation without heat shock in trial 1 generated more CFU/µg DNA than the transformation in trial 3 with heat shock. Is the conclusion correct?

Calculating the Mean

Since there are three trials, statistical analysis can be performed to determine how confident we can be in our conclusion. First, we calculate the mean ($\bar{x}$), or the average transformation efficiency, with and without heat shock. The mean is the midpoint in a distribution of data and is calculated by adding all the values ($x_1 + x_2 + x_3 + ...$) and dividing by the number (n) of values.

$$\bar{x} = \frac{x_1 + x_2 + x_3 + ...}{n}$$

In this example, there are three values for each condition (heat shock (+HS) or no heat shock (–HS)), so n = 3.

$$\bar{x}\ (+HS) = \frac{6,000 + 5,400 + 3,500}{3} = 4,967 \text{ CFU/µg DNA}$$

$$\bar{x}\ (-HS) = \frac{4,000 + 1,800 + 1,200}{3} = 2,333 \text{ CFU/µg DNA}$$

The average transformation efficiency with heat shock is much higher than the average transformation efficiency without heat shock, which suggests that heat shock results in higher transformation efficiency. However, the fact that one value with heat shock (in trial 3) was below one value without heat shock (in trial 1) has not been accounted for.

Calculating the Standard Deviation

Additional analyses are performed to determine the distribution of the data. If values are close together, this indicates higher reproducibility and the likelihood that the next time the experiment is performed the outcome will have a value close to the mean. This variance from the values to the mean is calculated as the standard deviation. To calculate standard deviation perform the following steps (see Table 9.8 for an example):

1. List the values and the mean.

2. Subtract the mean from each value, which gives the deviation from the mean.

3. Square each value (that is, multiply it by itself).

4. Add all these squares to obtain a total sum.

5. Divide the total sum by one less than the number of values (n – 1); this is the variance.

6. Calculate the square root of the value; this is the standard deviation.

Though standard deviations can be calculated using a scientific calculator or a computer program like Microsoft Excel or Google Sheets, which have statistical functions, the theory behind the value and how it is derived should be understood.

Standard deviations can help determine the level of confidence you can have that data are statistically different from others and whether the data support the hypothesis. For example, if the mean transformation efficiencies in the earlier experiment are one standard deviation apart, then the data would be truly different and the hypothesis would be supported in ~68% of cases, but ~32% of the time the hypothesis would not be supported. If the means are two standard deviations apart, then in ~95% of cases the hypothesis would be supported.

In general, means are required to be two standard deviations apart to be considered significantly different, and this is often shown in a graphical format as error bars. Each error bar on the graph in Figure 9.6 shows the standard deviation above and below the mean. If error bars on two bars of the graph overlap (as they do here), then they are less than two standard deviations apart and are not considered statistically significant.

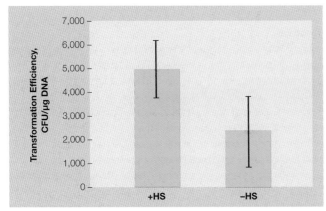

Figure 9.6. **The effect of heat shock on transformation efficiency (simulated data).** A histogram of the mean of transformation efficiencies with (+HS) and without (–HS) heat shock of three independent trials. Standard deviations are shown as error bars.

Drawing Conclusions from Statistical Data

The conclusion for the transformation experiment would be that, although the data suggest that transformation efficiency is reduced with no heat shock, they do not fully support the hypothesis. The recommendation for this research would be to repeat the experiment two or three more times. If the trend seen in the original data is reproduced, then the standard deviations will decrease upon having a larger sample size and will likely provide an acceptable level of confidence.

The level of data analysis you perform will depend on your level of statistical expertise. For example, the data in the transformation example can also be analyzed by a Student's t-test, which tests whether data support a null hypothesis (the hypothesis that there is no difference in the data; in this case, the hypothesis that heat shock has no effect on transformation). Different types of analyses may lead to different conclusions, which is why an understanding of statistical analyses is vital to data analysis.

Recruit mentors or math teachers to help with data analysis and to find the best way to interpret and present your data. Use online resources to help with statistical analyses. Be aware, however, that while scientific calculators and software can perform statistical calculations, the type of analysis chosen must be appropriate for the data set and this requires an understanding of the theory behind the statistics.

Funding Research Projects

In the research world, academic scientists write grant proposals to get research funded. These proposals describe the importance of the research to society and include a plan of the experiments and the materials required to perform the research. If the materials for your research project are not readily available, you may need to look for funding or other support, and this may need similar justification.

Table 9.8. **Standard deviation on transformation efficiency (TE) with/without heat shock (HS) (simulated data).**

	With Heat Shock			Without Heat Shock		
	TE, CFU/µg DNA	Deviation (Value – Mean)	Deviation Squared	TE, CFU/µg DNA	Deviation (Value – Mean)	Deviation Squared
Trial 1	6,000	1,033	1,067,089	4,000	1,667	2,778,889
Trial 2	5,400	433	187,489	1,800	–533	284,089
Trial 3	3,500	–1,467	2,152,089	1,200	–1,133	1,283,689
Average (mean)	4,967			2,333		
Sum of squared deviations			3,406,667			4,346,667
Variance (sum/(n – 1))			1,703,333			2,173,333
Standard deviation (√variance)			**1,305**			**1,474**

If the school does not have a budget to support student research projects, seek donations of supplies or funds. Scientific supply companies or local research institutions or hospitals may donate supplies. If your project is in the same area of interest as that of a local researcher, it is worth asking him/her for specific supplies such as enzymes or DNA. Specific requests to individuals are much more likely to yield success than generic requests to a general office. For nonscientific supplies, local businesses or big box stores will often make a donation to local schools. For funding, try presenting your research plan to a local nonprofit service group such as the Rotary, Kiwanis, Soroptimist, or Lions Club, or to local hospital foundations and businesses. Alternatively, try general fundraising methods like bake sales and car washes.

When requesting support, be sure to fully explain the research being carried out and how the donation will make the project possible. Be specific about your needs; include the quantities you need and a description of your planned experiments. Describe what you plan to do with the funds and how the research is relevant to society. Regardless of who helps you with your project, always write a letter of thanks and describe how their help contributed to your research. Not only is this polite but it helps to build your personal network and paves the way for students coming after you.

Table 9.9. **Outline of a scientific communication**.

Section	Details
Title	A succinct description of the research that is appropriate for the audience.
Abstract/ summary	A short summary of the research. This should be written after the report has been drafted and should give an overview of the purpose of the research, how it was done, the major findings, and the implications of the results.
Introduction	A summary of the background to the research and the overall research question explored, including any hypotheses.
Materials and methods	A review of all the procedures used in the research, including the reagents used.
Results	The data generated by the research shown as figures, tables, and graphs, and a narrative that shows how each figure contributes to the major findings.
Discussion/ conclusions	A summary of the data, a discussion of any issues with the data, a discussion of how the research answered the research question, any implications of the research, and suggestions for further research.
References/ bibliography	A list of the citations or references used in the communication.

9.3 Communicating Research

All the scientific advances in the world have been built on scientific discoveries of the past. The best science requires discussion and input from other scientists. Scientists constantly challenge and question each other's data, and this discourse results in high-quality research. Scientific discoveries are most reliable when different laboratories using various methods come to the same conclusions. Scientists share their discoveries with other scientists in many ways.

Once you complete your research project, you will need to communicate the results to the rest of the scientific community through poster presentations, research papers, or oral presentations. All presentations have a similar outline, which is summarized in Table 9.9. Posters and oral presentations provide an opportunity for direct feedback from an audience, while research papers include much more detail on the research and its findings.

Posters

Junior scientists frequently present their work as posters. Research posters include the sections discussed in Table 9.9 but in an abbreviated form highlighting the most important points.

The size and format of the poster are often mandated by the scientific meeting at which they are presented so check the rules on poster formats before beginning. Posters can be made using separate sheets of paper or cardstock mounted individually on a board or they can be printed as large-format, single sheets generated in a computer program like Microsoft PowerPoint or Google Slides (see Figure 9.7). The poster format depends not only on the rules of the meeting but also on the budget available since large-format prints can be expensive.

The text of the poster should be legible from at least 6' away so use a large and legible font. Generally, the title of the poster should have a font size of 80 point or more, headings should be 36 point, and text font should be 18–24 point. Also, use sans serif fonts like Helvetica or Arial, which are considered easier to read, instead of serif fonts like Times.

Organize the different sections of the poster so the reader's eyes will run up and down the poster in columns (which is how we read a newspaper) rather than left to right (which is how we read books). This allows readers to spend more time standing in one place while perusing the poster.

Before preparing the final layout, draw a rough draft of the poster on a sheet of paper to visualize how it will look when complete (see Figure 9.8). Maximize the use of graphics and minimize the use of text. For example, replace the procedures usually described in the materials and methods with a flowchart or tables. Populate the results section with tables and figures, giving each a title that describes the finding and a legend that describes the experimental details. Include short narrative summaries of the findings if necessary; the data in the

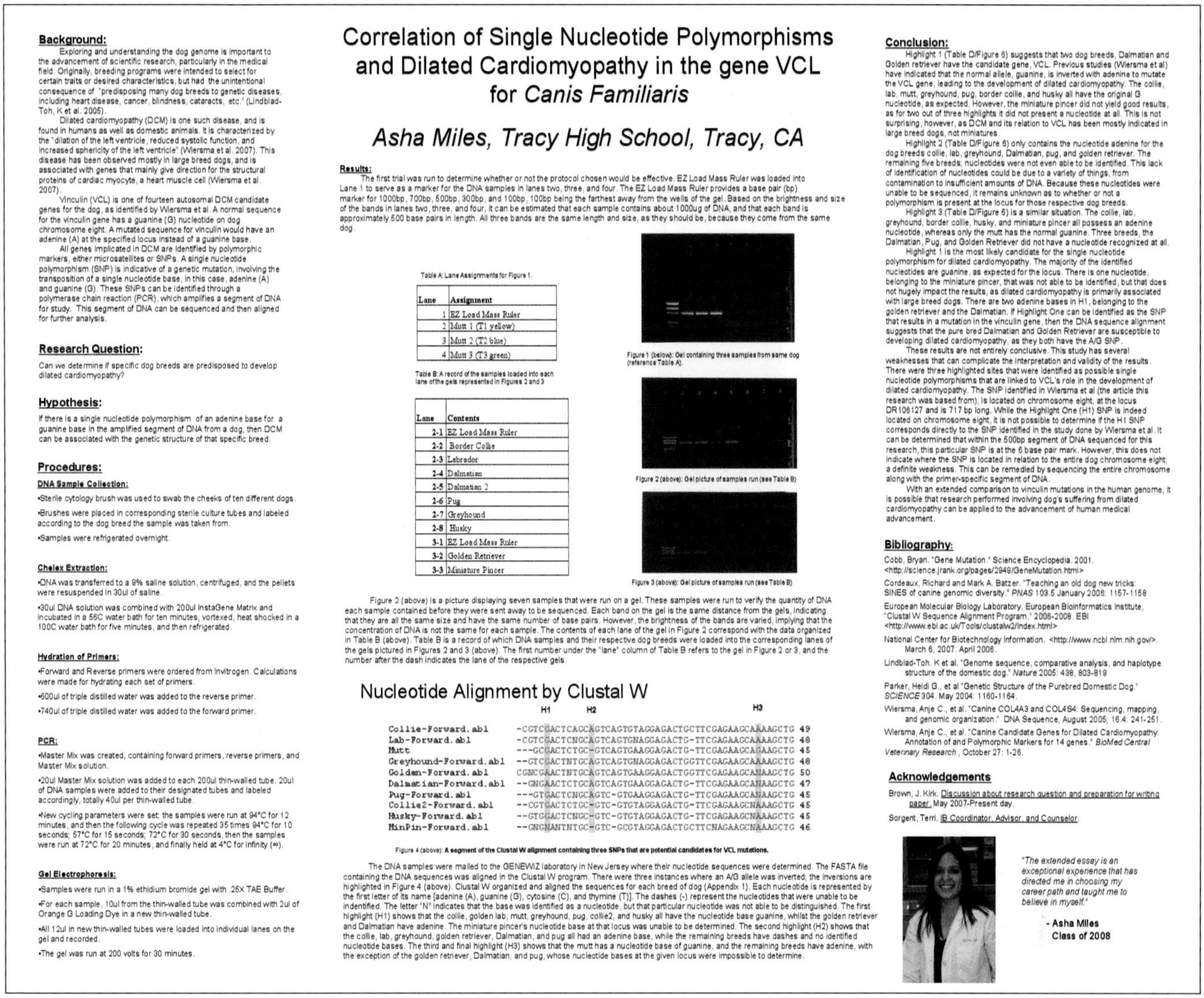

Correlation of Single Nucleotide Polymorphisms and Dilated Cardiomyopathy in the gene VCL for *Canis Familiaris*

Asha Miles, Tracy High School, Tracy, CA

Background:

Exploring and understanding the dog genome is important to the advancement of scientific research, particularly in the medical field. Originally, breeding programs were intended to select for certain traits or desired characteristics, but had the unintentional consequence of "predisposing many dog breeds to genetic diseases, including heart disease, cancer, blindness, cataracts, etc." (Lindblad-Toh, K et al. 2005).

Dilated cardiomyopathy (DCM) is one such disease, and is found in humans as well as domestic animals. It is characterized by the "dilation of the left ventricle, reduced systolic function, and increased sphericity of the left ventricle" (Wiersma et al. 2007). This disease has been observed mostly in large breed dogs, and is associated with genes that mainly give direction for the structural proteins of cardiac myocyte, a heart muscle cell (Wiersma et al. 2007).

Vinculin (VCL) is one of fourteen autosomal DCM candidate genes for the dog, as identified by Wiersma et al. A normal sequence for the vinculin gene has a guanine (G) nucleotide on dog chromosome eight. A mutated sequence for vinculin would have an adenine (A) at the specified locus instead of a guanine base.

All genes implicated in DCM are identified by polymorphic markers, either microsatellites or SNPs. A single nucleotide polymorphism (SNP) is indicative of a genetic mutation, involving the transposition of a single nucleotide base, in this case, adenine (A) and guanine (G). These SNPs can be identified through a polymerase chain reaction (PCR), which amplifies a segment of DNA for study. This segment of DNA can be sequenced and then aligned for further analysis.

Research Question:

Can we determine if specific dog breeds are predisposed to develop dilated cardiomyopathy?

Hypothesis:

If there is a single nucleotide polymorphism of an adenine base for a guanine base in the amplified segment of DNA from a dog, then DCM can be associated with the genetic structure of that specific breed.

Procedures:

DNA Sample Collection:
- Sterile cytology brush was used to swab the cheeks of ten different dogs.
- Brushes were placed in corresponding sterile culture tubes and labeled according to the dog breed the sample was taken from.
- Samples were refrigerated overnight.

Chelex Extraction:
- DNA was transferred to a 9% saline solution, centrifuged, and the pellets were resuspended in 30ul of saline.
- 30ul DNA solution was combined with 200ul InstaGene Matrix and incubated in a 56C water bath for ten minutes, vortexed, heat shocked in a 100C water bath for five minutes, and then refrigerated.

Hydration of Primers:
- Forward and Reverse primers were ordered from Invitrogen. Calculations were made for hydrating each set of primers.
- 600ul of triple distilled water was added to the reverse primer.
- 740ul of triple distilled water was added to the forward primer.

PCR:
- Master Mix was created, containing forward primers, reverse primers, and Master Mix solution.
- 20ul Master Mix solution was added to each 200ul thin-walled tube. 20ul of DNA samples were added to their designated tubes and labeled accordingly, totaling 40ul per thin-walled tube.
- New cycling parameters were set: the samples were run at 94°C for 12 minutes, and then the following cycle was repeated 35 times 94°C for 10 seconds; 57°C for 15 seconds; 72°C for 30 seconds, then the samples were run at 72°C for 20 minutes, and finally held at 4°C for infinity (∞).

Gel Electrophoresis:
- Samples were run in a 1% ethidium bromide gel with .25X TAE Buffer.
- For each sample, 10ul from the thin-walled tube was combined with 2ul of Orange G Loading Dye in a new thin-walled tube.
- All 12ul in new thin-walled tubes were loaded into individual lanes on the gel and recorded.
- The gel was run at 200 volts for 30 minutes.

Results:

The first trial was run to determine whether or not the protocol chosen would be effective. EZ Load Mass Ruler was loaded into Lane 1 to serve as a marker for the DNA samples in lanes two, three, and four. The EZ Load Mass Ruler provides a base pair (bp) marker for 1000bp, 700bp, 500bp, 300bp, and 100bp, 100bp being the farthest away from the wells of the gel. Based on the brightness and size of the bands in lanes two, three, and four, it can be estimated that each sample contains about 1000ug of DNA, and that each band is approximately 500 base pairs in length. All three bands are the same length and size, as they should be, because they come from the same dog.

Table A: Lane Assignments for Figure 1.

Lane	Assignment
1	EZ Load Mass Ruler
2	Mutt 1 (T1 yellow)
3	Mutt 2 (T2 blue)
4	Mutt 3 (T3 green)

Table B: A record of the samples loaded into each lane of the gels represented in Figures 2 and 3

Lane	Contents
2-1	EZ Load Mass Ruler
2-2	Border Collie
2-3	Labrador
2-4	Dalmatian
2-5	Dalmatian 2
2-6	Pug
2-7	Greyhound
2-8	Husky
3-1	EZ Load Mass Ruler
3-2	Golden Retriever
3-3	Miniature Pincer

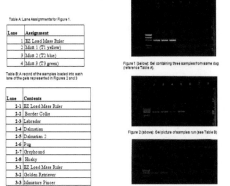

Figure 1 (below): Gel containing three samples from same dog (reference Table A)

Figure 2 (above): Gel picture of samples run (see Table B)

Figure 3 (above): Gel picture of samples run (see Table B)

Figure 2 (above) is a picture displaying seven samples that were run on a gel. These samples were run to verify the quantity of DNA each sample contained before they were sent away to be sequenced. Each band on the gel is the same distance from the gels, indicating that they are all the same size and have the same number of base pairs. However, the brightness of the bands are varied, implying that the concentration of DNA is not the same for each sample. The contents of each lane of the gel in Figure 2 correspond with the data organized in Table B (above). Table B is a record of which DNA samples and their respective dog breeds were loaded into the corresponding lanes of the gels pictured in Figures 2 and 3 (above). The first number under the "lane" column of Table B refers to the gel in Figure 2 or 3, and the number after the dash indicates the lane of the respective gels.

Nucleotide Alignment by Clustal W

```
                      H1        H2                 H3
Collie-Forward.ab1    -CGTCGACTCAGCTCAGTGTAGGAGACTGCTTCGAGAAGCAAAGCTG  49
Lab-Forward.ab1       -CGTCGACTCGCAGTCAGTGAAGGAGACTG-TTCGAGAAGCAGAGCTG  48
Mutt                  ---GCGACTCTGC-GTCAGTGAAGGAGACTG-TTCGAGAAGCAAAGCTG  45
Greyhound-Forward.ab1 --GTCGACTNTGCAGTCAGTGAAGGAGACTG-TTCGAGAAGCAAAGCTG  45
Golden-Forward.ab1    CGNCGAACTNTGCAGTCAGTGAAGGAGACTGGTTCGAGAAGCANAAGCTG 50
Dalmatian-Forward.ab1 -GNGAACTCGCAGTCAGTGAAGGAGACTG-TTCGAGAAGCANAAGCTG  47
Pug-Forward.ab1       ---GTCGACTCNGCAGTC-GTGAAGGAGACTG-TTCGAGAAGCANAAGCTG  45
Collie2-Forward.ab1   --CGTCGACTCTGC-GTC-GTGTAGGAGACTG-TTCGAGAAGCANAAGCTG  45
Husky-Forward.ab1     --GTCGAACTCNGC-GTC-GTGTAGGAGACTG-TTCGAGAAGCANAAGCTG  45
MinPin-Forward.ab1    --GTCGGANTNTGC-GTC-GCGTAGGAGACTCTTCNAGAAGCNAAAGCTG  45
```

Figure 4 (above): A segment of the Clustal W alignment containing three SNPs that are potential candidates for VCL mutations.

The DNA samples were sent to the GENEWIZ laboratory in New Jersey where their nucleotide sequences were determined. The FASTA file containing the DNA sequences was aligned in the Clustal W program. There were three instances where an A/G allele was inverted; the inversions are highlighted in Figure 4 (above). Clustal W organized and aligned the sequences for each breed of dog (Appendix 1). Each nucleotide is represented by the first letter of its name (adenine (A), guanine (G), cytosine (C), and thymine (T)). The dashes (-) represent the nucleotides that were unable to be identified. The letter 'N' indicates that the base was identified as a nucleotide, but that particular nucleotide was not able to be distinguished. The first highlight (H1) shows that the collie, golden lab, mutt, greyhound, pug, collie2, and husky all have the nucleotide base guanine, whilst the golden retriever and Dalmatian have adenine. The miniature pincer's nucleotide base at that locus was unable to be determined. The second highlight (H2) shows that the collie, lab, greyhound, golden retriever, Dalmatian, and pug all had an adenine base, while the remaining breeds have dashes and no identified nucleotide bases. The third and final highlight (H3) shows that the mutt has a nucleotide base of guanine, and the remaining breeds have adenine, with the exception of the golden retriever, Dalmatian, and pug, whose nucleotide bases at the given locus were impossible to determine.

Conclusion:

Highlight 1 (Table D/Figure 6) suggests that two dog breeds, Dalmatian and Golden retriever have the candidate gene, VCL. Previous studies (Wiersma et al) have indicated that the normal allele, guanine, is inverted with adenine to mutate the VCL gene, leading to the development of dilated cardiomyopathy. The collie, lab, mutt, greyhound, pug, border collie, and husky all have the original G nucleotide, as expected. However, the miniature pincer did not yield good results, as for two out of three highlights it did not present a nucleotide at all. This is not surprising, however, as DCM and its relation to VCL has been mostly indicated in large breed dogs, not miniatures.

Highlight 2 (Table D/Figure 6) only contains the nucleotide adenine for the dog breeds collie, lab, greyhound, Dalmatian, pug, and golden retriever. The remaining five breeds' nucleotides were not even able to be identified. This lack of identification of nucleotides could be due to a variety of things, from contamination to insufficient amounts of DNA. Because these nucleotides were unable to be sequenced, it remains unknown as to whether or not a polymorphism is present at the locus for those respective dog breeds.

Highlight 3 (Table D/Figure 6) is a similar situation. The collie, lab, greyhound, border collie, husky, and miniature pincer all possess an adenine nucleotide, whereas only the mutt has the normal guanine. Three breeds, the Dalmatian, Pug, and Golden Retriever did not have a nucleotide recognized at all.

Highlight 1 is the most likely candidate for the single nucleotide polymorphism for dilated cardiomyopathy. The majority of the identified nucleotides are guanine, as expected for the locus. There is one nucleotide, belonging to the miniature pincer, that was not able to be identified, but that does not hugely impact the results, as dilated cardiomyopathy is primarily associated with large breed dogs. There are two adenine bases in H1, belonging to the golden retriever and the Dalmatian. If Highlight One can be identified as the SNP that results in a mutation in the vinculin gene, then the DNA sequence alignment suggests that the pure bred Dalmatian and Golden Retriever are susceptible to developing dilated cardiomyopathy, as they both have the A/G SNP.

These results are not entirely conclusive. This study has several weaknesses that can complicate the interpretation and validity of the results. There were three highlighted sites that were identified as possible single nucleotide polymorphisms that are linked to VCL's role in the development of dilated cardiomyopathy. The SNP identified in Wiersma et al (the article this research was based from), is located on chromosome eight, at the locus DR106127 and is 717 bp long. While the Highlight One (H1) SNP is indeed located on chromosome eight, it is not possible to determine if the H1 SNP corresponds directly to the SNP identified in the study done by Wiersma et al. It can be determined that within the 500bp segment of DNA sequenced for this research, this particular SNP is at the 6 base pair mark. However, this does not indicate where the SNP is located in relation to the entire dog chromosome eight; a definite weakness. This can be remedied by sequencing the entire chromosome along with the primer-designated segment of DNA.

With an extended comparison to vinculin mutations in the human genome, it is possible that research performed involving dog's suffering from dilated cardiomyopathy can be applied to the advancement of human medical advancement.

Bibliography:

Cobb, Bryan. "Gene Mutation." Science Encyclopedia. 2001. <http://science.jrank.org/pages/2949/GeneMutation.html>

Cordeaux, Richard and Mark A. Batzer. "Teaching an old dog new tricks: SINES of canine genomic diversity." PNAS 103:5 January 2006: 1157-1158

European Molecular Biology Laboratory. European Bioinformatics Institute, "Clustal W Sequence Alignment Program." 2006-2008. EBI <http://www.ebi.ac.uk/Tools/clustalw2/index.html>

National Center for Biotechnology Information. <http://www.ncbi.nlm.nih.gov/>. March 6, 2007. April 2006.

Lindblad-Toh, K et al. "Genome sequence, comparative analysis, and haplotype structure of the domestic dog." Nature 2005: 438, 803-819

Parker, Heidi G., et al "Genetic Structure of the Purebred Domestic Dog." SCIENCE 304. May 2004: 1160-1164.

Wiersma, Anje C., et al. "Canine COL4A3 and COL4S4: Sequencing, mapping, and genomic organization." DNA Sequence, August 2005; 16.4: 241-251.

Wiersma, Anje C., et al. "Canine Candidate Genes for Dilated Cardiomyopathy: Annotation of and Polymorphic Markers for 14 genes." BioMed Central Veterinary Research, October 27: 1-26.

Acknowledgements

Brown, J. Kirk. Discussion about research question and preparation for writing paper. May 2007-Present day.

Sorgent, Terri. IB Coordinator, Advisor, and Counselor.

"The extended essay is an exceptional experience that has directed me in choosing my career path and taught me to believe in myself."

- Asha Miles
Class of 2008

Figure 9.7. **Research poster in a large-format print**. Poster courtesy of Asha Miles.

tables and figures can, however, often stand alone. Leave plenty of blank space in the poster so the reader does not feel overwhelmed.

Finally, check for spelling and grammatical errors. It is a good idea to show the poster to a fellow student, colleague, or instructor before printing it, as feedback from a peer can be quite helpful.

Research Papers

Research papers provide the most detailed information about a project and usually run several pages in length. Research papers cover each section described in Table 9.9 in detail.

In general, research papers are written in the third person and past tense because they describe research that has already been performed. The materials and methods are written as a narrative to describe how the experiments were performed rather than as a step-by-step protocol, and are written in sufficient detail so that others scientists can reproduce the work.

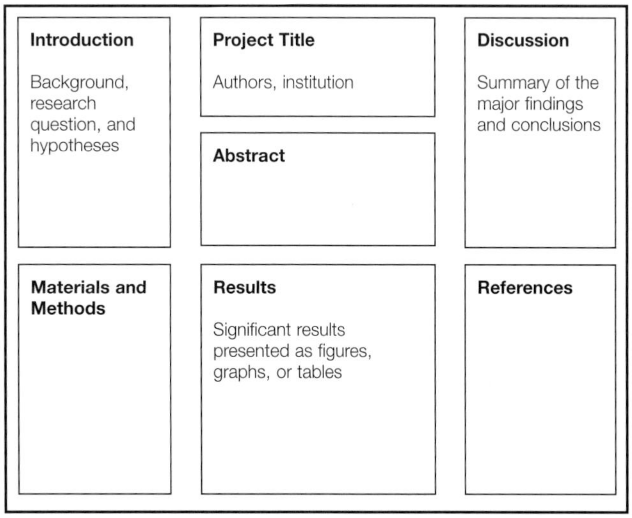

Introduction	Project Title	Discussion
Background, research question, and hypotheses	Authors, institution	Summary of the major findings and conclusions
	Abstract	
Materials and Methods	**Results**	**References**
	Significant results presented as figures, graphs, or tables	

Figure 9.8. **Sample layout of a scientific poster.**

When preparing papers for publication in a scientific journal, be sure to follow the journal's guidelines for submission, which specify, for example, the format and the maximum number of characters and figures. You can usually find the submission guidelines on the journal's website.

Oral Presentations

You may be invited to present your research at a team meeting, a formal scientific meeting, or a competition. Most research talks are organized in the same manner as written reports, with a title, introduction, materials and methods, and results and discussion. Audience members cannot go back to review the material, however, as they can with a written report, so be succinct and arrange the presentation in a logical fashion.

An oral presentation usually includes visual aids, such as a Microsoft PowerPoint or Keynote presentation. Since the time allotted for most oral presentations is 10–15 minutes, there is only a short period of time for the audience to read the slides. Therefore, instead of showing long paragraphs, encapsulate the salient points of each slide in brief sentences or bulleted lists. As with a poster, use a font size of at least 18 points and incorporate as many figures and tables as possible. Use a consistent format throughout the presentation.

Be concise and accurate, and give priority to the most important features of the research. There is not the same distinction between the results and discussion in an oral presentation as there is in a written report. When presenting a table or graph, it is acceptable to speculate on or discuss implications at that point, rather than waiting until later, since the audience will not be able to return to the data later. At the end of the talk, acknowledge everyone who assisted with the research and ask the audience for questions or comments. This is a good opportunity to get perspectives from other people about the research and presentation.

9.4 Project Ideas

Ideas for a research project can come from many sources, including documentaries, books, research articles, classes, previous students, and life events. The best research project usually stems from a topic you are naturally interested in and care about. Often an initial research question can be too broad to address in the time allowed for the project; however, such a question can often be broken down into progressively smaller questions that can be addressed in the scope of the course (see Figure 9.2).

This section lists project ideas to provide some direction if you do not have your own idea yet. It may also provide ideas on how to reduce the scope of a broad research question. Most of the project ideas have not been investigated, which means that results are not guaranteed. In addition, there may be issues with the experimental details that you may need to solve. This section does not provide protocols or background research.

The project ideas are categorized by the following topic areas:

- Research tools and techniques
- Agricultural and animal research
- Food science
- Humans and health
- Forensics
- Energy and the environment

Many projects address multiple topic areas (for example, an investigation into GMOs can be categorized as an agricultural, food science, environmental, or even a human health project depending on its specifics). Therefore, it is worth reading through all the project ideas, not just the section you think you find most interesting. Techniques that may be used for each project are listed, as is the chapter of this text with the most relevance to the project. For projects that use a Bio-Rad kit as a basis, request a copy of the kit instruction manual. This provides additional background information that may apply to your project. Also visit the Bio-Rad website (explorer.bio-rad.com) where you may find additional information on the kit, such as PowerPoint presentations, webinars, or extension activities.

In addition, projects are categorized by levels of complexity: ●, ■, and ◆ with ◆ being the most difficult. ● projects are great as a starting point for students new to research and biotechnology; ■ projects require more experience and usually involve techniques covered by the course; and many ◆ projects require a long period of time and significant background research to complete. These projects can also be performed over multiple years by different teams, with each team taking the research a step further. This is how research works in the real world; scientists entering a laboratory often continue the work of a scientist who has moved on.

Research Tools and Techniques

Developing or optimizing the research tools and techniques used to investigate a problem is fundamental to any research project. These types of projects can apply to many different topic areas. For example, if the effect of a specific recombinant protein on bacterial growth is the broad research question, then the path to answering that question may include optimizing transformation protocols that will help to meet the goal of expressing the protein in the bacteria. These projects can involve comparing different techniques or reagents, optimizing a technique to make it more efficient, using a tool in a new system, generating tools (such as plasmids), or identifying or separating molecules that could be needed for other research (see Table 9.10).

Table 9.10. **Research tools and techniques projects**.

Level	Project Idea	Potential Techniques	Chapter
●	**Optimize bacterial growth** Optimize the growth conditions of a strain of bacteria. Variables to test include type of growth media, additives to growth media, temperature, and oxygenation (through shaking).	Bacterial culturing and counting	3
●	**Investigate bacterial growth rates** Determine the growth rate and phases of a strain of bacteria. Bacteria grow at different rates and go through different phases as they grow (see Figure 5.13). Determine the number of bacteria at different time points.	Bacterial culturing and counting	3
●	**Optimize bacterial transformation** Optimize the modified calcium chloride transformation in Activities 5.A and 5.B. Variables to test include temperature and duration of heat shock, quantity of LB broth, incubation time with LB broth, growth phase of bacteria when transformed, starting bacteria from a broth culture compared to starting bacterial colonies on an agar plate, and the amount of bacteria or plasmid.	Bacterial culturing, transformation, and colony counting	5
■	**Develop chemicompetant bacteria** Develop a protocol for making chemically competent (chemicompetent) bacteria and test their transformation efficiency. Protocols for making competent cells can be found online and the cells can be stored in a −80°C freezer, if available. The cells can serve as a resource for the class. Bio-Rad also has reagents to make high-efficiency chemicompetent cells with a short and easy protocol; however, the cells cannot be stored (Transformation module (catalog #1665017EDU)). The instruction manual can be downloaded as part of the Ligation and Transformation module (catalog #1665015EDU).	Bacterial culturing and transformation	5
■	**Develop electrocompetent bacteria** Develop a protocol to make electrocompetent bacteria if an electroporator is available.	Bacterial culturing, transformation, and electroporation	5
■	**Compare transformation efficiencies** Compare the transformation efficiencies of homemade chemicompetent cells, purchased competent cells, and cells transformed using the modified calcium chloride protocol in Activities 5.A and 5.B.	Bacterial culturing and transformation	5
■	**Determine plasmid copy number** Compare different plasmids to determine whether they are high or low in copy number. Estimate the actual plasmid copy number based on bacterial number, plasmid size, and plasmid yield.	Bacterial culturing, transformation, minipreps, and DNA quantitation	5
■	**Compare miniprep plasmid yield** Compare plasmid yield from minipreps (see Activities 5.C and 5.D) when different strains of *E. coli* bacteria, such as HB101, DH5α, or JM109, or different growth media are used. For example, terrific broth (TB) is designed for higher plasmid yield.	Bacterial culturing, transformation, minipreps, and DNA quantitation	5
■	**Optimize miniprep protocols** Determine the optimal amount of bacteria for a specific miniprep protocol. Too many or too few bacteria can decrease the yield of plasmid DNA.	Bacterial culturing, transformation, minipreps, and DNA quantitation	5
■	**Evaluate miniprep protocols** Evaluate different miniprep protocols or kits to determine which provide the highest quality DNA, the highest yield, the fastest results, and which are the most cost-effective. Performing experiments that are properly controlled so that useful comparisons can be made requires special attention to detail and documentation.	Bacterial transformation, minipreps, and DNA quantitation	5

Table 9.10. **Research tools and techniques projects**. (continued)

Level	Project Idea	Potential Techniques	Chapter
●	**Generate stock bacterial cultures** Generate a library of stock bacterial cultures as a resource for the laboratory. These can be different bacterial strains that have interesting properties and morphologies for study by students, such as *Bacillus megaterium*, *Micrococcus luteus*, *Rhodospirillum rubrum*, and *E. coli* HB101. If a –80°C freezer is available, generate glycerol stocks (using protocols found online), or if no –80°C freezer is available, generate stab cultures (see Activities 5.A and 5.B). Stab cultures must be subcultured every few months to remain viable.	Bacterial culturing	3
●	**Bacterial viability in a stab culture** Determine the viability of different bacteria when stored as a stab culture. Different bacteria have different viabilities and respond differently to environmental conditions.	Bacterial culturing and stab cultures	3
◆	**Subclone a gene** Subclone a gene from one plasmid to another. Cut the gene out of the original plasmid with restriction enzymes (or use PCR) and ligate it into a new plasmid. Subcloning is performed when the gene needs a different regulation system, such as a different promoter, to answer a broader research question or if the gene must be expressed in a different host cell. Alternatively, a plasmid may need a different gene for antibiotic resistance.	Restriction digestion, PCR, ligation, transformation, miniprep, and agarose gel electrophoresis	5
■	**Optimize a PCR reaction** The amount of PCR product can usually be increased and the presence of background bands or primer-dimers can usually be decreased by changing PCR conditions. If a commercial master mix is being used, test variables such as primer concentration, annealing temperature, cycling times, and template concentration. A more sophisticated approach is to generate the master mix and test the variables described in Chapter 6. In this type of project, change one variable at a time and test the mix against stringent controls. The PCR reactions from the activities in Chapter 6 provide PCR reactions with which to start.	PCR	6
●	**Plan and execute a restriction digestion** Determine the restriction enzymes required to cut a specific segment of DNA, such as a gene, from a plasmid. Then determine the restriction enzyme digestion conditions for the reaction and perform the digestion to verify the prediction. Techniques used in Activities 4.D and 4.F should be applied to this project.	Restriction digestion and agarose gel electrophoresis	4
●	**Restriction enzyme plasmid mapping** Generate a restriction map of a plasmid. Digest the plasmid using two restriction enzymes. Use them separately and together in a double digest, then separate the digested plasmids on an agarose gel. The positions of the enzyme recognition sites on the plasmids can then be deduced using critical thinking and logic (see Activity 4.F). Challenge yourself by adding one or two additional enzymes and deducing their positions on the map.	Restriction digestion and agarose gel electrophoresis	4
●	**Effect of BSA on enzyme activity** Some restriction enzymes work better if bovine serum albumin (BSA) is added. The product information for each enzyme will indicate whether it requires BSA. For enzymes that require BSA, determine the effect of the presence and absence of BSA as well as different concentrations of BSA on a restriction enzyme digestion reaction. To test if a restriction digestion reaction is working as well as a control, you may need to reduce the concentration of enzyme or the time for the reaction. Compare the number of bands generated with those generated in a control under optimal conditions. If restriction enzymes are not working optimally rather than not cutting at all, they will cut some recognition sites but not others (a partial digest), creating a different banding pattern than a fully digested sample.	Restriction digestion and agarose gel electrophoresis	4

Table 9.10. **Research tools and techniques projects**. (continued)

Level	Project Idea	Potential Techniques	Chapter
●	**Optimize a restriction digest** Optimize a restriction digestion reaction on lambda DNA or a plasmid. Test variables such as time, temperature, salt concentration, enzyme concentration, DNA concentration, and additives like BSA. The reagents from Activity 4.D can provide a restriction digestion reaction with which to start.	Restriction digestion and agarose gel electrophoresis	4
■	**Restriction enzyme activity** The concentration of an enzyme is usually given in units of activity. Verify the published rates of reaction for an enzyme by conducting the test used to determine its activity. Test different experimental conditions to determine if the activity can be increased.	Restriction digestion and agarose gel electrophoresis	4
●	**Generate DNA size standards** Generate DNA size standards for different applications. Some experiments need large DNA fragments, while others need small fragments. One common standard is lambda phage digested with HindIII. Standards based on lambda digested with other enzymes, a different phage, or plasmid DNA could be generated.	Restriction digestion, restriction site mapping, and agarose gel electrophoresis	4
●	**Develop a restriction enzyme activity** Develop a restriction enzyme experiment for younger students and generate the required reagents for the activity. Conduct the activity with the students.	Restriction digestion and agarose gel electrophoresis	4
■	**Determine antibody limits of detection** Determine the detection limits for an antibody, for example utilizing the ELISA Immuno Explorer™ kit used in Activities 8.B and 8.C. Perform a serial dilution of either the antigen or the antibodies and find the concentration at which there is no color change.	ELISA	8
■	**Optimize an ELISA** Optimize an ELISA assay to use the smallest amount of antibody; this may include increasing antigen or secondary antibody concentration, increasing incubation times or temperatures, and changing wash conditions.	ELISA	8
◆	**Develop an ELISA** After gaining experience optimizing ELISAs, develop an ELISA using different antibodies. Refer to How To… Select an Antibody on page 301.	ELISA	8
◆	**Optimize a western blot** Optimize a western blot (such as that used in Activity 8.D) to have a single protein band on the gel of the expected size using the least amount of primary antibody. Determine which parts of the protocol must be changed to allow use of lower antibody concentrations (for example, increase the antigen or secondary antibody concentration, increase the incubation times or temperatures, and change the wash conditions).	Western blotting	8
◆	**Develop a western blot** After gaining experience optimizing western blots, develop a western blot protocol to detect a different antigen using a different primary antibody.	Western blotting	8

Table 9.10. **Research tools and techniques projects**. (continued)

Level	Project Idea	Potential Techniques	Chapter
◆	**Develop an immunocytochemistry assay** Develop or optimize an assay to investigate an antigen in a cell using fluorescent immunocytochemistry. This requires access to a fluorescence microscope. Use an existing protocol and positive control samples from a research laboratory as a starting point.	Immunocytochemistry and cell culture	8
●	**SDS-PAGE of green fluorescent protein (GFP)** Perform SDS-PAGE on GFP using Bio-Rad's pGLO™ kit SDS-PAGE extension (catalog #1660013EDU) to explore the effects of detergents and heat on protein conformation.	SDS-PAGE	7
■	**Time course of GFP induction** Induce the expression of GFP from a liquid culture of pGLO bacteria in LB/amp broth without arabinose by adding arabinose and taking samples of bacteria from 5 minutes out to 3 hours. Examine the induction of GFP by SDS-PAGE electrophoresis using Bio-Rad's pGLO kit SDS-PAGE extension (catalog #1660013EDU).	Protein expression and SDS-PAGE	7
■	**SDS-PAGE of chromatography fractions** Perform hydrophobic interaction chromatography on bacterial samples such as bacteria transformed with pGLO plasmid and use SDS-PAGE (see Activity 7.D) to analyze the level of purity and the different protein profiles generated from the fractions. Samples may need to have the salt in the elution buffer removed using a desalting column (such as Bio-Rad's Micro Bio-Spin® 6 columns (catalog #7326221EDU)) before they are electrophoresed.	Chromatography and SDS-PAGE	7
■	**Purify proteins with chromatography** Use multiple types of chromatography sequentially on a bacterial sample to gradually separate the mixture of proteins in the sample and examine the proteins using SDS-PAGE.	Chromatography and SDS-PAGE	7
●	**Bioinformatics and phylogenetic trees** Obtain sequences of a protein of interest from multiple species from GenBank and develop phylogenetic trees using Clustal Omega.	Bioinformatics	7

Agricultural and Animal Research

The drive in agricultural research is to improve farming efficiency, crop yields, and animal production and to lessen environmental impacts. Agricultural projects can be targeted to a local issue, such as a pest that is causing local farmers a problem, or to broader agricultural issues, such as GM crops and the effects of pesticides and herbicides on the environment (see Table 9.11).

Other important agricultural research includes investigating factors such as fertilizers, soil chemistry, and hydrology that affect crop yield or plant growth rates. The chemistry of hydroponics, plant bioremediation, and animal health are other potential areas for research. Research involving animals requires special precautions (see the Animal Research section later in this chapter).

Table 9.11. **Agricultural and animal research projects**.

Level	Project Idea	Potential Techniques	Chapter
◆	**Clone and sequence a plant gene** Clone the *glyceraldehyde 3-phosphate dehydrogenase* (*GAPDH*) gene from a plant of your choice using Bio-Rad's Cloning and Sequencing Explorer series (catalog #1665000EDU). This can be a project for the whole class. This series of modules takes 6–8 weeks to complete and guides students through the process of extracting DNA, amplifying a section of the *GAPDH* gene using nested PCR, ligating the gene fragment into a cloning plasmid, transforming the ligation into bacteria, performing minipreps, digesting the plasmids with restriction enzymes, sequencing the gene fragment, and performing bioinformatics analysis on the gene sequence.	DNA extraction, nested PCR, ligation, transformation, minipreps, restriction digestion, DNA sequencing, and bioinformatics	4/5/6

Table 9.11. **Agricultural and animal research projects**. (continued)

Level	Project Idea	Potential Techniques	Chapter
●	**Investigate plant pathogens** Investigate *Agrobacterium tumefaciens* and other bacteria from naturally occurring galls on plants or trees. Compare different galls on the same plant species, different areas of the same gall (such as close to the trunk or stem compared to the outer sections of the gall), or galls from different species. This requires research into the best and safest method to examine *A. tumefaciens*.	Bacterial cell staining	3
■	**Isolate plant bacteria** Isolate *A. tumefaciens* from naturally occurring galls on plants or trees. Methods to isolate the bacteria will need to be researched.	Bacterial culture and cell staining	3
◆	**Investigate plant susceptibility to infection** Isolate *A. tumefaciens* as above and then inoculate another plant with the bacteria in a controlled environment to test whether another plant species or cultivar is susceptible to the disease. Could susceptibility to infection be changed? Methods to perform this experiment will need to be researched and there may be local or state agricultural guidelines regulating handling of *A. tumefaciens*.	Microbiology and horticulture	3
●	**Identify soil bacteria** Bacteria in soil are vital for the health of the soil and its ability to support crops. Examine bacteria in different soil samples. Extend the project by trying to identify the isolated bacteria, which will require research into the morphologies of soil bacteria.	Bacterial staining	3
◆	**Detect bacteria with antibodies** Develop an ELISA test to detect *Bacillus subtilis* (or other bacteria) in soil.	ELISA	8
●	**Nitrogen-fixing bacterial counts** *Sinorhizobium meliloti* is a symbiotic bacterium that lives in the root nodules of alfalfa and fixes nitrogen for the plant. Use bacterial staining techniques (see Activity 3.D) to classify nitrogen-fixing bacteria from root nodules. Compare root nodules from the same plant, plants grown in different conditions, or different plant species. Extend the project by isolating and quantifying nitrogen-fixing bacteria from root nodules using techniques similar to Activity 3.E.	Bacterial cell culturing, serial dilutions, and plate counts	3
◆	**Nitrogen-fixing bacterial identification** Extend the previous project by using molecular biology techniques to amplify 16S rRNA regions of the bacterial genome and identify the species using PCR, and DNA sequencing, and searching bioinformatics databases.	PCR, DNA sequencing, and bioinformatics	6
●	**Antimicrobials in plants** Many plants have purported medicinal properties. Use the disk diffusion method to examine spices, herbs, or other plant products claimed to possess antibacterial properties (see Activity 3.B).	Microbiology and the disk diffusion assay	3
◆	**Plant DNA sequencing** Develop a method to differentiate plant species by comparing the DNA sequences of the gene for the 18S rRNA ribosomal subunit. This method could be used to look at evolutionary differences among plants or as a forensic tool to differentiate one plant from another. This will require research into previously used PCR primer sequences or developing new PCR primer sequences. Compare the DNA sequences using bioinformatics tools such as Clustal Omega (see Activity 7.F).	Isolating plant DNA, PCR, DNA sequencing, and bioinformatics	6
◆	**Plant sequence bioinformatics** Identify plants whose 18S rRNA sequences are in the NCBI GenBank database and perform bioinformatics analysis to identify differences in restriction enzyme recognition sites among different species. Propose a test based on these data to differentiate plants based on amplifying the region by PCR and then digesting the PCR products with restriction enzymes to identify the plant species. Test methods include bioinformatics analysis of GenBank DNA sequences, Clustal Omega, NEBcutter®, and PCR primer design. Test this method in the laboratory if there is sufficient time in the project or test the method as a second project.	Bioinformatics (optional: PCR and restriction digestion)	7

Table 9.11. **Agricultural and animal research projects**. (continued)

Level	Project Idea	Potential Techniques	Chapter
■	**Identify genetic modification in GM food** Identify genetically modified crops using the GMO Investigator™ kit and then test whether the crop contains the 35S CaMV promoter or the NOS terminator by using primers for these specific modifications to amplify the crop genomic DNA. Extraction of DNA from plants may require use of Bio-Rad's Nucleic Acid Extraction module (catalog #1665005EDU) rather than the InstaGene™ extraction method for food used in Activity 6.B.	DNA extraction and PCR	6
■	**Develop assay to test GM crops** Research genes used in genetic modifications, such as the *pat* gene (isaaa.org/gmapprovaldatabase/ hosts a database of all crop genetic modifications) and use PCR to determine if these are present in crops or snack foods.	DNA extraction and PCR	6
◆	**Horizontal transfer of genetic modifications** Test weeds in the vicinity of GM crops to learn if genetic modifications have been transferred (a process known as horizontal transfer).	DNA extraction and PCR	6
■	**Quantify the amount of GM material in snack food** Use Bio-Rad's GMO Investigator kit real-time PCR extension (catalog #1662560EDU) to quantify the level of GM material in snacks. This project requires access to a real-time thermal cycler.	DNA extraction and real-time PCR	6
●	**Test crops for genetic modifications with antibodies** Use a commercial ELISA or dipstick test kit to identify proteins expressed as a result of a genetic modification in a particular crop, for example the Cry1 endotoxin in BT corn.	ELISA	8
◆	**Develop an ELISA for genetic modifications** Develop an ELISA to identify a protein expressed as a result of a genetic modification in a particular crop by using a primary antibody to the recombinant protein.	ELISA	8
■	**Plant tissue culture** Induce callus formation from plant tissue. Order callus induction media from a science supplier. Although not ideal, if a tissue culture hood is not available create your own by spraying down a box with 70% ethanol or isopropanol, which will reduce the chance of contamination. Kits to perform plant tissue culture are also available from scientific suppliers and will make a plant tissue culture project easier since the reagents and protocols will be provided.	Plant tissue culture	3
◆	**Clone a plant** Grow a callus and transfer it to root and shoot induction media and eventually into potting soil. Then test the plants to determine whether they are clones.	Plant tissue culture	3
◆	**Genetically modify a plant** Perform genetic modification of plant tissue by using *A. tumefaciens* to insert genes of interest into a plant genome. This project requires use of a tissue culture hood and may be subject to local and state regulations on working with *A. tumefaciens* and the creation of novel organisms.	Plant tissue culture	3
●	**Milk protein levels through lactation** Compare the protein concentration of different cow breeds or from cows at different stages of their lactation cycles (time since birth) using methods described in Activity 7.A.	Protein concentration and standard curves	7
●	**Protein concentrations of animal milk** Compare the protein concentration of milk from different farm animals such as cows, sheep and goats using the Bradford assay used in Activity 7.A.	Protein concentration and standard curves	7
■	**Lactation profiles of mammals** Explore the protein profiles of milk from different animals or at different stages of lactation by SDS-PAGE or western blots using methods similar to those used in Activities 7.D and 8.D.	SDS-PAGE and western blotting	8

Table 9.11. **Agricultural and animal research projects**. (continued)

Level	Project Idea	Potential Techniques	Chapter
●	**Protein profiles of animals** Compare the protein profiles of meat from different species using SDS-PAGE in a method similar to that in Activity 7.D.	SDS-PAGE	7
■	**Protein profiles of animal organs** Compare the protein profiles from different animal or fish organs in the same species using SDS-PAGE using methods similar to those used in Activity 7.D.	SDS-PAGE	7
■	**Test meat with myosin antibodies** Extend the previous activities by performing a western blot on meat to compare myosin light chain proteins among organs or species as in Activity 8.D.	SDS-PAGE and western blotting	8
◆	**Develop a western blot to an animal protein** Develop a western blotting protocol to examine a protein of interest (different from myosin light chain) in meat to extend the previous activity.	SDS-PAGE and western blotting	8
◆	**Animal maternity tests** Identifying the parents of animals is important to animal breeders, whether to ensure offspring do not mate with their mothers or to verify an animal descended from a particular champion. Identify the maternity of livestock using PCR and DNA sequencing. Research into the best DNA sequences to use for testing maternity will be necessary (you will likely find that mitochondrial DNA sequences should be used). You will also need to determine which PCR primers to use. Test methods include isolating DNA from the animals (hair or saliva samples may provide sufficient DNA, see Activity 6.C), amplifying the region by PCR, and sequencing the DNA. You will then need to compare the animal to its mother. A less rigorous project may be to develop the PCR test for maternity and use the test as a project next year.	DNA extraction, PCR, DNA sequencing and bioinformatics	6
■	**Biofuel and symbiotic gut bacteria** Combine the areas of livestock farming and biofuel research by looking for microorganisms with cellulose-digesting abilities. Herbivores like cattle can digest cellulose due to symbiotic bacteria in their guts. Isolate and compare the cellulose-digesting abilities of cellulose from the guts of herbivores. Obtain samples from fresh dung or the chambered stomachs of the animals after slaughter. This will require development of a method for testing cellulase activity; methods from Activity 7.E could provide a starting point.	Microbiology and enzyme assays	7
◆	**Identify animal gut bacteria** Identify bacteria from the guts of herbivores by determining the sequence of the 16S rRNA ribosomal subunit gene. Extract DNA from fresh dung or the chambered stomachs of animals after slaughter. Develop a PCR assay to amplify the 16S rRNA region of the bacteria, clone the different PCR products into plasmids, and send the plasmids for sequencing. Use NCBI BLAST to identify the bacteria from the sequence.	PCR, DNA sequencing, and bioinformatics	6
◆	**Dog medical tests** Investigate medical conditions in animals. For example, many large dog breeds suffer from dilated cardiomyopathy (DCM), or enlarged hearts. A single change in one nucleotide in a gene called vinculin (VCL) (a single nucleotide polymorphism, or SNP) has been associated with DCM; many more genes are suspected of being associated with this condition. Find a SNP in these genes using PCR and DNA sequencing of DNA from different dog breeds.	PCR, DNA sequencing, and bioinformatics	6
◆	**Identify insect pests** Develop a method for differentiating insect species by comparing the DNA sequences of the gene for COI. Focus on identifying evolutionary differences among insects or developing a tool to identify or differentiate among particular insect species (DNA barcoding). Note that extracting DNA from insect tissues is challenging due to the number of compounds in the exoskeleton. One suggestion is to remove the muscle from the inside of the largest leg segment and use this tissue to extract DNA.	DNA extraction, PCR, DNA sequencing, and bioinformatics	6

Food Science

Biotechnology plays a significant role in various areas of the food industry, including evaluating food safety, improving production, and evaluating and improving nutritional values. Student projects can include evaluating food for its nutritional or GMO content; levels of pesticides, hormones, or allergens; or levels of contamination with other food stuffs or microorganisms (see Table 9.12). The use of microorganisms in the production of foods such as yogurt, cheese, or kimchi can be investigated, as can their use in the composting of food. Sanitation and cross-contamination in food preparation can be researched and food labeling can be checked by developing methods to identify food. When conducting food science research, remember that it is not safe to eat anything that has been in the science laboratory.

Table 9.12. **Food science projects**.

Level	Project Idea	Potential Techniques	Chapter
●	**Count yogurt bacteria** Quantify the number of bacteria in yogurt by performing a serial dilution and plate counts using techniques from Activity 3.E.	Bacterial culturing	3
●	**Investigate yogurt bacterial growth** Investigate the effects of different agar formulations and incubation temperatures on the growth of yogurt bacterial strains. Some bacterial strains may grow on one type of agar but not on another, and some yogurt strains are thermophilic (they grow better at higher temperatures ~50°C). The reagents and techniques from Activity 3.C could be applied to this project.	Bacterial culturing	3
●	**Optimize yogurt production** Optimize yogurt production using different strains of bacteria and different incubation conditions. Assess how the consistency of yogurt is affected by these factors. **Safety note:** Yogurt produced in the laboratory is not safe to eat.	Bacterial culturing and yogurt production	3
■	**Identify cheese molds** Investigate cheese molds (fungi), which are introduced into cheeses to change their consistency and flavor since they digest cheese proteins such as caseins. Soft cheeses like brie are encased in a layer of mold, while blue cheeses have molds inserted into the cheese. Use a microscope, stains, and information found through background research to identify different molds using techniques similar to those in Activities 3.D and 3.F.	Microscopy and mold staining	3
■	**Enzyme activity of cheese mold** Investigate the activity of enzymes from cheese molds on the degradation of casein. For example, use milk agar, which contains casein, peptone, and beef extract, to demonstrate clearing where enzymes (caseases) from mold (or bacteria) have degraded the casein.	Mold cultures and enzyme assays	7
●	**Cheese making** Investigate the cheese making process. Compare recombinant chymosin (rennin) and natural rennin (made from grinding cow stomach linings) on cheese production. Test the effects of other variables, such as molds, temperature, and formulations, important in cheese making to optimize the process. **Safety note:** Cheese made in the laboratory is not safe to eat.	Enzyme activity and cheese making	7
●	**Investigate dyes in candy** Separate the dyes in candy or other colored foods using electrophoresis as in Activity 4.C. Use deductive reasoning to identify which candy colors contains which food dyes based on the ingredients lists.	Dye electrophoresis	4
◆	**Test veggie burgers for meat** Verify that vegan or vegetarian food does not contain meat by performing PCR on samples using primers that specifically amplify animal DNA. Research will need to be performed to find animal specific primers.	DNA extraction and PCR	6
◆	**Test for insect contamination** Test for insect contamination in vegetarian or vegan food using PCR and primers specific to insects.	DNA extraction and PCR	6

Table 9.12. **Food science project ideas**. (continued)

Level	Project Idea	Potential Techniques	Chapter
■	**Confirm food is organic** Verify that organic food does not contain genetic modifications using PCR as in Activity 6.B.	DNA extraction and PCR	6
■	**Confirm fish identities** Verify the identity of fish served as sushi against whole fish that can be identified from the fish market by comparing their protein profiles using SDS-PAGE as in Activity 7.D.	SDS-PAGE	7
■	**Which fish are in imitation crab meat?** Imitation crab meat is made from fish, not crab. Narrow down the potential type of fish used by comparing the protein profile of imitation crab meat to known fish such as cod and pollock using SDS-PAGE as in Activity 7.D. Eliminate fish profiles that are different and more closely examine the profiles that are similar.	SDS-PAGE	7
■	**Survey foods for GMOs** Using PCR as in Activity 6.B, survey one food type, for example corn chips, to test how often it uses genetically modified material.	DNA extraction and PCR	7
●	**Food protein levels** Determine the protein concentrations in food and drinks using the Bradford assay as in Activity 7.A. Compare this to the protein content stated in the nutritional information on the package. Proteins will need to be extracted from solid food, and the appropriate dilution will need to be determined.	Protein concentration and standard curves	7
■	**Changes to protein profiles during cooking** Using SDS-PAGE, compare the protein profiles of meat at various levels of cooking: raw, rare, medium, and well-done. Determine which proteins appear and disappear in the profiles as the meat is cooked.	SDS-PAGE	7
◆	**Find food bacteria** Identify bacteria on food, such as meat from the grocery store, by performing PCR using primers that amplify only the DNA of certain types of bacteria. Background research will be required to find PCR primers that can identify bacteria.	DNA extraction and PCR	6
◆	**Identify bacteria on food** Identify bacteria on food by performing PCR to amplify the ribosomal RNA 16S gene, cloning the PCR products, and sending them for sequencing.	DNA extraction, PCR, ligation, transformation, minipreps, and sequencing	6

Humans and Health

Medical research is the largest research segment of the biotechnology industry, and a great deal of medical research is performed without using human subjects. Model systems such as mammalian cell cultures, yeast, and fruit flies are used to investigate genes or proteins involved in human disease and can be used in student research projects as well. Medicinal compounds can also be investigated in plants, and bioinformatics can be used to investigate human disease without the need to test human subjects. Direct research with human subjects requires special care and precautions (see Research on Human Subjects later in this chapter) but might include investigating the activities of human enzymes, such as amylase from saliva. Anthropological research into the origins of human populations is also a topical area and might involve investigations into human genetics or human lifestyles, such as the evolution of human food (see Table 9.13).

Table 9.13. **Humans and health projects.**

Level	Project Idea	Potential Techniques	Chapter
■	**Purify an enzyme that is targeted in cancer treatments** Dihydrofolate reductase (DHFR) is an enzyme involved in DNA replication. The chemotherapeutic drug methotrexate targets this protein to inhibit cancer growth. Use Bio-Rad's Protein Expression and Purification series (catalog #1665040EDU) to express this medically relevant protein in *E. coli*, purify it using affinity chromatography, and test its enzyme activity using a workflow similar to that used in pharmaceutical drug development. The series of modules takes 2–4 weeks to complete and guides students through the entire process.	Bacterial culturing, affinity chromatography, SDS-PAGE, and enzyme activity	7
●	**Antimicrobials in plants** Many plants have purported medicinal and antimicrobial properties. Use the disk diffusion method (see Activity 3.B) to examine the antimicrobial properties of spices, herbs, or other plant products that are claimed to possess them.	Disk diffusion assay	3
■	**Compare genetics of human populations** Compare the allelic frequency of the PV92 Alu insert in different populations, ethnicities, or backgrounds using protocols used in Activity 6.C. **Safety note:** handling saliva requires special precautions. Ideally, have individuals process their own samples until they have been heated to 95°C, which should kill any pathogens. **Privacy note:** Make sure samples are anonymous and randomized so that the genetic results cannot be tracked back to specific individuals.	DNA extraction and PCR	6
■	**Amplify the hypervariable region of human mitochondria** Sequence the hypervariable region 1 (HV1) on the D-loop of human mitochondria and compare it to the control sequence (the Cambridge reference sequence, or CRS) and other populations to investigate human origins. Primer sequences for the HV1 region are the forward primer L15971: 5'-TTAACTCCACCATTAGCACC-3' and the reverse primer H16391: 5'-GAGGATGGTGGTCAAGGGAC-3'. The PCR program is 94°C for 5 minutes, followed by 94°C for 30 seconds, 55°C for 30 seconds, 72°C for 30 seconds. Repeat the program for 35 cycles and then add a 72°C hold for 10 minutes. **Safety note:** Handling saliva requires special precautions. Ideally, have individuals process their own samples until they have been heated to 95°C, which should kill any pathogens. **Privacy note:** Make sure samples are anonymous and randomized so that the genetic results cannot be tracked back to specific individuals.	DNA extraction, PCR, and DNA sequencing	6
◆	**Amplify human or animal mitochondrial genes** Find or develop primer sequences to amplify and sequence mitochondrial regions or genes such as cytochrome b from classmates or from different animal species. Compare the sequences to each other and to those in databases. **Safety note:** Handling saliva requires special precautions. Ideally, have individuals process their own samples until they have been heated to 95°C, which should kill any pathogens. **Privacy note:** Make sure samples are anonymous and randomized so that the genetic results cannot be tracked back to specific individuals.	DNA extraction, PCR, and DNA sequencing	6

Table 9.13. **Humans and health projects.** (continued)

Level	Project Idea	Potential Techniques	Chapter
◆	**Mammalian cell tissue culture** Perform mammalian tissue culture using CHO cells (if biological safety cabinets and CO_2 incubators are available). Use SDS-PAGE to compare the protein profiles of cells grown under different growth conditions, for example, in the presence and absence of serum. Aseptic technique is vital for tissue culture and special training is required.	Tissue culture and SDS-PAGE	3 & 7
◆	**Investigate a medically important protein** Using PCR or antibodies in an ELISA or western blot, test for the presence of a medically important gene or protein in a model system such as yeast or fruit flies. Endeavor to characterize the gene or protein.	Model organisms, PCR, or immunoassays	6 or 8
◆	**Purify an animal protein** Perform animal cell tissue culture and chromatography to purify a protein of interest from animal tissue culture cells. Alternatively, extract proteins directly from an animal tissue sample by crushing or blending the tissue and extracting the proteins using a protein extraction buffer. You will need to develop the assays to track the protein of interest through chromatography fractions; these could be SDS-PAGE, enzyme assays, western blotting, or ELISA.	Tissue culture, protein extraction, chromatography, SDS-PAGE, enzyme assays, and immunoassays	7
●	**Protein level in breast milk** Quantify the protein concentration in breast milk using the Bradford assay, as in Activity 7.A. Compare protein levels at different stages of lactation or to milk from other mammals, such as cows and goats. **Safety note:** Handling bodily fluids is potentially hazardous and special precautions should be taken.	Protein quantitation	7
■	**Protein profiles of breast milk** Compare the protein profiles of human breast milk to cows' milk and baby formula using an SDS-PAGE method similar to Activity 7.D. **Safety note:** Handling bodily fluids is potentially hazardous and special precautions should be taken.	SDS-PAGE	7
●	**Proteins in saliva** Quantify the protein concentration in human saliva using the Bradford assay as in Activity 7.A and compare it to saliva from other animals, such as dogs. As an additional activity, compare the protein profiles of saliva using SDS-PAGE as in Activity 7.D. Investigate whether protein profiles change when foods with different smells are presented to a subject. **Safety note:** Handling saliva requires special precautions. Ideally, have individuals process their own samples until they have been heated to 95°C, which should kill any pathogens.	Protein quantitation and SDS-PAGE	7
●	**Enzymes in saliva** Investigate the enzyme amylase from saliva using an enzyme activity assay. Starch is the substrate for amylase and iodine turns blue/black in the presence of starch. Therefore, when starch is digested by amylase the color of iodine changes; this change can be measured by eye or by using a spectrophotometer. **Safety note:** Handling saliva requires special precautions. Ideally, have individuals process their own samples.	Enzyme activity	7

Forensics

DNA evidence is a vital tool in law enforcement and crime shows on television have made forensics a household term. Forensics is the use of science and technology to investigate crimes and it incorporates a number of the techniques used for DNA and protein analysis. Forensic analysis of human DNA has ethical and privacy implications that are discussed in Research on Human Subjects later in this chapter. However, a lot of forensic analysis is also performed on food, plants, and animals to identify biological material such as poached ivory or fish caught out of season. In addition, DNA and protein analysis can help match flora or fauna found at a crime scene to material found on a suspect. New methods for distinguishing different species for forensic analyses are therefore useful and necessary (see Table 9.14).

Table 9.14. **Forensics projects**.

Level	Project Idea	Potential Techniques	Chapter
●	**Stain crime scene bacteria** Investigate methods for locating a crime scene, such as evaluating the diversity of bacteria in the soil or environment by Gram staining, and use them to compare the types of bacteria present in various environments.	Bacterial staining	3
◆	**Fingerprint microorganism populations** A more advanced way to investigate the diversity of microorganisms in the soil or environment is through a random amplification of polymorphic DNA (RAPD) PCR assay. Develop and use a RAPD assay to fingerprint samples likely to be present at a crime scene, such as mud on a shoe. Use commercially available RAPD primers or sequences available in journal articles.	DNA extraction and PCR	6
◆	**Amplify crime scene bacteria** The presence of an uncommon microorganism on a suspect can be used as evidence that he/she was at a crime scene. Develop a PCR test using published primers to identify specific microorganisms whose presence or absence can differentiate among locations. Use controls to show the prevalence of the organisms in different locations; if the microorganism is found everywhere, it will not be useful as evidence.	DNA extraction and PCR	6
◆	**Plant DNA fingerprint** Develop a genetic fingerprint test for plants. The GMO Investigator kit used in Activity 6.B contains primers to the *PSII* gene. Sequence and compare the PCR product from different plants. Alternatively, identify restriction fragment length polymorphisms (RFLPs) in the PSII PCR product by digesting it with different restriction enzymes to find enzymes that differentially cut the product from different plants. Note that Bio-Rad's Nucleic Acid Extraction module, which was developed to extract DNA from plant tissue, may yield better results than the InstaGene matrix used in Activity 6.B. Also consider adding bioinformatics research into known plant *PSII* sequences for this project.	DNA extraction, PCR, restriction enzyme digestion or DNA sequencing, and bioinformatics	6
◆	**Animal DNA fingerprint** Develop a genetic fingerprint test for animals that could be used to identify animal tissue at a crime scene or to determine if meat or animal feed has been adulterated with animal tissue that should not be present. For example, due to the risk of BSE, or mad cow disease, cattle feed should not contain rendered cows. Use primers specific to different animal species, which can be found in published papers and usually target mitochondrial DNA.	DNA extraction and PCR	6
◆	**Animal paternity test** Develop a paternity test for cattle or other farm animals based on the sequences of the D-loop region in their mitochondrial DNA to help identify fraudulent animal breeders.	DNA extraction, PCR, DNA sequencing, and bioinformatics	6
■	**Protein fingerprint** Develop a protein fingerprint test using SDS-PAGE as in Activity 7.D for fish species that might help fish and game officials identify fish fillets suspected of being caught improperly or out of season. Alternatively, identify market substitution of meat or fish products.	SDS-PAGE	7
●	**Animal blood typing** Develop a test to identify the animal origins of blood samples from a crime scene using the Ouchterlony test as in Activity 8.A. Antisera for different animal species are commercially available.	Ouchterlony test	8

Energy and the Environment

The role of biotechnology in environmental research is growing. Environmental projects can include monitoring and sampling of organisms, pollutants, biofilms, or other indicators of environmental health, or the application of microorganisms for bioremediation (for example, oil-degrading bacteria). In addition, the use of organisms to generate energy is a major focus of the research community; this includes research into finding organisms and enzymes that can improve production of biofuels such as cellulosic ethanol and biodiesel. Projects can target global environmental issues or issues within your local area (see Table 9.15).

Table 9.15. **Energy and the environment projects**.

Level	Project Idea	Potential Techniques	Chapter
●	**Investigate biofilm formation** Investigate biofilms. Biofilms are thin layers of microorganismal communities whose secretions coat wood, rocks, plastic, or steel, and can cause damage to surfaces. Determine the conditions for biofilm growth by growing a biofilm using water from a natural resource such as a local pond. Determine which surfaces are more resistant than others and how long it takes for the film to form.	Bacterial culture	3
◆	**Identify biofilm bacteria** Perform PCR analysis on natural biofilms to identify which organisms are present. Use species-specific primers or clone 16S ribosomal RNA gene sequences into plasmids and have them sequenced.	DNA extraction, PCR, and DNA sequencing	6
■	**Biofilm protein profiles** Use SDS-PAGE to compare biofilms from different environments. If possible, also separate the extracellular secretions from the bacteria and analyze the proteins in the secretions.	SDS-PAGE	7
◆	**Identify proteins in biofilms** Identify specific proteins in biofilms using antibodies and either western blotting or ELISA.	SDS-PAGE and western blotting or ELISA	8
◆	**Test for environmental bacteria** Identify bacteria in water samples from local rivers or streams using PCR or ELISA. Perform background research into PCR primers specific for bacterial species or if ELISA is used, use commercial kits or antibodies specific to bacterial species.	PCR or ELISA	6 or 8
◆	**Identify environmental bacteria** Use PCR to amplify the 16S ribosomal RNA of random bacteria in a water sample from a local body of water or even a puddle. Clone the PCR products by ligating them into plasmids, transforming them into bacteria, and purifying the plasmids. Sequence the 16S rRNA and identify the bacteria using bioinformatics databases.	PCR, ligation, transformation, miniprep, DNA sequencing, and bioinformatics	6
◆	**Bacterial fingerprints with RAPD PCR** Compare the diversity of bacteria in different environments by developing a random amplification of polymorphic DNA (RAPD) PCR test to fingerprint the microorganisms present in an environment. Compare pristine environments to those that may have been changed due to human intervention or activity.	DNA extraction and PCR	6
◆	**Oil-degrading bacteria** Investigate oil remediation by sampling ocean water and looking for oil-degrading bacteria such as proteobacteria that are thought to help clean up oil spills. Investigate how to assay for oil-degrading abilities. Samples do not need to be from the ocean; oil-degrading bacteria could be found anywhere there is oil, even in parking lots.	Bacterial culturing	3
■	**Bioremediation abilities of bacteria** Test whether bacterial strains such as *E. coli* found in the laboratory have oil-degrading abilities.	Bacterial culturing	3

Table 9.15. **Energy and the environment projects**. (continued)

Level	Project Idea	Potential Techniques	Chapter
◆	**Bioremediation with recombinant bacteria** Researchers may have developed recombinant strains of *E. coli* to help with bioremediation, for example *E. coli* expressing enzymes that can degrade oil. If possible, obtain some of these bacteria and use them to investigate the projects above.	Bacterial culturing and enzyme assays	3 & 7
◆	**Metagenomic studies on bacteria** Collaborate with a university or institute to assist with a metagenomic study. Such studies attempt to identify all the microorganisms in a single sample using high-throughput sequencing or microarray technology. They look for differences among samples (for example, ocean water at different depths or with different levels of pollutants).	DNA sequencing and microarrays	6
◆	**Isolate and discover new bacterial viruses in soil** Recover and grow bacterial viruses from local soil samples. Visualize virus particles by microscopy (optional), perform DNA analysis, sequence the genomes, and perform bioinformatics to determine whether the samples contained previously undiscovered viruses. This can be a whole-class project. Visit seaphages.org for more information.	DNA extraction, restriction digestion, gel electrophoresis, plaque assay, DNA sequencing, and bioinformatics	4 & 6
◆	**Investigate local species biodiversity** Identify species present in a local environment by collecting samples, isolating DNA, and performing PCR to amplify the appropriate DNA barcode region. Then sequence the amplified DNA and perform bioinformatics analysis to determine the species as in Activity 6.D.	DNA extraction, PCR, gel electrophoresis, DNA sequencing, and bioinformatics	4 & 6
●	**Biofuel enzymes in mushrooms** Investigate biofuels. Test the cellobiase activity of enzymes in store-bought mushrooms to determine how temperature and pH affect their activity using the procedures in Activity 7.E. **Safety note:** Never touch wild fungi as they may be poisonous.	Enzyme activity	7
■	**Biofuel enzymes in decomposers** Biofuel researchers are seeking enzymes that can more efficiently release the energy stored in biological molecules. Decomposers like fungi and soil bacteria (for example, *B. subtilis*) break down biological molecules such as celluloses and starches. Investigate the enzymes present in decomposers using enzyme activity assays or microbiological techniques. Activity 7.E can act as a starting point. **Safety note:** Never touch wild fungi as they may be poisonous.	Enzyme activity	7
◆	**Biofuel enzyme purification** Use chromatography to purify cellulase enzymes from mushroom extracts. Try different chromatography columns and assess the fractions using SDS-PAGE and enzyme activity assays. Compare the activity to the purified enzyme used in Activity 7.E.	Chromatography, enzyme assays, and SDS-PAGE	7
■	**Biofuel bacteria in termite guts** Biofuel researchers are investigating bacteria found in the guts of termites, since these bacteria help termites digest wood. Develop a method to isolate these bacteria, stain them, and (time permitting) perform enzyme assays. Termites can be found in rotting wood and their digestive tracts can be extracted by placing the termite in a drop of water and pulling the head and abdomen apart using two needles, which should isolate the gut. Make a wet mount of the gut, and use a microscope to examine it and find bacteria. Perform a Gram stain as in Activity 3.D and test for cellulase activity.	Insect dissection, bacterial staining, and enzyme activity	3 & 7
●	**Fermentation for biofuel production** Develop a miniature fermenter (information on this and other projects on biofuels can be found on the website of the Department of Energy's Great Lakes Bioenergy Research Center, glbrc.org/outreach/classroom-materials).	Fermentation	7
■	**Algae as a source of biofuel** Biofuel researchers at Arizona State University are investigating blue-green algae (cyanobacteria) as a source of biofuel. Compare different growth conditions for cyanobacteria, including light and temperature, and determine which conditions generate the highest yields.	Bacterial culturing	3

9.5 Considerations and Tips for All Laboratory Projects

Microbiology

Microbiology projects can be simple and relatively inexpensive; however, microorganisms in the environment can be dangerous, particularly as the prevalence of antibiotic-resistant bacteria increases. Projects that have the potential to grow harmful bacteria, such as projects involving culturing bacteria from classroom surfaces, are not safe to perform in a school laboratory. Consider instead microbiology topics that explore safe bacteria, such as those found in food (for example, *Lactobacillus acidophilus* and *L. bulgaricus* from yogurt and acidophilus milk; *Bacillus cereus* from tofu and cocoa; *B. subtilis* from cocoa and natto; *Acetobacter aceti* from vinegar; or pure strains of non-pathogenic bacteria, such as *E. coli* HB101). A list of microorganisms that are safe for study in the classroom is available at sciencebuddies.org/science-fair-projects/references/microorganisms-safety (accessed April 10, 2018). Alternatively, consider projects where bacteria are explored but not cultured (for example, bacterial identification through Gram staining, PCR, and immunological methods such as ELISA and western blotting).

Review all projects involving microorganisms with a trained professional who has experience in microbiology to ensure that there is no risk of culturing hazardous organisms. In addition, follow all local safety regulations and dispose of any biohazardous waste properly.

Animal Research

Perform any research projects involving animals according to state and national guidelines. Such projects may require pre-approval by an institutional review board (IRB) and under the ordinances of the Institutional Animal Care and Use Committee (IACUC). Do not conduct any research that causes harm to or kills vertebrates (animals with a backbone) for the purpose of research unless you have justified the necessity of the research to and gained approval from the appropriate board and/or committee. You can obtain animal tissue from animals that have not been killed for the purposes of research from veterinarians, butchers, and slaughter houses. It is also likely that nonharmful research on live animals, such as animal behavioral studies or sampling of animal body fluids, will require approval, such as sign off by a veterinarian, to be acceptable for entry into a competition. In summary, animal research can be the basis of a great project but before starting the project, obtain the necessary approvals to ensure the safety and proper care of the animal subjects.

Research on Human Subjects

Research into people and their genetics or behaviors can make great projects; however, care must be taken on a few fronts.

First, diseases can be passed among individuals. Therefore, when collecting body samples, such as saliva, make sure each individual handles their own sample until it has been processed in a way that kills any pathogens (for example, mixing with alcohol, heating to 95°C, or bleaching). Never culture human bacteria as part of a student project (see Microbiology) and never use blood samples due to the risk of blood-borne pathogens.

Second, genetic studies should respect personal privacy. To ensure the privacy of your subjects, randomize the samples so that they cannot be traced back to the individual and never perform experiments that might reveal information about an individual's health. Let all subjects know of and consent to your plans for their sample and the information it generates. For competitions, you may need to document that you obtained their permission by using a consent form.

Third, genetic information can reveal details about familial relationships, which could be disruptive. Never test samples from relatives in the same family, even if you do not intend to compare relatives. Results may be misinterpreted by individuals, errors may be made in student sample preparation, or samples may be mixed up, all of which can lead to false assumptions about familial relationships.

In summary, conduct your research project safely and do not risk revealing information about a person that might affect their life.

Molecular Biology

Molecular biology projects involve manipulating DNA using techniques such as restriction enzyme digestion, electrophoresis, and PCR. Recombinant DNA can also be generated by ligation, and recombinant organisms can be generated, for example, by transformation of bacteria.

Guidelines for Working with Recombinant DNA

There are specific guidelines from the National Institutes of Health (NIH) with respect to working with recombinant DNA and recombinant organisms (NIH Guidelines for Research Involving Recombinant or Synthetic Nucleic Acid Molecules, April 2016, Department of Health and Human Services, National Institutes of Health). The Intel International Science and Engineering Fair provides similar guidelines in their international rules and guidelines. Take extra care when working with GMOs, as they may have unforeseen effects on human health and the environment. Also review the Governmental Regulation of Biotechnology section in Chapter 1.

Restriction Enzymes

The list of potential projects involving restriction enzymes is extensive. Information on restriction enzymes can be found in the catalogs of companies that sell them; for example, the New England BioLabs (NEB) catalog has an entire reference section on restriction enzymes, which is also available on its website (neb.com). Enzymes are usually purchased from a supplier such as NEB and the appropriate reaction buffer and any additives that may be required such as bovine serum albumin (BSA), are included in the purchase.

PCR

PCR can be the basis of many great research projects but it can also be challenging. Review the information in Chapter 6 before proposing a PCR-based project. Consider extending the PCR research of a previous student or extending a PCR activity performed in class. Such projects can yield more results than those requiring initiation of a new PCR project.

Extracting DNA

Extracting template DNA can be challenging. In fact, optimizing a DNA extraction method can be an entire project itself. Different methods may be used to extract DNA from bacteria, plants, and animals, and for each type of organism, there are many different options. In Activities 6.B and 6.C, InstaGene matrix is used to extract gDNA from human cheek or hair cells, plant seeds, and food. InstaGene matrix may also be used to extract gDNA from bacteria using a method similar to the cheek cell DNA extraction used in Activity 6.C. InstaGene matrix extraction works only for certain cell types; more extensive extraction procedures may need to be researched and tested. Plasmid DNA is extracted from bacteria in Activity 5.C using a commercial column purification protocol and similar kits for extracting gDNA exist, such as Bio-Rad's Nucleic Acid Extraction module. There are also many noncommercial laboratory protocols available in the literature and online. Many procedures involve hazardous chemicals, such as phenol, so review all safety information before use.

Primer Design

PCR primers determine the sequence of DNA that will be amplified. Ideally, sequences for primers that amplify a region of interest are available in a protocol or in the materials and methods section of a research paper. If this is not the case then primers may need to be designed, which can be a research project in itself. Background information on primer design is provided in Chapter 6; however, to design primers from scratch requires more in-depth, independent research and will likely involve using primer design software such as Primerfox and Primer3 as well as software such as Primer Show that can check the final primer sequences. (Internet searches will find these programs.) The National Center for Biotechnology Information also has primer designing tools. Take care that each primer sequence is complementary to the appropriate DNA strand and in the correct direction, 5' to 3' (primer checking software can help with this).

You can purchase primers as freeze-dried, desalted oligonucleotides (oligos) from a commercial company such as Eurofins MWG Operon or Integrated DNA Technologies, and they are relatively inexpensive (25 nmol is usually sufficient for a student project). Rehydrate the primers in 10 mM Tris, pH 8.0, or molecular biology grade water before use. The company will often provide information regarding the volumes necessary for rehydration to a stock concentration of 100 µM (a 200x stock if the final concentration of primer in the PCR reaction is 0.5 µM). Dispense (aliquot) this stock into multiple tubes using aseptic technique and store the tubes at –20°C. Use one tube at a time

for PCR experiments and mark it as the tube currently in use. This aliquoting procedure is vital to PCR studies to prevent contamination of primer stocks; if contamination occurs, discard the contaminated aliquot and use a new, uncontaminated aliquot.

Master Mixes

Review the Setting Up a PCR Reaction section in Chapter 6 and the suggested concentrations for the components. If possible, use a commercial master mix that requires only primers and template DNA, such as Bio-Rad's 2x master mix for PCR. Using commercial master mixes can reduce the variability among experiments that can be caused by pipetting errors and variability in raw components. If a commercial master mix does not work, is not appropriate for the experiment, or is not available, make a master mix from scratch using the various components described in Table 6.2.

Determining the Annealing Temperature

The melting temperatures (T_m) of primers, usually provided by the manufacturer, are used to determine the annealing temperature of the PCR reaction (see the Primer Design section of Chapter 6, page 196), which is critical to successful PCR. If the temperature is too low, the primer may bind nonspecifically (when it does not completely match the template). If the temperature is too high, the primer may not bind even if it is a complete match. The annealing temperature is often 2–3°C lower than the melting temperatures of the primers. If the primers have very different melting temperatures, the annealing temperature may be in between them. Because the T_m can be calculated using many different formulae, the T_m designated by the manufacturer may be different from the T_m calculated when designing the primers (though they should be close). You may need to determine the optimal annealing temperature yourself through experimentation and this may comprise a research project in itself.

Ligation and Transformation

Ligating DNA fragments into plasmids is an advanced technique. The DNA fragments can be generated from restriction enzyme digestions or by PCR. There are two main types of plasmids: those for housing DNA fragments (cloning plasmids) and those for expressing proteins of interest (expression plasmids). Cloning plasmids can be used to house PCR products for DNA sequencing or subcloning. Bio-Rad's Ligation and Transformation module contains the pJET1.2 cloning plasmid, and the Sequencing and Bioinformatics module contains sequencing primers for pJET1.2 (see DNA Sequencing on the next page). Other common cloning plasmids are pUC18 and pUC19. Cloning genes into expression plasmids to express proteins is quite an advanced technique and will likely require guidance from a mentor. For any project involving ligation, it is important to develop a cloning strategy to ensure that the correct enzymes and plasmids are used; if possible, review the strategy with an experienced researcher.

In contrast to transforming bacteria with plasmids as in Activities 5.A and 5.B, transformation protocols for ligation reactions require bacteria with much higher transformation efficiencies. You can use Bio-Rad's Ligation and Transformation module, which contains reagents to generate chemicompetent bacteria with high transformation efficiencies in a school laboratory in around 30 minutes. These work well for transforming ligations. Alternatively, use traditional protocols to generate competent cells or purchase competent cells from a supplier.

DNA Sequencing

DNA sequencing can provide data for many types of projects. Sequencing requires two things: a single DNA sequence and a sequencing primer. To perform the actual sequencing reactions, employ the help of a university DNA sequencing core facility or a DNA sequencing company. Each facility will have specific requirements and protocols for submitting samples, including specifications for the concentrations of template DNA and primers.

The DNA template used for sequencing can be either plasmid DNA or a PCR product but it must be free of any contaminating DNA sequences; otherwise, the sequence will be unreadable. Since plasmid DNA has been isolated and purified, it is almost always free of contaminating DNA. A PCR reaction, on the other hand, may contain nonspecific PCR products or primer-dimers that may interfere with sequencing.

To confirm that a PCR product is free of contaminating DNA sequences, separate it by agarose gel electrophoresis. The PCR product should migrate as a single band. Be aware, however, that even if the reaction shows a single band on a gel, this is not an guarantee that low level contamination is absent (a contaminating product of the same size, for example, may be present). If there are contaminating bands, to further isolate the PCR product, cut out the band from the gel and purify it to isolate the DNA from contaminating bands (a technique called gel purification). Commercial kits are available for this and protocols can be found online. Gel purification also removes dNTPs, enzymes, and primers, which is necessary before sequencing. If the PCR product is known to be a single band, however, you can remove the dNTPs, enzymes, and primers more simply by using size exclusion chromatography columns such as Bio-Rad's PCR Kleen™ spin columns.

If gel purification is not sufficient to purify the band (for example, if the PCR yields multiple products of similar sizes), then ligate the PCR products into plasmids (see Ligation and Transformation) and sequence the resultant plasmid.

The sequencing primer can be one of the PCR primers or, if the DNA is a plasmid, it may be a primer that matches a plasmid sequence on one side of the multiple cloning site (MCS). A typical sequencing reaction reads 600–800 bp, so if the DNA sequence is longer than 600 bp, use multiple sequencing primers. Also note that there is a region of at least 50 bp beyond the sequencing primer that does not yield a high-quality DNA sequence. Therefore, it is good practice to set up two sequencing reactions: one with a forward primer and one with a reverse primer. This ensures that a sequence is read twice, which may provide more confidence and clear up any ambiguities in the data.

Proteins
Obtaining Antibodies

Chapter 8 describes antibodies and how they are used. How To… Select an Antibody on page 301 provides guidelines on how to choose antibodies. The materials section of journal articles are also a great place to find antibodies that should work for your application and will often reference where to obtain the antibody. Various options exist for obtaining antibodies for use in research projects:

- Commercial sources — many vendors, including Bio-Rad, sell a wide range of primary and secondary antibodies. Commercial antibodies can be expensive, so be sure to check which applications a particular antibody is best suited for before purchase

- Government agencies — as an example, the Developmental Studies Hybridoma Bank was created by the NIH as a national resource and provides monoclonal primary antibodies at low cost

- Research labs — specific antibodies can be obtained by request from researchers if they have published information on the development of the antibody in a journal article. When requesting antibodies from a researcher, include a clear outline of the research question and make it very clear to the researcher that you have properly planned the experiment

Optimizing Immunoassays

When performing immunoassays, first optimize assay parameters, such as the type and concentration of antibodies and the incubation times and temperatures. Otherwise, the quality of the entire research project may be compromised. Assay optimization can also yield savings in terms of both cost and time.

The amount of antibody required will depend on the type of assay being performed. An Ouchterlony assay, for example, requires a higher concentration of antibody than a western blot, which usually needs a higher concentration than an ELISA. A starting concentration for each application is often provided with an antibody or in the research paper citing use of the antibody; however, the optimal concentration of antibody greatly depends on the concentration of the antigen. If the antigen is scarce, then a higher antibody concentration will be required to detect it.

Higher concentrations of antibody may cost more but in some cases, may yield time savings. Activities 8.B, 8.C, and 8.D have been optimized to work with very short incubation times; This was achieved by using higher antibody concentrations. More common incubation times for ELISAs and western blots are 30 minutes to

2 hours and require lower antibody concentrations, which lowers costs. If you use longer incubation times to improve detection, be sure to also increase number and duration of washing steps.

Antibody binding also depends on temperature. If the binding of an antibody is increased, then less antibody can be used, which can save resources. To increase the binding of an antibody, you can conduct immunoassays at a slightly elevated temperature of ~37°C, but make sure the assays do not dry out.

Enzymes and Substrates

When exploring enzyme activity, use control enzymes and substrates that have known concentrations. Many companies, including VWR and Sigma-Aldrich, sell enzymes such as proteases, cellulases, cellobiases, lysozymes, and lipases that can be used for projects. Enzyme concentration is frequently provided in activity units but you will need to determine what the units mean and how much is needed for your experiments.

When designing an enzyme-based project, one of the difficult questions may be how to monitor and quantify the progress of the enzyme-mediated reaction. The enzyme will change the substrate, but how will that be measured? In Activity 7.E, a synthetic substrate was used because the product of the natural substrate was invisible and hard to quantify.

Chromatography

Many chromatography columns can be reused if properly washed and stored, which can reduce the cost of a project. Size exclusion columns from Activity 7.B, for example, can be washed with 40 ml of distilled water to wash all the proteins out of the beads and then rinsed with 10 ml of column buffer to reequilibrate the column. Hydrophobic interaction columns from Activity 7.C can be washed with 20 ml of distilled water and then stored in 20% ethanol to reduce microbial contamination. Before use, equilibrate the column with equilibration buffer as described in part 3 of Activity 7.C.

Purifying a protein from a mixture can be challenging. You will need to use different types of chromatography columns to see separation from a mixture of proteins. Refer to the information in Chapter 7 on chromatography and consider what properties your protein has and what resources, such as different types of chromatography columns, are available for your project.

Chapter 9 Essay Questions

1. Discuss the differences between a scientific hypothesis, theory, and law and give examples.

2. Using a journal article you have read, discuss how statistical analysis was used in the paper and how it strengthened the data.

Additional Resources

Committee on Revising Science and Creationism: A View from the National Academy of Sciences, National Academy of Sciences and Institute of Medicine of the National Academies. (2008). Science, Evolution, and Creationism. (Washington, D.C.: The National Academies Press).

Cothron, J et al. (2005). Students and Research: Practical Strategies for Science Classrooms and Competitions. Fourth Edition. (Dubuque: Kendall/Hunt Publishing Company).

Primer design software:

Primer 3: Rozen S and Skaletsky HJ (2000). Primer3 on the WWW for general users and for biologist programmers. In Bioinformatics Methods and Protocols: Methods in Molecular Biology, Krawetz S, and Misener S, eds. (Totowa, NJ: Human Press), pp. 365–386.

Primer-BLAST: Ye J et al. (2012). Primer-BLAST: A tool to design target-specific primers for polymerase chain reaction. BMC Bioinformatics. 13:134. ncbi.nlm.nih.gov/tools/primer-blast/ (accessed June 1, 2018).

Appendix A:
Fast Gel Protocol

This protocol shortens the time required for agarose gel electrophoresis by using a higher voltage and a low-salt buffer. Agarose gels of any percentage are made with 1x TAE buffer and are electrophoresed at 200 V in one-fourth strength (0.25x) TAE buffer.

1. Follow Activity 4.B to make agarose TAE gels using 1x TAE buffer.

   Note: Make agarose gels with 1x (not 0.25x) TAE buffer or DNA bands will not resolve.

2. Make 0.25x TAE electrophoresis buffer. Use the equation $C_1V_1 = C_2V_2$ to calculate volumes. For example:

   - To make 300 ml of 0.25x TAE buffer from 50x TAE stock, add 1.5 ml of 50x TAE stock to 298.5 ml of distilled water

   - To make 300 ml of 0.25 x TAE buffer from 1x TAE buffer, add 75 ml of 1x TAE buffer to 225 ml of distilled water

3. Perform electrophoresis.

   a. Place the gel in the electrophoresis chamber and cover it with 0.25x TAE buffer; ensure the gel is submerged.

   b. Load the samples and run the gels at 200 V for no more than 20 min. Monitor the progression of the gel loading dye down the gel.

   Note: Never run an agarose gel in 1x TAE buffer at more than 100 V, as the heat generated will melt the gel. The lower salt concentration in 0.25x TAE buffer reduces heat generation, allowing use of the higher voltage.

Appendix B: Alternative Staining Methods for Agarose Gels

Option 1: Use 100x Fast Blast™ DNA Stain for Quick Staining (12–15 min)

Use this protocol if laboratory class time is sufficient to visualize the DNA bands after electrophoresis.

Note: Warm water (40–55°C) is essential for all destaining steps.

Safety reminder: Follow all laboratory safety protocols. Wear appropriate PPE. Read SDS before use.

1. Dilute the 500x concentrate to 100x by adding 24 ml of 500x concentrate to 96 ml of distilled water to make 120 ml of 100x concentrate. We recommend using 120 ml of 100x Fast Blast stain to stain two 7 x 7 cm or 7 x 10 cm gels in individual staining trays; each tray can accommodate two gels.

2. Carefully remove the gels from their gel trays and slide them into the staining tray. Add the 100x stain until the gels are completely submerged (about 120 ml of stain).

3. Stain for 2–3 min (maximum). Pour the 100x stain into a storage bottle and save for future use. The 100x stain may be reused at least seven times. Using a funnel, pour the 100x stain into a plastic storage bottle and store away from direct light for future use.

4. Rinse the gels for 10 sec in 500–700 ml of clean, warm (40–55°C) tap water.

5. Wash the gels for 5 min in 500–700 ml of clean, warm (40–55°C) tap water. Repeat this step. The bands may appear fuzzy immediately after the second wash but will become sharper within 5–15 min. To remove excess stain faster, lay 2-3 laboratory wipes on top of each other and roll them into a single roll. Add this roll to the wash tray to one side of the gel (but not on top of it) with the water. The laboratory wipes will bind dye molecules to help visualize the bands faster.

6. To obtain maximum contrast, additional washes in warm water may be necessary. Destain to the desired level but do not leave the gel in water overnight. If destaining cannot be completed in the time available, transfer the gel to a tray containing 1x Fast Blast stain for storage overnight.

7. The gel is ready to be documented and analyzed manually or using an imaging system.

Option 2: Use 1x Fast Blast DNA Stain for Overnight Staining

Use this method if there is not enough laboratory class time to allow for staining immediately after electrophoresis.

Safety reminder: Follow all laboratory safety protocols. Wear appropriate PPE. Read SDS before use.

1. Dilute the 500x concentrate to 1x by adding 0.24 ml of 500x concentrate to 120 ml of distilled water to make 120 ml of 1x concentrate. We recommend using 120 ml of 1x Fast Blast stain to stain two 7 x 7 cm or 7 x 10 cm gels in individual staining trays; each tray can accommodate two gels.

2. Carefully remove the gels from their gel trays and slide them into the staining tray. Add the 1x stain until the gels are completely submerged (about 120 ml of stain).

3. Place the staining tray on a rocking platform and gently shake overnight. If a rocking platform is not available, swirl the solution and gel a few times during the staining period. This is crucial because smaller fragments tend to diffuse without shaking. The bands will begin to develop after 2 hr, but at least 8 hr of staining is recommended.

4. The gel is ready to be documented and analyzed manually or using an imaging system.

Option 3: Use UView™ 6x Loading Dye and Stain before Electrophoresis for Immediate Gel Viewing

This protocol uses UView 6x loading dye and stain, a fast-acting loading dye that contains a nontoxic nucleic acid stain. UView 6x loading dye and stain is provided as a 6x concentrate that is added to DNA samples prior to loading on a gel. This replaces the need for separate loading dyes and DNA stains, and allows DNA to be visualized immediately after electrophoresis, without the need for subsequent staining and destaining steps. UView 6x loading dye and stain can be used with pre-poured or existing gels, as it does not need to be added during gel pouring.

Safety reminder: Follow all laboratory safety protocols. UV light is damaging to skin and eyes. Wear appropriate PPE. Read SDS before use.

1. Add UView 6x loading dye and stain to each electrophoresis sample in a 1:5 ratio and mix. For example, add 3 µl UView 6x loading dye and stain to a 15 µl sample for a total volume of 18 µl. Mix thoroughly and load samples onto gel.

2. After electrophoresis, visualize DNA bands using a gel imaging system with UV light. When using UView 6x loading dye and stain samples are best viewed with a 365 nm UV light.

Option 4: Use SYBR® Safe DNA Gel Stain Before Electrophoresis

In this protocol, concentrated SYBR® Safe stain is added directly to molten agarose prior to pouring the gels. Use this method to view and document gels immediately following electrophoresis using a digital imaging system with blue or UV light.

Safety reminder: Follow all laboratory safety protocols. UV light is damaging to skin and eyes. Wear appropriate PPE. Read SDS before use.

1. Follow the directions for preparing molten agarose. Combine agarose powder and running buffer in appropriate volumes and heat until the agarose powder is dissolved. Allow to cool to 50ºC.

2. Add SYBR® Safe stain to the molten agarose. Use 1 µl of 10,000x SYBR® Safe stain concentrate with 10 ml of agarose solution (thus, for 40 ml of agarose, add 4 µl of SYBR® Safe stain). Mix by swirling the molten agarose, then pour the gels.

3. To store gels containing SYBR® Safe stain, protect them from light by covering them with aluminum foil and store at room temperature overnight.

4. Run prestained SYBR® Safe gels in the same buffer system and electrophoresis conditions as the gels that are not made with SYBR® Safe stain.

5. View the gels directly on an imaging system with either blue light (480 nm) or UV light. Prestained SYBR® Safe gels do not need to be stained or destained.

   Note: Use of a blue light transilluminator minimizes UV exposure and damage to samples, an important factor in applications such as cloning.

Activity	DNA Marker Volume (µl)	UView 6x Loading Dye and Stain Added to DNA Marker (µl)	Sample Volume (µl)	UView 6x Loading Dye and Stain Added to Sample (µl)
4.D Restriction Digestion and Analysis of Lambda DNA	100	20	10	2
4.E Forensic DNA Fingerprinting	100	20	20	4
5.D DNA Quantitation	1 KB molecular ruler: 250 plus 165 µl water Precision molecular mass ruler: 250 plus 165 µl water	1 KB molecular ruler: 85 Precision molecular mass ruler: 85	10	2
6.A STR PCR Analysis	200	40	40	8
6.B GMO Detection by PCR	200	40	40	8
6.C Detection of the Human PV2 Alu Insert	100	20	Control: 10 Sample: 40	Control: 2 Sample: 8
6.D Fish DNA Barcoding	200	40	10	2

Option 5: Use SYBR® Safe DNA Gel Stain (0.5x Concentration) After Electrophoresis

This protocol is most appropriate when time is available for staining the gels and analyzing them using a digital imaging system with blue or UV light.

Safety reminder: Follow all laboratory safety protocols. UV light is damaging to skin and eyes. Wear appropriate PPE. Read SDS before use.

1. Prepare 0.5x SYBR® Safe DNA gel stain by adding 6 µl of 10,000x SYBR® Safe concentrate per 120 ml of running buffer. Alternatively, purchase the premade 0.5x concentration.

2. After electrophoresis, carefully remove the gel from its tray and slide it into the staining tray. Add enough stain to submerge the gels completely (about 120 ml of stain).

3. Cover the staining tray with aluminum foil to protect the gel from light. Place the staining tray on a rocking platform with gentle shaking, and leave it for 30 min at room temperature. If a rocking platform is not available, periodically swirl the solution and gel a few times to obtain thorough and uniform staining patterns.

4. Rinse the gel with water and place it into the imaging system. Illuminate with either blue light (480 nm) or UV light.

   Note: Use of a blue light transilluminator minimizes UV exposure and damage to samples, an important factor in applications such as cloning.

Option 6: Use Ethidium Bromide Before Electrophoresis

In this protocol, concentrated ethidium bromide is added directly to molten agarose prior to pouring the gels. Use this method immediately following electrophoresis using a digital imaging system with UV light.

Safety reminder: Follow all laboratory safety protocols. Ethidium bromide is a mutagen and suspected carcinogen. UV light is damaging to skin and eyes. Wear appropriate PPE. Read SDS before use.

1. Follow the directions for preparing molten agarose. Combine agarose powder and running buffer in appropriate volumes and heat until the agarose powder is dissolved. Allow to cool to 50°C.

2. Add ethidium bromide to the molten agarose. Use 0.5 µl of 10 mg/ml ethidium bromide for each 10 ml of agarose solution (thus, for 50 ml of agarose, add 2.5 µl of 10 mg/ml ethidium bromide). Mix by swirling the molten agarose, then pour the gels.

3. To store gels containing ethidium bromide, protect them from light by covering them with aluminum foil and store at room temperature overnight.

4. Run prestained ethidium bromide gels in the same buffer system and electrophoresis conditions as the gels not made with the ethidium bromide stain.

   Note: It is not necessary to add ethidium bromide to the buffer unless the gels will be electrophoresed for longer than 30 min at 100 V. If a long electrophoresis time is anticipated, ethidium bromide can be added to the running buffer prior to electrophoresis to a concentration of 0.5 µg/ml.

5. View the gels directly on an imaging system with UV light. Prestained ethidium bromide gels do not need to be stained or destained.

Option 7: Use Ethidium Bromide After Electrophoresis

This protocol is most appropriate when additional time is available for staining the gels and analyzing them using a digital imaging system with UV light.

Safety reminder: Follow all laboratory safety protocols. Ethidium bromide is a mutagen and suspected carcinogen. UV light is damaging to skin and eyes. Wear appropriate PPE. Read SDS before use.

1. Prepare 0.5 µg/ml ethidium bromide solution by adding 6 µl of 10 mg/ml ethidium bromide per 120 ml running buffer.

2. After electrophoresis, carefully remove the gel from its gel tray and slide it into the staining tray. Add enough stain to submerge the gels completely (about 120 ml of stain).

3. Cover the staining tray with aluminum foil to protect the gel from light. Place the staining tray on a rocking platform with gentle shaking and leave it for 30 min at room temperature. If a rocking platform is not available, periodically swirl the solution and gel a few times to obtain thorough and uniform staining patterns.

4. Rinse the gel with water and place it into the imaging system with UV light. If a background is evident, destain in water for 15 min and reimage.

Appendix C: Stain-Free SDS-PAGE Gel Imaging

This protocol allows for protein bands to be visualized on gels, such as the Bio-Rad Mini-PROTEAN® TGX Stain-Free™ Gels, and blots without the use of protein stains. Stain-free gels are used in the same way as standard SDS-PAGE gels, but they contain reagents that react with tryptophan amino acids with exposure to UV light. This causes the protein bands to fluoresce, making them visible without staining.

General protocol for imaging Bio-Rad Mini-PROTEAN® TGX Stain-Free gels

1. Immediately after electrophoresis is complete, remove the gel from the cassette and place it directly on a UV transilluminator or gel imaging system tray. No fixation or rinsing steps are required.

   Note: Do not allow gel to soak in water or other solution after electrophoresis and prior to activation and imaging. Soaking a gel before activation allows the stain-free reagent to diffuse out of the gel. Once activated, the gel can sit in buffer or water as usual.

   Note: Do not place any material, such as plastic film, between the gel and the imaging tray.

2. Expose the gel to 365 nm wavelength UV light for 1–5 minutes.

3. Document the gel using a camera or other gel documentation equipment.

Protocol for stain-free gel imaging using Bio-Rad's Image Lab™ software.

Image Lab software can be used to capture and visualize gel images and is available for download on the Bio-Rad website.

1. Immediately after electrophoresis is complete, remove the gel from the cassette and place it directly on a stain-free enabled Bio-Rad UV transilluminator (e.g. ChemiDoc MP,) or stain-free tray (Gel Doc EZ). No fixation or rinsing steps are required. To ease manipulation of the gel, spray the stain-free tray with water before positioning gel.

   Note: Do not allow gel to soak in water or other solution after electrophoresis and prior to activation and imaging. Soaking a gel before activation allows the stain-free reagent to diffuse out of the gel. Once activated, the gel can sit in buffer or water as usual.

   Note: Do not place any material, such as plastic film, between the gel and the imaging tray.

2. To acquire a stain-free gel image, double click the icon for Image Lab™ software on the computer connected to the imager. Follow the prompts on the start page under Gel Imaging Application: Select > Protein Gels > Stain-Free Gel

3. Select the gel activation, depending on the application.

   a. 1 min activation provides sufficient UV activation for gels used in western blots.

   b. 2.5 min activation provides a good balance between time required for UV activation and signal intensity. In general, this is a good default activation time.

   c. 5 min activation provides maximum signal intensity.

4. In the Imaging Area pane, select from the list of Bio-Rad gels, or enter the image area dimensions manually. This option is not necessary when using the Gel Doc EZ, as the entire area is imaged automatically.

5. In the Image Exposure pane, select from one of the following options:

   a. **Auto Exposure** — this setting estimates an optimal exposure time and ensures the best use of the dynamic range.
      i. **Intense bands** — optimizes exposure for all bands in the image area.
      ii. **Faint Bands** — a longer exposure time is used to make faint bands more visible, but more prominent bands might be overexposed.

   b. **Manual Exposure** — use this setting to manually override automated imaging. This setting is often used to duplicate an exposure time in order to compare band intensities from different gels.

6. In the Display Options window, keep the default settings.

7. To start the stain-free activation and image acquisition, select Run Protocol.

8. The Gel Image Preview window will now display the stain-free gel undergoing activation in real time.

9. Once activation is complete, an image of the stain-free gel is captured, based on the image exposure time selected.

10. The gel image can now be exported in a variety of image formats (TIFF, JPEG, Bitmap, or PNG).

Appendix D: Glossary

Abstract: Summary of a scientific article, usually appearing just after the title.

Adherent cultures: Cell cultures that adhere or stick to a solid surface such as a tissue culture dish or flask.

Adsorb: To take up and hold by adsorption (adhesion of molecules to the surface of a solid).

Aerosol: Suspension or dispersion of fine particles of solid, liquid, or gas.

Affinity chromatography: Chromatographic technique that separates molecules based on affinity for a specific binding partner.

Agar: Jelly-like substance made of linked sugars (polysaccharides) from seaweed; used in media for growing bacteria.

Agarose: Uncharged polymer typically used to make agarose gels for the separation by size of nucleic acids and other biomolecules through electrophoresis.

Allele: One of a pair or more alternative DNA sequences (usually genes) that occupy a specific position on a specific chromosome.

Allele frequency: Proportion of an allele in a population.

Alu: Short interspersed, nuclear elements (SINEs) of DNA around 300 bp found in the human genome that contain the AluI restriction site.

Amino acids: Molecules with a central carbon attached to an amino group, a carboxyl group, and a side chain (or R group); the building blocks of proteins.

Anneal: In PCR, to bind single-stranded DNA to complementary sequences. For example, oligonucleotide primers anneal to denatured, single-stranded template DNA.

Anode: Positive electrode in an electrophoresis chamber.

Antibiotic: Chemical that prevents or reduces the growth of microorganisms.

Antibody: Protein formed in response to a challenge of the immune system by a foreign agent. Antibodies bind to specific antigens.

Antigen: Foreign molecule or cell that elicits an immune response.

Aseptic technique: Set of methods that prevents microbial contamination of a specific environment.

Assay: Procedure that tests or measures a property of a sample.

Assay validation: Process of proving an assay or process is consistently performing correctly and giving the expected outcome.

Autoclave: Piece of equipment that sterilizes liquids, containers, and instruments by exposing them to high pressure and temperature for a defined period of time.

Bacteriophages (phages): Viruses that infect bacteria.

Base pairs: Complementary nucleotides held together by hydrogen bonds. In DNA, adenine is bonded by two hydrogen bonds with thymine (A–T) and guanine with cytosine by three hydrogen bonds (G–C).

Biochemistry: Study of the chemistry of biological molecules.

Biofuel: Fuel produced from renewable biological resources.

Bioinformatics: Application of information technology and computer science to biological research.

Biological safety cabinet (BSC): Ventilated cabinet that filters air to protect personnel, reagents, or the environment against particulates or aerosols from biohazardous agents.

Bioreactor: Large production tank used to grow bacterial cultures.

Bioremediation: Use of microorganisms for the removal of pollutants from the environment.

Biosafety levels (BSL): Classification system for biological laboratories, ranging from BSL-1 to BSL-4, with BSL-1 being the lowest level.

Biosensor: Device that uses biological materials to monitor the presence of chemicals in a sample or environment.

Biotechnology: Technological application that uses biological systems, living organisms, or derivatives thereof to make or modify products or processes for specific use.

Biuret test: Test for the presence of peptide bonds. Copper ions in a basic environment change from blue to purple in the presence of peptide bonds.

Blot: In molecular biology, a flexible, solid support such as nitrocellulose membrane that has had DNA, RNA, or proteins transferred onto it.

Bradford assay: Test that uses the Bradford reagent containing Coomassie (Brilliant) Blue dye to determine protein concentration using a spectrophotometer.

BSL: See Biosafety levels.

Calibration: Check or adjustment of an instrument to ensure it measures precisely and accurately.

Callus: Mass of undifferentiated plant tissue, often formed in plant tissue culture.

Cas9: Endonuclease that cuts double stranded DNA. The location of the cleavage site is determined by the guide RNA that is bound to the Cas9.

Cas genes: Encode enzymes of bacterial origin that cut DNA.

Cathode: Negative electrode in an electrophoresis chamber.

cDNA: See Complementary DNA.

Cell culture: Process of growing cells under controlled conditions.

Cell line: Immortalized animal or plant cells maintained for scientific purposes.

Chelating agent: Substance that binds metal ions in solution: examples include EDTA and Chelex® resin.

Chromatography: Set of techniques for separating molecules based on their physical and chemical characteristics.

Chromatography column: Plastic or glass cylinder designed to contain a chromatographic support.

Citation: Reference to a published source.

Cleanroom: Used in the biotechnology and pharmaceutical industries to control the exposure of raw materials and products to microbial contaminants.

Clinical trials: Regulated series of tests of pharmaceutical drugs in humans to evaluate their safety and effectiveness.

Clone: In cell biology, a cell or group of cells derived through cell division from the same parent cell, thus having identical genetic data.

Cloning: In molecular biology, to obtain a fragment of DNA such as a gene from a genome and ligate it into another piece of DNA such as a plasmid in order to replicate the original ligated DNA sequence.

Clustered regularly interspaced short palindromic repeats (CRISPR): DNA sequence repeated many times with unique sequences between the repeats that serves as an adaptive immune response in bacteria.

Codon: Set of three nucleotides (in DNA or mRNA) coding for a single amino acid. Some amino acids have multiple codons.

Colony: Group of bacteria on growth media, usually grown from a single bacterium; clone of identical organisms.

Colony forming unit (CFU): Single microorganism such as a bacterium with the potential to form a colony when grown on solid medium.

Commercialization: Stage of product development in which the details of marketing, manufacturing, and distribution are decided and product is prepared for sale.

Competent cell: Cell with the ability to take up extracellular DNA from the environment.

Complementarity: Property by which two strands of nucleotides have complementary base pairs such that they will anneal.

Complementary DNA (cDNA): Single-stranded DNA synthesized from an RNA template using a reverse transcriptase enzyme.

Conceptualization: Stage of product development in which marketing research identifies a need for a product and researchers confirm such a product can and should be developed.

Conjugation: Process by which bacteria shuttle DNA between one another using a bridge between them.

Constitutive: Constant expression (for example, of a gene).

CRISPR-Cas9 technology: A system derived from bacteria used by scientists to precisely manipulate genes and gene expression.

Deductive research: Research that transforms a general idea or theory into a testable hypothesis.

Deep tubes: Sterile test tube containing solid growth media such as LB agar used to grow stab cultures.

Denaturation: In PCR, template DNA is made single-stranded in the reaction by heating.

Deoxyribonucleoside triphosphate (dNTP): A building block used to make DNA or RNA composed of a nitrogenous base, a sugar, and three phosphates. The letter N is used in the abbreviation when the base in the molecule (typically, A, T, C, or G) is unspecified or unknown.

Development: Stage of product development that begins when a candidate product is in hand and includes refining the product, investigating the best manufacturing process, detailing specifications, and testing and efficacy.

Dideoxynucleotide triphosphates (ddNTPs): Specialized dNTPs that cannot be extended by a DNA polymerase used in a cycle DNA sequencing reaction to prematurely terminate a new DNA strand.

Digital PCR (dPCR): PCR technique that allows more sensitive detection and direct measurement of nucleic acid concentration. Samples are divided into many separate partitions that are small enough to contain either zero or one (or a few) template DNA molecules.

DNA barcoding: A short sequence of an organism's DNA that can be used for species identification.

DNA extraction: Removal of DNA from cells by cell lysis and purification.

DNA polymerase: Enzyme involved in DNA replication; links a new nucleotide to a growing strand of DNA.

DNA profiling: Comparison of individuals using differences in DNA sequences; often used in criminal investigations to exclude suspects.

DNA sequencing: Determination of the order of bases in an unknown DNA sequence.

DNA size standard: DNA fragments of known size used as a standard for comparison (for example, when estimating the size of DNA fragments following gel electrophoresis); also called markers, rulers, and ladders.

DNA stain: Colored or fluorescent compound used to visualize DNA, for example, Fast Blast™ DNA stain, ethidium bromide, UView™ 6x loading dye and buffer, and SYBR® Safe DNA stain.

dNTP: See Dexoynucleoside triphosphate.

Drug development: Process by which pharmaceutical and biotechnology industries test and develop candidate molecules to generate new drugs.

Drug discovery: Process by which pharmaceutical and biotechnology industries identify candidate molecules that have the potential to become new drugs.

Electroblotting: Use of an electrical field to transfer molecules from a gel onto a solid support or membrane.

Electrophoresis: Movement and separation of molecules in an electrical field.

Electroporation: Transformation of cells using an electrical field to increase permeability of cell membranes to allow DNA entry.

ELISA: See Enzyme-linked immunosorbent assay.

Embryonic stem cells: Stem cells derived from embryos.

Enzyme: Protein with catalytic activity.

Enzyme-linked immunosorbent assay (ELISA): Immunological assay that involves adsorption of proteins to multi-well polystyrene microplates and application of enzyme-linked antibodies to probe for specific proteins.

Epitope: Portion of an antigen that is recognized by an antibody. An antigen can have multiple epitopes.

Exon: Segment of a eukaryotic gene that is transcribed into mRNA, is retained after RNA processing, and encodes protein. Present in either the DNA sequence or the RNA transcript. Exons are separated in DNA and in the primary RNA transcript by introns (non-coding sections).

Extension: In PCR, DNA polymerase extends the primers by reading the complementary strands and adding matching nucleotides to the 3' end of the primer.

Facultative: Only when required (for example, genes that are transcribed only when needed).

Fluorescence-activated cell sorting (FACS): Method in which cells labeled with different fluorescent markers are passed through a fluorescence-activated cell sorter and separated into fractions based on the type of marker; also called flow sorting.

Fraction: With respect to chromatography, sample of material eluted from a chromatography column at a particular stage of the process. Each chromatography protocol involves collecting multiple fractions.

gDNA: See Genomic DNA.

Gel documentation systems: Systems used to acquire an image of a gel, usually with an inbuilt camera and often with a computer and imaging and analysis software.

Genetic code: A biological code consisting of three nucleotide units (codons) encoded by mRNA that map to a specific amino acid in the translated protein. There are 64 codons in the genetic code.

Genetic engineering: Manipulation of DNA of an organism; addition of new DNA sequences or deletion of endogenous DNA sequences.

Genetically modified crop (GM crop): Crop containing genetically engineered DNA.

Genetically modified organism (GMO): Organism whose genome has been altered in a way that does not occur naturally through mating or natural recombination.

Genetics: Study of genes and inheritance.

Genome: Complete sequence of the DNA (or RNA in some viruses) of an organism.

Genomic DNA (gDNA): Full complement of DNA within a cell that makes up its genome.

Genomics: Study of the genomes of organisms.

Genotype: Genetic composition of a cell, organism, or individual.

GLP: See Good laboratory practice.

GM: Genetically modified.

GMO: See Genetically modified organism.

GMP: See Good manufacturing practice.

Good clinical practice (GCP): Quality system used by governments to establish their own regulations for clinical trials involving human subjects.

Good laboratory practice (GLP): Quality system concerned with the organizational process and conditions under which non-clinical health and environmental studies are planned, performed, monitored, recorded, archived, and reported.

Good manufacturing practice (GMP): Quality system for ensuring the quality and safety of manufactured products used in health care, such as therapeutic drugs and diagnostic and medical devices.

Gram-negative: Classification of bacteria whose cell walls contain a second lipid membrane on the outside of a thin layer of peptidoglycans that does not stain with crystal violet dye in Gram stain (bacteria are pink after Gram staining).

Gram-positive: Classification of bacteria whose cell walls contain only one lipid membrane surrounded by a thick layer of peptidoglycans that sticks to crystal violet dye in Gram stain (bacteria are dark purple after Gram staining).

Gram stain: Combination of stains used to classify bacteria by their type of cell wall.

Guide RNA (gRNA): RNA approximately 100 nucleotides long that can form a complex with Cas9. A 20 nucleotide region at the end of the gRNA contains a spacer sequence complementary to the target DNA sequence.

High-throughput immunoassay: Assay capable of processing or screening a large number of compounds, substances, genes, or organisms.

Horizontal gel electrophoresis chamber: Plastic box or tank designed for agarose gel electrophoresis that can be filled with electrophoresis buffer and connected to a power supply.

Humanized antibody: Antibody of nonhuman origin that has been genetically modified to more closely resemble antibodies expressed in humans.

Hydrophobic interaction chromatography: Chromatographic technique that uses a hydrophobic matrix to separate proteins based on their degree of hydrophobicity.

Hypothesis: See Scientific hypothesis.

Immunity: Ability of an organism to resist disease.

Immunization: Induction of acquired immunity through stimulation of an immune response; when done deliberately, this is called vaccination.

Immunoassay: Test that uses the specificity of antibodies to detect or measure specific molecules in samples.

Immunocytochemistry: Use of antibodies for detection of a molecule of interest in a cell.

Immunoglobulin (Ig): Class of proteins that includes antibodies.

Immunohistochemistry: Use of antibodies for detection of a molecule of interest in a tissue.

Immunotherapy: A type of personalized medicine that uses parts of a person's immune system to fight diseases.

Induced pluripotent stem cell (iPSC): Cells artificially induced to have pluripotent properties in common with an embryonic stem cell.

Inducer: A molecule such as lactose or arabinose that binds to a gene repressor allowing for transcription.

Inductive research: Research that progresses from specific observations to patterns, to broader generalizations, and then to theories.

Intron: Eukaryotic gene segments that are interspersed between exons and are non-coding regions of DNA. They are transcribed into mRNA but are subsequently spliced from the primary transcript during RNA processing.

In vitro: Performed (as in an experiment) outside of an organism, such as a cell culture experiment or an assay in a test tube; literally "within the glass" (Latin).

In vivo: Performed (as in an experiment) within a living organism; literally "within the living" (Latin).

Ion exchange chromatography: Chromatographic technique that separates proteins by charge using a support with either a positive or negative charge.

Isoelectric point (pI or IEP): pH at which a molecule carries no net electric charge.

Kirby-Bauer, or disk diffusion, test: Test for microbial antibiotic resistance in which antibiotic disks inhibit the growth of a bacterial lawn in a measurable zone.

Lysogeny broth or Luria-Bertani broth (LB): Growth medium containing yeast extract, tryptone, and sodium chloride used to culture bacteria.

Ligase: Enzyme that repairs single-strand breaks in double-stranded DNA.

Liquid cultures: Large quantities of bacteria suspended in liquid media or broth.

Liquid media: Broth used to produce large quantities of bacteria, usually for protein production or for isolating plasmid DNA.

Locus (plural loci): Location of a genetic marker (that may or may not be a gene) on a chromosome.

Lowry assay: Copper-based protein assay that uses the Folin-Ciocalteau reagent to determine protein concentration.

Master mix: Mixture of components common to all reactions in an experiment; may contain all components needed for PCR except template DNA, for example.

Media: Liquid or gel that supports the growth of microorganisms, cells, or plants.

Melting temperature (T_m): Temperature at which 50% of a double-stranded DNA fragment is denatured into single strands; heavily dependent upon the DNA length and GC content.

Meniscus: Curve in the upper surface of a standing body of liquid (for example, liquid in a graduated cylinder).

Messenger RNA (mRNA): RNA that has been transcribed from DNA and serves as template for translation into a polypeptide by ribosomes; intermediary between DNA sequence and protein synthesis.

Metabolomics: Systematic study of the collection of all metabolites or metabolic processes in a biological cell, tissue, organ, or organism.

Microarray: Solid support, such as a microscope slide or chip, that hosts an array of microscopic spots of material to be assayed, including DNA, proteins, and antibodies. Microarrays are used in high-throughput assays to identify molecules that bind to the spotted material.

Microbiology: Study of microorganisms.

Microbiomics: Study of all microorganisms living in a place at the same time (microbiome).

Microcentrifuge: Instrument that spins microcentrifuge tubes at high speeds to separate substances of different densities.

Microorganism: Small (microscopic) organism.

Miniprep: Method used to extract and purify plasmid DNA from bacterial cells from a small culture volume that involves culturing bacteria, lysing the cells, and separating the plasmid from the cellular debris.

Model system: Simplified, idealized system that is accessible and easily manipulated.

Molarity (M): Moles per liter of solution, commonly stated as moles/L.

Mole: Unit of measurement for a chemical or substance; the number of atoms, molecules, or other elemental entity equal to the number of atoms in 12 grams of carbon-12, which is equivalent to the amount of the substance that has a mass in grams equal to its molecular mass.

Molecular biology: Study of genes and the flow of genetic information from DNA to RNA to proteins.

Molecular mass ruler: DNA standard with specific masses of DNA in each band that can be used to estimate DNA quantities on an agarose gel.

Monoclonal antibody: Antibody produced from a hybridoma cell line that recognizes a single epitope (hybridoma cell lines are produced by fusing a B cell to a specific type of tumor cell).

mRNA: See Messenger RNA.

Multiple cloning site (MCS): Short engineered DNA sequence in a DNA vector (often a plasmid) featuring multiple restriction enzyme recognition sites used for ligation of DNA fragments into the vector.

Myosin: Major muscle protein that is a component of thick filaments.

Nanobiotechnology: Area of study that applies nonbiological nanoscale devices to solve biological problems, and biologically based nanoscale devices to solve any problem.

Nanoparticles: Particles that are 1–100 nanometers in size.

Nanotechnology: Engineering of systems at the nanometer or molecular scale.

Negative control: Sample that generates a negative result and trust in any positive results observed when used according to the experimental conditions.

Normality (N): Molarity of a solution multiplied by the number of equivalents per mole; often used to describe the concentration of protons (H^+) or hydroxide ions (OH^-) in an acid or base.

Nucleotide: Fundamental unit of DNA and RNA comprising a sugar, phosphate group, and one of four bases: adenine, guanine, cytosine, and thymine (DNA) or uracil (RNA).

Operon: Functional unit of DNA containing a cluster of genes under the control of a single promoter.

Optical density (OD): In a solution, the amount of light of a specific wavelength that is absorbed.

Origin of replication (Ori): Sequence of DNA that specifies the location where DNA replication starts.

Ouchterlony, or double diffusion, assay: Test that uses antibodies and antigen in wells that both diffuse toward each other. If they interact, they will form a precipitate indicating recognition.

Pathogens: Infectious agents including bacteria, viruses, fungi, infectious proteins called prions, and parasites that cause disease in a host organism.

PCR: See Polymerase chain reaction.

PCR product: Target DNA amplified in a PCR reaction; also called an amplicon.

Pellet: Solid portion of a centrifuged sample (the liquid portion, over the pellet, is the supernatant).

Peptide bond: Bond between two amino acids that holds polypeptide chains together.

Peptidoglycan: Sugar-peptides that are components of bacterial cell walls.

Personalized medicine: Practice of medicine in which therapies are developed for, or directed at, the patients most likely to benefit from them.

Personal protective equipment (PPE): Protective clothing, helmets, goggles, or other garment designed to protect the wearer's body from injury by blunt impacts, electrical hazards, heat, chemicals, and infection; for job-related occupational safety and health purposes.

Petri plate: Round, flat container made of glass or plastic used to hold media for microbial and tissue culture.

Phenotype: Observable traits or characteristics of an organism, for example, hair color.

Plasmid: DNA molecule (usually circular) a few thousand base pairs in length capable of independent replication.

Plate count: Number of colonies of microorganisms that grow on a petri plate.

Polyacrylamide gel electrophoresis (PAGE): Electrophoresis technique used to separate mostly proteins but also nucleic acids, that uses a gel made from polyacrylamide. The gels run vertically and are very thin.

Polyclonal antibodies: Population of antibodies produced in response to immunization with a particular antigen; contains various antibodies that recognize multiple epitopes of the same antigen.

Polymerase chain reaction (PCR): Technique that uses multiple cycles of defined temperature changes to copy (amplify) specific DNA sequences.

Polypeptide: Molecule comprising two or more amino acids; the chain of amino acids that will be folded to make a protein.

Positive control: Sample that generates a positive result and trust in any negative results observed when used according to the experimental conditions.

Posttranslational modification: Chemical modifications made to a protein after translation (for example, glycosylation).

Power of discrimination: The level of ability to discriminate between any two genotypes; increases as more loci are analyzed.

Power supply: Instrument used to supply power for experiments requiring an electrical current.

PPE: See Personal protective equipment.

Primary antibody: In an immunoassay, the antibody that binds a target antigen, conferring specificity to the assay.

Primary cells: Animal cells cultured directly from tissue.

Primary structure: Level of protein structure that describes the specific order of amino acids in a polypeptide.

Primer: Short nucleic acid sequence (oligonucleotide) that binds to a single-stranded nucleic acid template and serves as the starting point for strand elongation.

Product: With respect to enzyme activity, the resulting molecules from enzymes acting on substrates. With respect to biotechnology, a wide variety of materials made from the use of modern molecular and microbial techniques.

Project proposal: Document describing and justifying a planned research project. It includes background information, objectives, methods, anticipated results, budget, and timeline.

Promoter: Region of DNA near a gene that facilitates or regulates its transcription.

Protease: Enzyme that breaks proteins at predictable places within a peptide sequence.

Protein: Functional assembly of one or more polypeptide chains of amino acids.

Proteomics: Study of the proteome (the whole complement of proteins) in cells, tissues, organs, organ systems, or organisms during a specific time period.

Quality assurance (QA): Oversees regulatory compliance and the company quality system including document control, internal and external audits, and record review.

Quality control (QC): Inspects and tests materials at all critical stages of the manufacturing process from raw materials to finished product.

Quantitative PCR (qPCR): See Real-time PCR.

Quaternary structure: Level of protein structure that describes the binding of multiple protein subunits to form the functional protein.

Real-time PCR: Technique that uses fluorescently labeled molecules to track the accumulation of amplified products with each cycle of PCR.

Recombinant DNA: Genetically engineered DNA made by linking fragments of DNA from different organisms.

Recombinant protein: Protein derived from genetically engineered DNA.

Replication: Process by which a DNA molecule is copied into a new strand of DNA.

Repressor: Protein that binds to the operator region near the operon and blocks RNA polymerase from transcribing.

Research: In product development, the stage in which scientists create initial products.

Research question: Question that forms the basis of an experiment or body of research.

Resin: With respect to column chromatography, one type of material that is contained within the column. Each resin has specific physical or chemical properties that bind substances with different specificities and allow separation of mixtures.

Restriction enzyme: Enzyme that cuts DNA at a specific sequence called a restriction enzyme recognition site.

Restriction fragment length polymorphism (RFLP): Difference in DNA sequence between highly similar DNA molecules where one sequence has a restriction enzyme recognition site that is not present in the other molecules; used to fingerprint DNA.

RFLP: See Restriction fragment length polymorphism.

Ribosomal RNA (rRNA): Structural and catalytic RNA that complexes with proteins to form a ribosome.

Ribosome: Non-membranous organelle composed of protein and rRNA and that plays a role in translation by reading the mRNA and translating that into a polypeptide with the aid of tRNA molecules that carry amino acids.

rRNA: See Ribosomal RNA.

RNA polymerase: Enzyme that produces RNA through transcription of DNA.

Safety data sheet (SDS): Document containing information on storage, handling, safety, and disposal of a chemical.

Sample loading buffer: High-density, colored solution that is mixed with samples before they are loaded on a gel to provide visibility and to help samples fall into the wells of the gel.

Scientific fact: An observation about the world that has been repeatedly confirmed.

Scientific hypothesis: Testable explanation of observed phenomena.

Scientific law: A detailed description of a natural phenomenon that is based on repeated experimental observations.

Scientific notation: Method used to write very large or very small numbers using a form where the number is multiplied by 10 raised to a positive or a negative number, for very large or very small numbers, respectively.

Scientific theory: Explanation of multiple related observations or events based on multiple hypotheses and verified many times by many independent researchers.

SDS: See Safety data sheet.

Secondary antibody: In an immunoassay, the antibody that recognizes the primary antibody.

Secondary structure: Level of protein structure due to hydrogen bonding between amino acids. There are two types of secondary structure: alpha helices and beta pleated sheets.

Selection: Process used to single out an organism with the necessary adaptation to survive under a particular condition. For example, antibiotic selection isolates cells or organisms that have an antibiotic resistant gene, often on a plasmid.

Serial dilution: Technique in which a substance in solution is diluted step-wise in a series of proportional amounts. For example, a bacterial culture can be diluted such that each tube has ten times less bacteria than the previous tube.

Serum (plural sera): Clear fluid obtained when the solid components (for example, red and white blood cells) are removed from whole clotted blood.

Short tandem repeat (STR): Class of DNA polymorphisms in which two or more nucleotides are repeated and occur directly adjacent to each other. The number of repeats can be different among individuals, so STRs are used to create DNA profiles.

Shuttle plasmid: Plasmid that can propagate in two host species, for example, bacteria and yeast.

Significant figures: The digits of a number that contribute to its precision; also called significant digits.

Size exclusion chromatography: Chromatography technique that uses a porous resin to trap small molecules but allow larger molecules to pass through, thereby separating a mixture by size.

SOP: See Standard operating procedures.

Slants: Sterile preparation of solid media, such as agar in a test tube, that has been solidified at an angle to provide an expanded surface area for plating microorganisms.

Southern blotting: Process in which DNA is separated on an agarose gel, denatured to make it single-stranded, and then transferred to a membrane. Blotted DNA is usually probed with labeled pieces of single-stranded DNA to find the location of specific DNA sequences.

Spectrophotometer: Instrument for measuring the absorbance of light by a solution at a particular wavelength.

Stab culture: Culture of microorganisms that has been stabbed into a deep tube of agar.

Standard operating procedures (SOP): Controlled document or record detailing how to perform a process or procedure.

Stem cells: Cells that have the potential to differentiate into multiple cell types.

Sterile technique: See Aseptic technique.

STR: See Short tandem repeat.

Substrate: Target molecule for an enzyme.

Supernatant: Liquid portion of a centrifuged sample (the solid portion at the bottom of the vessel is the pellet).

Suspension cultures: Cell culture in which the cells are suspended in liquid media.

Systems biology: Systematic study of the complex interactions in biological systems.

Template DNA: DNA sample copied in a PCR reaction.

Terminator: DNA sequence at the end of a gene that signals RNA polymerase to stop transcription.

Tertiary structure: Level of protein structure in which a single polypeptide chain is fully folded.

Theory: See Scientific theory.

Thermal cycler: Instrument used in PCR that automates the repeated cycles of heating and cooling.

T_m: See Melting temperature.

Transcription: Process by which DNA is copied into RNA.

Transcriptomics: Study of the transcriptome, every RNA transcript expressed in a cell or organism.

Transduction: Process by which DNA is transferred from one bacterium to another by a virus.

Transfection: Deliberate introduction of nucleic acids into cells (most often used for eukaryotic cells).

Transfer RNA (tRNA): Non-coding RNA that carries an amino acid for use in translation.

Transformation: Uptake of DNA from the outside environment, usually by bacteria; can be naturally or artificially induced.

Transformation efficiency: Measurement of the number of successful transformations per microgram of DNA used.

Translation: Process by which ribosomes translate mRNA sequences into chains of amino acids (polypeptides).

tRNA: See Transfer RNA.

Variable: In a scientific experiment, the part of the experiment that is subject to change.

Western blotting: Process by which proteins separated on a polyacrylamide gel are transferred to a membrane or blot. The protein on the blot is then probed with antibodies to locate specific proteins.

Appendix E:
Laboratory Skills
Assessment Rubric

Activity	Skill	Novice	Developing	Proficient
2.A 2.B 2.C 2.D 2.E and all activities	Follow laboratory protocols	Student may not understand the importance of following proper laboratory procedures. Procedure is performed out of order or is missing steps, or the methods recorded in the laboratory notebook are incomplete.	Student understands the importance of following proper laboratory procedures. Procedure is performed in the appropriate order but one or more procedural steps are missing. The methods recorded in the laboratory notebook are missing one or two steps.	Student understands the importance of following proper laboratory procedures. Procedure is performed in the appropriate order with no steps missed. The methods are clearly and completely recorded in the laboratory notebook.
2.A and all activities	Select and wear proper PPE	Student may not understand the purpose of PPE. Student may need to be reminded to wear PPE, is missing critical PPE, or does not verify with the instructor that the PPE is appropriate.	Student understands the purpose of PPE. Student remembers to wear the most critical PPE but may be missing PPE that protects clothing. Student may not have verified with the instructor that the PPE is appropriate.	Student understands the purpose of PPE. Student wears PPE appropriate to the task and PPE is worn correctly. Student asked the instructor for information on appropriate PPE when unsure.
2.A	Extract DNA from cells	Student may not understand the purpose behind DNA extraction. Student performs the procedure incorrectly, misses steps, or performs steps out of order, resulting in no visible DNA.	Student understands the purpose behind DNA extraction. Student performs procedure but performs one or more steps incorrectly. DNA is visible but may be broken up (flocculant) or present in small quantities, making collection difficult.	Student understands the purpose behind DNA extraction and follows procedures correctly. DNA is easily visible and may be present in strands or clumps that can be transferred to another container.
2.A	Precipitate DNA	Student may not understand the principles of DNA precipitation. Student may perform the protocol incorrectly and DNA is not visible.	Student understands the principles of DNA precipitation. Student may handle the sample roughly, leading to DNA that is broken up or flocculent.	Student understands the principles of DNA precipitation, performs the protocol carefully, and obtains white, thread-like pieces of DNA.
2.A 2.B 2.C 2.D 2.E 2.F and all activities	Maintain a laboratory notebook	See Laboratory Notebook Rubric (Appendix F).	See Laboratory Notebook Rubric (Appendix F).	See Laboratory Notebook Rubric (Appendix F).
2.B 2.C	Use a serological pipet with a pipet pump or filler	Student may not demonstrate the ability to use a serological pipet correctly. Student may not use a pump or filler, or inserts the pipet loosely into the pump or filler such that liquid does not remain in the pipet and pours out upon transfer. The cotton plug becomes wet or volumes transferred are inaccurate.	Student demonstrates the ability to use a serological pipet correctly. Student inserts the pipet into the pump or filler correctly but liquid leaks. Student does not read the volume from the bottom of the meniscus or transfers a slightly inaccurate volume.	Student demonstrates the ability to use a serological pipet correctly. Student inserts the pipet into the pump or filler correctly and no liquid escapes. Student reads the volume from the bottom of the meniscus and transfers an accurate volume.

Activity	Skill	Novice	Developing	Proficient
2.B	Use a transfer pipet	Student may not demonstrate the ability to use a transfer pipet correctly. Student may not squeeze the pipet bulb prior to inserting the pipet into the liquid (creating bubbles), does not apply consistent pressure, or transfers an incorrect volume.	Student demonstrates the ability to use a transfer pipet correctly. Student squeezes the pipet bulb prior to inserting the pipet into the liquid but does not draw liquid up slowly or to the appropriate volume. The volume of liquid transferred was not correct.	Student demonstrates the ability to use a transfer pipet correctly. Student squeezes the pipet bulb gently prior to inserting the pipet into the liquid, draws the liquid slowly up to the appropriate volume, maintains the volume during transfer, and transfers the correct volume of liquid.
2.B and many activities	Select an appropriate size of micropipet	Student does not know the different ranges of micropipets or selects an inappropriate size for his/her desired volume.	Student knows the different ranges of micropipets and selects a pipet with a range that covers the desired volume. If two types of pipet cover the desired volume (for example, 20 µl), student may not choose the pipet with the middle to upper range closest to the desired volume.	Student knows the different ranges of micropipets and selects a pipet with a range that covers the desired volume. If two types of pipet cover the desired volume (for example, 20 µl), student chooses the pipet with the middle to upper range closest to the desired volume (for example, a 2–20 µl pipet over a 20–200 µl pipet).
2.B 2.C and many activities	Use an adjustable-volume micropipet	Student does not demonstrate the ability to set the pipet volume and transfers an incorrect volume. Student selects correct tips but may not apply them firmly or student pushes the pipet plunger to the second stop when drawing in liquid.	Student demonstrates the ability to set the pipet volume but transfers an incorrect volume. Student selects correct tips, applies them firmly, and pushes the pipet plunger to the first stop, however pipet is depressed while inserting it into the liquid, creating bubbles, or student releases the plunger too quickly, sucking back liquid.	Student demonstrates the ability to correctly set the pipet volume and the volume transferred is correct. Student selects correct tips and applies them firmly. Student pushes the pipet plunger to the first stop prior to inserting the pipet into the liquid and draws the liquid slowly into the tip. Student expels the liquid by pushing the plunger to the first stop followed by the second stop.
2.B	Calculate percent error	Student may not understand how to calculate percent error. An incorrect answer is obtained, calculations are not completely written out, or units of measurement are not included.	Student understands how to calculate percent error. Calculations are partly written out, units of measurement may be missing, or answers are mostly correct but may not be presented as a percentage.	Student understands how to calculate percent error. Calculations are fully written out, units of measurement are present, and answers including percentages are correct.
2.B 2.E	Calculate mean and standard deviation	Student may not understand how to calculate mean and standard deviation. An incorrect answer is obtained with no units of measurement identified, the answer is incomplete, and data are inaccurate or not displayed.	Student understands how to calculate mean and standard deviation. The correct answer is obtained but units of measurement are not consistently or clearly identified. Calculations are difficult to follow.	Student understands how to calculate mean and standard deviation. The correct answer is obtained and units of measurement are clearly and consistently identified. Calculations are easy to follow.
2.C 2.D	Perform calculations using the $C_1V_1 = C_2V_2$ equation	Student may not understand how to calculate dilutions using the equation $C_1V_1 = C_2V_2$. An incorrect answer is obtained, calculations are not completely written out, or units of measurement are not included.	Student understands how to calculate dilutions using the equation $C_1V_1 = C_2V_2$. Calculations are partly written out, answers may be correct, but appropriate units of measurement are missing.	Student understands how to calculate dilutions using the equation $C_1V_1 = C_2V_2$. Calculations are fully written out, units of measurement are included, and answers are correct.

Activity	Skill	Novice	Developing	Proficient
2.C 2.D	Wash glassware properly	Student may not demonstrate the ability to thoroughly wash glassware. Student washes glassware with tap water but does not rinse them completely. Residue remains when solutions or liquids are added.	Student demonstrates the ability to thoroughly wash glassware. Student washes glassware thoroughly with soap and tap water but does not rinse properly with distilled water.	Student demonstrates the ability to thoroughly wash glassware. Student washes glassware thoroughly with soap, tap water, and a brush and rinses glassware 3x with tap water and then twice with distilled water.
2.C	Select and use appropriate equipment to measure volumes	Student may not demonstrate the ability to select the appropriate piece of equipment to measure a specific volume and so delivers a volume that is inaccurate.	Student demonstrates the ability to select the appropriate piece of equipment to measure a specific volume but delivers a volume that is inaccurate.	Student demonstrates the ability to select the appropriate piece of equipment to measure a specific volume and delivers a volume that is accurate.
2.C	Make percent solutions	Student may not demonstrate the ability to make percent solutions. Calculations may be incorrect, units of measurement may be missing, or the incorrect measuring equipment is used. The final solution cannot be used.	Student demonstrates the ability to make percent solutions. Calculations are correct and units of measurement are present but one or more units are missing. Volumes are measured but with inappropriate equipment. The final solution is useful.	Student demonstrates the ability to correctly make percent solutions. Calculations are correct, units of measurement are clearly and consistently identified, and volumes are measured accurately with the appropriate equipment. The final solution is useful.
2.C 2.D	Write a label for a reagent bottle	Student does not demonstrate the ability to clearly and properly label a reagent bottle. Key information, such as contents or concentration, is missing from the label.	Student demonstrates the ability to clearly and properly label a reagent bottle. Key information, such as contents, concentration, date, and initials are present but a hazard warning may be missing.	Student demonstrates the ability to clearly and properly label a reagent bottle. The label contains contents, concentration, date, initials of who made the solution, and any hazards.
2.C	Use a Sep-Pak C18 cartridge and syringe	Student does not demonstrate the ability to use a Sep-Pak cartridge and syringe correctly. Student dismantles the syringe incorrectly, leading to the wash flowing through the cartridge in the wrong direction. The fractions do not elute as expected.	Student demonstrates the ability to use a Sep-Pak cartridge and syringe correctly. Student dismantles the syringe mostly in the proper order prior to adding the next wash. The wash mostly flows through the cartridge in the same direction and may exit as a stream rather than droplets. The fractions elute as expected.	Student demonstrates the ability to use a Sep-Pak cartridge and syringe correctly. Student dismantles the syringe in the proper order prior to adding the next wash. The wash always flows through the cartridge in the same direction and exits as droplets. The fractions elute as expected.
2.C 2.D 2.E	Use a graduated cylinder	Student may not demonstrate the ability to use a graduated cylinder. Student may select an inappropriate cylinder, does not read volume markings correctly, or does not read the volume from the bottom of the meniscus. The volume transferred is not accurate.	Student demonstrates the ability to use a graduated cylinder. Student selects an appropriate cylinder and reads volume markings accurately but does not read the volume from the bottom of the meniscus. The volume transferred is mostly accurate.	Student demonstrates the ability to use a graduated cylinder correctly. Student selects an appropriate cylinder, reads volumes accurately, and reads the volume from the bottom of the meniscus. The volume transferred is accurate.
2.C and most activities	Interpret results and draw conclusions from data	See Laboratory Notebook Rubric (Appendix F).	See Laboratory Notebook Rubric (Appendix F).	See Laboratory Notebook Rubric (Appendix F).

Activity	Skill	Novice	Developing	Proficient
2.D	Calculate molarity	Student may not understand how to calculate molarity. An incorrect answer is presented, calculations are not completely written out, or units of measurement are not included.	Student understands how to calculate molarity. Calculations are partly written out and answers may be correct but they are missing appropriate units of measurement.	Student understands how to calculate molarity. Calculations are fully written out, units of measurement are present, and answers are correct.
2.D	Use a balance	Student does not demonstrate the ability to use a balance correctly. Student may not zero or tare the balance with the weigh boat or weighing paper prior to use, takes the reading prematurely, may not record units, or returns excess material to the storage container.	Student demonstrates the ability to use a balance correctly. Student zeroes or tares the balance with the weigh boat or weighing paper prior to use but takes the reading prematurely, does not record units, or returns excess material to the storage container.	Student demonstrates the ability to use a balance correctly. Student zeroes or tares the balance with the weigh boat or weighing paper prior to use, takes the reading only when the balance is steady, records the units, and does not return excess material to the storage container.
2.D	Make dilutions from concentrated solutions	Student may not demonstrate the ability to make a dilution from a concentrated stock solution. Calculations may be incorrect or missing units of measurement, solutions may be incorrectly labeled, or the incorrect measuring equipment may be used. The final solution cannot be used.	Student demonstrates the ability to make a dilution from a concentrated stock solution. Calculations are correct, solutions are correctly labeled, and units of measurement are present but one or more units are missing. Student measures volumes but uses equipment that may not be the most accurate. The final solution is useful.	Student demonstrates the ability to make a dilution from a concentrated stock solution. Calculations are correct, solutions are correctly labeled, and units of measurement are clearly and consistently identified. Student measures volumes accurately with the appropriate equipment. The final solution is useful.
2.D	Make molar solutions using a volumetric flask	Student may not demonstrate the ability to make a molar solution using a volumetric flask. Calculations may be incorrect, units of measurement may not be present, and labels may be missing vital information. Solution may not be made in a working volume and final volume may not be accurate. The final solution cannot be used.	Student demonstrates the ability to make a molar solution using a volumetric flask. Calculations are correct, though some units of measurement may be missing. Solutions are correctly labeled but may be missing some information. Solution is made in a working volume. When solid is dissolved, the final volume is added up to the marking on the flask but the volume may not be read from the bottom of the meniscus. The final solution is useful.	Student demonstrates the ability to make a molar solution using a volumetric flask. Calculations are correct and include units of measurement. Solutions are correctly labeled and made in a working volume. When solid is dissolved, the final volume is added up to the marking on the flask and is read from the bottom of the meniscus. The final solution is useful.
2.D	Make compound molar solutions	Student may not demonstrate the ability to make compound molar solutions. Calculations may be incorrect, units of measurement may not be present, and labels may be missing vital information. Student measures volumes inaccurately or with inappropriate equipment. The final solution cannot be used.	Student demonstrates the ability to make compound molar solutions. Though calculations are correct, some units of measurement may be missing and though solutions are correctly labeled, some information may be missing. Student measures volumes accurately and with the appropriate equipment. The final solution is useful.	Student demonstrates the ability to make compound molar solutions. Calculations are correct and include units of measurement, and solutions are correctly labeled. Student measures volumes accurately and with the appropriate equipment. The final solution is useful.

Activity	Skill	Novice	Developing	Proficient
2.D	Use and calibrate a pH meter	Student may not demonstrate the ability to use and calibrate a pH meter. Student chooses an incorrect pH standard solution for calibration of the instrument, reads the pH value too early, uses an incorrect acid/base for adjusting pH, or badly overshoots the pH. The final pH is incorrect and the solution cannot be used.	Student demonstrates the ability to use and calibrate a pH meter. Student chooses correct pH standards for calibration of the instrument and the correct acid/base for adjusting the pH but reads the pH value before the reading is stabilized or slightly overshoots the end point. The pH of the final solution is slightly off but is adequate and the final solution can be used.	Student demonstrates the ability to use and calibrate a pH meter. Student chooses correct pH standards for calibration of the instrument, reads the pH value when the reading is stabilized, and uses the correct acid/base for adjusting pH. The pH of the final solution is correct and the solution can be used.
2.E	Titrate a solution	Student may not demonstrate the ability to titrate a solution. Student performs the procedure out of order, misses steps, or does not swirl solutions. Student adds either too much or too little solution to obtain the endpoint. Two out of three trials significantly overshoot the end point and the correct concentration cannot be calculated from the data.	Student demonstrates the ability to titrate a solution. Student performs the procedure in the appropriate order but makes one or more mistakes. Student does not swirl flasks sufficiently. One trial out of three is overshot but the correct concentration can be calculated from the data.	Student demonstrates the ability to titrate a solution. Student performs the procedure correctly, completely, and in the appropriate order. Student swirls flasks after every addition until the color disappears and reduces additions to single drops before the end point is reach. Student stops titration as soon as the end point is reached. The correct concentration can be calculated from the data.
2.E	Use a burette	Student may not demonstrate the ability to use a burette properly. Student interprets volume incorrectly, does not read the volume at the start of the experiment or at appropriate times during the course of the experiment.	Student demonstrates the ability to use a burette properly. Student reads the volume at the start of the experiment and at appropriate times during the course of the experiment but interprets the volume incorrectly.	Student demonstrates the ability to use a burette properly. Student interprets volume correctly and reads volumes at the start of the experiment and at appropriate times during the course of the experiment.
2.E	Calculate unknown molarity	Student may not understand how to determine the molarity of a solution. Student does not use the equation $M_1V_1 = M_2V_2$, uses the equation incorrectly, or does not write out calculations completely.	Student understands how to determine the molarity of a solution. Student uses the equation $M_1V_1 = M_2V_2$ but does not substitute values correctly or otherwise obtains an incorrect answer.	Student understands how to determine the molarity of a solution. Student correctly determines the molarity of a solution using the equation $M_1V_1 = M_2V_2$ with correct values substituted and calculations completely written out.
2.F	Write an SOP	Student may not understand how to write an SOP. Header is inaccurate, some required sections are not addressed, or the procedure is not a numbered step-by-step list. The procedure cannot be carried out by following the SOP.	Student understands how to write an SOP. Header and some sections are missing some minor information but otherwise all required sections are present and addressed, and all procedural steps are numbered. The procedure needs some additional explanation to be carried out properly.	Student understands how to write an SOP. Header is completed accurately, all required sections are present and addressed, and all procedural steps are numbered. The procedure can be carried out by following the SOP without additional explanation.

Activity	Skill	Novice	Developing	Proficient
3.A	Perform calculations necessary for growth media production	Student may not demonstrate the ability to perform calculations for preparing growth media. Calculations do not have methods written out, units of measurement are missing, or most answers are incorrect.	Student demonstrates the ability to perform calculations for preparing growth media. Calculations are partially written out, units of measurement may be missing, but most answers are correct.	Student demonstrates the ability to perform calculations for preparing growth media. Calculations are fully written out with units of measurement and the answers are correct.
3.A	Label agar plates and broth	Student may not understand the importance of properly labeling agar plates and broth. Media are missing vital information or information is written in the incorrect location.	Student understands the importance of properly labeling agar plates and broth. Media have partial information or information is written in the incorrect location (on the tube cap, plate lid or covering the bottom of the plate).	Student understands the importance of properly labeling agar plates and broth. Media have media name, date, and initials. Plates are labeled properly on the bottom in an arc around the plate. Tubes are labeled on the side.
3.A	Make LB, LBS, LB/amp, and LB/amp/ara agar	Student may not demonstrate the ability to properly make agar. If agar cooled, it may be lumpy, not set, or obviously contaminated. Student may add ampicillin or arabinose prior to initial boiling rather than to the molten agar at 55°C or may add the wrong supplements. Agar cannot be used for experiments.	Student demonstrates the ability to make agar. When agar sets, it is smooth but may be uneven in color or appear very dark, indicating a long heating time. If ampicillin or arabinose is required, student adds them when the molten agar is above 55°C and may not properly mix them into the agar. Agar may be contaminated and does not perform correctly, indicating supplements are incorrect (can be assessed only once agar is used).	Student demonstrates the ability to make agar. When agar sets, it is smooth, light brown, and uniform in color. If ampicillin or arabinose is required, student adds them when the molten agar is around 55°C and swirls the agar to mix just prior to pouring the agar. Upon usage, agar is not contaminated and performs as expected (can be assessed only once agar is used).
3.A	Use the autoclave to sterilize media (all autoclave use must be under direct supervision of instructor)	Student may not demonstrate knowledge of proper autoclave use. Student may fill bottles more than halfway, screw lids on tightly, not apply autoclave tape, or not label containers correctly. Student may not know how to set the autoclave or programs the autoclave incorrectly. Student may start autoclave without consulting the instructor.	Student demonstrates knowledge of proper autoclave use. Student fills bottles less than halfway and prior to starting the autoclave, loosens caps on all bottles, and applies autoclave tape but labels may be incomplete. Student still needs assistance with autoclave settings but consults instructor before starting the autoclave.	Student demonstrates knowledge of proper autoclave use. Prior to starting the autoclave, student loosens caps on all bottles (which are less than half full), labels containers correctly, and applies autoclave tape. Student is confident and can correctly program the autoclave. Student consults instructor before starting the autoclave.
3.A 3.B 3.C 5.A 5.B 5.C	Use aseptic technique	Student may not demonstrate aseptic technique. Student may not sterilize loops at the appropriate time, may not use plate lids to shield plates, may place lids on tables or leave them off containers for extended periods of time, or may not flame the mouths of glass containers. After transfer of bacteria from a plate or culture to another container, contamination occurs.	Student demonstrates aseptic technique but with some minor technical errors that have the potential for yielding contamination. Occasionally, student does not sterilize a loop prior to use, does not shield some plates with their lids, or does not flame the mouths of glass containers. After transfer of bacteria from a plate or culture to another container, contamination does not occur.	Student demonstrates aseptic technique. Student sterilizes loops at appropriate times, shields plates, never places plate lids on the counter, and flames the mouths of glass containers when appropriate. After transfer of bacteria from a plate or culture to another container, contamination does not occur.

Activity	Skill	Novice	Developing	Proficient
3.A	Pour agar plates	Student may not demonstrate the ability to properly pour agar plates and does not use aseptic technique. Student may make plates that are uneven, contain bubbles, or have areas that are partly uncovered or that have agar splashed onto the lid. Plates may have lumps due to inadequate melting or to cooling below optimal temperature prior to pouring. Upon use, plates may be contaminated or cannot be used for experiments.	Student demonstrates the ability to properly pour agar plates and tries to use aseptic technique but may miss some steps. Student does not shield plates with their lids while pouring or pours plates that have different depths, uneven surfaces, or rough surfaces. Upon use, some plates are contaminated, but most can still used for experiments.	Student demonstrates the ability to properly pour agar plates and uses aseptic technique. Student shields plates from contamination with their lids while pouring and pours plates to even depths (around one third the depth of plate) with smooth surfaces. Upon use, no plates are contaminated.
3.A	Make LB, LB/amp, and LB/amp/ara broth	Student may not demonstrate the ability to properly prepare LB broth and may not use aseptic technique. Student may add ampicillin or arabinose prior to initial boiling or autoclaving of the broth. Broth may be contaminated (cloudy when not inoculated) or broth sets because LB agar was used instead of LB broth.	Student demonstrates the ability to properly prepare LB broth and tries to use aseptic technique but may miss some steps. Student adds ampicillin or arabinose when broth is over 55°C. Broth may not perform correctly, indicating that supplements are incorrect (assessed once broth is used for experiments). Broth may be dark brown in color rather than light brown/yellow but can still be used for experiments.	Student demonstrates the ability to properly prepare LB broth using aseptic technique. Broth is light brown/yellow and has no contamination. Student adds ampicillin or arabinose when broth is less than 55°C. Broth performs as expected.
3.A	Make LB agar deep tubes	Student may not demonstrate the ability to properly prepare LB agar deep tubes and may not use aseptic technique. Tubes may have different volumes, and agar may be lumpy, have bubbles, or is obviously contaminated. Upon use, most tubes are contaminated and cannot be used.	Student demonstrates the ability to properly prepare LB agar deep tubes and tries to use aseptic technique but may miss some steps. Some tubes have different volumes but are near one third the volume of the tube. Upon use, some tubes are contaminated but most can be used.	Student demonstrates the ability to properly prepare LB agar deep tubes using aseptic technique. Tubes are one third full and not contaminated.
3.B	Perform a modified Kirby-Bauer test	Student may not understand the purpose behind the Kirby-Bauer test. Plates may be poured incorrectly, contain lumps, be contaminated, or have bacteria on the surface. Disks may not be placed in the correct position, or zones are not correctly labeled.	Student understands the purpose behind the Kirby-Bauer test. Plates are poured properly and bacteria are spread evenly but the disks are not evenly spaced or plates have minor labeling errors. Plates still can be used for the experiment.	Student understands the purpose behind the Kirby-Bauer test. Plates are poured properly, bacterial lawn is even, and disks are placed correctly in labeled quadrants. Experiment works properly.
3.B	Inoculate an agar plate for a bacterial lawn	Student may not demonstrate the ability to correctly inoculate an agar plate with a bacterial lawn. Student may not use aseptic technique and may gouge the agar surface through incorrect use of the inoculation loop. The plate may have bacteria on only part of the plate and cannot be used for the experiment.	Student demonstrates the ability to correctly inoculate an agar plate with a bacterial lawn. Student uses aseptic technique and the plate has a bacterial lawn but the edges of the plate are sparse and the lawn is streaky. The plate can be used for the experiment.	Student demonstrates the ability to correctly inoculate an agar plate with a bacterial lawn. Student uses aseptic technique, a bacterial lawn covers the entire surface area of the plate, and the plate can be used for the experiment.

Activity	Skill	Novice	Developing	Proficient
3.C 5.A 5.B	Streak bacteria to isolate single colonies	Student may not demonstrate the ability to streak bacteria to isolate a single colony. Student may not use aseptic technique, may not hold the lid over the plate while streaking, and may gouge the surface of the agar through incorrect use of the inoculation loop. There may be no evidence of dilution of bacteria on the plate. Plate may be contaminated or cannot be used to obtain a single colony.	Student demonstrates the ability to streak bacteria to isolate a single colony. Student uses aseptic technique and the plate has distinct colonies but the streaking pattern is incorrect. Plate may have some contaminated colonies but can be used to obtain a single colony.	Student demonstrates the ability to streak bacteria to isolate a single colony. Student uses aseptic technique. Plate has no contamination, has distinct colonies, and the streaking pattern is correct. Student demonstrates the ability to aseptically inoculate a plate with a liquid culture.
3.C 3.E 5.A 5.B	Aseptically inoculate a plate with a liquid culture	Student may not demonstrate the ability to aseptically inoculate a plate with a liquid culture. Student may not sterilize loops or the mouths of containers prior to use or misses other steps, leading to contamination.	Student demonstrates the ability to aseptically inoculate a plate with a liquid culture. Student flames the metal loop and mouths of glass containers before but not after use. Student uses fresh, sterile plastic loops for each inoculation. Student may touch the sides of containers with the loops and the plate grows with minor contamination.	Student flames a metal loop and mouths of glass containers before and after use or uses fresh, sterile plastic loops for each inoculation. Student does not allow loops to touch the sides of containers. The plate grows successfully without contamination.
3.C 5.C 7.C	Aseptically inoculate a liquid culture from a plate	Student may not understand the reason behind aseptically inoculating a single colony and does not isolate a pure colony to obtain a pure culture of bacteria. Student may not sterilize a loop prior to use, may not cool a flamed metal loop before use, resulting in a burned and inviable bacterial colony, or may not select a single colony. Student may not flame the mouths of containers prior to use or may miss other steps, leading to contamination.	Student understands the reason behind aseptically inoculating a single colony but does not isolate one colony to obtain a pure culture of bacteria. Student flames the metal loop and mouths of glass containers before but not after use. Student uses fresh, sterile plastic loops for each inoculation. Student may touch the sides of containers with the loops and may select a single colony but places the plate lid on the counter or does not use it as a shield. The liquid culture grows successfully.	Student understands the reason behind aseptically inoculating a single colony and isolates one colony to obtain a pure culture of bacteria. Student flames a metal loop and mouths of glass containers before and after use and uses fresh, sterile plastic loops for each inoculation. Student does not allow loops to touch the sides of containers. Student selects a single colony using the lid to shield the plate. The liquid culture grows successfully.
3.C 7.E	Organize experimental data	Student may not organize data logically in tables, lists, or charts. Tables, lists, and charts are inappropriate for the dataset being analyzed, and labels, headings, or units of measurement are misidentified. Description of the data set is incomplete and requires extensive explanation to be understood by another individual.	Student places data in tables, lists, or charts but is missing appropriate labels, headings, or units of measurement. Description of the data set would require some explanation to be understood by another individual.	Student organizes data logically in tables, lists, or charts with clear labels, headings, and units of measurement. Description of the data set is organized and requires no additional explanation to be understood by another individual.

Activity	Skill	Novice	Developing	Proficient
3.C 3.F	Make a wet mount	Student may not demonstrate the ability to prepare a wet mount. Student places a drop of liquid on the slide but may lower the coverslip in such a way that many bubbles are generated. Water may not completely cross the coverslip, making it difficult to view the slide.	Student demonstrates the ability to prepare a wet mount. Student places a drop of liquid on the slide and lowers the coverslip onto the slide but uses excess water or creates one or two bubbles. Water may extend onto the coverslip but the slide is still usable.	Student demonstrates the ability to prepare a wet mount. Student places a drop of liquid on the slide and lowers the coverslip onto the slide. There are no air bubbles or excess water.
3.C 3.D	Observe bacteria using a microscope	Student may not demonstrate the ability to use a microscope properly. Student begins on low or high power and needs help focusing the microscope. Student is unable to find color or locate bacteria.	Student demonstrates the ability to use a microscope properly. Student begins on low power with the objective as close to the slide as possible. Student focuses away and sees color but needs help locating the bacteria.	Student demonstrates the ability to use a microscope properly. Student begins on low power with the objective as close to the slide as possible. Student focuses away until they see color, uses the fine focus to locate areas with scattered bacteria, then changes to high power and successfully focuses on an individual bacterium.
3.D	Perform Gram staining of bacteria	Student may not demonstrate the ability to Gram stain bacteria. Student may not heat fix bacteria properly and subsequently washes them off during Gram staining. Student may not perform procedure correctly or uses stains out of order.	Student demonstrates the ability to Gram stain bacteria. Student fixes bacteria onto the slide and performs staining in the correct order and at proper times but overloads bacteria on the slide.	Student demonstrates the ability to Gram stain bacteria. Student fixes bacteria onto the slide, performs staining in the correct order and at proper times. The bacteria are clearly visible, correct in shape, and either gram-positive (purple) or gram-negative (pink).
3.D	Heat fix bacteria to a slide	Student may not demonstrate the ability to heat fix bacteria onto a slide. Student may not draw a circle on the slide with a wax pencil, use a large droplet of water, or add a large amount of bacteria. Student may pass bacteria through the flame too quickly or for too long, resulting in burnt bacteria or bacteria not properly fixed onto the slide. The following Gram staining does not work.	Student demonstrates the ability to heat fix bacteria onto a slide. Student draws a circle using a wax pencil for each type of bacteria added to the slide and adds a small droplet of water aseptically but may add an entire bacterial colony to the droplet. Student may allow water containing the bacteria to evaporate only partially and passes bacteria through the flame but at varying times. After Gram staining, a majority of the bacteria are stained and remain on the slide.	Student demonstrates the ability to heat fix bacteria onto a slide. Student draws a circle using a wax pencil for each type of bacteria added to the slide, adds a small droplet of water aseptically, and touches a single bacterial colony to the droplet. Student allows the water containing the bacteria to evaporate completely and passes the dried bacteria through the flame three times for approximately 1 second per pass. After Gram staining, the bacteria are stained and remain on the slide.
3.D	Differentiate gram-positive bacteria from gram-negative bacteria	Student may not demonstrate the ability to differentiate gram-positive and gram-negative bacteria. Student does not choose gram-positive and gram-negative based on the color of the stained bacteria or mixes them up and misidentifies them.	Student demonstrates the ability to differentiate gram-positive and gram-negative bacteria. Student identifies gram-positive or gram-negative bacteria correctly when staining is optimal but when staining is suboptimal has difficulty differentiating the two.	Student demonstrates the ability to differentiate gram-positive and gram-negative bacteria. Student identifies gram-positive bacteria by their purple color and gram-negative bacteria by their pink color.

Activity	Skill	Novice	Developing	Proficient
3.D	Identify bacteria from cell shape	Student may not demonstrate the ability to identify bacteria from cell shape. Student identifies bacteria incorrectly based on their shape.	Student has difficulty identifying bacteria according to their cell shape, confuses cell shapes, misidentifies bacteria, or is not familiar with the formal names for bacterial shapes.	Student demonstrates the ability to correctly identify bacteria from their cell shape using the formal names.
3.D 3.F	Sketch microscopic details	Student may not understand the importance of properly sketching microscopic details. Sketch shows the basic shape of a cell but is drawn quickly and inaccurately, and is missing labels or size estimation.	Student understands the importance of properly sketching microscopic details. Sketch shows the overall shape of a cell and is drawn accurately but some labels may be incorrect. Shading is included and size is noted.	Student understands the importance of properly sketching microscopic details. Sketch shows the overall shape of a cell and is drawn and labeled accurately. Shading is done using dots and the cell size is shown using a scale bar.
3.D 3.F	Estimate the size of cells using a microscope	Student may not demonstrate the ability to correctly estimate the size of cells using a microscope. Scale is missing or inaccurately calculated.	Student demonstrates the ability to correctly estimate the size of cells using a microscope. Student estimates size using information provided by the instructor (for example, the diameter of the field of view at each magnification divided by the number of cells that fit across the field). Scale is shown using a size bar but the proper number of significant figures is not used.	Student demonstrates the ability to estimate the size of cells using a microscope. Student estimates size using information provided by the instructor (for example, the diameter of the field of view at each magnification divided by the number of cells that fit across the field). Scale is shown using a size bar and the proper number of significant figures is used.
3.E	Perform a serial dilution of a bacterial culture	Student may not understand why serial dilutions are used. Student does not use aseptic technique, sets pipets incorrectly, or does not mix media and cells completely between dilutions or prior to spreading. Plates do not display a proportional decrease in bacterial colonies or are contaminated, or colonies are unevenly distributed and cannot be used to calculate CFU/ml.	Student understands why serial dilutions are used. Student uses aseptic technique but does not mix the media and cells completely between dilutions or prior to spreading. Plates generally display a proportional decrease in bacterial colonies correlating with the dilution factor but one or more plates may not align. Plates may have some contaminating colonies, or colonies may not be evenly distributed but plates can be used to calculate CFU/ml.	Student understands why serial dilutions are used. Student uses aseptic technique and mixes the media and cells completely between dilutions and prior to spreading. All plates display a proportional decrease in bacterial colonies correlating with the dilution factor. Plates are not contaminated and bacterial colonies are evenly distributed on plates and can be used to calculate CFU/ml.
3.E	Calculate concentration of bacteria in a culture	Student may not demonstrate the ability to calculate the concentration of bacteria from a culture. The calculations may not have methods written out and units of measurement may not be included or may be incorrect. Incorrect values may be used for calculations and the answer is incorrect.	Student demonstrates the ability to calculate the concentration of bacteria from a culture. Calculations are written out and units of measurement are correct but they may not be added at each step. One or more steps are missing. The dilution factor and the fraction of the culture that was inoculated have been accounted for. The answer is almost correct but may be off by an order of magnitude.	Student demonstrates the ability to calculate the concentration of bacteria from a culture. Calculations are written out with full method and units of measurement (CFU/ml). The correct dilution factor was used and the fraction of the culture that was inoculated has been accounted for. The answer is correct.

Activity	Skill	Novice	Developing	Proficient
3.F	Observe eukaryotic cells using a microscope	Student may not demonstrate the ability to properly use a microscope. Student begins on low or high power and needs help focusing the microscope. Student is unable to find color or see any cells using the microscope.	Student demonstrates the ability to properly use a microscope. Student begins on low power with the objective as close to the slide as possible. Student focuses away and sees color but needs help locating cells.	Student demonstrates the ability to properly use a microscope. Student begins on low power with the objective as close to the slide as possible. Student focuses away until they see color and then moves to high power and successfully focuses on an individual cell.
4.A 7.F	Use NEBcutter software to predict restriction sites	Student may not demonstrate the ability to use NEBcutter® software to predict restriction sites. Student may not paste a nucleotide sequence into the proper window, may have the sequence in an incorrect format, may not select the sequence correctly by shape (linear or circular) based on its source, or may not select a restriction enzyme. The resulting virtual gel is inaccurate for the assigned restriction enzyme. Restriction sites are incorrectly identified.	Student demonstrates the ability to use NEBcutter software to predict restriction sites. Student successfully pastes a nucleotide sequence into the proper window in the correct format but may need assistance selecting its shape (linear or circular) or the correct restriction enzyme. The resulting virtual gel is accurate and the list of restriction sites is shown. Student may require assistance interpreting results.	Student demonstrates the ability to use NEBcutter software to predict restriction sites. Student successfully pastes a nucleotide sequence into the proper window in the correct format, correctly selects its shape (linear or circular) based on its source, and selects the correct restriction enzyme. The resulting virtual gel is accurate and the list of restriction sites is shown. Restriction sites are correctly identified and the student requires no assistance interpreting results.
4.A 7.F	Record detailed software instructions into a laboratory notebook	Student may not understand the importance of recording detailed software instructions into a laboratory notebook. Student may not record steps for using software in the methods sections of the laboratory notebook and methods cannot be replicated.	Student understands the importance of recording detailed software instructions into a laboratory notebook. Student records steps for using software in the methods section of the laboratory notebook but the steps would require some explanation to be replicated by another individual.	Student understands the importance of recording detailed software instructions into a laboratory notebook. Student records steps for using software in the methods section of the laboratory notebook. Steps require no additional explanation to be understood by another individual.
4.B 4.C 4.D 4.E 5.D 6.A 6.B 6.C	Prepare agarose gels	Student may not demonstrate the ability to prepare an agarose gel. Student may calculate or measure volumes and masses incorrectly, allow transparent particles to remain in the molten agarose solution, or obtain a volume after heating that is significantly different than the original volume, significantly changing the percentage of agarose in the gel. Student may allow agarose to start to set prior to pouring or does not notice if the agarose leaks. Student may position the gel comb incorrectly or remove it prematurely, causing collapse of the wells. When the gel sets, it may lumpy, contain visible flecks of undissolved agarose, or is otherwise not usable.	Student demonstrates the ability to prepare an agarose gel. Student calculates and measures volumes and masses correctly and accurately and boils the agarose solution until no transparent particles are visible. The volume after heating is close to the original volume. Student does not cool molten agarose to around 55°C before pouring but notices any agarose leaks from the gel tray and, if necessary, prepares a new gel. The depth of the gel is not ideal but the gel is usable.	Student demonstrates the ability to prepare an agarose gel. Student calculates and measures volumes and masses correctly and boils agarose solution until no transparent particles are visible. The volume after heating is close to the original volume. Student cools molten agarose to around 55°C before pouring and places the comb in the correct position. Agarose does not leak from the gel tray and the set gel has a smooth surface with uniform texture and is 0.5–1.0 cm in depth. Student removes the comb once the gel is opaque.

Activity	Skill	Novice	Developing	Proficient
4.C 4.D 4.E 5.D 6.A 6.B 6.C 6.D	Load an agarose gel	Student may not demonstrate the ability to correctly load an agarose gel. Student may puncture the gel with the pipet tip, miss the well with the sample, or release the pipet plunger prior to tip removal, thereby sucking sample out of the well and yielding inaccurate data. The data from the gel cannot be used.	Student demonstrates the ability to correctly load an agarose gel. Student delivers samples into the wells with minor technical errors (such as depressing the plunger too quickly). Uneven loading is evident from unevenly sized bands of sample loading dye during electrophoresis but data from gel can be analyzed.	Student demonstrates the ability to correctly load an agarose gel. Student holds the pipet beneath the surface of the buffer and over the well of the gel, depresses the pipet plunger slowly to the first stop, and gently delivers sample into the well. Student does not release the pipet plunger until the pipet is lifted from the buffer. During electrophoresis, bands of sample loading dye are evenly sized.
4.C 4.D 4.E 5.D 6.A 6.B 6.C 6.D	Perform agarose gel electrophoresis	Student may not understand the purpose of agarose gel electrophoresis. Student may use buffers composed of different salts (TAE vs. TBE) for the gel and the running buffer, may not cover the gel in the chamber with buffer, orient the gel in the wrong direction, load samples unevenly or in the wrong order, or may not record changes in the notebook. Student may not set the power supply correctly, resulting in the gel melting or running too slowly. After gel electrophoresis, the gel may be stained incorrectly and no distinct bands are visible.	Student understands the purpose of agarose gel electrophoresis. Student loads samples unevenly or in the wrong order but records loading errors and the actual order in the notebook. Student sets voltage or time on the power supply slightly incorrectly but the gel still runs. After electrophoresis, staining of the gel is not optimal but the results can be analyzed.	Student understands the purpose of agarose gel electrophoresis. Samples are loaded properly and the power supply is set with the correct voltage and time. After gel electrophoresis, staining of the gel produces distinct bands that can be analyzed.
4.C 4.D 4.E 5.D 6.A 6.B 6.C 6.D 7.D 8.D	Use a power supply	Student may not demonstrate the ability to properly operate a power supply. Student uses an incorrect voltage or switches the electrodes. Experimental results are not useable.	Student demonstrates the ability to properly operate a power supply. Student uses the correct voltage for the task and matches the electrodes correctly but minor errors (such as timing) affect experimental results.	Student demonstrates the ability to properly operate a power supply. Student uses the correct voltage for the task, matches the electrodes correctly, and runs the gel for the appropriate length of time.
4.D 4.E	Perform a restriction digest	Student may not demonstrate the ability to perform a restriction digest. The concentrations of reagents may not be appropriate to digest the DNA sample. Only a single band is visible, indicating that the plasmid was not digested.	Student demonstrates the ability to perform a restriction digest. The concentrations of reagents are appropriate to digest the DNA sample. Samples are incubated at the correct temperature but the timing may be incorrect. Fragments are visible by electrophoresis but may not match predicted sizes, thereby indicating incomplete digestion.	Student demonstrates the ability to perform a restriction digest. The concentrations of reagents are appropriate to digest the DNA sample. Samples are incubated at the correct temperature and duration. Fragments are visible by electrophoresis and match predicted sizes.

Activity	Skill	Novice	Developing	Proficient
4.D 4.E 4.F 5.D 6.A 6.B 6.C 6.D	Analyze an agarose gel	Student may not understand how to analyze gel electrophoresis results and may not record obvious bands or estimate DNA sizes incorrectly.	Student understands how to interpret an agarose gel. Student requires minor assistance from the instructor in DNA analysis. Student is able to decipher bands but small or faint DNA bands are missed. Student may make minor errors in DNA size estimations.	Student understands how to interpret an agarose gel and is able to determine the number of DNA bands as well as the size of each band using the DNA size standard for comparison.
4.D 4.E 4.F	Generate a standard curve using a DNA size standard	Student does not understand the purpose of generating standard curves. Student plots data on the wrong axes, uses improper labels or units, plots data points inaccurately, or does not draw the line of best fit through the data points. The graph cannot be used to determine the sizes of unknown DNA bands.	Student understands the purpose of generating standard curves. Student plots the graph correctly on semilogarithmic axes with minor errors in labeling. The line of best fit may include the 23,130 bp point from the lambda HindIII size standard. The sizes of unknown DNA bands can be determined from the graph but may be slightly inaccurate.	Student understands the purpose of generating standard curves. Student plots the graph on semilogarithmic axes with the distance migrated in mm on the x-axis and the size of DNA in bp on the semilog y-axis. The line of best fit is drawn through the data points and the sizes of unknown DNA bands can be determined from the graph.
4.D 4.E 4.F	Determine DNA fragment sizes using a standard curve	Student may not demonstrate the ability to determine the size of DNA from a standard curve. Estimates of fragment sizes are unreasonable and the line of best fit or the equation of the line was not used properly to determine the sizes of DNA fragments.	Student demonstrates the ability to determine the size of DNA from a standard curve. The line of best fit or the equation of the line is used to determine the DNA fragment length. The answer is reasonable but incorrect due to a calculation error or misinterpretation of the graph.	Student demonstrates the ability to determine the size of DNA using a standard curve. The line of best fit or the equation of the line is used to determine the DNA fragment length and the answer is correct.
4.F	Construct a plasmid map from restriction digestion data	Student may not demonstrate the ability to construct a plasmid map. The number of cuts may be incorrectly determined or the restriction sites may be inaccurately placed on the map, and the reasoning for the positioning may not be given.	Student demonstrates the ability to construct a plasmid map. The number of cuts is determined by the number of fragments visualized when a plasmid is digested with one enzyme. A plasmid map is drawn with incorrect placement of some restriction sites and justification for the decision, or the map may show correct placement of the sites but have only partial justification for the decision.	Student demonstrates the ability to construct a plasmid map. The number of cuts is determined by the number of fragments visualized when a plasmid is digested with one enzyme. A plasmid map is drawn with correct placement of all restriction sites with justification using both single and double digest data.
5.A 5.B	Transform bacteria by calcium chloride transformation	Student may not understand the theory behind calcium chloride transformation. Student does not demonstrate proper aseptic technique or does not perform important procedural steps, resulting in no transformants.	Student understands the theory behind calcium chloride transformation. Student demonstrates proper aseptic technique and follows the procedure properly but conducts some steps at the wrong temperature, resulting in no transformants.	Student understands the theory behind calcium chloride transformation. Student demonstrates proper aseptic technique, follows the procedure properly with appropriate steps conducted on ice, and transformation results are as expected.

Activity	Skill	Novice	Developing	Proficient
5.A 5.B	Calculate transformation efficiency	Student may not demonstrate the ability to calculate transformation efficiency. Transformation efficiency is not based on the determination of transformants per microgram of DNA.	Student demonstrates the ability to calculate transformation efficiency. Transformation efficiency is based on the correct determination of transformants per microgram of DNA. Student uses the concentration of DNA and the fraction plated to calculate an answer; however, minor errors in fraction or concentration are evident.	Student demonstrates the ability to calculate transformation efficiency. Transformation efficiency is based on the correct determination of transformants per microgram of DNA. Student uses the concentration of DNA and the fraction plated to calculate the correct answer.
5.C 6.D	Use a microcentrifuge	Student may not demonstrate proper operation of a microcentrifuge. Student may not balance the tubes, resulting in vibrations or excess noise.	Student demonstrates proper operation of a microcentrifuge. Student places tubes in balanced positions with equal volumes. The hinges of the microcentrifuge tubes may not all point outward, the inner lid may not be used, or the speed and time may be set incorrectly.	Student demonstrates proper operation of a microcentrifuge. Student places tubes in balanced position with equal volumes. The hinges of the microcentrifuge tubes all point outward, and the speed and time are set correctly.
5.C	Purify plasmid DNA (perform a miniprep)	Student may not understand the purpose of plasmid purification. Student may or may not demonstrate proper aseptic technique, uses the microcentrifuge incorrectly, or does not perform one or more steps (or performs them out of order), resulting in no plasmid purified (assessed by further analysis).	Student understands the purpose of plasmid purification. Student demonstrates proper aseptic technique, uses the microcentrifuge correctly, and performs the procedure correctly but the final volume of the purified plasmid is higher or lower than expected; yield is also lower than expected (assessed by further analysis).	Student understands the purpose of plasmid purification. Student demonstrates proper aseptic technique, uses the microcentrifuge correctly, and follows the procedure. The final volume and yield of purified plasmid are as expected (assessed by further analysis).
5.D	Quantitate DNA using gel analysis	Student may not understand the theory of using gel analysis to quantitate DNA. Student does not use the mass ruler's intensity as a standard for comparison to estimate the concentration of unknown bands on a gel. The concentration given is not logical or no units are provided.	Student understands the theory of using gel analysis to quantitate DNA. Student uses the mass ruler's intensity as a standard for comparison to estimate the concentration of unknown bands on a gel. The concentration is given in ng of DNA but the estimation is different from the expected concentration.	Student understands the theory of using gel analysis to quantitate DNA. Student uses the mass ruler's intensity as a standard for comparison to estimate the concentration of unknown bands on a gel. The correct concentration is given in ng of DNA.
5.D 7.A 7.E	Use a spectrophoto-meter	Student may not understand the theory behind spectrophotometry or may not know how to operate the instrument. Student sets the program incorrectly, uses the wrong wavelength, or does not blank the instrument before taking readings. The data obtained are incorrect.	Student understands the theory behind spectrophotometry, sets the program correctly, and knows how to operate the instrument. Student may not be comfortable with the automated features of the instrument, uses an inappropriate sample to blank, leaves fingerprints on the cuvette, or forgets to remove the cuvette after assay reading. Results are close to expected values.	Student understands the theory behind spectrophotometry, sets the program correctly, uses an appropriate sample to blank the instrument, demonstrates proper operation of the spectrophotometer, and uses automated features. Student inserts the cuvette in the correct orientation and removes it after assay reading. Results are as expected.

Activity	Skill	Novice	Developing	Proficient
5.D	Quantitate DNA using a spectrophoto-meter	Student may not demonstrate the ability to quantitate DNA using a spectrophotometer. Student may not use quartz or other UV transparent cuvettes, may set the spectrophotometer incorrectly, may dilute the samples incorrectly, or may not mix them thoroughly. Concentration may not be calculated correctly. An incorrect value is obtained.	Student demonstrates the ability to quantitate DNA using a spectrophotometer. Student uses quartz or other UV transparent cuvettes, sets the spectrophotometer correctly, and dilutes the samples properly. Student may not mix the samples thoroughly, leading to inconsistent readings. There may be minor errors in the calculations or the units may not be included in the calculations. The value obtained is close to correct.	Student demonstrates the ability to quantitate DNA using a spectrophotometer. Student uses quartz or other UV transparent cuvettes, sets the spectrophotometer to read at 260 nm, mixes the samples thoroughly, and determines the concentration by multiplying absorbance x 50 µg/ml x dilution factor with the correct units. The correct value is obtained.
5.D	Determine DNA purity by the A_{260}:A_{280} ratio	Student may not understand the theory behind using the A_{260}:A_{280} ratio for determining DNA purity. Absorbance values at 260 nm and 280 nm are not recorded and calculations are incorrect.	Student understands the theory behind using the A_{260}:A_{280} ratio for determining DNA purity. The calculation for determining purity is correct but pertinent information or conclusions are missing.	Student understands the theory behind using the A_{260}:A_{280} ratio for determining DNA purity. Student divides the A_{260} value by the A_{280} value and a ratio greater than 1.8 is noted as a pure sample, while a ratio less than 1.7 is noted as a sample contaminated with protein. Absorbance values at 260 nm and 280 nm are recorded.
6.A 6.B 6.C 6.D	Set up PCR reactions	Student may not understand the theory behind PCR. Student may not set up PCR reactions properly on ice, close tubes as soon as reagents are added, or use filter tips. Tubes may not contain equal volumes and may not be appropriately labeled. PCR results are not as expected and may be contaminated, resulting in unexpected PCR products.	Student understands the theory behind PCR. Student sets up PCR reactions properly on ice but leaves tubes open longer than necessary when reagents are added and may not use filter tips. Tubes contain equal volumes and are appropriately labeled. PCR results are as expected but may be contaminated, resulting in unexpected PCR products.	Student understands the theory behind PCR. Student sets up PCR reactions properly on ice, opens and closes tubes quickly when reagents are added, and uses filter tips. Tubes contain equal volumes and are appropriately labeled. PCR results are as expected and are not contaminated, resulting in expected PCR products.
6.A 6.B 6.C 6.D	Use a thermal cycler	Student does not understand the theory behind thermal cycling and can use only a pre-existing program.	Student understands the theory behind thermal cycling but may not be familiar with how the instrument can be used for different types of PCR. He/she can edit a pre-existing program but needs assistance programming from scratch.	Student understands the theory behind thermal cycling including how the instrument can be used with different types of PCR. He/she can program the instrument from scratch.
6.B	Use a mortar and pestle	Student may not demonstrate proper use of a mortar and pestle. Student does not clean the mortar and pestle before and after use or apply firm pressure while grinding. Water may splash out of the mortar and sample may not be ground to a fine paste.	Student demonstrates proper use of a mortar and pestle. Student cleans mortar and pestle prior to use but may not clean it afterward. Student may not apply firm pressure while grinding or water may splash out of the mortar. Sample is ground to a fine paste.	Student demonstrates proper use of a mortar and pestle. Student thoroughly cleans the mortar and pestle prior to and after use, uses a stirring motion, and applies firm pressure while grinding. Sample is ground to a fine paste.

Activity	Skill	Novice	Developing	Proficient
6.B 6.C	Perform DNA extractions using InstaGene™ matrix	Student may not understand the theory behind DNA extractions. Student may not add correct volume to the InstaGene matrix or mix samples before, during, and after heating. If the pellet is disturbed, he/she may not centrifuge the sample again. When DNA is added to the PCR tube, matrix may be transferred.	Student understands the theory behind DNA extractions. Student adds the correct volume to the InstaGene matrix but may not mix samples before, during, and after heating. If the pellet is disturbed, he/she may not centrifuge the sample again. When DNA is added to the PCR tube, no matrix is transferred.	Student understands the theory behind DNA extractions. Student adds correct volume to the InstaGene matrix, mixes samples before, during, and after heating, and if the pellet is disturbed, centrifuges the sample again. When DNA is added to the PCR tube, no matrix is transferred.
6.C	Calculate allele frequency	Student may not understand the principles behind calculating gene frequencies. Major errors are made in the calculation, and the method is incorrect or not shown. The answer is not correct.	Student understands the principles behind calculating gene frequencies. Minor errors are made, such as forgetting to multiply the possible alleles by two or mixing up the numerator and denominator. The answer is not correct; however, the method was mostly correct and all work is shown.	Student understands the principles behind calculating gene frequencies. The total number of possible alleles and the actual number of alleles in the population (class) are determined and used to calculate the allele frequency. The answer is correct and all work is shown.
7.A	Perform a Bradford assay	Student may not understand the purpose behind the Bradford assay. The standards may not be mixed properly, shown by the solution in the top of the cuvette being a different color than the bottom, or may not be incubated for 5 min prior to reading. The series of standards may not show a gradation of color or may not have equal volumes. The spectrophotometer may be used incorrectly. The standards do not generate a linear standard curve and the protein concentration cannot be calculated correctly.	Student understands the purpose behind the Bradford assay. Each standard may not be mixed properly, shown by the solution in the top of the cuvette being a different color than the bottom, or may not be incubated for 5 min prior to reading. The series of standards has a gradation of color and equal volumes. The spectrophotometer is used correctly. The standard curve is mostly linear and the protein concentration is calculated correctly.	Student understands the purpose behind the Bradford assay. Each standard is mixed and incubated for 5 min prior to reading. The series of standards has a gradation of color and equal volumes. The spectrophotometer is used correctly. The standards generate a linear standard curve and the protein concentration is calculated correctly.
7.A 7.E	Generate a standard curve using absorbance values	Student may not understand the purpose of generating standard curves, graphs data incorrectly, plots data on the wrong axes, or uses improper labels or units. Data points are plotted inaccurately and the line of best fit is not drawn through the data points. The linear equation cannot be used to determine unknown values.	Student understands the purpose of generating standard curves, may graph data with the variables on the incorrect axes, and may not label axes properly. Data are plotted correctly, and a line of best fit is drawn. The equation of the line and/or r^2 value may not be shown. The graph or linear equation can be used to determine unknown values.	Student understands the purpose of generating standard curves, graphs data with the dependent variable (absorbance) on the y-axis and the independent variable on the x-axis, and labels axes with units. Data are plotted correctly, and a line of best fit is drawn. The equation of the line and/or r^2 value is shown. The graph or linear equation can be used to determine unknown values.
7.A 7.E 8.C	Determine experimental values using a standard curve	Student may not understand how to use the line of best fit or the linear equation to determine the values for experimental samples. The values may not have appropriate units and are incorrect.	Student understands how to use the line of best fit or the linear equation to determine the values for experimental samples but may not be familiar with both methods. The values may not have appropriate units but the values are correct.	Student understands how to use both the line of best fit and the linear equation to determine the values for experimental samples. Values are given with appropriate units and are correct.

Activity	Skill	Novice	Developing	Proficient
7.B	Perform size exclusion chromatography	Student may not understand the purpose of size exclusion chromatography and performs the technique incorrectly. Student may not collect fractions properly or collects fractions of unequal volumes.	Student understands the purpose of size exclusion chromatography but misidentifies the order in which molecules elute from of the column. Student properly collects fractions of equal volumes.	Student understands the purpose of size exclusion chromatography and the order in which molecules elute. Student properly collects fractions of equal volumes.
7.C	Make a bacterial lysate from a liquid culture	Student may not understand the purpose of lysing bacteria. Bacterial cells are not lysed correctly and the pellet is green while the supernatant is clear.	Student understands the purpose of lysing bacteria. Bacterial cells are partially lysed, and the supernatant is light green in color with some green remaining in the pellet.	Student understands the purpose of lysing bacteria. Bacterial cells are lysed correctly, and the supernatant is green in color, while the pellet is white.
7.C	Perform hydrophobic interaction chromatography (HIC)	Student may not understand the concept of HIC. Student performs the technique improperly, fractions do not elute as expected, and errors are made in the order in which buffers are added to the column.	Student understands the concept of HIC. Student performs the technique properly but fractions are collected incorrectly or the GFP collected is dilute.	Student understands the concept of HIC. Student performs the technique properly and collects fractions correctly. GFP elutes after addition of TE buffer.
7.D 8.D	Extract proteins using Laemmli sample buffer	Student may not understand the purpose of extracting proteins. An improper amount of tissue may be used or samples may be incubated for the wrong duration. The final samples may contain pieces of tissue.	Student understands the purpose of extracting proteins. The tissue may be cut irregularly or is the wrong size. The samples are incubated for the correct duration and no tissue is in the final samples.	Student understands the purpose of extracting proteins. The tissue is cut into pieces of the correct size and they are incubated for the correct duration. No tissue is in the final samples.
7.D 8.D	Perform SDS-polyacrylamide gel electrophoresis (SDS-PAGE)	Student may not understand the purpose of SDS-PAGE. Student may incorrectly assemble the gel and the electrophoresis chamber, prepare or load samples incorrectly, or load samples in the wrong order without recording the changes. The buffer may leak without the problem being noticed or fixed. The resulting gel cannot be used to analyze the samples.	Student understands the purpose of SDS-PAGE. Student may need assistance assembling the gel and the electrophoresis chamber. Student prepares samples properly but may lose some sample when loading (or may load samples into the wrong well) but records errors, which do not affect results. Buffer may leak but is corrected by filling the inner and outer chambers. The bands of loading dye may be uneven or may smile or frown as the gel runs. The gel can be used for analysis.	Student understands the purpose of SDS-PAGE. Student properly assembles the gel in the electrophoresis chamber. Student prepares samples properly and loads them into their assigned wells without loss of sample. The buffer does not leak and the gel runs as expected. The bands of loading dye are evenly sized and run in a horizontal line down the gel.

Activity	Skill	Novice	Developing	Proficient
7.D	Stain polyacrylamide gels	Student may not understand why polyacrylamide gels need to be stained. The gel is broken, stained with the wrong volume of stain, stained with the wrong type of stain, or improperly destained. The gel cannot be analyzed.	Student understands why polyacrylamide gels need to be stained. The gel is broken but stained with the appropriate volume of protein stain. The gel is incompletely destained, possibly in too small a container, resulting in a blue background, but with visible darker blue protein bands that can be analyzed.	Student understands why polyacrylamide gels need to be stained. The gel is intact and stained with the appropriate volume of protein stain. The gel is completely destained, resulting in dark blue bands on a clear background.
7.D 8.D	Generate a standard curve using a protein size standard	Student does not understand the purpose of generating standard curves. Student may plot data on the wrong axes or use improper labels or units. Data points may be plotted inaccurately and the line of best fit may not be drawn through the data points. The graph cannot be used to determine the sizes of unknown protein bands.	Student understands the purpose of generating standard curves. Student plots the graph correctly on semilogarithmic axes with minor errors in labeling. The line of best fit is generated. The sizes of unknown protein bands can be determined from the graph but may be slightly inaccurate.	Student understands the purpose of generating standard curves. Student plots the graph on semilogarithmic axes with the distance migrated in mm on the x-axis and the size of protein in kD on the semilog y-axis. The line of best fit is drawn through the data points and the sizes of unknown protein bands can be determined from the graph.
7.D 8.D	Determine protein molecular masses using a standard curve	Student may not demonstrate the ability to determine the size of proteins from a standard curve. Estimates of protein masses are unreasonable and the line of best fit or the equation of the line is not used properly to determine the sizes of proteins.	Student demonstrates the ability to determine the size of proteins from a standard curve. The line of best fit or the equation of the line is used to determine the protein mass. The answer is reasonable but incorrect due to a calculation error or misinterpretation of the graph.	Student demonstrates the ability to determine the size of proteins using a standard curve. The line of best fit or the equation of the line is used to determine the protein mass, and the answer is correct.
7.D	Generate a character matrix and a cladogram	Student may not demonstrate knowledge of the importance of cladograms in evolutionary studies. Student cannot interpret the gel data to generate a shared character matrix or the data from the shared matrix cannot not be transferred to a cladogram.	Student demonstrates knowledge of the importance of cladograms in evolutionary studies. Student requires assistance to interpret the data from the gel to generate a shared character matrix or a cladogram. The outlying sample and the samples with the most in common are correctly identified and placed on the cladogram; however; the remaining samples may be placed incorrectly.	Student demonstrates knowledge of the importance of cladograms in evolutionary studies. Data from the gel are correctly interpreted and used to generate a shared character matrix. The character matrix are then used to generate a cladogram. The outlying sample and the samples with the most in common are correctly identified and placed on the cladogram, and the remaining samples are placed correctly.

Activity	Skill	Novice	Developing	Proficient
7.E 8.C	Perform a serial dilution	Student may not understand why serial dilutions are used. Student may not mix samples thoroughly before each dilution and may pipet volumes incorrectly. Final volumes are inconsistent. A useable linear curve is not obtained.	Student understands why serial dilutions are used. Student may not mix samples thoroughly before each dilution but does pipet correct volumes. Final volumes are consistent. A useable linear standard curve is obtained but with one or two data points slightly off.	Student understands why serial dilutions are used. Student mixes samples thoroughly before each dilution and pipets correct volumes. Final volumes are consistent. A linear standard curve is obtained.
7.E	Use a thermometer	Student may read the thermometer before it has equilibrated and may read the scale incorrectly. The reading is incorrect and may not have the correct units.	Student may read the thermometer before it has equilibrated but interprets the scale correctly. The reading is slightly incorrect and may not have the correct units.	Student reads the thermometer after it has equilibrated and interprets the scale correctly. Temperature is recorded with the correct units.
7.E	Perform time course experiments	Student may not demonstrate the ability to take measurements at the appropriate times, fails to record new time points, or may not take into account new time points in the analysis.	Student mostly demonstrates the ability to take measurements at the appropriate times but may miss or be late on a time point. If a time point is missed or late, the student records the new time point and takes the change into account when analyzing data.	Student demonstrates the ability to take measurements at the appropriate times and all data are recorded accurately. Any differences are taken into account when analyzing data.
7.E	Determine the initial rate of an enzyme reaction	Student may not understand how to determine the rate of an enzyme reaction. A standard curve may not be generated or may not be used to determine the amount of product generated at each time point. The initial rate of reaction cannot be determined correctly.	Student understands how to determine the rate of an enzyme reaction. A standard curve is generated and used to determine the amount of product at each time point. The initial rate of reaction can be determined by plotting the data on a graph but not by directly calculating the rate of change. Some labels or units may be missing but the answer is correct.	Student understands how to determine the rate of an enzyme reaction. A standard curve is generated and used to determine the amount of product generated at each time point. The initial rate of reaction can be determined both by plotting the data on a graph and by directly calculating the rate of change. All graphs and calculations have proper labels and units, and the answer is correct.
7.F	Search GenBank for DNA sequences using accession numbers	Student may not demonstrate the ability to access the NCBI website, retrieve information using a GenBank accession number, or independently obtain the FASTA format or DNA sequence.	Student demonstrates the ability to access the NCBI website and is able to retrieve information using a GenBank accession number but may need help in locating the DNA sequence or obtaining the FASTA format.	Student demonstrates the ability to access the NCBI website, is able to retrieve information using a GenBank accession number, and independently locates the DNA sequence and FASTA format.
7.F	Use NEBcutter to identify ORFs	Student demonstrates the ability to access the NEBcutter website and paste the pBAD-GFPuv sequence into the correct text box. The sequence is in the incorrect format, the open reading frames are not correctly identified, or the gene name is not identified.	Student demonstrates the ability to access the NEBcutter website and paste the pBAD-GFPuv sequence in the correct format. The open reading frames are correctly identified but the student requires assistance with sequence analysis and with identifying the gene name.	Student demonstrates the ability to access the NEBcutter website and paste the pBAD-GFPuv sequence in the correct format. The open reading frames are correctly identified and the student requires no assistance with sequence analysis or with identifying the gene name correctly.

Activity	Skill	Novice	Developing	Proficient
7.F	Use BLAST to identify proteins from amino acid sequence	Student does not demonstrate the ability to use BLAST or interpret BLAST results to identify proteins.	Student demonstrates the ability to use BLAST but may need help interpreting BLAST results to identify proteins.	Student demonstrates the ability to use BLAST and interpret BLAST results to identify proteins.
6.D	Use bioinformatics tools to analyze DNA quality, assemble a consensus sequence, and search BOLD	Student does not demonstrate the ability to analyze DNA quality, assemble a sequence, search BOLD, or interpret results from BOLD to determine the best match for sample.	Student may demonstrate the ability to analyze DNA quality, assemble a sequence, and search BOLD, but may need help interpreting the results from BOLD to determine the best match for sample.	Student demonstrates the ability to analyze DNA quality, assemble a sequence, search BOLD, and interpret BOLD results to determine the best match for sample.
7.F	Use PDB identifiers to retrieve and view 3-D protein structures in Jmol	Student needs a lot of assistance accessing the Protein Databank, retrieving the protein structure, visualizing the 3-D structure using Jmol, or changing the rendering to show helices and pleated sheet structures.	Student needs minor assistance with one aspect of accessing the Protein Databank, retrieving the protein structure, visualizing the 3-D model using Jmol, or changing the rendering to show helices and pleated sheet structures.	Student demonstrates the ability to access the Protein Databank, retrieves the protein structure successfully, visualizes the 3-D model using Jmol, and can change the rendering to show helices and pleated sheet structures.
7.F	Create a phylogram using Clustal Omega	Student may have trouble accessing the NCBI website, retrieving protein sequence information, and generating a text document with sequences in FASTA format. Assistance is required to use Clustal Omega and a phylogram is not obtained.	Student demonstrates the ability to access the NCBI website, retrieve protein sequence information, and generate a text document with sequences in FASTA format but needs some assistance. Clustal Omega is used successfully to align the sequences and generate a phylogram but assistance is required.	Student demonstrates the ability to access the NCBI website, retrieve protein sequence information, and generate a text document with sequences in FASTA format. Clustal Omega is used successfully to align the sequences and generate a phylogram.
8.A	Perform Ouchterlony double immunodiffusion assay	Student may not understand the theory behind the Ouchterlony assay. Student performs procedure incorrectly, misses multiple steps, or performs steps out of order. Holes may be unevenly distributed. The control precipitate is not visible.	Student understands the theory behind the Ouchterlony assay. Student follows the procedure but performs one step incorrectly. Holes may be unevenly distributed but the control precipitate is visible.	Student understands the theory behind the Ouchterlony assay. Student follows procedures correctly. Holes are evenly distributed and the control precipitate is visible.
8.B	Perform a qualitative ELISA	Student may not understand the theory behind the ELISA. Student may make multiple pipetting or washing errors or may add samples out of order. Triplicate wells may not be identical or controls are inaccurate. Results cannot be interpreted or are interpreted incorrectly.	Student understands the theory behind the ELISA. Student may pipet samples into one or two incorrect wells, wash inadequately, or splash samples into adjacent wells. Triplicate wells may not be identical but results can be interpreted using results from two of the three wells. Results are interpreted correctly.	Student understands the theory behind the ELISA. Student pipets and washes wells carefully without contamination. Positive control wells are blue and negative control wells are clear. Triplicate wells give identical results. Results can be interpreted and are interpreted correctly.

Activity	Skill	Novice	Developing	Proficient
8.C	Perform a quantitative ELISA	Student may not understand the theory behind a quantitative ELISA. Student may perform the procedure incorrectly, miss steps, or perform steps out of order. Wells for the standard do not display a proportional decrease in sample concentration correlating with the dilution factor. Concentration of the antigen cannot be determined.	Student understands the theory behind a quantitative ELISA. Student follows the procedure but performs one or more procedural steps incorrectly. Wells for the standard generally display a proportional decrease in sample concentration correlating with dilution factor but one or more wells may not align due to incomplete mixing between dilutions or inaccurate pipetting. Concentration of the antigen can be determined.	Student understands the theory behind a quantitative ELISA. Student follows procedures correctly. Wells for the standard display a proportional decrease in sample concentration correlating with the dilution factor. Concentration of the antigen can be determined.
8.C	Use a microplate reader	Student may not demonstrate the ability to use a microplate reader. The wavelength or program is set incorrectly, blanks are not identified accurately, and results are not as expected.	Student demonstrates the ability to use a microplate reader. Wavelength or program is set correctly and blanks are identified accurately but automated features are not used. Results are close to expected values.	Student demonstrates the ability to use a microplate reader. Wavelength or program is set correctly, blanks are identified accurately, and results are as expected. Automated features are used correctly.
8.D	Perform western blotting	Student may not understand the theory behind western blotting. Student may perform SDS-PAGE incorrectly. Student may assemble the blot in the wrong order or with air bubbles, or use the wrong buffer for transfer. Protein standards may not be visible on the blot. Immunodetection may be performed incorrectly. There may be few or no bands on the blot or the bands may be too faint for analysis. The data cannot be analyzed.	Student understands the theory behind western blotting. Student performs SDS-PAGE correctly. Student assembles the blot in the correct order but air bubbles may be present. Student uses the correct buffer for transfer. The protein standards are visible on the blot but may be faint. Immunodetection yields bands; however, there may be background on the blot and some bands may be weak, faint, or incomplete (due to the presence of air bubbles). The results can be used for data analysis.	Student understands the theory behind western blotting. Student performs SDS-PAGE correctly. Student assembles the blot in the correct order and takes care to remove air bubbles. Student uses the correct buffer for transfer. The protein standards are visible on the blot. Immunodetection yields clear, strong bands against a white background that can be used for data analysis.

Appendix F:
Laboratory Notebook
and Project Rubrics

Laboratory Notebook Rubric

Student: *Date:*
Experiment: *Evaluator:*

Objective	Novice (1)	Developing (2)	Proficient (3)
Organization of entries	Few pages are numbered, dated, or signed.	Most pages are numbered, dated, and signed.	All pages numbered and signed, and every section is dated.
	There are no or few titles or headings.	A few titles and headings are missing or unclear.	Each experiment has a title and the appropriate headings: goal or purpose, background, method, results, and conclusions.
	Table of contents is missing, very incomplete, or out-of-date.	Table of contents is mostly up-to-date and complete.	Table of contents lists date, title, and page numbers for each experiment.
	Most experiments that run onto multiple pages of the notebook do not have "from page" and "to page" information provided.	Most but not all experiments that run onto multiple pages of the notebook have "from page" and "to page" information provided.	All experiments that run onto multiple pages of the notebook have "from page" and "to page" information provided.
Materials and methods	Methods and materials are incompletely described.	Methods and materials are described in most experiments but another person may have trouble repeating the experiment.	The methods and materials are completely described for all experiments and the experiment could easily be repeated by another person.
Results and analysis	Figures and tables are not included when appropriate.	Appropriate figures and tables are mostly included but may not be labeled properly.	All appropriate figures and tables are included and properly labeled with titles and legends.
	Observations are not recorded.	Observations are noted but not in sufficient detail.	Observations are carefully recorded.
	Data analysis is not described or included.	Data are analyzed but not thoroughly. Methods or calculations used may not be fully explained or written out.	Data analysis is complete, and analysis methods and calculations are written out in full.
	No error analysis is included.	Error analysis is qualitative, incorrect, or incomplete.	Error analysis is qualitative; quantitative estimates are given if appropriate. Suggestions for changes to methods or additional experiments to clarify results are provided.
Interpretation	Conclusions are not documented.	Immediate thoughts are recorded for most experiments.	Results are interpreted in the context of the hypothesis being tested or technique being conducted.

Project Progress Report Rubric

Student:		*Progress check number:*	
Project title:		*Evaluator:*	

Objective	Novice (1)	Developing (2)	Proficient (3)
Background knowledge	It is evident that the student is not familiar with the background necessary for the project. Experimental results are not aligned to what is already known about the topic.	There is some evidence that the student is familiar with the project topic but some experiments do not relate results to background knowledge.	It is evident that the student is familiar with the necessary background for the project. Experiments and interpretations are based on background knowledge that is referenced in the experimental writeups.
Organization of laboratory notebook	Notebook is disorganized and difficult to follow.	Notebook is fairly organized but some headings or sections are missing.	Notebook is organized well, and experiments are easy to follow.
Content	Materials, methods, and results are incompletely described.	Materials, methods, and results are described in most experiments but may be difficult for another person to follow.	Materials, methods, and results are completely described.
Analysis	Data analysis is not described or included.	Data are analyzed, but not correctly or completely.	Data analysis is complete, appropriate, and correct.
Interpretation	Conclusions have not been documented.	Immediate thoughts have been recorded for most experiments.	Results are interpreted in the context of the research question. The next experiment is suggested and justified using data.
Experimental errors and improvements	Experimental errors are not addressed.	Experimental errors are mentioned, but corrections or suggestions for improvement are not given or have not been made.	Experimental errors are addressed, and suggestions for improvement or experiments to clarify the data are provided and acted upon.
Research process	It is not evident how each experiment contributes to the research question. Experiments may be illogical or not thought out.	Most experiments contribute to answering the research question but some are unnecessary or not well thought out.	It is fully evident how each new experiment contributes to answering the research question.
Schedule	Project is not on schedule, very little progress has been made, and there is no evidence of efforts being made to get the project back on track.	Project is not on schedule. Progress has been made and efforts are being made to get back on track but schedule or plan has not been revised.	Project is on schedule or schedule has been revised to account for experimental results or unforeseen delays.
Team work (if appropriate)	Team is not aligned, responsibilities are not clear, and not all team members are contributing.	Team is aligned and each team member is contributing but responsibilities are not clearly defined and communication could be improved.	Team is aligned to the same goal and is communicating well. Each member has a contributing role and is clear on their responsibilities.
Progress	Very little progress has been made from the last check.	Intermediate progress has been made and there is evidence of inactivity.	Satisfactory progress has been made and evidence of constant research is clear.

Project Proposal Rubric

Student:
Project title: *Evaluator:*

Component	Novice (1)	Developing (2)	Proficient (3)
Background/ introduction	Background is irrelevant or does not support the proposed project.	Background is relevant but not enough is provided for the proposed project.	Background is relevant and sufficient for the proposed project.
	Research questions/hypotheses are vague or don't relate to the proposed experiments.	Research questions/hypotheses are clear but do not properly relate to the proposed experiments.	Research questions/hypotheses are clear and relate to the proposed experiments.
	Citations used are minimal and may come from unvalidated sources.	Citations used are substantial but some are from unvalidated sources.	Citations used are substantial and from valid, relevant sources.
	Overall idea is not well justified.	Overall idea is justified but not well organized.	Overall idea is justified and well organized.
Experimental methods	Some protocols are vague and/or not included.	Protocols are included but minor sections need to be improved for clarity.	Protocols are included and could be followed.
	Protocols may not be aligned to the research questions.	Protocols are aligned to the research questions but may be unnecessary.	Protocols are aligned to the research questions/hypotheses.
	Timeline given is not realistic.	Timeline is realistic but no room is given for experimental variability.	Timeline is realistic and room is given for experimental variability.
Materials required	A limited list of supplies needed is provided.	A complete list of supplies may be given but the suppliers or specific details may be missing.	A complete detailed list of supplies is provided with the suppliers named.
	Quantities are missing or limited.	Quantities are provided but they do not align with the proposed experiments.	Quantities are provided and align with the proposed experiments.
Data analysis plan	A data analysis plan is not provided, is very minimal, or lacks detail.	A data analysis plan is provided but could be improved based on the student's background.	A realistic data analysis plan is provided and is appropriate to the level of the student's background.
	Statistical methods are not proposed.	Statistical methods are proposed but not consistently applied.	Statistical methods are proposed and cover all the proposed work.
Safety	Minimal issues are listed as potential safety hazards.	Major issues are listed as potential safety hazards but minor hazards are missing.	All the safety hazards in the protocols have been addressed.
	No plan for hazardous waste disposal is provided.	A plan for hazardous waste disposal is given but is incomplete.	A comprehensive hazardous waste plan is provided that covers all the proposed experiments.
	PPE is not addressed.	PPE is addressed but not listed for each experiment.	PPE is addressed and appropriate for all proposed experiments.
Expected results	Expected results are not provided or would not answer the research questions.	Limited expected results are provided and may not completely answer the research questions.	Expected results are provided for all proposed experiments and would answer the research questions.
	Controls are missing.	Controls are included but may not be appropriate or sufficient.	Controls are in place and include both positive and negative controls where possible.
	No plan for experimental modification is given.	A minimal attempt at experimental modification is provided.	A solid plan of modification is provided that relates to the planned experimentation.
Budget	A budget is not provided or is not useable.	Prices with suppliers are provided but may not align with materials list. A total project cost may not be provided.	Prices with suppliers are provided and align with the materials list. A total project cost is provided.

Presentation Content Rubric

Score	1	2	3	4	5
Presentation visuals	An attempt was made but there is a general lack of visual support and there may be errors.	Some visuals support the material and only some aspects of the materials to be covered were represented.	Most visuals supported the material. Most of the material to be covered were represented.	All visuals supported the material. Most aspects of the material to be covered were represented.	All visuals supported the material. All aspects of materials to be covered were represented.
Background information	The background information did not relate to the research and little knowledge of the subject area was shown.	A small amount of background information was provided or relevant to the research.	Some of the information was relevant to the research but too much extra information or too little background was given.	The majority of the information was appropriate and a good link was made to the research conducted.	The background information was appropriate for the presentation, explained well, and relevant to the research.
Research questions and hypotheses	The research questions did not relate to the hypotheses and the hypotheses were poorly designed.	The research questions were not very clear and did not relate well to the hypotheses. Hypotheses had errors in design.	The research questions were generally clear but they may have minor errors and their relation to the hypotheses may have been unclear.	The research questions were clearly explained, and the majority of hypotheses related to the research questions.	The research questions were clearly explained, and hypotheses related directly to the research questions.
Materials and methods	The methods were not appropriate for answering the research question or experiments were not controlled.	The methods were only slightly appropriate for answering the research question or multiple controls were missing.	The methods were somewhat appropriate for answering the research question but one important control was missing.	The methods were generally appropriate for answering the research question and all variables were controlled.	The methods were appropriate for answering the research questions and controls were appropriate for the variables.
Results/ conclusions	Major errors were present in the data presented. Data were not displayed properly and major errors were made in the conclusion.	Data were displayed, but there were errors in presentation. Some major mistakes were made in the conclusions.	Data were generally displayed in ways that helped interpretation, some inconsistencies were noted but some inaccurate conclusions were drawn.	Data were generally displayed in ways that helped interpretation and most inconsistencies were noted. All the data were explained but not every conclusion was drawn from the data.	Data were displayed in ways that helped interpretation, and inconsistencies were noted. All the data were explained and appropriate conclusions were drawn.
Analysis of experimental errors	Very few errors were explored or explained.	Some major errors were not identified, but explanations with plans for how to correct them were given for those presented.	Most but not all errors were identified and explained with some plans for how to correct them.	All errors were identified and explained with plans for how to correct them but some explanations were not clear.	All errors were identified and explained completely with plans for how to correct them in the future.
Further experiments	No future experiments were presented.	Potential future experiments were mentioned but without reasoning.	Potential future experiments were presented but did not align with the research question.	Potential future experiments were presented but reasoning for the experiments was not clear.	Potential future experiments were presented with reasoning aligned with the research question.
Questions	No questions could be answered and no mastery of the subject was displayed.	Many questions could not be addressed and many answers were inaccurate.	Not all the questions were addressed or the student could not answer some questions they should have had answers for.	All the questions were addressed but some inaccurate answers were given. The student did not make it clear that they were guessing at an answer.	All the questions were addressed appropriately. The student was clear when they did not know the answer or were guessing at an answer.

Presentation Style Rubric

Score	1	2	3	4	5
Presentation visuals	An attempt was made but there was a general lack of visual support.	Some visuals supported the material and only some aspects of the materials to be covered were represented.	Most visuals supported the material and most of the material to be covered was represented.	All visuals supported the material and most aspects of the material to be covered were presented.	All visuals supported the material and all aspects of materials to be covered were represented.
Delivery	The presenters had no voice inflection and made major speaking errors. Presentation was very difficult to follow.	The presenters had little voice inflection and made some major speaking errors or presentation was difficult to follow.	The presenters had some voice inflection and made an effort to present clearly but made a few speaking errors.	The presenters generally had good voice inflection and made very few speaking errors when delivering the presentation.	The presenters had good voice inflection and did an outstanding job delivering the presentation.
Teaching style	The presenters did not use any teaching strategies.	The presenters used a couple of teaching strategies but they were not effective.	The presenters used some teaching strategies but they may not have been appropriate for the audience.	The presenters used a wide range of teaching strategies but some were not as effective as others.	The presenters effectively used a wide range of teaching strategies to help the audience learn the important concepts.
Content coverage	Very few aspects of subject were covered and very few objectives were met.	Some aspects of the subject were covered but some major objectives were not met.	Most aspects of the subject were covered but at least one major objective was not met.	Most aspects of the subject were covered and most of the objectives were met.	All aspects of the subject were covered and all objectives were met effectively.
Preparedness	Most of the presentation was not well prepared or not understood.	Some of the presentation was not well prepared or much was not understood by the presenters.	Most of the presentation was well prepared but some important aspects were not understood by the presenters.	Most but not all of the presentation was well prepared or only minor aspects were not understood by the presenters.	The presentation was well prepared and completely understood by the presenters.
Group coordination	The group had major delays or problems working as a group.	The group had errors in coordination that delayed the presentation and some members did not cooperate and contribute.	The group had errors in coordination that delayed the presentation or some members did not cooperate and contribute.	The group had slight errors in coordination but there were no delays and all members cooperated and contributed.	The group was well coordinated and all members cooperated and contributed.
Written or support material	The material did not support the presentation and had major errors.	The material barely supported the presentation.	The material mostly supported the presentation but contained distracting errors.	The material supported the presentation fairly well.	The material supported the presentation appropriately.

Appendix G: Trademarks and Legal Notices

Trademarks and Legal Notices

The following are either registered trademarks or trademarks of Bio-Rad Laboratories Inc. in the U.S. and/or other countries. For the most up-to-date and comprehensive list of Bio-Rad trademarks, visit www.bio-rad.com.

Aurum™
BioLogic™
Bio Plex®
Bio-Rad®
Bio-Safe™
Bio-Rad Explorer™
Chelex®
ChemiDoc™
Crime Scene Investigator PCR Basics™
DC™
ELISA Immuno Explorer™
EZ Load™
Fast Blast™
Genes in a Bottle™
Genie HIV1/HIV2™
GMO Investigator™
Helios®
iMark™
InstaGene™
Kaleidoscope™
Micro Bio-Spin™
MicroPulser™
Mini-PROTEAN®
Mini-Sub®
MRSASelect™
PCR Kleen™
PDS-1000/He™
pGLO™
Poly-Prep®
PowerPac™
Precision Plus Protein™
Profinity™
Prot/Elec™
T100™
TGX™
ThINQ!™
Trans-Blot®
Turbo™
UView™

The following trademarks are the property of the companies listed.

23andMe: 23andMe, Inc.
454 SEQUENCING: Roche Diagnostics GmbH
Accumet Research: Fisher Scientific
AAAS: American Association for the Advancement of Science
AncestryDNA: Ancestry.com Operations, Inc.
ATryn: GTC Therapeutics Inc.
Chy-Max: Chr. Hansen, Inc.
Claritin: Schering Corporation
EvaGreen: Biotium, Inc.
Excel: Microsoft Corp.
Ficoll: GE Healthcare Bio-Sciences AB
GenBank: U.S. Dept. of Health and Human Services
GloFish: Yorktown Technologies L.P.
Google Sheets: Google LLC
Herceptin: Genentech Inc.
Hoechst: Hoechst GMBH
HuCAL: MorphoSys AG
Humulin: Eli Lilly and Company
HydroShear: Digilab, Inc.
Intel: Intel Corporation
Jell-O: Kraft Foods Global Brands LLC
Keynote: Apple Computer Inc.
Kool-Aid: Kraft Foods Global Brands LLC
Kymriah: Novartis AG
Logger Pro: Vernier Software & Technology
MyHeritage: MyHeritage LTD
NanoDrop: Thermo Fisher Scientific Inc.
NEBcutter: New England Biolabs Inc.
Novozymes: Novozymes A/S LLC
OraQuick: Orasure Technologies, Inc.
Orthoclone: Johnson & Johnson
Parafilm: Bemis, Inc.
PowerPoint: Microsoft Corp.
Protropin: Genentech Inc.
Provenge: Dendreon Corporation
Pyrex: Corning Inc.
Rexin-G: Epeius Biotechnologies Corporation
Rituxan: Idec Pharmaceuticals Corporation
Roundup Ready: Monsanto Technology LLC
Science: American Association for the Advancement of Science
Sep-Pak: Waters Technologies Corporation
SmartStax: Monsanto Technology LLC
Spectronic: Spectronic Instruments Inc.
SYBR: Molecular Probes Inc.
Syngenta: Syngenta Participations AG
Vectibix: Immunex Corporation
Wheaton: Wheaton Industries Inc.
Xenomouse: Xenotech Inc.

Legal Notices

Notice regarding Bio-Rad Bio-Plex® system and assay: The Bio-Plex suspension array system includes fluorescently labeled microspheres and instrumentation licensed to Bio-Rad Laboratories, Inc. by the Luminex Corporation.

Notice regarding Bio-Rad Precision Plus Protein™ standards: Precision Plus Protein standards are sold under license from Life Technologies Corporation, Carlsbad, CA, for use only by the buyer of the product. The buyer is not authorized to sell or resell this product or its components.

Bio-Rad thermal cyclers and real-time thermal cyclers are covered by one or more of the following U.S. patents or their foreign counterparts owned by Eppendorf AG: U.S. Patent Numbers 6,767,512 and 7,074,367.

Notice regarding Primer3 software: Primer3 includes software developed by the Whitehead Institute for Biomedical Research.

Notice regarding Bio-Rad products that contain SYBR Green: SYBR is a trademark of Molecular Probes, Inc. Bio-Rad Laboratories, Inc. is licensed by Molecular Probes, Inc. to sell reagents containing SYBR Green I for use in real-time PCR, for research purposes only.

Notice regarding Bio-Rad products that EvaGreen® Supermix: EvaGreen is a trademark of Biotium, Inc. Bio-Rad Laboratories, Inc. is licensed by Biotium, Inc. to sell reagents containing EvaGreen dye for use in real-time PCR, for research purposes only.

Index

Note: *f* denotes figures; *t* denotes tables; *s* denotes sidebars, **bold** denotes defined/glossary terms

C

Index

Index

S

Index

Short tandem repeat (STR). 202, 206–210, 206*f*, 210*f*, **378**
Shuttle plasmids, 151–152, **378**
Significant figures, 41, **378**
Size exclusion chromatography (SEC), 248, 248*f*, 258–260, 363, **378**, 396
Sketching microscopic details, 110, 389
Slants, 73, 73*f*, **378**
Small volume dilutions, 45
SmartStax corn, genetically modified, 8
Sodium chloride (salt), bacterial environment, 72, 72*f*
Sodium dodecyl sulfate (SDS), 244–245, 244*f*, 245*f*
Solid media, 73
Solubility, proteins in biology, 239
Solutions, preparing. *See* Preparing solutions.
SOPs (standard operating procedures)
    definition, **378**
    laboratory skills assessment, 384
    uses for, 23
    writing, 22, 64–65, 384
Southern, Edwin, 125
Southern blotting, 125, **378**
Spectrophotometer
    components of, 242, 242*f*
    definition, 4, **378**
    spectrophotometric quantitation, 159–160, 159*f*
    laboratory skills assessment, 393, 395
Spirillum, 71, 71*f*
Spreaders, 74
Stab culture, 73, 73*f*, **378**
Stain-free gel imaging, 371
Staining
    agarose gels, 124, 367–370
    eukaryotic cells, 108–110
Standard curves, 254–256, 317, 328, 395, 397
Standard deviation, calculating, 342–343, 343*f*, 343*t*, 381
Standard operating procedures (SOPs). *See* SOPs (standard operating procedures).
Staphylo prefix, 71
Starch hydrolysis test, 78, 78*f*
Start-ups, biotechnology companies, **15**
Stem cell research
    cell differentiation, 82*f*
    eukaryotic cells used for, 81
    iPSC (induced pluripotent stem cells), 81, 84*f*
Stem cell research, hESC (human embryonic stem cells), 81, 82*s*
Stem cells, 81, 82*f*, 82*s*, 84*f*, **378**
Sterile technique. *See* Aseptic technique.
Sterile transfer, 75–76, 75*f*
Sterility, eukaryotic cell culture, 84
Sterilizing laboratory equipment. *See* Laboratory equipment, disinfecting and sterilizing.
Sticky ends, 117, 117*f*, 118, 149
Stock solutions in terms of "x," 43
STR (short tandem repeat), 202, 206–210, 206*f*, 210*f*, **378**
Streak plate techniques, 76, 76*f*, 387
Strepto prefix, 71

Student research projects
    abstract of an article, 339, **372**
    assay validation, 341, **372**
    choosing a project, 336–337
    collaborating with scientists, 335
    competitions, 334*t*, 335
    controls, 341
    experimental design, 341–342
    funding, 343–344
    group projects, 335
    high school research, 335–336
    individual projects, 335
    internet as a source, 339
    libraries, 339
    mentors, 339, 341
    methods, 341
    negative controls, 341
    peer-reviewed journal articles, 339, 340*t*–341*t*
    performing background research, 339, 341
    planning a project, 339
    positive controls, 341
    primary research articles, 339
    Regeneron Science Talent Search, 334, 338*s*
    resources, 339*t*
    review articles, 339
    sample project proposal, 338*f*
    scoping a project, 337–338
    trials, 341–342
    undergraduate research, 336, 337*t*
    venues for research or competitions, 335–336, 336*t*
    whole class projects, 335
Substrate
    activity, 276–286
    definition, **379**
    proteins in biology, 239, 239*f*
    tips for, 364
Supernatant, 157, **379**
Suspension cultures, 82, **379**
Suttle, Curtis, 200*s*
Swanson, Robert, 147
Synthesizing protein. *See* Protein synthesis.
Synthetic biology, 83*s*, 241
Systems biology, 5, **379**

T

T100™ thermal cycler, 192
TA (Topo) cloning, 118
Targeted gene editing, 120,120*f*. *See also* CRISPR (clustered regularly interspaced short palindromic repeats).
Tatum, Edward, 147
TB (tuberculosis), 70, 156
Temperature, used in PCR, 192–194, 194*f*, 195–197
Template DNA, 188–189, 196, **379**
Termination, protein synthesis, 236, 237*f*
Terminator, 149, **379**
Tertiary structure, 238, 238f, **379**
Testing
    for HIV, 8*f*, 306*s*
    new drugs. *See* Drug testing.
    paternity, 203, 203*f*
    pregnancy tests, 300, 300*f*
    starch hydrolysis test, 78, 78*f*
Testing, Kirby-Bauer, or disk diffusion test
    activity, 93–95
    culturing bacteria, 74, 74*f*
    definition, **375**
    laboratory skills assessment, 386
TGS (Tris/glycine/SDS) running buffer, 245
TGX™ Precast gel, 245, 246*s*
TGX Stain-Free™ gel, 371
Theory. *See* Scientific theory.
Thermal cyclers
    definition, **379**
    uses for, 4
    laboratory skills assessment, 394
Thermal gradient, optimizing PCR, 193, 197, 197*f*
Thermometers, 38, 38*f*, 398